Second Edition

SUSTAINABLE WORLD

APPROACHES TO ANALYZING & RESOLVING WICKED PROBLEMS

Sonya Remington-Doucette

Cover images © Shutterstock, Inc.

Kendall Hunt
publishing company

www.kendallhunt.com
Send all inquiries to:
4050 Westmark Drive
Dubuque, IA 52004-1840

Copyright © 2013, 2017 by Kendall Hunt Publishing Company

ISBN 978-1-5249-1236-9

All rights reserved. No part of this publication may be reproduced, stored in a retrieval system, or transmitted, in any form or by any means, electronic, mechanical, photocopying, recording, or otherwise, without the prior written permission of the copyright owner.

Published in the United States of America

DEDICATION

To my son, born since the publication of the first edition of this book. With his future in mind, he sustained my energy and motivation, as a tired new mom torn in many different directions, throughout those late night work sessions during which the majority of this second edition was written.

CONTENTS

CHAPTER 1	**Introduction to Sustainability**		1
	Core Questions and Key Concepts		3
	Key Terms		4
	Section 1.1	People and Nature, Today and throughout Human History	6
	Section 1.1.1	Recent Human Transformation of Natural Systems	7
	Section 1.1.2	Human Interaction with Natural Systems in the Past	21
	Section 1.2	Understanding Human Transformations Using the I = PAT Model	23
	Section 1.2.1	Present-Day Application of I = PAT	23
	Section 1.2.2	Comparing Human Transformations Then and Now Using I = PAT	25
	Section 1.3	Natural Systems and Human Well-Being	31
	Section 1.3.1	Connections between Natural Systems and Human Well-Being	31
	Section 1.3.2	Defining Human Well-Being	34
	Section 1.3.3	Aspects of Human Well-Being Affected by Natural System Transformations	36
	Section 1.4	Origins of Sustainability	39
	Section 1.4.1	Environment and Natural Capital	40
	Section 1.4.2	Economy and Development	43
	Section 1.4.3	Society and Equity	44
	Section 1.4.4	Birth of Sustainability	50
	Bibliography		57
	End-of-Chapter Questions		60
CHAPTER 2	**Wicked Problems and Their Resolution**		63
	Core Questions and Key Concepts		65
	Key Terms		65

		Section 2.1	Understanding Sustainability Problems	66
		Section 2.1.1	Introduction to Wicked Problems	67
		Section 2.1.2	Wicked Problems, Socioecological Systems, and Complexity	74
		Section 2.2	Resolving Sustainability Problems	99
		Section 2.2.1	Overview of the Transformational Sustainability Research Framework	99
		Section 2.2.2	Making Tradeoffs for Sustainability	102
		Section 2.2.3	Participatory and Transdisciplinary Approaches	106
		Bibliography		111
		End-of-Chapter Questions		114
CHAPTER 3	**Current State Analysis**			**119**
	Core Questions and Key Concepts			121
	Key Terms			122
	Section 3.1	Defining the System and Classifying Drivers		124
		Section 3.1.1	Defining the System	125
		Section 3.1.2	Classifying Driver Scale	131
		Section 3.1.3	Classifying Driver Influence	138
	Section 3.2	Causal Chain Analysis		146
		Section 3.2.1	Assessing Causality	147
		Section 3.2.2	Extent of Influence, Current Trends, and the Importance of Context	151
		Section 3.2.3	Drivers and Causal Chains: A Western Indian Ocean Case Study	155
	Section 3.3	Stakeholder Analysis		161
		Section 3.3.1	Identifying Stakeholders	162
		Section 3.3.2	Analyzing Stakeholder Behavior, Interest, and Influence	165
		Section 3.3.3	Investigating Stakeholder Relationships	173
	Bibliography			174
	End-of-Chapter Questions			176
CHAPTER 4	**Indicators of Sustainability**			**185**
	Core Questions and Key Concepts			187
	Key Terms			187
	Section 4.1	Introduction to Sustainability Indicators		189
		Section 4.1.1	Sustainability Indicators versus Traditional Indicators	190
		Section 4.1.2	Characteristics of Effective Sustainability Indicators	194

Section 4.2	Sustainability Indicators, a Case Study from the Primorska Region of Slovenia	219
Section 4.2.1	The Participatory Process of Indicator Development	220
Bibliography		226
End-of-Chapter Questions		227

CHAPTER 5 Resilience and Patterns of Change — 233

Core Questions and Key Concepts — 235
Key Terms — 236

Section 5.1	Regime Shifts, Thresholds, and Resilience	238
Section 5.1.1	Regimes	238
Section 5.1.2	Thresholds and Regime Shifts	241
Section 5.1.3	Resilience	250
Section 5.2	Feedbacks and Resilience	261
Section 5.2.1	Shifting Dominance of Stabilizing and Reinforcing Feedbacks	265
Section 5.2.2	System Fluctuations and Feedbacks	281
Section 5.3	Adaptive Cycles and Resilience	286
Section 5.3.1	Introduction to Adaptive Cycles	287
Section 5.3.2	Adaptive Cycle Application	289

Bibliography — 298
End-of-Chapter Questions — 300

CHAPTER 6 Complex Adaptive Systems — 307

Core Questions and Key Concepts — 309
Key Terms — 310

Section 6.1	Emergent Feature and Behaviors	312
Section 6.1.1	Sophisticated Properties from Simple Individual Interactions	312
Section 6.1.2	Emergent Properties versus Collective Properties	316
Section 6.2	Interactions between the System and External Conditions	323
Section 6.2.1	Disturbance and Patterns of Fluctuation	324
Section 6.2.2	Disturbance, Internal Dynamics, and Stability Landscapes	332
Section 6.3	Adaptation	362
Section 6.3.1	Biological Evolution by Natural Selection	363
Section 6.3.2	Cultural Evolution by Learning	364

| | Bibliography | 369 |
| | End-of-Chapter Questions | 371 |

CHAPTER 7 Thinking about the Future — 379

Core Questions and Key Concepts — 381
Key Terms — 382

Section 7.1 Scenarios and Future Thinking — 383
- Section 7.1.1 Challenges to Future Predictions about SESs — 385
- Section 7.1.2 Introduction to Future Scenarios — 391
- Section 7.1.3 Scenario Typology for Future Thinking — 394

Section 7.2 Visioning and Future Thinking — 431
- Section 7.2.1 Visions for Sustainability — 432
- Section 7.2.2 The Visioning Process — 437
- Section 7.2.3 Quality Criteria for Visioning — 445

Bibliography — 450
End-of-Chapter Questions — 452

CHAPTER 8 Sustainability Transitions — 461

Core Questions and Key Concepts — 463
Key Terms — 464

Section 8.1 Understanding Transitions — 465
- Section 8.1.1 Niches, Regimes, and Landscapes — 470
- Section 8.1.2 Transition Pathways — 477
- Section 8.1.3 Tying It All Together — 489

Section 8.2 Guiding Sustainability Transitions — 494
- Section 8.2.1 Multi-Phase Concept — 496
- Section 8.2.2 Building a Transition Strategy — 502
- Section 8.2.3 Intervention Points — 509

Bibliography — 521
End-of-Chapter Questions — 522

CHAPTER 9 Governing the Commons — 533

Core Questions and Key Concepts — 535
Key Terms — 536

Section 9.1 Introduction to Tragedy of the Commons — 537
- Section 9.1.1 Excludability and Rivalry — 537
- Section 9.1.2 The Market, the State, or Communities? — 539

	Section 9.2	Characteristics of Successful Common Property Regimes	542
	Section 9.3	Evaluating the Tragedy of the Commons Using Essential Ingredients	566
	Bibliography		570
	End-of-Chapter Questions		572

Glossary   575

Index   585

FOREWORD

Textbooks in sustainability are often patchwork-like compilation of chapters that address different facets of sustainability. Yet, they rarely offer a practical perspective that helps students and professionals integrate and apply sustainability theories and tools. This textbook by Sonya Remington-Doucette is different in several respects: First, it offers a holistic approach to sustainability challenges from a complex systems perspective; second, it combines theoretical perspectives with practical case studies; and third, it aims at building students' capacity to not only analyze sustainability challenges, but also to develop viable solutions to these challenges. What the field of sustainability needs is more guidance on how to distinguish and link concepts, methods, and topical domains. This is one of the few attempts to provide such an orientation; and it does so in a compelling and accessible manner.

Dr. Remington-Doucette deserves credit for reviewing and synthesizing a great deal of material relevant to sustainability students and professionals, ranging from the theory of social-ecological systems to transition management. Readers become familiar with key questions of sustainability, including: How did the idea of sustainability arise in human societies? What are key features of wicked sustainability problems? How do you adopt a systems perspective to analyze them? What is the role of sustainability indicators in this process? How can we think in systematic ways about possible future pathways of problems and solutions? What are the critical ingredients for transitions toward sustainability? All of these questions are addressed from a perspective that integrates material from various disciplines and links qualitative and quantitative approaches.

Dr. Remington-Doucette draws on the concept of transformational sustainability research (TSR) and makes it accessible to students and professionals. My research lab at the School of Sustainability and colleagues from around the world have contributed to the development of this paradigm. The core idea is to enhance our capacities in initiating and supporting transformations toward sustainability that go beyond describing and analyzing sustainability problems. Over the last 10 years, this important shift from problems to solutions has been inaugurated in many fields that deal with societal problems such as social works and public health. It is timely to follow this lead and shift the focus in sustainability efforts, too. The problems—such as the overexploitation of natural resources, violent conflicts, or inequitable access to education and health services—are numerous, complex, and urgent. Simple technical fixes are insufficient; we need instead novel solutions that aim at long-term impacts and account for unintended consequences. Many obstacles in academia and society hinder us from developing and delivering such solutions. A major one is the lack of capacity in advanced sustainability thinking—a particular type of thinking that integrates systems thinking, goal formation, anticipation, and strategy building, favorably undertaken in teams and with public engagement.

This textbook embraces this perspective and prepares students and professionals not only to better understand the challenges of sustainability societies face around the world, based on many illustrative case studies, but also offers tools and concepts to develop viable solutions to these challenges. The book is structured accordingly. It starts with a concise introduction to the history of sustainability thinking and the idea of transformational sustainability research and practice. The remainder of the book walks the reader through the different elements of analyzing and resolving wicked sustainability problems, from current state analysis with informative excursions on indicators, resiliency, and complex adaptive systems, through future thinking to transition management and governance. Each chapter provides a solid overview of the relevant concepts and tools, offers illustrative examples and case studies, and refers the interested reader to relevant literature.

Dr. Remington-Doucette provides here a textbook for students and professionals that addresses and integrates concepts and tools of sustainability with all the breadth and depth necessary for a sound educational experience. Beyond this, I appreciate how she conveys three key messages that support life-long learning offered to everyone who is seriously interested in building his/her sustainability capacity.

Sustainability knowledge requires new ways of thinking and exploration. It needs to continuously grow and evolve (we need more and more solution ideas)—and, maybe surprisingly, it can be generated anywhere, in universities as well as in government units, businesses, and nonprofit organizations. Therefore, we should embrace the practice of sustainability *research*, in the pragmatic meaning of innovative knowledge generation that helps us to resolve wicked sustainability problems. Sustainability needs to bridge the divide between academic research and education on the one hand, and professional practice and training on the other. It is a societal enterprise and needs quality support from all professions and sectors.

Also, research products such as books and journal articles are nothing to be afraid of; they should be considered opportunies for personal growth. Dr. Remington-Doucette does a great job in making these research products accessible. Researchers increasingly try to communicate their findings and ideas more clearly and with less jargon. So, the references provided here are an invitation to continue reading and exploring the different domains of sustainability.

And finally, this book is an enthusiastic call for applying the presented concepts and methods and doing good things in the world. The material synthesized by Dr. Remington-Doucette is not meant to be just learned and used in examinations—it is meant to be used for changing the course of governments, corporations, and our society at large. Don't hesitate—put them to work!

Arnim Wiek
Associate Professor, School of Sustainability,
Arizona State University, Tempe AZ, USA

PREFACE

This textbook's content is based on an undergraduate introductory sustainability course (SOS 110: Sustainable World) offered in the School of Sustainability (SOS) at Arizona State University. It evolved from a need, perceived by me while teaching this class, for a single classroom resource that would provide content for the course. This resource needed to be written in language accessible by the introductory reader and it needed to use clear and consistent terminology. The chapters were reviewed by the following renowned sustainability scientist research faculty, who are experts in the content areas contained in the chapters of this book and who also teach in SOS: Dr. Charles Redman (Chapters 1 and 2), Dr. David Manuel-Navarrete (Chapters 3 and 4), Dr. Sander van der Leeuw (Chapters 5 and 6), Dr. Cynthia Selin (Chapter 7), Dr. Arnim Wiek (Chapter 8), and Dr. Michael Schoon (Chapter 9). These researchers have been engaged in sustainability research for years, or even decades, of their professional lives, as explained in the sustainability profiles included in this textbook. The SOS 110 course is taken by a diverse group of undergraduate students majoring in a wide range of disciplines, including sustainability, business, ecology, architecture, engineering, and many more. Thus, the book is appropriate for any college undergraduate, advanced high school student, or as an introduction to the content areas for any interested reader. It is perfect for an intensive stand-alone course focused on sustainability problem solving, whether inside the classroom through case studies or engaged in actual problem solving in real-world contexts. The book is a reasonable length, such that it can be used as supplemental reading for any course, in any discipline, focused on resolving real-world sustainability problems.

Most textbooks used in environmental science and environmental studies courses, and even those used in courses about sustainability, are focused on analyzing problems in natural and human systems. Some books present solutions options. However, no textbook at present, in a language and format accessible at the introductory level, presents the methodologies that underlie the actual problem solving *process*. How can we actually go about resolving real-world sustainability problems as change-makers or transition managers? This textbook is intended to provide a broad introduction to the knowledge base that will allow you to do just that. From this perspective, sustainability is not only about understanding the *why* of real-world problems through analysis, but it is also about learning *how to* actually resolve those problems.

This book presents a problem-solving approach drawn from ideas in the emerging field of sustainability science. Sustainability science adopts concepts and methods from many different disciplines, including policy studies, planning, business, ecology, sociology, and many others. The framework that ties together much of this textbook, presented in Chapters 2, 3, 4, 7, and 8, is drawn from more than a decade of research carried out by Professor Arnim Wiek and his graduate student team in the Sustainability Transition and Intervention Research Lab in SOS. A good share of his work was carried out in collaboration with Professor Daniel Lang, who is a Professor for Transdisciplinary Sustainability Research at Leuphana University of Lüneburg in Germany. Their framework, which is known as *Transformational Sustainability Research* (*TSR*), is first introduced at the end of Chapter 2 and includes methodologies and tools for analyzing the current state (Chapter 3), developing indicators of sustainability (Chapter 4), thinking about the future using future scenarios and visioning (Chapter 7), and devising transition strategies (Chapter 8).

The word *research*, which is the third word in *TSR*, is by no means restrained to research in the academic sense. Instead, research in the *TSR* context is defined in a general sense as *knowledge generation*. Knowledge generation aimed at resolving sustainability problems through *TSR* is and *must be* carried out by both academic researchers and practitioners if we are to resolve the sustainability challenges of the 21st century. Practitioners include those in government, business, and nonprofit careers, as well as those involved in local grassroots activities. In short, knowledge generation through *TSR* can happen anywhere and, in fact, *is* happening in many places around the world today. Many of these instances involve joining expert academic knowledge with the traditional, local, or on-the-ground knowledge of practitioners through collaborative and participatory activities. Wherever *TSR* occurs, the generation of knowledge through research should follow certain mechanisms and rules for quality control. This is what *TSR* provides for any academic, professional, or grassroots activity aimed at transformational endeavors meant to resolve sustainability problems. This book provides only a very broad introduction to *TSR*. The interested reader requiring additional detail should consult the references provided at the end of each chapter, and the references within those references, and also the 2016 book edited by Harald Heinrichs, Pim Martens, Gerd Michelsen, and Arnim Wiek titled *Sustainability Science: An Introduction*.

In addition to a broad introduction to *TSR* (end of Chapter 2), and an introduction of the methodologies that compose the *TSR* framework (Chapters 3, 4, 7, and 8), the book begins with an introduction to sustainability (Chapter 1) and wicked problems (beginning of Chapter 2) and also delves into the concept of complexity, with specific reference to how this concept relates to the socio-ecological systems (SESs) of concern to sustainability (Chapters 5 and 6). Finally, the book ends with a special focus on *who* should be involved with the process of resolving sustainability problems (Chapter 9). Government and market solutions have been given the lion's share of attention. Thus, much accessible literature can be found as introductions to these two areas and they are not included in this book. With increasing recognition of the importance of the late Dr. Elinor Ostrom's work, for which she won the Nobel Prize in Economics with Dr. Oliver E. Williamson, a third general area from which solutions have been drawn has arisen: resource management activities of local, and often informal, community institutions. Dr. Ostrom's 1990 book *Governing the Commons* provides a good book-length introduction to these ideas. In this textbook, I have attempted to distill these ideas into an accessible format as a broad introduction to Governing the Commons (Chapter 9). Again, the reader interested in more detail should delve into the references provided at the end of Chapter 9. Each chapter contains an opening vignette about *Students Making a Difference*, which tells the story of a former or present student in SOS who has taken action on a real-world sustainability problem.

Each chapter is broken down into Core Questions, which head each section, and Key Concepts at the beginning of each subsection. Key terms are bolded within the text and defined in the margins. Separate boxes are included periodically throughout the chapters to demonstrate the application or meaning of certain concepts. Questions for discussion and concept application are included at the end of each chapter. These End of Chapter questions are broken down into two types—General Questions and Project Questions—with the exception of Chapter 1, which contains only General Questions. The purpose of the General Questions is to allow for concept application and to stimulate discussion based on the specific case studies, or hypothetical scenarios, described in each question. Project Questions are designed for application to a specific case study, which is a sustainability challenge researched by students or other readers of this book, either within the classroom or while engaged in actual problem solving activities or initiatives in the real world. The Project Questions were used in the SOS 110 course, as students engaged in real-world problem solving activities. As such, they have been revised and refined over the past several years with the input of SOS 110 undergraduate students and SOS Graduate Teaching Assistants. The Project Questions are designed to guide you through your own problem solving process for a specific sustainability problem of your choice. To answer these questions, you will need to do your own research and active problem solving based on a real-world issue of concern to you.

From an educational perspective, this book's focus is on contributing to the development of six sustainability competencies: *systems thinking, normative (values thinking), anticipatory (futures thinking), strategic thinking (action-oriented), interpersonal (collaboration),* and *integrated problem-solving*. While sustainability education is focused on competencies deemed important in traditional disciplines, such as critical thinking and problem-solving skills, the sustainability competencies are unique to sustainability education. Initially, five competencies were synthesized from a literature review of competencies in sustainability education conducted by Professor Arnim Wiek, Lauren Withycombe Keeler, and Professor Charles Redman who is the Founding Director of SOS. This work was published in *Sustainability Science* (2011, Volume 6: 203–218). A sixth competency—integrated problem-solving—was added later, as described in an article by Arnim and many others in the 2015 book titled *Handbook of Higher Education for Sustainable Development* (pp. 241–260). The sustainability competencies are meant to foster the knowledge, skills, and attitudes needed by change agents or transition managers to resolve sustainability problems in the real world. They are closely tied to the *TSR* framework.

The information presented in this book is not new. Rather, it was drawn from the innovative ideas and hard work of many academic researchers and practitioners who have spent years or decades focused on resolving sustainability problems. By reading widely, talking to specialists in the topic areas covered in this textbook, and sitting in on several graduate and undergraduate level courses in SOS, I have attempted to compile, synthesize, and present this large body of work in a format accessible to the introductory reader who is new to sustainability science. Sustainability science, or simply sustainability, is variously, and sometimes contentiously, considered a discipline, field of study, or focus area. Whatever its label as a body of knowledge, a huge amount of information drawn from a wide variety of disciplines has contributed to the information presented in this book. I have done my best to credit all of the important thinkers who have contributed to the body of knowledge presented in this textbook. However, due to the large number of contributors and disciplines involved, I have surely missed some. I sincerely apologize for this and welcome comments for improvements to the next edition of this textbook. I view this textbook as a work in progress that will be continually revised and welcome suggestions for revision!

Many theories and concepts presented in this book are very young, active areas of research and continually evolving. How they translate into practice is a big unknown in sustainability.

Theories are generalizations about how the world works or ought to work. However, as will be emphasized many times in this book, the specific context in which a problem is embedded, in terms of place, time, culture, and many other factors, is extremely important to pay attention to when attempting to resolve sustainability problems. As such, the theories and concepts described in this book should be viewed as general places to start. They will need to be refined and altered for each specific situation.

Without the help, collaboration, and generous support of many, the completion of this textbook would have been impossible. I would like to thank the undergraduate students who took my SOS 110 course. During the course, they read and commented on (albeit, for extra credit points!) early and continually evolving draft chapters of this textbook. This book was created with them in mind, so I very much took into account their valuable comments. I would like to thank the forward-thinking SOS Graduate Teaching Assistants who taught SOS 110 alongside of me and contributed valuable insight to the structure and clarity of this textbook during our weekly meetings. In alphabetical order, they are Shirley-Ann Augustin, Edgar Cardenas, Tamsin Connell, Nigel Forrest, Felicia French, Nonso Gbemudu, Auriane Koster, Nelson Mandrell, Genevieve Metson, Michael Moreau, Sheryl Musgrove, Ann Marie Raymondi, Nivedita Regarajan, Benjamin Warner, and Lea Wilson. The six previously mentioned faculty researchers who reviewed the chapters of this book offered valuable insight and suggestions on draft chapters. I very much appreciate that they generously volunteered time from their busy schedules for quality control of the information presented here. Dr. Sander van der Leeuw, the SOS Dean during my time as a Senior Lecturer when I wrote this book, encouraged innovation in education and always enthusiastically supported my many ideas and projects. SOS generously provided a summer stipend to support the writing of this book. They also funded my attendance at several national and international sustainability conferences, which allowed me to bounce ideas for this book off of many sustainability researchers and educators. I would like to also thank Kimberly Grout, the Sustainability Concierge at the Global Institute of Sustainability (GIOS). Her humorous personality and warm friendship made the overwhelming writing of this textbook enjoyable, when I came up for air several times a day to pass by her desk on the way to the break room. In general, I would like to thank all of the amazing students, staff, and faculty in GIOS and SOS.

I would also like to thank those outside of SOS and GIOS. The staff at Kendall Hunt publishing company made the publication of this textbook possible. I would especially like to thank Samantha Smith and Beth Trowbridge for their patience when my deadline for the completion of the second edition was extended several times over the past year. Kendall Hunt also generously agreed to donate 2% of the profits from this textbook to sustainability education. Jean MacGregor and her collaborators at the Curriculum for the Bioregion at the Evergreen State College in Olympia, Washington, provided me with the financial support and intellectual resources to first become involved with sustainability education more than a decade ago. I would like to thank my colleagues at Bellevue College, Washington for their patience and generosity during the writing of the second edition. Finally, I want to thank my family and friends for their encouragement, support, and patience, particularly in the final days of completing this book!

Sonya Remington-Doucette
June 2017

REVIEWER SUSTAINABILITY PROFILES

The reviewers of this textbook are renowned Sustainability Scientist researchers in Arizona State University's (ASU) Global Institute of Sustainability (GIOS) and teachers in the School of Sustainability (SOS). SOS is educating a new generation of leaders through collaborative real-world learning experiences, transdisciplinary approaches, and problem-oriented training to address the environmental, economic, and social challenges of the 21st century. SOS offers BA and BS degrees in Sustainability, as well as graduate degrees.

Dr. Charles Redman, Chapter 1 and 2 Reviewer

Dr. Redman is an Anthropologist and Founding Director of SOS. He has been committed to interdisciplinary research since, as an archaeology graduate student, he worked closely in the field with botanists, zoologists, geologists, art historians, and ethnographers. Dr. Redman received his BA from Harvard University, and his MA and Ph.D. in Anthropology from the University of Chicago. He taught at New York University and at SUNY-Binghamton before coming to ASU in 1983. At ASU, he was Chair of the Department of Anthropology for nine years, Director of the Center for Environmental Studies for seven years, and was chosen to be the Julie Ann Wrigley Director of GIOS in 2004. From 2007–2010, he served as Founding Director of SOS. Dr. Redman's interests include human impacts on the environment, sustainable landscapes, rapidly urbanizing regions, urban ecology, sustainability education, and public outreach. He has authored or co-authored 14 books including *Explanation in Archaeology, The Rise of Civilization, People of the Tonto Rim,* and *Human Impact on Ancient Environments.* Most recently, he co-edited four books: *The Archaeology of Global Change; Applied Remote Sensing for Urban Planning, Governance and Sustainability; Agrarian Landscapes in Transition;* and *Polities and Power: Archaeological Perspectives on the Landscapes of Early States.* Dr. Redman is currently a researcher in GIOS and teaches in SOS.

Dr. David Manuel-Navarrete, Chapter 3 and 4 Reviewer

Dr. Manuel-Navarrete is an Assistant Professor in SOS, where he teaches an undergraduate Capstone course focused on real-world problem solving. Manuel-Navarrete received his BA in Environmental Sciences and MS in Ecological Economics from the Autonomous University of Barcelona and his Ph.D. in Geography from the University of Waterloo. His research focuses on tourism sustainability, climate change adaptation, social inequalities, and socio-ecological boundaries. He combines ethnographic and participatory methods with systems and complexity frameworks to analyze human-environmental interactions. He has worked as a researcher in an international multi-site project on knowledge systems for sustainability funded by NOAA and coordinated by Harvard University. Employment in the United Nations Economic Commission for Latin America and the Caribbean provided him with practical experience in environmental policy, as well as access to national and international decision-making arenas. While at King's College London, Dr. Manuel-Navarrete investigated the role of governance arrangements and development visions in climate change adaptation. He is a member of desiguALdades.net, an interdisciplinary, international, and multi-institutional research network on social inequalities in Latin America based at the Free University of Berlin. Dr. Manuel-Navarrete has published numerous books, chapters, and journal articles about sustainable food systems, adaptation to climate change, ecological integrity and more.

Dr. Sander van der Leeuw, Chapter 5 and 6 Reviewer

Dr. van der Leeuw is the 2012 United Nations Champion of the Earth for Science and Innovation and Co-Director of ASU's Complex Adaptive Systems Initiative. He served as Dean of SOS from 2010 to 2013. Dr. van der Leeuw received his BA in History, ADM/MLitt in Medieval History/Prehistory, and Ph.D. at the University of Amsterdam. His expertise lies in the role of invention, sustainability, and innovation in societies around the world. He and his research team investigate how invention occurs, what the preconditions are, how the context influences it, and its role in society. An archaeologist and historian by training, Dr. van der Leeuw has studied ancient technologies, ancient and modern man-land relationships, and Complex Systems Theory. He has done archaeological fieldwork in Syria, Holland, and France, and conducted ethno-archaeological studies in the Near East, the Philippines, and Mexico. Since 1992 he has coordinated a series of interdisciplinary research projects on socio-natural interactions and modern environmental problems. The work spans all the countries along the northern Mediterranean rim. Dr. van der Leeuw is an External Professor of the Santa Fe Institute and a Corresponding Member of the Royal Dutch Academy of Sciences. He teaches courses on the ethnography of innovation.

Dr. Cynthia Selin, Chapter 7 Reviewer

Dr. Selin is an Assistant Professor in SOS and the Consortium for Science, Policy and Outcomes. She teaches the upper-division undergraduate course *Future Thinking and Strategies*, which offers 5 students hands-on experience with developing future scenarios focused on real-world sustainability problems. Selin graduated with a BS in American Studies from University of California-Santa Cruz in 1996, where she wrote her senior thesis on the promise and peril of nanotechnology. In 2000, she completed a MA in Science, Technology, and Society from Roskilde University in Denmark, where her research focused on technology assessment, and, more specifically, scenario planning. She further pursued these interests at Copenhagen Business Schools' Institute for Management, Politics, and Philosophy for her doctoral project, completed in 2006. The Ph.D. dissertation, entitled *Volatile Visions: Transactions in Anticipatory Knowledge* explores three interwoven research areas—foresight methodologies, the sociology of expectations, and the emergence of nanotechnology—to understand the development of new technologies and to explore the tools and methods used to grasp their emergence. Dr. Selin has published in *Science, Technology and Human Values, Futures,* and *Time and Society*.

Dr. Arnim Wiek, Chapter 8 Reviewer

Dr. Wiek is an Associate Professor in the School of Sustainability, where he teaches the upper-division undergraduate course *Sustainability and Society* as well as the graduate course *Transformational Sustainability Research*. He is the director of the *Sustainability Transition and Intervention Research Lab* and an international leader in the emerging field of Sustainability Science. Dr. Wiek received an MA in Philosophy at the Free University in Berlin, an MS in Environmental Sciences at the University of Jena in Germany, and a Ph.D. at the Swiss Federal Institute of Technology in Zurich. He had research and teaching engagements at the Swiss Federal Institute of Technology Zurich, the University of British Columbia, Vancouver, and the University of Tokyo. Dr. Wiek has developed the *transformational sustainability research framework* and has published (together with Lauren Withycombe Keeler and Charles Redman) a widely cited article on *key competencies in sustainability* as a reference framework for the design of sustainability curricula and courses. Both concepts serve as a basis for this textbook. Dr. Wiek's research group has been involved in solution-oriented sustainability research on emerging technologies, urban development, resource governance, climate change, and public health in the United States, Canada, different European countries, Sri Lanka, Mexico, and Costa Rica. The group carries out all research in close collaboration with governmental agencies, businesses, nonprofit organizations, and community groups.

Dr. Michael Schoon, Chapter 9 Reviewer

Dr. Schoon is an Assistant Professor in SOS, where he teaches an undergraduate Capstone course focused on sustainability problem solving in the Metro Phoenix Area. He earned two BS degrees (Mechanical and Aerospace Engineering) from the University of Arizona, and a MBA and Ph.D. in Environmental Policy, Public Policy, and International Relations from Indiana University. Schoon's research focus is policy and governance in sustainable systems. His dissertation work occurred at Indiana University's (IU) Ostrom Workshop in Political Theory and Policy Analysis, a research center headed by the late Noble Prize winner Dr. Elinor Ostrom. His focus on transboundary protected areas (Peace Parks) in southern Africa won the American Political Science Association's best dissertation award. He joined the Center for the Study of Institutional Diversity at ASU to continue research on collaboration across borders of all types, ranging from those between landowners to nation-states. His research on collaborative, cross-border institutional arrangements covers a range of environmental issues, including biodiversity conservation, water sharing, and fire management, and combines multiple methodological approaches. Dr. Schoon is active in international research communities through the Resilience Alliance, the Beijer Institute of Ecological Economics, and ASU's Complex Adaptive Systems group. He serves on the board for IUCN's Transboundary Conservation Specialist Group and as co-Editor-in-Chief of the *International Journal of the Commons*.

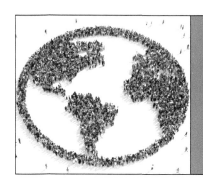

Chapter 1
INTRODUCTION TO SUSTAINABILITY

Students Making a Difference

Creating a Just Transition to Sustainability Through Activism

Sarra Tekola is a graduate student working on her Ph.D. She is also an activist fighting for social justice and against climate change. She works to educate people about these issues, and participates in protests and acts of civil disobedience to bring attention to them. She was once awoken at 4:00 a.m. by SWAT team police officers holding a gun to her face. When asked why she takes such risks, she responds by saying that is it her responsibility as a Ph.D. student in America: "My father is from Ethiopia.

He left because of political disorder that happened after the famine. I'm sitting in a position of privilege in America, even having less privilege being a black person in America, and it is due to that privilege that I feel a duty to use it and risk arrest, because in Ethiopia you can't risk arrest." In Ethiopia, as in many other countries around the world, the penalties that come with arrest can be much steeper, such as severe physical abuse and other inhumane acts. Sarra has never been arrested, but she feels that risking arrest through civil disobedience is an American privilege that should be used to advance sustainability.

In 2016, Sarra camped in a tent on the train tracks leading to two oil refineries near Anacortes, WA. These refineries are the largest point sources of carbon pollution in Washington state and also where workers have died due to unsafe working conditions. This act of civil disobedience was coordinated by Break Free, an organization that supports bold actions to break free from fossil fuel use. The aim was to stop trains, carrying crude oil, from reaching the refineries where the oil is processed. Sarra and her fellow activists from Break Free occupied the train tracks for 3 days, stopping about 600 trains from reaching the refineries, until SWAT teams arrived to remove them.

Sarra, along with the members of Break Free and many other sustainability organizations, wants a just transition away from fossil fuels. In a just transition, it is important to prioritize the needs of people who could be harmed by a transition. In the case of Anacortes, this meant caring for the workers at the refinery and the surrounding communities. The end goal, if fossil fuels are to be kept in the ground, is to close down the refinery. However, if done carelessly and with haste, this would hurt oil refinery workers, who would lose their jobs, and the overall local economy of the surrounding communities. Among those who could be harmed, many are from communities of color and other marginalized groups. Sarra and the members of Break Free who were involved in the Anacortes protest followed these principles by involving local communities and oil refinery workers during the planning phase that led up to their occupation of the train tracks. In this chapter, you will learn about the birth of the sustainability movement, and how sustainability is just as much about social justice and economic vitality as it is about environmental health.

Core Questions and Key Concepts

Section 1.1: People and Nature, Today and throughout Human History

Core Question: How have humans transformed natural systems today and in the past?

Key Concept 1.1.1—The five central natural system transformations caused by present-day human activities are habitat degradation, climate change, invasive species, biogeochemical cycle alterations, and overexploitation.

Key Concept 1.1.2—Contrary to some beliefs, humans did noticeably transform natural systems in the past in many ways but at smaller scales and slower rates than today.

Section 1.2: Understanding Human Transformations Using the I = PAT Model

Core Question: How can present and past human transformations of natural systems be understood?

Key Concept 1.2.1—Differences in human population growth (P), affluence (A), and technology (T) explain why developed countries have a greater impact (I) on natural systems than developing countries.

Key Concept 1.2.2—Differences in human population growth (P), affluence (A), and technology (T) explain why overall human impacts (I) on natural systems are greater today than in the past.

Section 1.3: Natural Systems and Human Well-Being

Core Question: How do changes in natural systems affect human well-being?

Key Concept 1.3.1—The human activities that have transformed natural systems have provided society with many benefits, but at the same time have degraded the free services that ecosystems provide to humans.

Key Concept 1.3.2—In sustainability, the vitality-integrity framework is used to define human well-being.

Key Concept 1.3.3—The five major aspects of human well-being affected by natural system degradation are basic material needs, health, safety, good social relations, and freedom of choice and action.

Section 1.4: Origins of Sustainability

Core Question: How did the idea of sustainability arise in human societies?

Key Concept 1.4.1—Increasing environmental degradation led to concern about loss of natural capital, which supports prosperous human societies.

Key Concept 1.4.2—The idea of human progress and increasing poverty in underdeveloped regions around the world highlighted the need for economic development to ensure human well-being and advancement.

Key Concept 1.4.3—Growing inequity due to increasingly scarce resources and uneven economic growth led to concern about equity among people living today and in the future.

Key Concept 1.4.4—Increasing global concern for ecosystems, economies, and people led to publications and international meetings that defined sustainability as we know it today.

Key Terms

- socioecological systems (SESs)
- Millennium Ecosystem Assessment (MEA)
- Intergovernmental Panel on Climate Change (IPCC)
- ecosystem services
- adaptation
- mitigation
- transpiration
- eutrophication
- troposphere
- megafauna
- I = PAT model
- per capita
- Gross Domestic Product (GDP)
- vitality
- integrity
- stakeholders
- livelihood
- social capital
- natural capital
- economic development
- social equity
- intragenerational equity
- intergenerational equity
- vulnerability science
- change agents

> "Sustainable development is development that meets the needs of the present without compromising the ability of future generations to meet their own needs."
>
> —*Brundtland Report, World Commission on Environment and Development, 1987*

Sustainability is fundamentally about preserving the conditions necessary for the survival of prosperous human societies. A common misunderstanding is that sustainability is about environmental conservation only. However, the environment will continue to exist if our human societies disappear. Thus, sustainability includes but goes far beyond conservation of natural systems. Sustainability involves the complex and interconnected natural and human systems, often called **socioecological systems (SESs)**, which comprise our world today. The goal of sustainability is to simultaneously promote healthy ecosystems (environment), human well-being (society), and viable economies (economy) (Figure 1.1). Each of these three pillars of sustainability is equally important. If only one or two pillars are promoted and others are disregarded, then the outcome will not be sustainable. Sustainability addresses the compromises and tradeoffs among the three pillars, which are often necessary for resolving real-world sustainability challenges. Creating a sustainable world requires a mix of many things, including gaining accurate knowledge about natural and human systems; altering individual behavior and consumption patterns; engaging values and changing social norms; promoting technological advances; and fostering the social, political, and economic institutions necessary to guide the actions of individuals and societies toward sustainable behavior.

socioecological systems (SESs) an integrated conceptual model for understanding how energy, matter, and information are transferred among the different components of natural and human systems and how these exchanges impact the long-term development of both systems.

To help you understand how the concept of sustainability came to be, this chapter includes a brief introduction to several of the challenges faced by human societies of the past and present. These challenges are generally grouped into two categories. First, *natural systems are becoming increasingly degraded* due to excessive resource extraction and pollution. Although human societies are shielded from this degradation to varying degrees by technology and culture, natural systems provide us with vital life support services such as clean air to breathe and water to drink, food to eat, and a stable climate in which to live. The survival of human societies, and of other species, ultimately depends on the health of natural systems. Second, there is *growing inequity in human systems*. The unequal distribution of wealth and power intensifies with each passing year in societies around the world. As a result, some people receive an unfair share of the resources provided by natural systems and also bear an unfair burden of the pollution produced by others. This creates problems, such as poverty and social conflict, which lead to unstable human systems. Both the root causes and many of the solutions

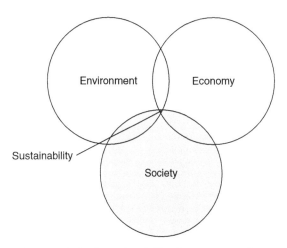

Figure 1.1 **Sustainability is concerned with society and economy, as well as the environment.**

to these two major challenges lie within our economic systems. If economic growth continues unchecked under status quo conditions, the integrity of natural and human systems will continue to be compromised. On the other hand, economic development that improves the quality of life for everyone is a major pathway for lessening growing inequity and preserving natural systems. Finding the right balance is a major sustainability challenge.

Section 1.1: People and Nature, Today and throughout Human History

Core Question: How have humans transformed natural systems today and in the past?

> **Millennium Ecosystem Assessment (MEA)** a five-year project (2001–2005) based on the work of more than 1,300 international experts that assessed the present conditions and future trends of ecosystem services worldwide, the effects of changes in the conditions of these services on human well-being, and the options for action that would promote the conservation and sustainable use of these services.

> **Intergovernmental Panel on Climate Change (IPCC)** a scientific body established in 1988 by the World Meteorological Organization and the United Nations Environmental Programme to regularly review the most recent work of thousands of climate scientists from 195 countries around the world and assess the potential environmental, social, and economic impacts of climate change based on this review in order to provide the most up-to-date scientific information for decision makers.

We live today in human-dominated ecosystems. Over the past 60 years, ecosystems around the world have been more significantly transformed by human action than at any other time in human history. This section begins with a brief overview of ecosystem transformations caused by human activities, including habitat degradation, anthropogenic (human-caused) climate change, introduction of invasive species, pollution by disruption of biogeochemical cycles, and overexploitation of species through hunting, fishing, logging, and other extraction activities. Much of the information presented in this section is drawn from two recent reports called the **Millennium Ecosystem Assessment (MEA)** and the **Intergovernmental Panel on Climate Change (IPCC)** assessment. Both reports were written by more than 1,000 experts from around the world and present the current changes to natural systems that have resulted from human actions, project further changes in these systems under different future scenarios, and evaluate the consequence of these changes for human well-being. The MEA assesses the consequences of ecosystem change and biodiversity loss for human well-being and provides a scientific basis for actions needed for sustainable use of ecosystems. It is focused specifically on **ecosystem services**, which are defined as the benefits provided to humans for free by ecosystems that make life both possible and worth living. These services include clean air, climate regulation, and flood mitigation as well as aesthetic, spiritual, and recreational uses of ecosystems. The IPCC assessment report is specifically focused on climate change, its causes and current effects, and future projections of the impact of further changes on natural and human systems. Specifically, the report focuses on **adaptation** and **mitigation** options for different sectors such as water, agriculture, forestry, tourism, infrastructure and human settlements, waste management, human health, transportation, and energy. Following an overview of human transformations of natural systems, a short account of the interactions between humans and natural systems over the course of human history is provided.

Section 1.1.1—Recent Human Transformation of Natural Systems

Key Concept 1.1.1—The five central natural system transformations caused by present-day human activities are habitat degradation, climate change, invasive species, biogeochemical cycle alterations, and overexploitation.

Habitat Degradation. Habitat degradation occurs in all ecosystems from terrestrial and fresh water to coastal and open oceans. In terrestrial ecosystems, the single largest change has been habitat degradation, ranging from fragmentation to complete destruction. About 40% of all terrestrial habitats on Earth, from tropical forests to the Arctic tundra, have been converted from their natural states (Figure 1.2). Twenty-five percent of all terrestrial habitat has been converted to agricultural land for crop production and raising livestock (Figure 1.3). Only regions with land not fit for agriculture remain mostly unchanged, such as deserts, tundra, and boreal forests. By 2050, almost 50% of all terrestrial habitat may be converted. Fresh water habitats have been largely transformed through dam construction and withdrawal of water for human uses. In fact, three to six times more water is stored in reservoirs today than flows in natural river channels. Sixty percent of the large river systems of the world are affected, including the Colorado River in North America, the Yellow River in Asia, and the Nile in Africa. Some of these large rivers no longer flow into the ocean. Most of this water (70%) is used for agriculture. Modification of fresh water flow is one of the three major human activities threatening fresh water systems. The other two (invasive species introductions and excessive nutrient inputs) are discussed below. Major areas of habitat in the ocean have also been lost or severely degraded worldwide, including at least 35% of mangrove swamps and 40% of coral reefs. The greatest threat to these coastal areas is habitat conversion due to activities such as industrialization, aquaculture, port development, tourist resorts, and urban sprawl. Today, the most rapid habitat degradation is occurring in developing nations. In temperate and boreal habitats of developed nations,

---

ecosystem services the benefits afforded by ecosystems to human systems that make human life both possible and worth living, including goods such as food, fiber, fuel, and fresh water and services such as soil formation, nutrient cycling, crop pollination, climate regulation, and spiritual values and inspiration.

---

adaptation actions aimed at adjusting human or natural systems to minimize the harmful impacts of hazards, such as by changing agricultural practices in ways that will allow food production systems to continue to function in a different climate.

---

mitigation actions intended to permanently eliminate or lessen long-term hazards or risks to human systems, such as by ceasing or reducing anthropogenic CO_2 emissions contributing to climate change.

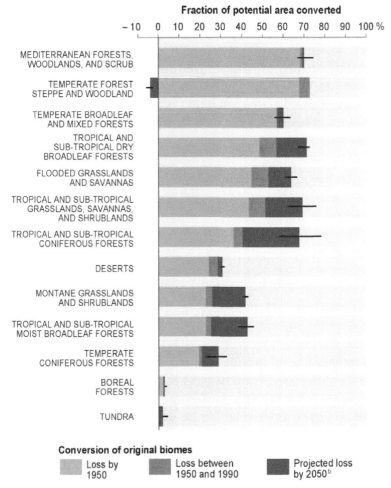

Figure 1.2 Different terrestrial habitats have been altered from their natural state.

habitat degradation has slowed from previous rates or has even been reversed, allowing forests to grow back.

Habitat fragmentation is a more subtle form of habitat degradation, compared to complete destruction of habitat (Figure 1.4a and 1.4b). When habitat is fragmented, pieces of the original habitat remain, such as the fragments of forest shown in Figure 1.4a. All animals need to move around to find food, habitat, and suitable mates for reproduction, or for seasonal migration to more habitable climates. However, when the pieces

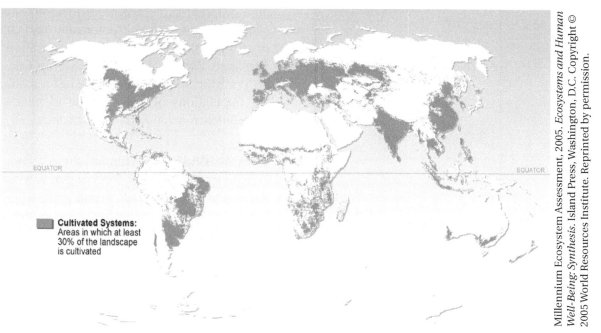

Figure 1.3 A large amount of terrestrial habitat has been converted to agricultural land worldwide.

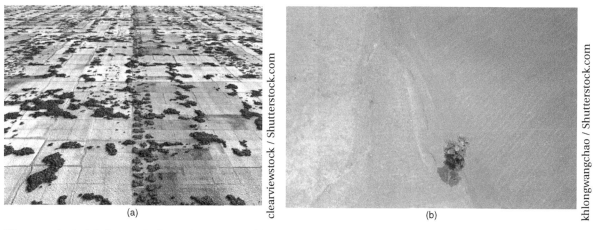

Figure 1.4 (a) Forest habitat fragmented by farmland; (b) Only one tree remains in a completed destroyed forest in Thailand.

of original habitat are too small, there is not enough room for animals to carry out these life-sustaining activities, and the population of organisms dependent on the fragmented habitat will likely not survive. One solution to habitat fragmentation may be to connect pieces of habitat using wildlife corridors, such as those shown in Figure 1.5a and 1.5b. Wildlife corridors are being used around the world to help animals move around more freely through human-built environments, including farmland, urban areas, suburban housing developments, and major highways and rail lines. Endangered tigers, elephants, and rhinoceros use corridors connecting nine different protected areas in India and Nepal to move through deforested habitat. In Botswana, the nonprofit organization, *Elephants Without Borders,* has set up wildlife corridors to guide one of the last remaining populations of free-roaming elephants through encroaching urban landscapes.

Climate Change. Humans are affecting climate stability. This is occurring mostly by disruption of the global carbon cycle when human activities add excess carbon dioxide (CO_2) and methane to the atmosphere, but other greenhouses gases such as nitrous oxide are also increasing and contributing to climate change (Figure 1.6). Of all the greenhouse gases emitted into the atmosphere from human activities, carbon dioxide is currently the major contributor to human-caused climate change. At the beginning of the Industrial Revolution in 1750, the concentration of CO_2 in the atmosphere was 280 ppm. Today concentrations are over 400 ppm, with the major sources of CO_2 into the atmosphere being fossil fuel burning and deforestation. According to many climate scientists, the safe upper limit for CO_2 concentrations is 450 ppm, and many advocate for an even lower safe limit of 350 ppm. Human-caused climate change is projected to result in rising sea levels, increased intensity and frequency of drought, flooding, and wildfires, and a generally less predictable climate. A major concern with changes in climate is that the agricultural systems on which humans depend will break down under these conditions. Ever since the establishment of agriculture about 10,000 years ago, when the climate warmed and stabilized after the last ice age, human societies have enjoyed a relatively stable climate with a global mean temperature of about 15°C (59°F). For the past 10,000 years, these temperatures have fluctuated by only about 0.5°C (0.9°F) or less around this mean. Over the last century alone, global temperatures have risen by about 0.7°C (1.3°F). According to future climate scenarios developed by the Intergovernmental Panel on Climate Change in its 2000 *Special Report Emissions Scenarios* (IPCC-SRES), global mean temperatures are expected to increase by 2.0 to 6.4°C (3.6 to 11.5°F) above preindustrial levels by 2100. How human societies

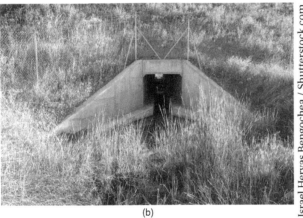

Figure 1.5 (a) This highway overpass in Banff National Park (AB, Canada) allows wildlife to cross highways and move safely across the mountainous landscape; (b) This wildlife passage allows wetland species to move underneath a high-speed railway in Leon Province, Spain.

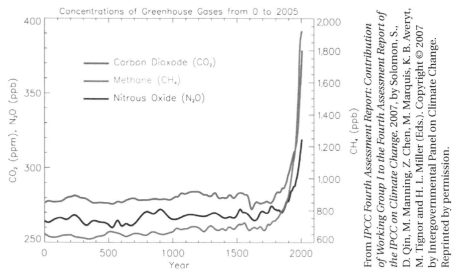

Figure 1.6 Carbon dioxide is one of many green house gases that is causing global climate change.

will adapt to these new conditions is uncertain and of high concern.

Solutions to the climate crisis fall into two categories: mitigation and adaptation. Attempts at mitigation range from new technologies such as carbon capture and sequestration (CCS), to economic solutions such as carbon cap-and-trade programs, to international agreements to reduce greenhouse gas emissions such as the most recent 2015 Paris Agreement. All of these mitigation efforts, and many more, are aimed at eliminating or lessening the severity of climate change. However, climate change is already well underway and, even if greenhouse gas emissions stopped today, climate change would continue for many years. Therefore, mitigation alone is not enough. Figuring out ways to adapt to our new global climate is necessary and the economic costs of this can be high. In general, developing countries will experience the most harm as the climate changes, due to limited resources and, therefore, a lower capacity for adaptation. Despite these limitations, creative low-cost strategies for adaptation are being implemented in developing countries around the world.

Figure 1.7 People living on the shores of the Tonle Sap Lake in Cambodia build their homes on stilts to protect them from monsoon season flooding, during which the lake rises by several meters.

In the coastal city of Da Nang, Vietnam, homes are being retrofitted with stilts to adapt to sea level rise and flooding. Building houses on stilts is a technology adopted from regions that experience natural annual flooding, such as in villages along the Amazon River in Brazil and Tonle Sap Lake in Cambodia (Figure 1.7). In Lagos, Nigeria, another coastal

Figure 1.8 A rural Asian farmer talks on his smartphone while tending water buffalo.

city, instead of stilts floating homes are being built with discarded barrels made of plastic. Smaller rural farmers are another group heavily impacted by climate change. Heavy rain and severe flooding attributed to climate change can destroy farms and homes in these regions. If farmers had advanced warning, then they could prepare their homes and farms for potential flooding, but such remote regions lack basic communication services, such as radio, television, and internet. The use of cell phones, however, is widespread (Figure 1.8) and rapid alert networks are being set up to notify farmers by this means. In other regions of the world, a combination of drought and flooding can reduce crop yields. One adaptation is the use of a simple greenhouse, which allows farmers to grow crops year-round on flat lands in less time, rather than only seasonally on flood- and erosion-prone terraces, and crops that require less water.

Invasive Species. The introduction of invasive species to ecosystems around the world has resulted in a more uniform distribution of species, extinction of native species, and overall biodiversity loss. Although humans have been moving plants and animals into nonnative habitats for millennia, over the past 50 years more efficient transportation systems and international trade have resulted in greater rates of invasive species introduction than ever before. Invasive species introductions are not always harmful, but when they are, they result in damage to economies and ecosystems.

Figure 1.9 A woman in Bor, South Sudan prepares cassava, an important staple of the African diet.

In addition to biodiversity loss, invasive species threaten food security in developing countries. For example, in rural regions of Africa, agriculture supports 80% of a largely undernourished and impoverished population. Cassava is one of the most important crops in Africa (Figure 1.9), but invasive species, such as the cassava mealybug and the grain borer, have caused dramatic reductions in cassava crop yields, which directly affect the survival and livelihoods of millions of people. Ironically, international aid organizations have introduced some invasive species to Africa. For example, during the Ethiopian drought of the 1980s, wheat grains were imported as food aid. However, the wheat is thought to have been contaminated by seeds of congress weed, an invasive species that now dominates Ethiopian pastures but that livestock do not like to eat, resulting in grazing shortages. Owing to resource limitations, invasive species are incredibly difficult to combat and eradicate in developing countries.

Where resources are available, solutions to invasive species problems are being implemented. For example, at the turn of the twenty-first century, the invasive lionfish began to appear in warm water reefs of the Atlantic Ocean, the Caribbean Ocean, and the Gulf of Mexico (Figure 1.10). Scientists suspect that people released these exotic tropical fish into the ocean when they no longer wanted them as pets in their home aquariums. Lionfish do not have many natural predators in Atlantic Ocean regions, due to their poisonous spines, which harm potential predators and humans alike. In addition, they reproduce quickly and are voracious predators, having stomachs that can expand up to 30 times following a meal. The lionfish will likely damage local economies, as they harm the coral reef ecosystems desirable for tourism. They also prey on economically important commercial fish species, such as grouper and snapper. Despite the lionfish's destructive capabilities, there is hope. Reef Environmental Education Foundation (REEF) is a Florida nonprofit organization compiling a database of lionfish sightings, and working to increase public education and awareness through research trips and workshops. Perhaps the most exciting event held by REEF is an annual Lionfish Derby, where SCUBA divers compete to catch the most lionfish in one day (Figure 1.11). Between 2009 and 2015, 16,134 lionfish have been successfully removed from the ocean as a result of the derby. What happens to the lionfish caught during the derby? First, scientists collect data to learn more about the fish. Then, lionfish are filleted and cooked for a free public tasting. REEF

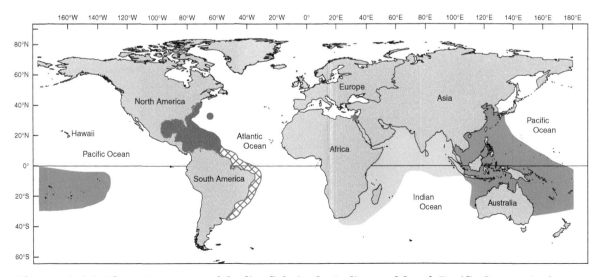

Figure 1.10 The native range of the lionfish, in the Indian and South Pacific Oceans, is shown in blue. The regions of the Atlantic Ocean recently invaded by the lionfish are shown in red.

14 *Sustainable World: Approaches to Analyzing and Resolving Wicked Problems*

Figure 1.11 Lionfish are highly prized by SCUBA divers and other underwater enthusiasts.

hopes this will help to create a market for lionfish cuisine. *The Lionfish Cookbook* by Tricia Ferguson and Lad Akins, published in 2016, is an attempt to turn a destructive pest into a gastronomic delicacy.

Biogeochemical Cycle Alteration. Biogeochemical cycles have been altered on regional to global scales. These cycles are the conduits by which chemical elements (such as carbon, nitrogen, phosphorus, hydrogen, oxygen, and sulfur) move through the biosphere, lithosphere, atmosphere, and hydrosphere over time. They are necessary for maintenance of Earth's natural systems and have been disrupted by humans in many ways. Hydrogen and oxygen travel through natural systems together in the form of water molecules (H_2O) by way of the hydrologic cycle (**Figure 1.12**). The other major biogeochemical cycles—carbon, sulfur, phosphorus, and nitrogen—are tightly linked to the hydrologic cycle because these chemical elements dissolve in river, lake, and ocean water or are washed out of the atmosphere by rain. Water also makes up about 60% of the human body and, in some organisms, as much as 90%.

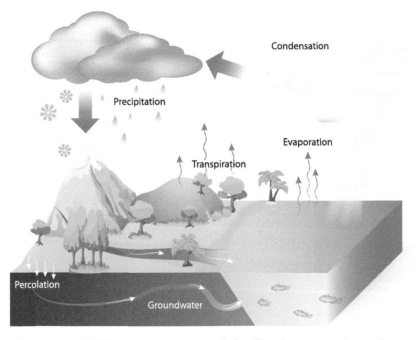

Figure 1.12 Water moves around the Earth system through many physical processes, which include evaporation, precipitation, percolation, and many others.

Human impacts on the hydrologic cycle have major consequences for both water quantity and quality. In addition to changes in the quantity of water flowing in rivers, due to dam construction and withdrawal for agriculture as discussed above, there have been major impacts on groundwater and on water in the atmosphere. Humans are using supplies of groundwater faster than they can be replenished by natural processes, which are mainly infiltration and percolation through soils and bedrock. Humans are also decreasing replenishment by increasing the area of impermeable surfaces on land, including compacted soils and concrete parking lots, through which water cannot flow back into groundwater basins. River water stored in reservoirs and groundwater pumped to the Earth's surface have increased the quantity of water vapor in the atmosphere. This water would not enter the atmosphere in large quantities when flowing in rivers or stored underground, but becomes available for evaporation when stored or pumped. Deforestation has the opposite effect of reducing atmospheric water vapor because trees return water back to the atmosphere through a process called **transpiration**. When trees are removed, less of this occurs. Changes in the amount of water vapor can affect local weather patterns and contribute to longer-term climate change through feedbacks. Water quality changes are also a concern and are often coupled with other biogeochemical cycles, such as nitrogen, phosphorus, and sulfur. However, water quality can also be degraded when synthetic chemicals, such as pesticides and pharmaceuticals, enter drinking water supplies.

transpiration the process by which plants absorb water in liquid form, typically through their roots, and release it into the atmosphere as water vapor, primarily through openings on their leaves called stomata.

One consequence of human transformation of the carbon cycle—climate change—was discussed above, but ocean acidification is equally threatening to human and natural systems. Not as well-known as climate change, ocean acidification is often referred to as "The Other CO_2 Problem." When CO_2 dissolves in ocean water, carbonic acid forms. The more acidic water causes problems for organisms that build shells or other body parts using the carbonate ion found in seawater. This is analogous to a carbonate-based antacid tablet dissolving in your acidic stomach to alleviate heartburn. Like the antacid tablet, carbonate-based shells of organisms will become damaged in acidic ocean water (**Figure 1.13**). These organisms include many species in coral reefs, which harbor more than 25% of the ocean's biodiversity and support many economic activities ranging from commercial fishing to tourism. Since the time that humans began releasing CO_2 into the atmosphere, the global ocean has experienced a 30% increase in acidity, on average.

Figure 1.13 The corrosive seawater resulting from ocean acidification causes pitting and other damage to shells of ocean creatures, such as on this sea snail shell.

Alterations of the sulfur cycle have also made ecosystems more acidic, but for a different reason. Sulfur is emitted naturally into the atmosphere from volcanoes as sulfur dioxide gas, but humans have been releasing much larger quantities of sulfur dioxide from coal-fired power plants. This has resulted in air pollution and acid rain, harming both human health and ecosystems.

Oyster farming is responsible for millions of jobs worldwide, especially in rural communities, and contributes millions of dollars to local economies (Figure 1.14). Like other shelled creatures in the ocean, the carbonate shells of oysters are harmed by ocean acidification. In the shellfish industry, the problem is not for adult oysters on farms, but for the tiny microscopic oyster larvae raised in hatcheries, which supply the farms with oysters. With the help of scientists and funding from local governments, hatcheries have devised ways to adapt to more acidic oceans. At the Whiskey Creek hatchery along the western coast of the United States, scientists helped hatchery owners set up monitoring programs to identify the times at which ocean water would not harm oyster larvae. With this knowledge, the owners only pull water into the hatchery during those times, or add neutralizing chemicals to the water when it is too acidic. As a result, hatchery production improved. Although this is a short-term solution, as the only real solution is reducing global CO_2 emissions, the model established at Whiskey Creek is spreading to other hatcheries, ranging from the eastern coast of the United States to the English Channel.

Phosphorus is a nutrient present in fertilizers, animal manure and human sewage, and mining waste. Flows of phosphorus to fresh water and coastal ecosystems have tripled over the past 50 years due to human activities. As a result, low oxygen "dead zones" where fish and other aquatic organisms cannot survive have appeared. Dead zones are created by a process called **eutrophication**, which occurs when rates of primary productivity by aquatic organisms (such as algae) increase due to excess nutrient inputs (Figure 1.15). This additional productivity creates extra biomass, which is eaten by aquatic primary consumer organisms. Consumers use oxygen for respiration when they eat biomass. The higher respiration rates of consumers, resulting from the extra biomass, deplete oxygen concentrations in the water.

Of all the biogeochemical cycles, the nitrogen cycle has been most intensively transformed by humans. It is also the most complex cycle to understand. Like phosphorus, nitrogen is

> **eutrophication** the process by which a body of fresh water or saltwater, such as a lake or coastal area, receives excessive quantities of nutrients that lead to algae blooms and low oxygen conditions.

Figure 1.14 **Shellfish farmers working at low-tide in the English Channel off the northern coast of France.**

also a nutrient and excessive nitrogen inputs to fresh water and coastal ecosystems cause eutrophication. Unlike phosphorus, which has its sources on land, nitrogen comes from a gas in the atmosphere. In natural ecosystems, nitrogen-fixing bacteria perform the special function of converting nitrogen gas (N_2) from the atmosphere into types of nitrogen that can be used by all other organisms in an ecosystem, such as ammonia and nitrate (Figure 1.16). Humans developed the industrial Haber-Bosch process to make the same chemical conversion for the production of synthetic nitrogen fertilizers. As a result of this process, which began in the early 20th century, humans have doubled the rate of natural nitrogen fixation. In addition to causing eutrophication when excess nitrogen fertilizers run off into water bodies, overuse of synthetic nitrogen fertilizers also results in the release of extra nitrous oxide gas (N_2O) into the atmosphere. N_2O is a greenhouse gas and the third largest contributor to climate change after CO_2 and methane. Humans have also transformed the nitrogen cycle by burning fossil fuels, which release a suite of nitrogen gases (often denoted as NO_X) into the atmosphere (Figure 1.16). Like SO_2, NO_X gases combine with water in the atmosphere to produce acid rain. These gases also react with hydrocarbons in the **troposphere**, in the presence of sunlight, to form photochemical smog or ozone (O_3). This tropospheric O_3 causes human health problems, including impaired respiratory function and increased incidence of asthma and allergies. Higher up in the next layer of the atmosphere (the stratosphere), nitrogen gases destroy the beneficial O_3 that blocks harmful, cancer-causing UV radiation from reaching the Earth's surface.

Figure 1.15 Excess nutrient inputs to water bodies cause algae and other photosynthetic organisms to produce excess biomass, which appears as a green scum or film on lakes, rivers, and coastal waters.

troposphere the layer of the atmosphere closest to the Earth's surface that extends from 4 miles elevation at the poles to 11 miles at the equator and is characterized by temperature decreases with increases in altitude of about 6.5°C per kilometer.

Overexploitation. Overexploitation of any species leads to a reduction in the number of individuals in a population of that species and can eventually lead to complete species extinction. Extinction is something that happens naturally. In fact, over 99% of all species that have ever lived on Earth are extinct. The problem today is that extinction rates are up to 1,000 times higher than natural extinction rates and, if current trends continue, they could increase another tenfold in the future (Figure 1.17). On land, 23% of mammals, 12% of birds, and 32% of amphibians are threatened with extinction. Species living in fresh water are even more at risk than those on land, although exact percentages are

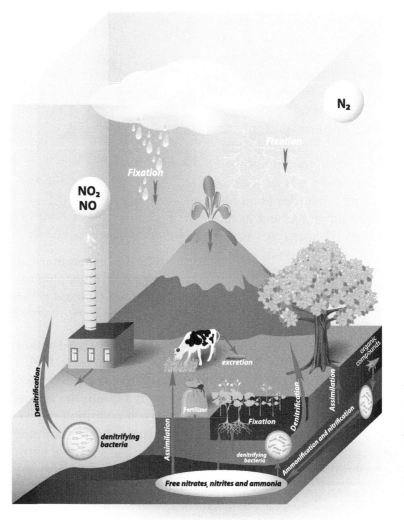

Figure 1.16 The transformation of the nitrogen cycle by human activities has contributed to eutrophication, global climate change, acid rain, and ozone destruction.

not as well known. Other human impacts, such as habitat destruction and eutrophication, harm species on land and in fresh water. However, in ocean ecosystems, the single greatest threat is overexploitation of species through fishing. Fishing pressure has increased with the rise of industrial fishing over the past 50 years. Over that time period alone, the total biomass in the world's oceans (especially of large commercially fished species) has been reduced by 10%. More than 25% of commercial fish populations are harvested unsustainably and about 50% are fished at the maximum sustainable level. Fishing has disproportionately affected coastal areas in the past, but today it is also affecting the open ocean. Much overfishing has been driven by advances in fishing technologies (**Figure 1.18**).

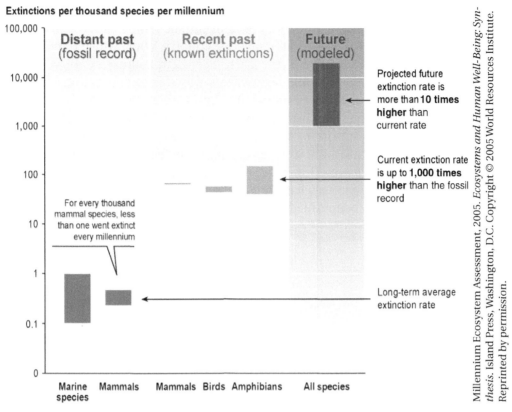

Figure 1.17 Extinction rates today are much higher than in the past and are expected to increase into the future.

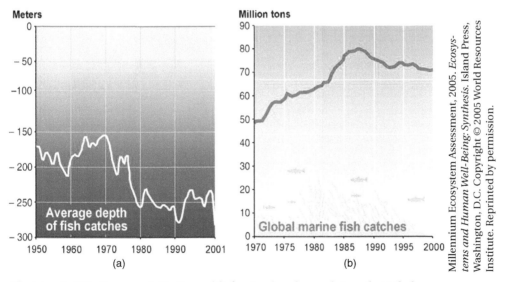

Figure 1.18 Overexploitation of fisheries has been driven largely by advances in fishing technologies.

Protecting marine life in international waters is a challenge, as a governing body that effectively creates and enforces laws to regulate overexploitation is still lacking. For example, unlike species such as oysters, lobsters, and other relatively stationary ocean organisms, whales roam the entire global ocean. As such, their protection cannot fall under the jurisdiction of any one country. When whale populations declined dramatically in the early twentieth century century, a 1946 treaty to sustainably manage whales was signed and nations became members of the International Whaling Commission (IWC). By 1986, enough nations supporting whale conservation had joined, and a controversial moratorium on whaling was instated. Exceptions to the whaling moratorium, such as for aboriginal sustenance and scientific research, have acted as loopholes for some IWC nations. For example, Japan infamously has granted permits for scientific research, where whales have been killed for "research" and then leftover whale meat sold to consumers (Figure 1.19a and 1.19b).

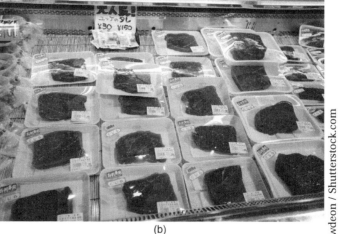

Figure 1.19 (a) The harpoon used to kill whales on the Japanese whaling ship *Yushin Maru;* (b) Whale meat is sold in Japanese markets and used to prepare special cuisines, such as raw whale meat sashimi.

Section 1.1.2—Human Interaction with Natural Systems in the Past

Key Concept 1.1.2—Contrary to some beliefs, humans did noticeably transform natural systems in the past in many ways but at smaller scales and slower rates than today.

The trends in ecosystem transformation described above began on a global scale at the dawn of the Industrial Age about 300 years ago and have accelerated over the past 60 years. These global-scale alterations of natural systems are unprecedented in human history and are fundamentally connected to exponential human population growth, current consumptive behavior, and technological developments. Over the past 50 years alone, human populations have more than doubled from 3 billion in 1960 to 7 billion in 2011. This exponential growth has led to increased demand for food, water, fuel, and many other resources. Resource use, and also the production of pollution, has intensified with advances in technology that allow humans to extract resources more efficiently than ever. Things were not always this way, even though the modern human species (*Homo sapiens*) has been living on Earth for about 200,000 years. It is not that humans did not affect natural systems in the past. They definitely did. Over the course of human history, there has always been a dynamic interaction between human and natural systems. As you begin your study of sustainability, it is important to acknowledge these deep and long-standing interactions. The notion of indigenous people as gentle stewards of the Earth and the existence of vast pristine wilderness devoid of human transformations are largely misconceptions. Indigenous people of the past (and present) altered the landscape through practices such as herding, fire management, and shifting cultivation systems.

One example of past human transformations of natural systems is the worldwide extinction of charismatic **megafauna**. The dominant hypothesis explaining why megafauna disappeared attributes extinction to human overexploitation by hunting. The megafauna are composed of large animals including the woolly mammoth of Eurasia, the saber tooth tiger of North America, and the giant monitor lizard of Australia (**Figure 1.20**). According to this hypothesis, prehistoric humans hunted these animals to extinction on several continents around the world. Whether hunting is the primary cause of extinction is confounded by the fact that an increase in the global mean annual temperature of 6°C (11°F) occurred between 15,000 and 10,000 years ago at the end of the Pleistocene glaciation. This is about the same time that these animals went extinct in North America. A change in temperature could cause extinction because the large cold-adapted animals

> **megafauna** a word used, especially by paleontologists and archeologists, to denote any large mammal with a weight in excess of 100 pounds.

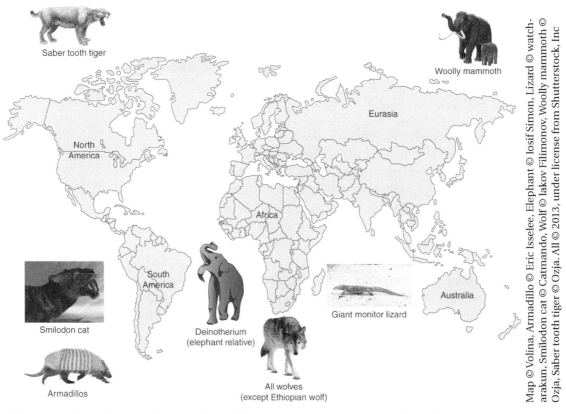

Figure 1.20 Charismatic megafauna have gone extinct on several continents.

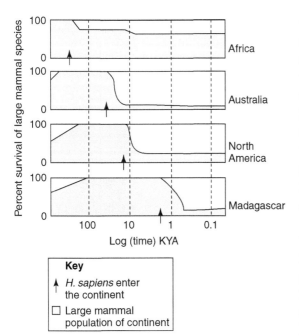

Figure 1.21 Extinctions of megafauna occurred on many continents just after humans arrived.

were not accustomed to a hot climate. However, extinction of these animals occurred on other continents at times that do not coincide with this global temperature change. For example, extinctions in Australia occurred from 40,000 to 30,000 years ago before the end of the last ice age. In South America, extinctions occurred less than 10,000 years ago, which was well after the global temperature change. The only consistent factor for all continents is the arrival of *Homo sapiens* just before extinctions occurred (**Figure 1.21**). Why did humans hunt these large animals to extinction? One suggestion is that these animals were a food source and that humans targeted bigger animals, rather than smaller ones, first. This was an efficient way to maximize the amount of edible biomass acquired with each hunting effort. Other reasons might be that these large mammals were threats as predators of humans or nuisances as competitors for valuable food sources.

What is the difference between overexploitation of species, and other human transformations of natural systems, in the past and those occurring today? After all, humans in the past extracted natural resources and emitted waste into natural systems just as we do today. The major difference is that the overall effect of humans on natural systems today is much *more widespread* and it occurs at *faster rates*. This difference can be best explored in more detail by using the I = PAT model.

Section 1.2: Understanding Human Transformations Using the I = PAT Model

Core Question: How can present and past human transformation of natural systems be understood?

> **I = PAT model** an equation used to convey the impact (I) of both resource use and pollution generation due to human activities in terms of the number of people (population, P), the level of consumption (affluence, A), and the means by which resources are used or pollution is generated (technology, T).

The I = PAT model defines the impact (I) of humans on natural systems as the total quantity of resources used and pollution generated by human activities. The total impact, I, depends on three factors: population (P), affluence (A), and technology (T). Population is simply the number of people using resources and generating waste. Affluence is essentially monetary wealth and is an indicator of **per capita** consumption of resources and production of waste. Technology is the quantity of resources used and pollution generated *per unit of production* with a given technology. In simple terms, these three factors point to the number of people (P) using resources and generating waste, the quantity or magnitude (A) of their resource use or waste generation, and the intensity of resource use and waste generation resulting from use of a certain type of technology (T).

> **per capita** a term that literally means "by head" in Latin and is used to denote the average per person, such as the quantity of resources used or pollution generated by the average person in a certain context.

Section 1.2.1—Present-Day Application of I = PAT

Key Concept 1.2.1—Differences in human population growth (P), affluence (A), and technology (T) explain why developed countries have a greater impact (I) on natural systems than developing countries.

Today, people living in developed countries generally have a greater impact (I) than those living in developing countries. This idea can be illustrated by comparing a wealthy American family living in the suburbs to a poor Tanzanian family living in a rural area using the I = PAT model (Figure 1.22). Today, the average household size in the United

Sustainable World: Approaches to Analyzing and Resolving Wicked Problems

Figure 1.22 (a) American family home in the suburbs; (b) Tanzanian family home in a mountainous rural area.

States is two to three people, whereas the average household size Tanzania is about five people. Therefore, in terms of population only, the Tanzanian family (P = 5) has a larger impact than the American family (P = 2 – 3) because there are more people using resources and generating waste.

In terms of affluence (A), the American family has the larger impact. The American family has more affluence, or wealth, in terms of income and therefore has higher rates of consumption. The annual household income of the Tanzanian family is less than $400. The American family earns at least $60,000 per year. Because the American family is more affluent, it can purchase more market goods and services. For example, the American family might own two cars and have a heating-cooling system in their home to keep them cool in the summer and warm in the winter. Both the car and the heating-cooling system use resources and generate waste. Cars are constructed with metals and other nonrenewable mineral resources. They also require gasoline, which is not only a resource that must be extracted, but burning it in a car creates CO_2, which contributes to climate change. Cars also emit other exhaust gases that increase air pollution, affecting human health. The same is true for the electricity required for the heating-cooling system. If electricity is generated in a coal-fired power plant, as almost half of the electricity in the United States is, then the coal must be mined. CO_2 and other air pollutants are also emitted when it is burned. The Tanzanian family, on the other hand, can afford neither a car nor a home heating-cooling system. Although P is greater for their family, they have a much smaller A ($400) than that of the American family ($60,000). Therefore, accounting for P and A, the overall impact (I) of the American family on natural systems is much larger.

This brings us to technology. Both families have vegetation outside of their homes that they need to maintain in some way. The American family has a grass lawn (Figure 1.22a), which is maintained each week by mowing, and applying water and fertilizer. The Tanzanian family has native vegetation surrounding their home (Figure 1.22b), which only needs occasional maintenance by trimming or completely removing vegetation. The families use two different types of technology (T) for lawn maintenance. The American family has a gas-powered lawnmower, synthetic nitrogen fertilizer, and a hose supplying water from a residential water service. The Tanzanian family removes rain-fed vegetation by hand, using simple tools for cutting when needed. For the sake of concept illustration, let's say that both families maintain the same area of land around their homes and, therefore, have the same *unit of production*. (Recall that the technology component of I = PAT is concerned with the quantity of resources used and pollution generated *per unit of production* with a given technology.) A larger quantity of resources is required *per unit of production* for the technologies used by the American family compared to the Tanzanian family: more materials are needed to construct the lawn mower, the hose, and the physical infrastructure supplying water to residents of the area; more water and fertilizer are used to maintain a type of vegetation (grass) that does not thrive on its own in the environment in which it was planted; synthetic nitrogen fertilizer is produced by the industrial Haber-Bosch process, which requires large energy inputs, likely in the form of fossil fuels. Pollution production is also considered in the I = PAT equation. More pollution is generated by the American family: the gas-powered lawnmower and the Haber-Bosch process emit CO_2, and other pollutants, when gasoline and other fossil fuels are burned; when lawns are watered, fertilizer flows into central drainage systems and ultimately ends up in local or regional water bodies, contributing to eutrophication problems; grass waste is sent to a landfill or municipal composting facility, which requires energy (and more pollution) for transportation and processing. In contrast, the Tanzanian family either leaves vegetation waste on the ground relatively close-by to naturally compost, or uses the materials for building structures, making fires for cooking, or other uses. Burning wood does produce CO_2, and removing vegetation can result in nutrient loss from soils, but the amounts are minuscule compared to the CO_2 emissions of the American family.

Section 1.2.2—Comparing Human Transformations Then and Now Using I = PAT

Key Concept 1.2.2—Differences in human population growth (P), affluence (A), and technology (T) explain why overall human impacts (I) on natural systems are greater today than in the past.

Based on the comparison above, the American family has a larger overall effect on natural systems according to the I = PAT model. This model will now be used to compare modern-day humans with those who lived in the past to understand why human transformations of natural systems today are so different from those of the past. This comparison is summarized in **Table 1.1**.

Population Then and Now. The first major difference is human population size (Figure 1.23). The capacity of Earth's natural systems to supply humans with the ecosystem services required for human well-being is limited. For a long time this limitation was not a concern because, at any one time, there were fewer than 10 million humans inhabiting the entire earth. In addition to low population numbers, most populations had low densities and were widely dispersed. Under these conditions, there were plenty of resources to go around, and the Earth was able to assimilate the relatively small amounts of waste generated by human activities. For much of human history, our species lived as hunter-gatherer bands, and transformations of natural systems were very local. If a band of hunter-gatherers depleted the natural resources in a given area, or created so much waste that a site became unsanitary, then the group could pick up and move to another site. By doing this, they would gain access to new resources and escape

Table 1.1 Human Transformations of Natural Systems throughout Human History Using the I = PAT Framework

	Hunter-Gatherer Societies (200,000–12,000 years ago)	Agriculture and Urbanization (12,000–300 years ago)	Industrialization and Globalization (Early 1700s–present day)
Population	Small populations, low-density, dispersed and disconnected	Populations begin to steadily increase, densities increase, less dispersed and more connected	Exponential population growth begins in earnest, urbanization and densities intensify, deeply interconnected
Affluence (Consumption)	Concept of private property, and therefore excessive consumption, not pervasive; bartering dominates exchange	Wealth accumulation begins as people settle down to live in one place, trade routes open, and monetary systems arise	Rise of modern capitalist economy, growth in international trade, rise of banking systems and financial market
Technology	Tool-making for hunting and gathering, some clothing, fire	Agricultural production and animal husbandry, written language	Mechanization, modern medicine, sanitation, Green Revolution, paper and electronic currency, rise of green technologies, Internet

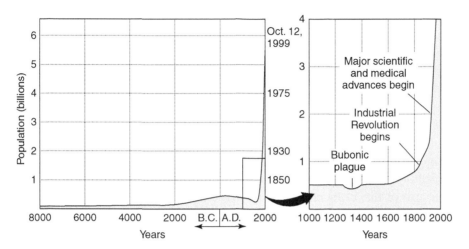

Figure 1.23 The human population has grown exponentially only recently.

their waste. Natural systems in the old site would have time to regenerate and recover before another band moved in. As human populations increased, bands began to come into contact with each other and conflict often resulted. To avoid this conflict, bands often practiced sustainable resource use and waste disposal so that they did not have to pick up and move into already occupied territory. It is important to note that there are exceptions to the hunter-gatherer society during this time period, such as tribes of the Pacific Northwest Coast in North America and the Natufians during the Levant of the eastern Mediterranean. These societies were more socially complex than hunter-gatherers and formed permanent agricultural settlements prior to the worldwide rise of agricultural about 12,000 years ago. Some of these complex societies collapsed because they failed to live sustainably.

With the rise of agriculture about 12,000 years ago, natural landscape transformation really began in earnest as farming and herding activities became more common. Humans switched from a hunter-gatherer lifestyle to a settled agricultural lifestyle once they figured out that they could cultivate certain plants and domesticate certain animals. Humans began to grow and raise more of their own food. With more food available, population numbers began to steadily increase. People became less dispersed and more connected as small villages formed. These villages morphed into towns and eventually full-blown cities. Higher population densities and increased connectivity among human settlements created problems, especially with the opening of trade routes. One major problem was disease. The most famous example of this is the bubonic plague, also called the Black Death, which rapidly spread through Europe and killed 25% to 50% of the population in the 14th century. This disease likely arose in the Gobi Desert of Asia and was transmitted by flea or rodent bites. These organisms are thought to have traveled with cargo along trade routes from Asia to Europe, carrying the disease with them. Once in Europe, the conditions of high human population density and lack of modern medicine were ripe for disease spread. Although premodern waste disposal systems existed, sanitation was also a major issue. Human and animal waste from households, agriculture, and industry contaminated water used for drinking, cooking, and bathing.

Exponential human population growth began about 300 years ago (Figure 1.23). The population of our species was less than 1 billion for much of human history. According to United Nations population data, at the turn of the 19th century in 1804, the human population had reached 1 billion. Exponential growth led to 2 billion by 1927, 3 billion by 1960, 4 billion by 1975, 5 billion by 1987, 6 billion by 1999, and 7 billion by 2011. Technological change was a major driver of this growth pattern. Major medical and scientific advances led to antibiotics, vaccines, and sanitation systems that greatly reduced the incidence of disease. (Something as basic as soap was not widely used in Europe until about 1000 A.D.) In the 1950s, the intensification of food production to feed ever-increasing human populations led to even more food and even more people. Technology, such as air conditioning and heating systems, promoted growth and expansion of human populations into new climatic zones where they previously could not thrive. Exponential population growth is expected to continue into the future. Based on various population growth scenarios, the world will have somewhere between 7 and 11 billion people by 2050.

Affluence Then and Now. Population numbers are only part of the picture, and affluence must also be considered. In hunter-gatherer societies, people would take with them only what they could carry and leave the rest behind. In fact, the concept of private property did not really exist and permanent personal possessions were few. There was no affluence, or wealth accumulation, as we think if it today in terms of money stored in a bank. Barter, or the exchange of one good for another, was the dominant form of commerce. As a result, the resource use and waste generation that goes along with consumption was very low. With the rise of agricultural settlements and urbanization, in addition to systems of commerce and trade, people began to accumulate wealth and acquire many more material goods than before. Part of this was practical. It was simply easier to accumulate personal possessions when settled in one place. Another factor was the ability to accumulate wealth more easily in the form of currency. Unlike bartering, currency systems allow goods to be exchanged more easily (Figure 1.24). For example, in a bartering system, if you have a goat and you would like to acquire some wheat to make bread, then you need to find someone else who wants a goat and who also has some wheat he or she is willing to give up. In a currency system, if you want some wheat, all you need to do is find someone willing to

(a) (b)

Figure 1.24 (a) The author barters with a vendor in a Tanzanian market, as she exchanges her wallet and a few AAA batteries for a decorative wall hanging; (b) A man uses cash, a form of currency, to buy food at a supermarket.

accept your money. This person can then use that money to buy whatever he or she wants, whether that is a goat, a bushel of corn, or a sack of flour. Money is also more easily stored for long periods of time, as opposed to wheat, which will eventually go bad or a goat, which will eventually die. Systems of currency arose at around the same time as agriculture and the physical unit of exchange has varied through time, ranging from shells to metal coins. In addition to settlement in one place and the rise of currency systems, trade also increased consumption. Resources began to flow from one far-reaching place to another. Overall, increased affluence led to more resource use and waste generation. Inequalities also arose between people who had wealth and those who did not.

Just as population growth accelerated with the start of the Industrial Revolution, unprecedented and sustained income growth led to increased standards of living for ordinary people previously unseen in human history. In fact, income levels worldwide grew faster at this time than population, with the average per capita income increasing by an order of magnitude between 1,800 and 2,000. Improved transportation networks and technological developments, such as the steam engine for railways, led to the expansion of international trade. Transition from an agricultural economy to one rooted in machine-based manufacturing increased production rates of goods and services. Currency systems continued to improve with more easily transportable paper bank notes, compared to heavy metal coins, followed eventually by electronic means of exchange. New financial markets in this modern capitalist society provided incentives for more production and consumption. All of this increased affluence, accompanied by a surge in resource use and waste generation. Some of the most harmful impacts on human systems were for the lower classes, which experienced harsh factory working conditions and severely decreased living standards. Most material benefits flowed to developed nations, while people in developing countries suffered as the gap between wealthy and poor societies expanded.

Technology Then and Now. Changes in technology have also shaped human transformation of natural systems through time. Technologies developed and used by hunter-gatherer societies had a relatively minimal effect, *per unit of production*, on natural systems. Humans at this time used stone tools for hunting and butchering animals. After figuring out how to create and manipulate fire, they could make ceramics such as pottery fired from clay. People started to wear very basic clothing made from animal fur and skin (leather) or woven from leaves, grass, and other vegetation. With the rise of agricultural settlements, smelting was developed and metal tools made from copper, bronze, tin alloys, and eventually iron replaced stone tools. Written language emerged. Countless technologies were developed all over the world ranging from the iron plough in China to the aqueducts of the Roman Empire to the printing press of the Renaissance.

The next major technological leap for humans was a shift from manual labor powered by humans and draft animals in agricultural systems to the mechanized technologies of the Industrial Revolution. Major sectors of technological innovation intensified resource use and waste generation during this time to levels higher than any previously experienced, including textiles, mining, metallurgy, and the steam engine. The use of fossil fuels as a source of dense, concentrated energy began and drove many industrial processes. Advances in science and technology spurred the Green Revolution in the 1950s, which transformed our food production systems. The birth of green technologies, such as renewable and clean energy alternatives to fossil fuel and nuclear energy, has occurred in the most recent past.

Summary of Impacts (I) Then and Now. The end result of increases in population, affluence, and technology over the course of human history has given birth to a global society living on a planet covered with human-dominated natural systems. Today, there are more than 7 billion people on Earth and exponential population growth continues. Most growth is occurring in developing nations, where poverty and inequity contribute to social problems. Global **Gross Domestic Product (GDP)** is projected to continue rising well into the future, contributing to even more affluence. By 2050, as projected by the MEA, the world economy is expected to quadruple—this will further increase pressure on natural and human systems. Global consumption continues to rise, especially in the developing world as people living in these places strive to meet both basic material needs and go beyond

---

Gross Domestic Product (GDP) the total market value of all goods and services produced in a certain period of time, usually a year.

basic needs to emulate material living standards of wealthy developed nations. (These two types of needs are defined as vitality and integrity in Section 1.3.2.) Humans today are using natural systems near or beyond capacity. If we degrade these systems, there is no new place to move for additional resources and to escape our waste. People no longer extract, process, and consume resources locally. A salmon fillet that you buy in your local supermarket may come from a fish caught in Alaska, shipped to China to be filleted and packaged, and then shipped back to the United States for consumption. The average human today does not grasp the quantity of resources used or waste generated to bring a product to the local supermarket. We are disconnected from the effects of our actions on natural systems and the consequences of damaged ecosystem services to people who are directly dependent on them. Although globalization has distanced us from the effects of our actions, it has also made our societies more interconnected. As a result, if we continue to live unsustainably, we face a collapse of our entire global society rather than localized "pockets of collapse" characteristic of the past such as the Mayan Empire in Central America and the Easter Islanders in the eastern Pacific Ocean.

Societies of the past that did collapse inflicted detrimental impacts on natural systems that caused problems, but that is not the only factor contributing to the decline of a society. In his book *Collapse: How Societies Choose to Fail or Succeed*, scientist Jared Diamond illustrates, through a multitude of case studies, that the problems faced by societies can also be ascribed to natural climate changes, hostility or friendliness of neighboring societies, and a society's response to these problems. The last factor listed—*a society's response—always* appears to be significant in determining whether a society survives or collapses. As a global society, we have recognized and begun to acknowledge the detrimental human-driven transformations of natural systems. Our response thus far has led us to the idea of sustainability.

Section 1.3: Natural Systems and Human Well-Being

Core Question: How do changes in natural systems affect human well-being?

Section 1.3.1—Connections between Natural Systems and Human Well-Being

Key Concept 1.3.1—The human activities that have transformed natural systems have provided society with many benefits, but at the same time have degraded the free services that ecosystems provide to humans.

The changes made to natural systems described in Sections 1.1 and 1.2 of this chapter have indeed resulted in degradation of these systems, but the human activities causing this degradation provide society with many benefits. Improvements in human health have been made by using nitrogen fertilizers, which have helped to increase agricultural yields and provide food for the hungry. Disruption of the hydrological cycle by extracting groundwater has provided households with clean water to drink. Poverty has been alleviated through economic activities such as tree harvesting and fishing, which has provided income to people living in newly industrializing countries. Advances in medicine have added synthetic pharmaceutical chemicals to drinking water, but these drugs also combat deadly diseases and make us more comfortable when we are sick. Many animal species have been exploited through subsistence hunting and fishing to provide a protein source for many people around the world. Fossil fuels provide us with an energy source to fuel our cars and to heat our homes, but these nonrenewable resources will someday run out and in the process contribute to climate change and pollution. In all of these cases, there is a trade-off between benefits to humans and preservation of natural systems. Sustainability is about finding a balance and a way to simultaneously promote healthy natural systems, human well-being, and economic activities.

The effects are not one-way: Humans change natural systems, but changes in natural systems also have consequences for human well-being. It is difficult to disentangle effects of natural system degradation on human well-being from the other social, cultural, political, and economic factors that also influence human well-being (**Box 1.1** gives an example of this.). Attempting to disentangle the multiple and interacting factors underlying a sustainability problem will be explored in more detail in Chapter 3 with casual chain analysis and stakeholder analysis. Despite these multiple and complicating factors, there is substantial evidence that natural system degradation affects human well-being.

BOX 1.1 DID ANTHROPOGENIC CLIMATE CHANGE CAUSE THE SYRIAN CIVIL WAR?

The Syrian civil war, which began in 2011 and is ongoing, has been referred to as "the worst humanitarian crisis since the Second World War" (Andrew Tabler, Washington Institute for Near East Policy). In 2015, a controversial scientific study claimed that human-caused climate change has made severe drought two to three times more likely to occur in Syria, and that the 2006–2009 drought may have triggered the war. The drought was deemed the worst drought in 900 years. It preceded the start of intense conflict in 2011, and resulted in water scarcity and crop failures, followed by unemployment and famine in rural areas, and finally mass migration of farmers into Syrian cities between 2006 and 2011. Overcrowded urban areas, already under stress from an influx of refugees from the war in Iraq, became even more stressed.

Despite the focus on drought and climate change, even the scientists who published the study admit that the problem is more complicated. Government policies are another contributing factor. Even before the drought, water resources were used unsustainably as a result of Syrian government policies. Migration into cities from rural areas began prior to the drought as a result of new free market policies, which eliminated government subsidies that farmers depended on, such as for fertilizers and diesel fuel. Unemployed farmers, who lost their livelihoods due to water scarcity, were also not provided adequate aid by the government.

Population growth, and its accompanying problems, is not the only factor that caused conflict in cities, as political unrest existed in Syrian cities even before migrants moved in. Before conflict and civil war erupted in 2011, peaceful protests of the oppressive government of President Bashar al-Assad were common. Protestors were often jailed, and even tortured, by the oppressive government, which led to more intense uprisings and eventually civil war. Despite all of the factors that may have caused the conflict, the central focus of the civil war is regime change toward a more democratic government.

As a result of the war, many Syrians are fleeing their country to become refugees in nearby nations. Many have migrated from Syria, then across Turkey to embark on a dangerous and sometimes deadly voyage across rough seas, then to refugee camps on the Greek island of Lesbos, and then on to seek asylum in European Union countries (Figure 1.25). These refugees are part of one of the most significant migrations in recorded human history. They have been referred to by some as climate refugees because of the connections among the drought, climate change, and the Syrian conflict.

Did the degradation of natural resources and ecosystem services affect human well-being in this case? Did climate change, drought, and water scarcity cause the Syrian civil war? Drought certainly played a part, but bad government policies, a corrupt government, and population growth also played large roles.

Figure 1.25 (a) Syrian refugees travel by land from Syria to the coast of Turkey and then into the European Union by sea.

(Continued)

Figure 1.25 (b) Travel by land through Turkey, toward the Aegean Sea, is by foot across a dry dusty landscape and migrants bring only what they can carry; (c) Syrian refugees arrive in an overcrowded inflatable raft on the Greek island of Lesbos after a cold, rough journey across the Mitilini Strait from Turkey.

Section 1.3.2—Defining Human Well-Being

Key Concept 1.3.2—In sustainability, the vitality-integrity framework is used to define human well-being.

vitality basic needs for survival that make life possible.

integrity human desires beyond basic needs that make life worth living.

Before delving into the ways that natural system degradation affects human well-being, the meaning of the term itself must first be explored. *Human well-being* has been defined in many different ways. In the context of sustainability, human well-being is often defined in terms of both vitality and integrity (**Figure 1.26**). **Vitality** includes basic needs for survival, such as food, water, and shelter. **Integrity** is concerned with human desires beyond basic needs that make life worth living. In simple terms, vitality can be thought of as a "need" and integrity as a "want."

The relationship of human well-being to vitality is often evident. For example, human well-being is clearly determined by the amount of food we have to eat and whether we have a secure shelter in which to sleep at night. It is clear that food and shelter are basics needs for survival. However, the exact meaning of vitality is not universal for all people in all places and at all times because it can depend on location, culture, age, gender, socioeconomic status, and many other factors. The context-dependent meaning of vitality is illustrated with a case study on human well-being and air conditioning during the 1995 Chicago heat wave (**Box 1.2**). What is needed

Figure 1.26 Vitality and integrity overlap with the five major aspects of human well-being, as defined by the Millenium Ecosystem Assessment.

BOX 1.2 AIR CONDITIONING DURING THE 1995 CHICAGO HEAT WAVE

Air Conditioning and Vitality. Is air conditioning a basic need for survival? In his book *Heat Wave*, sociologist Eric Klinenberg describes how a nationwide heat wave in 1995 disproportionately affected elderly poor males living in Chicago. This sector of the population had a very high mortality rate. In this case, location mattered because the city of Chicago is hotter than surrounding areas. The urban heat island effect, which was first discovered by British meteorologist Luke Howard in the early 19th century, results in metropolitan areas that are several degrees hotter than surrounding rural areas. This is primarily due to building materials used in cities that retain heat (Figure 1.27). Emission of low-quality waste heat during energy use also contributes. If these same elderly poor males lived in rural areas, it would have been cooler and they may have been less affected by the heat wave. Age and socioeconomic status are important factors as well. Young people are generally healthier and less susceptible to fatal heat-related health problems. Wealthier people in the city could afford air conditioning, whereas many poor people could not. Klinenburg even suggested that gender played a role because elderly poor women were more socially engaged than their male counterparts and, therefore, banded together to live in one apartment during the heat wave and collectively paid for air conditioning. Given the specific conditions of location, age, socioeconomic status, and gender, air conditioning was a basic need for survival (vitality) for some people during the 1995 Chicago heat wave. The major point to take away from this case study is that what is needed for human well-being in terms of vitality is not always as clear-cut as you might think.

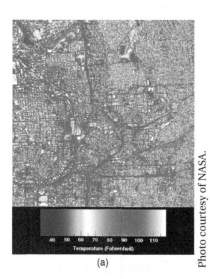
(a)

(b)

Figure 1.27 (a) Satellites can be used to detect the temperature in different regions of a city, ranging from relatively cool blue areas to very hot white areas; (b) Roofs of buildings are among the hottest areas in an urban heat island and green roofs composed of vegetation, such as this one shown on the U.S. Coast Guard Headquarters in Washington, D.C., can help reduce elevated temperatures.

(Continued)

> **Air Conditioning and Integrity.** Burning fossil fuels, such as coal, to produce electricity to power an air conditioner, emits CO_2 that contributes to climate change. Changes in climate negatively impact natural and human systems in many ways. If we use less air conditioning, we will use less electricity and therefore will burn less coal and emit less CO_2. Let's say that, in the summer, the temperature outside of your office building averages 90°F. Your company decides to stop using air conditioning to become more sustainable. You will not die from exposure to these temperatures, so your basic needs for survival (vitality) will be met, but the quality of your life beyond basic needs (integrity) may decline for many reasons. Instead of watching an extra hour of baseball on the weekend, you will have to sacrifice this time to make more frequent trips to the dry cleaner because you sweat more in your business shirts during the week. You may be cranky from being in the heat all day. This may affect interactions with your spouse when you get home and, in the long term, might be detrimental to your relationship.
>
> The point here is that whether you have air conditioning in the office where you work might affect the integrity aspects of your well-being. You may think that zero air conditioning is an extreme example and that using *less* air conditioning is more realistic. But where do we drawn the line and with respect to whom? If you are going to use air conditioning, should office temperatures be 65°F, 75°F, or 85°F? Your coworker might be comfortable at 85°F, whereas you might be sweating buckets. One worker might value sustainability more than another, such that he or she is okay with incurring some inconvenience if that means helping to mitigate human-caused climate change. Your boss might value personal comfort and employee productivity above all else. The specific aspects of human well-being affected by air conditioning can be very different for a variety of people and this type of disagreement can lead to much contention when defining and resolving sustainability challenges.

stakeholders individuals, organizations, or other entities that benefit or are harmed by other's actions, or who carry out the actions causing the benefits or harms, and as a result have an interest in some policy, conflict, organizational goal, or other issue that will influence their future actions.

for human well-being in terms of integrity is even less clear and this is one of the major reasons sustainability challenges are so difficult. When multiple people (**stakeholders**) come together to discuss a sustainability challenge, there is often much disagreement about the integrity aspect of human well-being. Thus, the relationship of integrity to human well-being is ill defined and can be the source of much disagreement in the face of sustainability challenges. This point is illustrated with an air-conditioning example in Box 1.2.

Section 1.3.3—Aspects of Human Well-Being Affected by Natural System Transformations

Key Concept 1.3.3—The five major aspects of human well-being affected by natural system degradation are basic material needs, health, safety, good social relations, and freedom of choice and action.

Although the exact meaning of human well-being is neither universal nor clearly defined, it is affected by the degradation of natural systems in several ways. The MEA conducted assessments at 33 locations around the world to survey how alterations in ecosystem services influence five major aspects of human well-being: *basic material for a good life*, *health*, *security*, *good social relations*, and *freedom of choice and action*. None of these five aspects are mutually exclusive, and they can be generally mapped onto the vitality-integrity framework as ranging from most basic to higher-order needs (Figure 1.26). To illustrate what is meant by these five major aspects, results from one of the 33 MEA assessments focused on the Laguna Lake Basin located in the Philippines (Figure 1.28) are presented here. Two major ecosystems services contributing to human well-being within this basin are provision of food (fish) and the water supply.

The *basic material for a good life* includes having enough food, water, shelter, and other basic needs for survival. It also means a secure and sufficient **livelihood** by which these basic needs are attained. Physical deterioration of Laguna Lake through eutrophication, which is driven by untreated sewage inputs containing large quantities of phosphorus and nitrogen, has resulted in a long-term decline in fishery

> **livelihood**
> the continuous means by which a person adequately secures life's necessities to meet the needs of the self and household in a dignified way, such as through earning wages or self-employment.

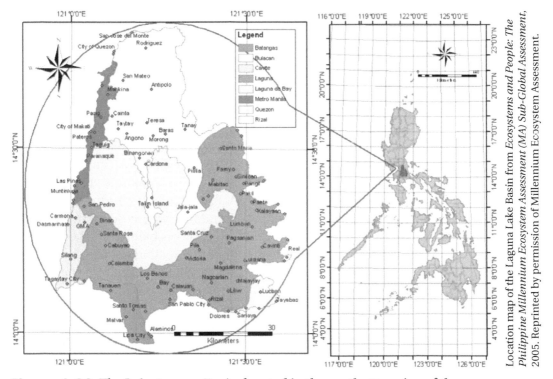

Figure 1.28 The Lake Laguna Basin, located in the northern region of the country, contains both the lake itself and many cities located on the land surrounding the basin.

productivity. This threatens the provision of fish as a food source and also the livelihood of fishermen who rely on catching fish to earn an income. Changes in the quality of the water supply have affected the *health* of local communities. Many rivers flowing into Laguna Lake are contaminated with household sewage and waste from livestock and food processing plants. As a result, the lake water contains high concentrations of total coliform that renders it unsafe for human consumption. This presents a continual health hazard to local people who are dependent on this water for drinking and other activities requiring a sanitary water supply.

The *security* aspect of human well-being includes consistent, dependable, and adequate access to the resources required to meet basic needs and also security from threats of human-caused and natural disasters. Agricultural development in the uplands of the San Cristobal River basin, which is one of 22 major rivers flowing into Laguna Lake, has led to increased flash flood hazards for communities located downstream along the lake's shoreline. This is because forest ecosystems, especially in the upper elevations of a river basin where the landscape is steeper, regulate water supply by providing the important ecosystem service of flood control. Compared to an agricultural field, a forest has relatively abundant vegetation and organic-rich soils that together absorb a larger quantity of precipitation. This helps to reduce the intensity of flooding in lowland regions. Increased occurrence of flash floods, especially during the wet season from May to October, threatens the security of downstream residents.

Good social relations in the area have been compromised due to the fishery decline. Laguna Lake is located in a region of the Philippines with a very high population density. As a result, there are many people competing to use limited resources and social conflict is common. Before fishery productivity declined, open water catch fisheries dominated (Figure 1.29a). Once productivity began to decline in the 1960s and early 1970s, aquaculture became more common in an attempt to return productivity to original levels (Figure 1.29b). Open water fishermen clashed with aquaculture operations, whose dense networks of fishpens in the lake both created navigation problems for fishermen and contributed additional sewage pollution, threatening further eutrophication. Fishermen destroyed fishpens, and aquaculture operations employed armed guards for protection. This type of conflict destroys the social cohesion and mutual respect necessary for good social relations. Good social relations, or **social capital**, is a required element for building a more sustainable world and depends on trust, understanding, and mutual respect among different stakeholder groups. A group of stakeholders with high social capital have the capacity to find common ground, talk and listen to one another, and generally work together to address a sustainability challenge. Good social relations can be fragile

---
social capital
the productive social relationships grounded in mutual trust and understanding by which diverse groups of people have the capacity to communicate with one another, find common ground, and work together to resolve sustainability problems.

Figure 1.29 (a) A fisherman in the Philippines navigates the open water, looking for fish to catch. (b) Fish pens from an Asian aquaculture operation clutter the open water, making it difficult for fishing boats to navigate.

because they are generally slow to build but quick to deteriorate in the face of conflict.

The *freedom of choice and action* is the final aspect of human well-being explored by the MEA. In the case of Laguna Lake, if open water catch fishermen lose the option to pursue what they value doing for a living, then their well-being is compromised. Decreases in open water fishery productivity and encroachment on the lake by aquaculture operations led to a more than 50% reduction in the number of employed fishermen. Freedom of choice and action can be thought of as the highest-order aspect of well-being, such that the other four aspects are often prerequisites to fully achieving this level.

Section 1.4: Origins of Sustainability

Core Question: How did the idea of sustainability arise in human societies?

"[S]ustainability is like truth and justice—concepts not readily captured in concise definitions. We all want truth and justice; but what these mean can also vary greatly from individual to individual and between societies. My justice may be your exploitation, and my truth may be your lies!"

— *Simon Bell and Stephen Morse, Sustainability Indicators: Measuring the Immeasurable?*

> *"Defining sustainability is ultimately a social choice about what to develop, what to sustain, and for how long."*
>
> — Thomas Parris and Robert Kates, Characterizing a Sustainability Transition

Historically, humans have not used the exact word *sustainable* to describe how they did or did not live, but the idea of sustainability is not new. The impact of humans on the environment on which they depend has been a persistent issue for human societies. The sustainability idea can be traced back to ancient civilizations, such as in Egypt, Mesopotamia, Greece, and Rome, where environmental degradation due to soil salinization and deforestation were prominent concerns. With the publication of *Our Common Future* by the international Brundtland Commission in 1987, the word *sustainability* has come into common usage. However, the definition of sustainability is still up for debate. The words *sustainability* and *sustainable development* are often used interchangeably and will be in this book. At the beginning of this chapter, sustainability was described as fundamentally concerned with preserving the conditions necessary for the survival of prosperous human societies. However, what these specific conditions are will depend on who you ask, where you are, and during what time period you ask. In general, we can say that something is sustainable if it does not diminish the well-being of people living today or in the future and preserves the Earth's natural systems. In order to understand the present-day meaning of the term *sustainability*, it is useful to explore the historical development of the term.

Sustainability is about simultaneously promoting healthy ecosystems, human well-being, and viable economies (refer back to Figure 1.1). This three-pillar model of environment-society-economy is the most commonly used starting point for describing what is meant by sustainability. This section provides a brief history of how we came to be concerned with all three. To understand the ideas underlying the three-pillar model, the evolution in thinking about each of the pillars and how they eventually came together to define sustainability will be presented here. The ideas, theories, and social movements that underlie environmentalism, achievement of societal equity, and economic development have long and diverse histories. The next subsections scratch the surface of these histories, focusing on the parts that led to present-day sustainability.

Section 1.4.1—Environment and Natural Capital

Key Concept 1.4.1—Increasing environmental degradation led to concern about loss of natural capital, which supports prosperous human societies.

The environment pillar of sustainability (Figure 1.1) is concerned with natural capital, which includes all natural resources (e.g., fresh water, forests, and fish) and ecosystems services (e.g., climate regulation, flood mitigation, and crop pollination) that support human systems. Although the modern-day environmental movement began in the 20th century, people have been concerned with the degradation of natural systems for thousands of years. People living in ancient agricultural societies were concerned with salinization of the soils on which they depended for farming. Human concern with natural systems was not always for practical reasons related to food production, but also for spiritual, emotional, and cultural reasons. In the late 12th century, Catholic friar Saint Francis of Assisi spoke about a deep spiritual connection to nature and humankind's obligation to protect it. He eventually became known as the Patron Saint of the Environment (Figure 1.30). In the 18th century, during the Enlightenment movement happening in Europe and America, German philosopher Jean-Jacques Rousseau called for a "return to nature" and, as a result, advocated for its preservation. This call was in response to the inequities and general deterioration of human well-being generated by the economic interests of industrialized society. He is famous for his idea of the noble savage and believed that humans experienced less suffering when they lived close to nature. Similar thinking was promoted by 19th-century American poets and writers of the Transcendentalist movement, such as Ralph Waldo Emerson and Henry David Thoreau. George Perkins Marsh, an American lawyer and politician during this time period, is often referred to as the first modern-day environmentalist due to his 1864 book on ecology called *Man and Nature*. At the dawn of the 20th century, rising concern about environmental degradation in the United States led to the creation of federal agencies to manage public lands such as the Forest Service (USFS), Bureau of Land Management (BLM), Fish and Wildlife Service (USFWS), and National Park Service (NPS).

By the early 20th century, the US environmental movement had divided into two groups based on conflicting ideas about how humans should interact with nature (Figure 1.31). The conservationists, such as the first head of the USFS, Gifford Pinchot, believed that nature should be used for human benefit but that this use should be regulated to prevent serious degradation. The preservationists, such as naturalist and Sierra Club cofounder John Muir, and ecologist and author of *A Sand County Almanac*, Aldo Leopold, thought nature needed to be protected for its own sake by establishing wilderness areas. Regardless of specific ideas about environmental protection, by the 1960s a greater public environmental awareness than ever before had emerged. Much of this was the product of two decades of unprecedented economic growth, accompanied by equally rapid rates

natural capital the resources and services provided by natural systems that support and sustain human activities.

Figure 1.30 Paintings of Saint Francis of Assisi show him surrounded by scenes of nature to reflect his role as Patron Saint of the Environment.

Figure 1.31a Gifford Pinchot, known as "Father of the Forest Service," attempted to find a balance between human use of nature and protection of natural resources; (b) John Muir believes nature should be protected from any use by humans and pushed for the establishment of Yosemite National Park in the late nineteenth century.

Figure 1.32 Rachel Carson was a scientist whose work led the United States to ban the pesticide DDT in 1972.

of environmental degradation, driven by reconstruction efforts after the end of World War II. Scientist Rachel Carson's 1962 book *Silent Spring* increased concern about the impact of pesticide use on ecosystems (Figure 1.32). This shifted society's unwavering faith in technology as a way to solve our problems to a realization that technology also causes problems. This book was also a turning point in many ways. It greatly increased public awareness, and led to the creation of the US Environmental Protection Agency (EPA), as well as the federal ban on the pesticide DDT in 1972. Ecologist Paul Ehrlich's 1968 book *The Population Bomb* warned of the long-term ecological consequences of continued exponential human population growth and accompanying food shortages. The first Earth Day was held on April 22, 1970. During the Apollo space mission, the first photo of Earth was taken from space on December 7, 1972. With a new awareness of human global environmental damage, this photo solidified people's perception of a fragile and finite Earth. The major laws underlying US environmental regulation today were passed during this time, including the Clean Air Act, the Clean Water Act, and the Endangered Species Act.

The environmental movement in the United States is only one brush stroke in a large global picture. In many developing countries around the world, such as in Africa, South America, and parts of Asia, different histories of the environmental movement exist. For example, the environmental movement in India is a social justice movement, focused on people and their rights to the land. In 1991, Sunita Narain, a change maker at the Center for Science and the Environment in New Delhi, India, referred to climate change as "environmental colonialism." The ultimate goal of these movements is similar to that in the United States: to preserve the land. However, the focus is different: to protect human rights, rather than the rights of plants and animals.

Section 1.4.2—Economy and Development

Key Concept 1.4.2—The idea of human progress and increasing poverty in underdeveloped regions around the world highlighted the need for economic development to ensure human well-being and advancement.

While there was great concern about environmental protection and restraining the economic activities that led to environmental degradation in developed parts of the "North" (Europe, North America, Japan), poorer regions of the world such as South America, parts of Asia, and Africa (the "South") were more focused on promoting **economic development**. In these regions, development would ensure access to basic resources required for survival and human well-being. In order to understand the issues surrounding sustainability and economic development, it is useful to look back a few centuries and explore the assumptions and ideas that have driven development.

In general, the idea of progress is the notion that human societies will advance through time and this has driven development. The ancient Greeks and Romans first proposed the advancement of human societies through successive stages. The goal of this advancement in Western societies evolved over the next few centuries from the religious purpose of moral advancement to achieve salvation after death to the secular aim of a better material life on Earth. Although the end goal changed, the means for progress remained consistent: The human condition would improve as society modernized through advancements in science and technology. Thinking during the Industrial Revolution added to this list by including economic growth, increased material wealth, and human domination over nature to the definition of progress. During this time, the demand for natural resources required to fuel economic growth increased. These resources were acquired largely by colonizing underdeveloped regions of the world. A major consequence of colonization was that wealth flowed from poor developing nations of the South to wealthy developed nations of the North. From an economic growth perspective, this unequal distribution of wealth among nations was a roadblock to continual enlargement of global markets. In the 1950s, two

> **economic development**
> a process by which the quality of life is improved for everyone without damaging the resources and services provided by natural systems.

development theories—modernization and dependency—emerged as competing solutions to this problem and still underlie thinking about international development to alleviate poverty today.

Modernization theory argues that developing countries should imitate the development model of industrialized countries of the North. This would involve establishing a market economy in countries of the South and promoting high levels of material goods consumption to increase wealth. The basic idea is that developing nations should foster technological advances, open up their domestic markets to multinational corporations, focus on producing goods for export to other countries, and move away from government financial support and toward privatization. The overall result would be to extend the wealth and natural resources of their countries to the entire world. As the world as a whole became more affluent, wealth would "trickle down" to poorer countries and alleviate poverty. At the 1944 Bretton Woods Conference, held in the wake of World War II, several international agencies were established to promote this type of development. With the goal of reducing poverty in developing nations, these agencies would provide loans for capital improvement projects (World Bank) and stabilize international exchange rates (International Monetary Fund).

Dependency theory claims that development by modernization has actually widened the wealth gap between rich and poor countries and increased poverty. The North has exploited the natural and human resources of the South, first by colonization and more recently through multinational corporations. Some supporters of this theory claim that development of countries in the North has depended on intentional underdevelopment of the South, so that their resources could be more easily exploited. Under these assumptions, the solution to poverty in developing nations is for these countries to become independent and self-reliant by closing their markets to the North and transferring market control to governments that can use the profits to alleviate poverty.

Section 1.4.3—Society and Equity

Key Concept 1.4.3—Growing inequity due to increasingly scarce resources and uneven economic growth led to concern about equity among people living today and in the future.

social equity equal opportunity for all people through fair access to education, livelihoods, and other resources, democratic participation, and self-determination.

The society pillar of sustainability is largely concerned with **social equity**. It involves cultivating a society that respects human rights, promotes justice and fairness, and maintains human well-being. Equity is not the same thing as equality. Equality means that each person receives exactly the same thing, whereas equity is more concerned with each person getting the same quality. For example, if we wanted to ensure that no one goes hungry, we would promote equality by giving everyone the same type and amount of food. Let's say we give each person one peanut butter and jelly sandwich. Some people may be happy and well fed, but others may have an allergic reaction to nuts. If we wanted to promote equity, we would provide a variety of sandwich options. As long as all people get the same quality of outcome from the sandwich, such as being happy and well fed after eating, then we have equity.

Sustainability is concerned with two major types of equity. **Intragenerational equity** is concerned with equity among all people living today. For example, we want to ensure that each person has an adequate livelihood to meet his or her basic needs. **Intergenerational equity** is concerned with equity between people living today and those of future generations. For example, if we use up all natural resources today, then there will be no resources remaining for people living in the future. To understand how the idea of equity contributes to sustainability, two ideas about how to ensure social equity will be explored, followed by a brief review of the modern-day social justice movement.

intragenerational equity social equity among all people living on Earth today.

intergenerational equity social equity among people living today and those who will live in the future.

The two ideas about equity roughly align with the two development theories described in Section 1.4.2. In the 18th century, Scottish philosopher Adam Smith published *The Wealth of Nations*. This is the first modern piece on economics and, as a result, Smith became known as the Father of Economics (Figure 1.33). A major idea in his book was that of "an invisible hand" that guides the self-interested actions of each individual to a result that unintentionally serves the public interest, even though each individual is concerned only with benefits to him- or herself and not to society as a whole. For example, a farmer earns an income by growing food. People buy this food and, as a result, they do not go hungry. The farmer is concerned only with the benefits to himself (monetary profit from selling food), but he also unintentionally provides a service to society (provision of food). Smith believed that when an individual pursues self-interest only, he or she also promotes societal well-being. He thought that equity in society would be promoted through self-interested actions of individuals in markets. This would result in competitive low prices, and therefore affordability, and also a variety of goods and services to satisfy everyone's needs. Based on this reasoning, he advocated for a market economy based on private property. Today, his ideas are often misconstrued by supporters of free markets as they advocate for zero government regulation. However, this model of economic development has proven to generate inequity in some instances rather than promoting it. Although it is not often acknowledged, Smith was aware of this. He was suspicious of business interests and claimed that they would form monopolies, leading to high

Figure 1.33 This statue in Edinburgh, Scotland memorializes Adam Smith, whose ideas guide thinking about free markets to this day.

Figure 1.34 The epitaph on Karl Marx's grave in Highgate Cemetery, London echoes his belief that the working class should revolt against capitalism.

prices for the public, and manipulate politicians to pass policy serving their own self-interests rather than those of society. Smith's ideas predicted how unrestrained economic growth would lead to social inequity. He recognized the need for governments to ensure fairness by assigning property rights and providing social services. In Smith's mind, the right mix of markets and governments would result in equity. **Box 1.3** provides an example of how markets and governments can work together to promote sustainability.

Not everyone agreed with this model of how a society would achieve equity. In the 19th century, German philosopher Karl Marx arrived on the scene as a major voice on social equity (**Figure 1.34**). Marx was highly critical of free market capitalism, the dominant economic system of his time that continues today, because it creates inequity between a wealthy class that owns the means of production and a poorer working class that serves the wealthy class as a source of labor. Because of the tensions between these two classes, Marx believed that a free market capitalist society would eventually collapse and evolve into a society with less inequity through collective ownership of the means of production. He believed that only collective ownership and control of resources by communities of people, rather than the individual private ownership advocated for by Smith, would prevent the social inequities generated by markets. Overall, he imagined an egalitarian society in which equity would triumph over the societal problems caused by free markets. Both Adam Smith and Karl Marx had their own ideas about how equity in a society could be achieved. Smith is associated with capitalism and Marx with socialism, which are the two dominant economic systems today.

Regardless of how social equity is achieved, it is a major issue for sustainability. The society pillar started to make its way into the mainstream sustainability movement as the importance of considering human well-being, in addition to environmental health, was recognized. For example, in the 1970s the major famines in the Sahel region of Africa made people realize the environment is not the only problem. People who do not have their basic needs met cannot afford to think about sustainability. They are more concerned with the short-term, urgent challenges of how they will feed their families from one day to the next. This does not allow much time or energy to dedicate to long-term thinking about environmental health. Today, the environmental justice movement has made efforts to integrate sustainability. Environmental justice emerged in the 1980s from earlier social justice activism during the civil rights movement in the United States. It is concerned with how power and privilege lead to both unfair access to

BOX 1.3   HOW GOVERNMENTS CAN ALLOW MARKETS TO WORK TOWARD CONSERVING FISH AND REDUCING POLLUTION

In his 2011 book *But Will the Planet Notice? How Smart Economics Can Save the World*, environmental economist, Gernot Wagner, describes how markets can solve many sustainability problems. He gives two examples related to fisheries and pollution.

Transferrable fishing quotas: The Alaskan halibut fishery. Halibut is a large fish that roams in the Pacific coast of North America. It is valued by fish-eating consumers, and its fisheries provide employment and commercial profits in the billions of dollars (Figure 1.35). In the 1980s and 1990s, halibut populations began to noticeably decline and restrictions were put on fishing. These restrictions allowed fisherman only a few 24-h periods each year to catch halibut. These "derby"-style fishing periods resulted in dangerous working conditions for fisherman, as they rushed to catch as many fish as possible, often equal to what they would normally catch over the course of an entire year, despite sometimes stormy weather and rough seas. These restrictions resulted in profit as well as loss. Fishing gear would often be lost, as fishermen deployed more gear than usual, to catch as many fish as possible. Of course, fish populations were also harmed more than necessary during these "derbies," as some fish were caught and lost with fishing gear, whereas the fish caught could not be processed quickly enough.

The "derby"-style fishing restrictions, which limited the number of days that fishermen could

(a)

(b)

Figure 1.35 (a) A fisherman weighs his catch at a harbor in Kachemak Bay, Homer, Alaska; (b) A boat harbor in Homer, Alaska, often referred to as "the Halibut fishing capital of the world," is teeming with fishing boats.

(Continued)

spend at sea, did not achieve the goal of reducing the number of fish caught. In 1995, a more successful policy was introduced that aimed directly at the goal. The policy put a limit on the number of fish caught each year by each fisherman, regardless of the number of days taken to catch the fish. This new policy, known as an individual fishing quota (IFQ) program, handed out a certain number of "quota shares," which allotted a specific quantity of fish to each fishermen. The number of quota shares available to all fisherman was based on the total annual sustainable catch, which was predetermined by fishery scientists. The 1995 IFQ program gave fishermen more flexibility regarding when to fish and how to fish. It even allowed them to decide whether to even fish at all, as they could sell their shares to another fishermen if they decided that selling shares, during a given year, would be more profitable than fishing. In addition to greater flexibility for fishermen, the IFQ program brought higher profits, safer working conditions, and protection of halibut populations. With benefits to society, economy, and environment, the Alaskan IFQ program can be considered a sustainability success story!

Cap-and-trade permits: Lead and sulfur dioxide pollution in the United States. Cap-and-trade is a similar market-based government program that results in sustainability. In these programs, the government puts a "cap," or limit, on the total amount of pollution that can be emitted in any given year. The annual limit is then reduced over a defined time period in order to reduce total pollution emitted over time. This type of program has successfully curbed lead and sulfur dioxide emissions in the United States, and many hope it can be used to reduce CO_2 emissions globally.

Until the 1970s, tetraethyl lead was added to gasoline as an anti-knocking agent, which protected car engines from damage (Figure 1.36). However, it was soon discovered that lead exposure led to high blood pressure and lowered IQs in children. These health effects, in addition to damage to catalytic converters, led the U.S. Congress to instruct the U.S. Environmental Protection Agency (EPA) to reduce lead pollution. To do this, the EPA created a market for lead, in which each company had a permit to pollute the environment with a certain amount of lead each year. If a company found it more profitable, in a given year, to reduce their lead emissions, such as by using new technologies, then they could sell their pollution permits to competing companies for a profit.

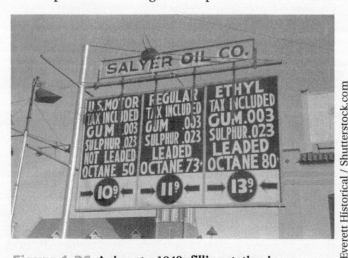

Figure 1.36 A sign at a 1940s filling station in Oklahoma City, USA shows prices for unleaded and leaded gasoline options.

From 1974 through 1982, the EPA set increasingly lower limits on the total amount of lead that could be released to the environment. By the mid-1990s, less than 1% of U.S. gasoline sales involved leaded fuel.

A similar program helped reduce U.S. sulfur dioxide emissions, beginning in the 1990s. Sulfur dioxide pollution causes respiratory problems in humans, and results in acid rain that damages both ecosystems and historical treasures (Figure 1.37a). The program was aimed at coal-fired power plants, which emit sulfur dioxide when coal is burned (Figure 1.37b). Each power plant was given a certain number of pollution permits. As with lead, power plants could choose to reduce emissions, such as through use of new technologies, and sell their permits, if this was more profitable in any given year. The sulfur dioxide program was much more widespread than the lead program, with over 3,500 power plants involved. Since this program was implemented in the 1990s, sulfur dioxide emissions have been almost cut in half.

(a)

(b)

Figure 1.37 (a) The ancient faces on stone temples in Angkor Wat, Cambodia, a UNESCO World Heritage site, show pock marks and other damage from acid rain; (b) Sulfur dioxide is a by-product of burning coal, which is done to generate electricity, such as in this power plant near Joseph City, Arizona, USA.

environmental resources (such as clean air, clean water, city parks) and unequal exposure to pollution (such as toxic waste). Although sustainability will not have transformative power without incorporating environmental justice, the tension between meeting the short-term needs of people in the present and the longer-term focus of sustainability

vulnerability science an emerging field that seeks to understand socio-ecological systems with a focus on lessening or eliminating harm to human systems and, to a lesser extent, the supporting natural systems.

has made this challenging. The relatively new field of **vulnerability science** seeks to unite environmental justice with the sustainability movement.

Section 1.4.4—Birth of Sustainability

Key Concept 1.4.4—Increasing global concern for ecosystems, economies, and people led to publications and international meetings that defined sustainability as we know it today.

With the recognition that we needed to preserve natural capital, promote social equity, and allow economic development so that people living in poverty can meet their basic needs, the three pillars of environment, society and economy converged, and the sustainability movement began to take shape. The worldwide sustainability movement began in earnest in the 1970s. The first widely known publication to promote the idea of sustainability was *Limits to Growth*, which was published by the Club of Rome in 1972. The authors were a young group of MIT scientists who used models to demonstrate the consequences of unregulated human population growth on a planet with finite natural capital. They predicted that humans would reach the limits to growth in the next 100 years and offered solutions.

The first international sustainability meeting was the United Nations (UN) Stockholm Conference on the Human Environment in 1972. The conference provided an impetus for the creation of the UN Environment Program (UNEP), which was the first international environmental agency. Many other similar national agencies were established in countries around the world. The year after the 1973 global oil crisis, which raised awareness of nonrenewable resource shortages, the second major UN meeting (UNCTAD) was held in Cocoyoc, Mexico. An important advancement for sustainability occurred at this meeting when environmentalists joined together with advocates of economic development to alleviate poverty. Together, they issued a declaration claiming that current development models were increasing social inequity and environmental degradation, which is especially apparent and distressing in the urban areas of developing nations (**Figure 1.38**).

Prior to this time, advocates for development and those for the environment were in conflict. As a result of the new alliance between advocates for the environment and those for social equity, several environmental organizations resolved to work with local communities around the world to align development needs with conservation issues. In the 1980s, organizations such as the International Union for the Conservation of Nature (IUCN), the World Wildlife Fund (WWF), and UNEP committed to this common goal. Also, during this time period, another unifying idea was brought to the forefront: Environmental degradation would eventually deplete natural capital to the point that opportunities for development would become limited. Previously, only the impacts of development on the environment had been acknowledged, not the impacts of environmental degradation on development.

In 1983, the World Commission on Environment and Development (WCED) was created by the UN General Assembly. WCED was made up of policymakers, planning officials, finance experts, and other stakeholders from more than 20 countries and was considered independent of the influence of any government or the UN. After three years of public hearings in countries around the world, the WCED published *Our Common Future*. WCED is also known as the Brundtland Commission, and *Our Common Future* as *The Brundtland Report*, because it was chaired by Norwegian Prime Minister Gro Harlem Brundtland. This effort embedded the idea of sustainable development into the minds of millions of people worldwide as "development that meets the needs of the present without compromising the ability of future generations to meet their own needs." It also made sustainability a political issue in most countries. The report claimed that development was not at odds with environmental protection and that it was needed to eliminate poverty, which limited the ability of people to orient toward the type of long-term thinking essential to sustainability. The report called for increased democratic participation by all people, although it was criticized for this because in practice the Brundtland Commission excluded local communities and some nongovernmental organizations (NGOs) from the process. Nonetheless, the publication of *Our Common Future* brought sustainability to the forefront as a global issue and catalyzed a large number of UN meetings over the next several decades that continue to the present day (**Table 1.2**).

Figure 1.38 (a) Favela da Rochina, the largest slum in Brazil, shows the severe inequity between wealth city dwellers living in high-rise apartments and people living in poverty; (b) Houses in Rochina are located one on top of another, in some cases on a steep hillside, which is dangerous during heavy rains and flooding.

Table 1.2 United Nations Meetings Related to Sustainability

Year	Meeting
1972	UN Stockholm Conference on the Human Environment
1974	UN Conference on Trade and Development (UNCTAD) in Cocoyoc, Mexico
1980	World Conservation Strategy
1983	World Commission on Environment and Development (WCED) formed
1987	*Our Common Future* published by WCED (Brundtland Commission)
1990	First Intergovernmental Panel on Climate Change (IPCC) Assessment Report
1992	United Nations Conference on Environment and Development (UNCED) in Rio de Janeiro, Brazil (Earth Summit)
1993	World Conference on Human Rights, Vienna, Austria
1994	World Conference on Natural Disaster Reduction, Yokohama, Japan; World Conference on Population and Development, Cairo, Egypt
1995	World Summit for Social Development, Copenhagen, Denmark; Second IPCC Assessment Report
1996	UN Conference on Trade and Development, Midrand, South Africa Habitat II, Istanbul, Turkey
1997	Earth Summit+5, New York, USA (Rio+5)
2001	Third IPCC Assessment Report
2000	UN Millennium Summit and the *Millennium Development Goals*
2002	Johannesburg World Summit on Sustainable Development (WSSD, Rio+10)
2005	Millennium Ecosystem Assessment (MEA)
2007	Fourth IPCC Assessment Report
2009	UN Copenhagen Climate Change Conference
2012	Rio+20
2014	Fifth IPCC Assessment Report
2015	UN Sustainable Development Summit and the *Sustainable Development Goals*

The first of these was UNCED, also known as the Earth Summit, held in 1992 in Rio de Janiero, Brazil, to commemorate the 20th anniversary of the 1972 Stockholm conference. With the publication of *Agenda 21* (agenda for the 21st century) and the *Rio Declaration on Environment and Development*, this meeting was the first international meeting to lay out the general principles and an action plan for sustainable development. It was the largest conference the UN had ever held at that time with over 20,000 heads of state and delegates, NGOs, members of the press, and other stakeholders in attendance. During UNCED, an agreement was made between wealthy developed countries and developing countries who could not afford to focus on sustainability. Developed countries agreed to allocate 0.7% of their GDP per year as development aid for developing countries. Three additional entities to aid in international environmental governance were also established: the UN Framework Convention on Climate Change (UNFCCC), the Convention on Biological Diversity (CBD), and the Statement of Forest Principles. The Commission on Sustainable Development (CSD) was created by the UN General Assembly a year later. Politically, the 1992 Earth Summit was a success in that it was attended by heads of state from almost every nation and increased overall global commitment to sustainability. Increased stakeholder participation, including NGOs and the business community, over the next 20 years was also counted as a success. However, there was still too much focus on the environment pillar and not enough on the social equity issues related to the society pillar.

Throughout the 1990s, meetings that focused on various aspects of sustainability were held in countries around the world (Table 1.2). Rio+5 in New York, USA, and Rio+10 (or WSSD) in Johannesburg, South Africa, were intended as progress reviews of the goals and action plans laid out during the 1992 Earth Summit in Rio. The overall conclusion of both of these meetings was that progress on sustainability was too slow and that the major problem was actual, on-the-ground implementation of sustainability in specific countries around the world. This is a major challenge of international commitments such as *Agenda 21* that, in reality, have to be carried out in specific countries with diverse institutional capacities and stakeholder interests. The devil is in the details. At the WSSD, progress was made by shifting focus away from the environment pillar and toward the development and equity issues related to the society pillar. This was largely a result of the 2000 Millennium Development Goals, which are focused primarily on human well-being in terms of poverty and disease reduction, education, and reproductive health. However, many analysts claim that the 2002 WSSD failed to consider environmental issues, where the 1992 Earth Summit succeeded, and that the Earth Summit failed at social equity and development issues, where the WSSD succeeded. A holistic focus on the three sustainability pillars still proves challenging.

We have come a long way since 1972, and rays of hope exist to show that progress toward sustainability has been made (see **Box 1.4**). Sustainability has moved from being an interesting and controversial idea to a guiding principle accepted by citizens, businesses, and governing bodies around the world. International organizations such as

---

BOX 1.4 AN UNSEEN GLOBAL MOVEMENT TOWARD SUSTAINABILITY?

In his 2007 book *Blessed Unrest*, Paul Hawken describes a dispersed and unnamed, yet self-organizing, worldwide movement that is healing ecosystems and people. This humanitarian movement is composed of one to two *million* nonprofit and nongovernmental organizations working on social and environmental issues. The tens of millions of diverse people working and volunteering for these organizations are not high-profile government scientists, politicians, or religious leaders, but instead are everyday people: students, retired professors, businesspeople, members of tribes, architects and engineers, and grandmothers, worried parents, and their teenagers. These organizations, and their inspiring stories of positive change in the world, are ignored by the mainstream media in favor of broadcasts about the latest murder, abduction, political scandal, or scary prediction about the future from the scientific community. As a result, most people don't hear of positive changes being made, and the dominant attitude toward the future is pessimistic and filled with doom-and-gloom thinking.

The diverse type of work carried out by the members of this global grassroots movement is illustrated in a passage from Hawken's book (p. 12):

Clayton Thomas-Müller speaks to a community gathering of the Cree nation about waste sites on their native land in northern Alberta, toxic lakes so big you can see them from outer space. Shi Lihong, founder of Wild China, films documentaries with her husband on migrants displaced by construction of large dams. Rosaline Tuyuc Velasquez, a member of the Maya-Kaqchukel people, fights for full accountability for tens of thousands of victims of death squads in Guatemala. Rodrigo Baggio retrieves discarded computers from New York, London, and Toronto, and installs them in the favelas of Brazil, where he and his staff teach computer skills to poor children. Biologist Janine Benyus speaks to 1,200 executives at a business forum in Queensland about biologically inspired industrial development. Paul Sykes, a volunteer for the National Audubon Society, completes his fifty-second Christmas Bird Count in Little Creek, Virginia, joining fifty thousand others who tally 70 million birds on one day. Sumita Dasgupta leads students, engineers, journalists, farmers, and Adivasis (tribal people) on a ten-day trek through Gujarat exploring the rebirth of ancient rainwater harvesting and catchment systems that bring life back to drought-prone areas of India. Silas Kpanan'Ayoung Siakor exposes links between the genocidal policies of President Charles Taylor and illegal logging in Liberia, resulting in international sanctions and the introduction of certified, sustainable timber policies.

This is a quick sampling of the work of only eight people committed to change. There are tens of millions of additional people, out there in the world, working on their own sustainability causes. The powerful, and hopeful, message in Hawken's book is that these millions of people self-organize to protect humanity in a similar manner that our immune system protects our body. Like the millions of molecules, cells, and organs that collectively protect our bodies from disease, toxins, and other harmful invaders, millions of people protect humanity from social injustices, political corruption, and environmental degradation. Our immune system preserves our bodies over time. The work of millions of people ensures the continued existence of humanity. Our immune system is dispersed throughout our entire body, just as these people are scattered across the globe (Figure 1.39). Our immune system recognizes and eliminates invaders to our body, just as millions of people recognize what is sustainable, and what is not, and work to eliminate unsustainable situations. Like our immune system, these people function independently of a central controller or coordinator. The immune system cannot be stopped: it will protect your body from harm no matter what you want it to do. Might it also be true that the global sustainability movement cannot be stopped from protecting humanity?

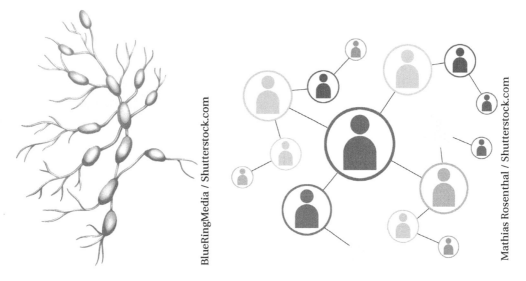

Figure 1.39 A network of lymph nodes, a major component of our immune system, is dispersed throughout our body, just as a network of people working toward sustainability is dispersed throughout the globe.

the World Bank, the IMF, and the World Trade Organization (WTO) have made formal commitments to sustainability. Many in the business and private sectors have adopted corporate and social responsibility (CSR) initiatives and voluntarily established organizations such as the World Business Council for Sustainable Development (WBCSD). International and local NGOs across the globe are incorporating sustainability principles into their agendas and activities in increasingly refined ways. At smaller local scales, thousands of cities, regions, and even countries around the world have implemented concrete and lasting change toward sustainability. New indicators of human progress, such as the Human Development Index (HDI) to measure social equity and the Happy Planet Index (HPI) for environmental health, are increasingly used alongside the traditional indicator of Gross Domestic Product (GDP), which measures only economic growth. Climate change has proved to be a successful link to sustainability that has carried political weight and gained media attention around the world. The assessment reports published by the Intergovernmental Panel on Climate Change (IPCC) since 1990 have played a role in raising awareness about the issue.

Despite this progress, vast challenges remain and some momentum has been lost. Global trends in environment degradation and social equity have not reversed and have, in fact, become worse. Economic growth over the past 20 years has not slowed down, which has led to unprecedented environmental degradation and increased resource use. Global biodiversity, ocean fish stocks, fresh water resources, and many vital ecosystem services continue to decline. Despite overall increases in consumption, the gap between rich and poor continues to grow. In 2009, 80% of natural resources were consumed by 20% of the world's people. Unlike the 1992 Earth Summit, Rio+10 in 2002 was not as politically successful with world leaders because of a change in focus on terrorism and national security after September 11, 2001, and the subsequent military conflicts in Iraq and Afghanistan. The ongoing global financial crisis that began in 2007 also shifted concern toward jobs and the economy and away from sustainability.

The current challenges for sustainability have to do with its ambiguity and current focus, competing agenda and stakeholder conflicts, and the persistent use of economic growth as the central indicator of human progress. The definition of sustainability remains vague. This is advantageous because it gives stakeholders around the world the flexibility to adapt sustainability to their specific time and place, but it also causes confusion regarding what to do and how. Although changes have been made at the local scale around the world, lasting worldwide societal transformation has not yet occurred. The focus of sustainability remains on the environment pillar, which hinders progress and real societal transformation. Without holistically considering the agendas of the powerful economic forces that shape our world, sustainability will not have long-term staying power. Social equity must also be part of the picture. In addition to the moral argument for equity, equitable societies have greater commitment to environmental protection.

Along these same lines, global stakeholders often have conflicting agendas and priorities. Specifically, developed countries are more concerned with environmental protection, whereas developing countries need economic development to meet basic human needs. Underlying both of these issues is the persistent definition of economic development as growth for profit, rather than development that increases human well-being. This has led to a singular focus on creating green technologies that will allow us to continue growth rather than a focus on fundamental behavior change needed to reduce consumption patterns. Social capital between developed and developing countries is eroding and, along with it, the ability to work together toward a global sustainability transition. Current international trade agreements leave developing countries at a disadvantage. Developed countries have not fulfilled their promise for 0.7% of GDP as development aid, as agreed upon at the 1992 Earth Summit. The level of mistrust has grown so high that some developing nations are suspicious of international sustainability agendas, which are viewed as secret efforts to suppress development so that people and natural resources in developing countries can continue to be exploited. Finally, there is an overall lack of political will. This is largely because economic institutions have had greater success in influencing government policies than those institutions promoting social equity and environmental health.

Some might argue that, given all of the effort exerted since 1972 and the relatively small victories, the global transition toward sustainability is a failure. However, in the long-term picture of human history, 40 years is a relatively short time. A global-scale societal sustainability transition is no small task. The problems are complex, wide-ranging, and embedded in complex dynamic systems. This book will not teach you exactly how to solve sustainability challenges in the same way that a chemistry textbook teaches you how to perform laboratory experiments. Sustainability challenges are not that simple. What this book will do is equip you with a knowledge base to approach these challenges as a **change agent** capable of identifying and analyzing sustainability problems, anticipating and imagining a more sustainable future, and ultimately managing transitions to sustainability. It will provide you with a framework for approaching complex sustainability problems, and some tools necessary to do this.

> **change agent** an individual who uses his or her knowledge, skills, and attitudes to directly or indirectly serve as a catalyst for change.

Bibliography

Agyeman, J. 2005. *Sustainable Communities and the Challenge of Environmental Justice*. New York: New York University Press.

Barlow, E.D. and A.N. Blake. *Managing Alaska's halibut: observations from the fishery*. Accessed August 16, 2016. https://www.mydigitalchalkboard.org/cognoti/content/file/resources/documents/6a/6a5ab6f6/6a5ab6f61b385311bdfe903cdf6c46dfdaddb343/downloaded file_3137954163774196498_489_halibut.PDF.

Boone, C. G. 2010. "Environmental Justice, Sustainability and Vulnerability." *International Journal of Urban Sustainable Development* 2 (1–2): 135–40.

Carson, Rachel. *Silent Spring*. Greenwich, Conn., Fawcett, 1962.

Daily, G. C., 1997. *Nature's Services: Societal Dependence on Natural Ecosystems*. Washington, DC: Island Press.

Diamond, J. 2005. *Collapse: How Societies Choose to Fail or Succeed*. New York: Penguin Group.

Drexhage, J., and D. Murphy. 2010. *Sustainable Development: From Brundtland to Rio 2012*. Background paper prepared by the International Institute for Sustainable Development for consideration by the High Level Panel on Global Sustainability at the September 19, 2010 meeting. Accessed May 25, 2012: http://sanidadambiental.com/wp-content/uploads/2011/02/GSP1-6.pdf.

Du Pisani, J. A. 2006. "Sustainable Development: Historical Roots of the Concept." *Environmental Sciences* 3 (2): 83–96.

Ehrlich, Paul. *The Population Bomb*. San Francisco: Sierra Club, 1969.

Gilsinan, K. 2015. T*he confused person's guide to the Syrian civil war: a brief primer, The Atlantic*. Accessed August 15, 2016. http://www.theatlantic.com/international/archive/2015/10/syrian-civil-war-guide-isis/410746/.

Global Invasive Species Programme (GISP). 2007. *Invasive Species and Poverty: Exploring the links*. Accessed August 8, 2016. http://www.issg.org/pdf/publications/gisp/resources/invasivesandpoverty.pdf.

Goodier, R. 2015. *Ten technologies developing countries are using now to adapt to climate change, Engineering for Change*. Accessed August 8, 2016. https://www.engineeringforchange.org/ten-technologies-developing-countries-are-using-now-to-adapt-to-climate-change/.

Gupta, A. 2009. *Invasion of the lionfish. Smithsonian Magazine*. Accessed July 28, 2016. http://www.smithsonianmag.com/science-nature/invasion-of-the-lionfish-131647135/?no-ist.

Harlow, J., A. Golub, and B. Allenby. 2011. "A Review of Utopian Themes in Sustainable Development Discourse." *Sustainable Development* 21: 270–80. doi:10.1002/sd.522.

Hawken, P. 2007. *Blessed Unrest: How the Largest Social Movement in History is Restoring Grace, Justice, and Beauty to the World*. London, England: Penguin Books, 342 pp.

Hopwood, B., M. Mellor, and G. O'Brien. 2005. "Sustainable Development: Mapping Different Approaches." *Sustainable Development* 13: 38–52.

Intergovernmental Panel on Climate Change. 2007. *Intergovernmental Panel on Climate Change Synthesis Reports*. Accessed July 17, 2013. http://www.ipcc.ch/.

Kelley, C.P., S. Mohtadi, M.A. Cane, R. Seagar, and Y. Kushnir. 2015. "Climate change in the Fertial Crescent and implications of the recent Syrian drought." *Proceedings of the National Academy of Sciences* 112 (11): 3241–3246.

Klinenberg, Eric. *Heat Wave: a Social Autopsy of Disaster in Chicago*. Chicago: University of Chicago Press, 2002.

Lumley, S., and P. Armstrong. 2004. "Some of the Nineteenth Century Origins of the Sustainability Concept." *Environment, Development and Sustainability* 6: 367–78.

Marsh, George P. *Man and Nature*. Cambridge: Harvard University Press, 1965.

Meadows, Donelle H. et al. *Limits to Growth; a Report for the Club of Rome's Project on the Predicament of Mankind*. New York: Universe Books, 1972.

Millennium Ecosystem Assessment. 2003. *Ecosystem and Human Well-Being*. Washington, DC: Island Press.

Millennium Ecosystem Assessment. 2005. *Millennium Ecosystem Assessment Synthesis Reports*. Accessed July 17, 2013. http://www.unep.org/maweb/en/index.aspx.

Mitlin, D. 1992. "Sustainable Development: A Guide to the Literature." *Environment and Urbanization* 4 (1): 111–24.

National Oceanographic and Atmospheric Administration. 2016. *Why are lionfish a growing problem in the Atlantic Ocean?* Accessed July 28, 2016. http://oceanservice.noaa.gov/facts/lionfish.html.

O'Connor, M. 2007. "The 'Four Spheres' Framework for Sustainability." *Ecological Complexity* 3: 285–92.

Ott, K. 2003. "The Case for Strong Sustainability." In *Greifswald's Environmental Ethics*, edited by K. Ott and P. Thapa, 59–64. Greifswald, Germany: Steinbecker Verlag Ulrich Rose. Accessed May 25, 2012. www.scribd.com/doc/30376702/The-Case-for-Strong-Sustainability.

Pezzoli, K. 1997. "Sustainable Development: A Transdisciplinary Overview of the Literature." *Journal of Environmental Planning and Management* 40 (5): 549–74.

Philippine Sub-Global Assessment Team. 2005. *Ecosystems and People: The Philippine Millennium Ecosystem Assessment (MA) Sub-Global Assessment*. Accessed March 22, 2013. www.unep.org/maweb/documents_sga/Philippine%20SGA%20Report.pdf.

Rittel, H. W. J., and M. M. Webber. 1973. "Dilemmas in a General Theory of Planning." *Policy Sciences* 4: 155–69.

Skaburskis, A. 2008. "The Origin of 'Wicked Problems.'" *Planning Theory & Practice* 9 (2): 277–80.

Smith, Adam. 1776. *The Wealth of Nations*. [S.l.] : Simon & Brown, 2012.

Steyn, P. 2013. *Urban wildlife corridors could save Africa's free-roaming elephants, National Geographic.* Accessed August 8, 2016. http://voices.nationalgeographic.com/2013/12/12/elephant-crossing-urban-wildlife-corridors-could-save-africas-free-roaming-elephants/.

Think Progress. 2015. *How Washington transformed its dying oyster industry into a climate success story.* Accessed August 11, 2016. https://thinkprogress.org/how-washington-transformed-its-dying-oyster-industry-into-a-climate-success-story-334f5ed3717c#.1xwbm0krc.

Wagner, G. 2011. *But Will the Planet Notice? How Smart Economics Can Save the World.* NY, USA: Hill and Wang, 258 pp.

Whyte, K. P., and P. B. Thompson. 2011. "Ideas for How to Take Wicked Problems Seriously." *Journal of Agriculture and Environmental Ethics* 25 (4): 441–45. doi:10.1007/s10806-011-9348-9.

World Commission on Environment and Development. *Our Common Future.* Oxford: Oxford, 1987.

World Wildlife Federation. 2010. *Terai Arc Landscape – securing corridors, curbing poaching and mitigating* HWC. Accessed August 8, 2016. http://wwf.panda.org/who_we_are/wwf_offices/india/?uProjectID=IN0961.

End-of-Chapter Questions

1. Use the I = PAT model to explain which of the following situations results in the larger human impact on natural systems in terms of resources used and pollution generated. For each example, explain why in terms of the three components of the model: Population, Affluence, and Technology.

    a. Which has a larger impact, pre- or post-1950s food production systems? Pre-1950s Food Production: Irrigation systems were not prevalent, and agriculture was limited to regions that received enough rainfall to support crops. Fields were worked largely by horse and plow, and fertilizer was primarily in the form of composted crop residues and animal waste. Before 1950, world population was less than 2.5 billion people and world GDP per capita was less than $2,000.

    Post-1950s Food Production: The Green Revolution of the 1950s resulted in an almost 90% increase in global food production to meet growing human populations, which by the mid-1970s exceeded 4 billion people. This increase in food production was largely a result of more intensive farming methods, such as use of high-yielding monoculture plant varieties, multicropping techniques, and higher inputs of fertilizer, pesticides, and water by irrigation. By 1980, the use of fossil-fuel intensive synthetic fertilizer (produced by the industrial Haber-Bosch process) exceeded crop residue and animal waste fertilizer use for the first time. By the mid-1970s, world population exceeded 4 billion people and world GDP per capita was almost $4,000.

b. **Which has a larger impact, 21st-century water use by an American family or a family living in Haiti?**

American family: The average household size of an American family is two to three people and each person uses over 500 liters of water per day. Amenities in the average American household include flush toilets, dishwashers, washing machines, and pools. Per capita income in the United States is almost $50,000.

Haitian family: The average household size of a Haitian family is four to five people, and each person uses fewer than 25 liters of water per day. In Haiti, many families do not have flush toilets, they wash dishes and clothes by hand, and they do not have pools in their yards. Per capita income in Haiti is less than $2,000.

c. **Which has a larger impact, energy consumption by a resident of Phoenix, Arizona, or the Navajo nation?**

Phoenix resident: Almost all homes have electricity and air conditioning to make the 100°F to 120°F summer temperatures livable. Average household size is 2.8 people and per capita income is almost $20,000.

Navajo nation: The Navajo nation covers about 27,000 mi^2 of land spanning the borders of Utah, Arizona, and New Mexico. About one-third of all households lack electricity and over 85% do not have access to natural gas. Average household size is 3.8 people and per capita income is slightly more than $7,000.

d. **Which had a larger impact, waste generation in 1960s America or in 2010?**

1960s America: Per capita waste generation was less than 3 pounds per person per day with a recycling rate of less than 7%. Of all waste produced per capita, 40% was food and yard waste, whereas 60% was product waste, including disposable products, packaging, and plastics. In 1960, per capita income was just over $15,000 (in 2005 dollars) and there was a population about 170 million people.

2010 America: Per capita waste generation was almost 4.5 pounds per person per day with a recycling rate of more than 30%. Of all waste produced per capita, 25% was food and yard waste, whereas 75% was product waste, including disposable products, packaging, and plastics. In 2010, per capita income was just over $40,000 (in 2005 dollars) and the population had reached 300 million people.

    e. Which had a larger impact, transportation systems in 19th-century America or in the 20th?

    19th century America: Transportation systems were dominated by horse and carriages. For much of the 19th century, the human population was less than 50 million people and per capita income was $5,000 or less (in 2005 dollars).

    20th-century America: By the late 20th century, transportation systems were dominated by gasoline-powered automobiles. By 1970, the population reached 200 million people and continued to grow to over 270 million by the year 2000. Over this same time period, per capita income doubled from $20,000 to $40,000.

2. Sustainability is fundamentally about preserving the conditions necessary for the survival of prosperous human societies. However, defining the exact conditions needed for this is far from straightforward and will vary depending on whom you ask. One way to think about this is to describe human well-being in terms of vitality and integrity. Decide whether each of the following contributes to human well-being in terms of vitality or integrity and explain why. Be specific about the place, time, and context because this will affect your answers and explanations.

    a. Basic sustenance, such as food and water

    b. Having clothing to wear

    c. Owning a car

    d. Possessing a cell phone or Smartphone

    e. Having a computer

Chapter 2
WICKED PROBLEMS AND THEIR RESOLUTION

Students Making a Difference

Moving Toward Zero Waste

Daniel Velez is a student with a vision. He wants the entire Phoenix metropolitan area to achieve zero waste. This will require regional improvements in recycling and composting, but he knows he needs to start small. He cofounded a company, Circle Blue, which has worked with a school in Tempe and an apartment complex in Phoenix to achieve zero waste. His work is rewarding, and he has created real change, but he explains how it is also challenging: "You envision what the end state will be, but getting to that end state is really, really challenging. Changing things is very difficult, so you have to be positive every day, but some days just feel like they suck. You definitely have your highs and lows in doing this type of work."

Early on, Daniel realized that achieving zero waste in the middle school and in the apartment complex would require different approaches. There is no one-size-fits-all solution. A lot of effort and time was required to adapt approaches to each particular context. Initial waste assessment surveys had to be kid-friendly for the school and tenant-friendly for the apartment. Different solutions were also needed. At the school, he easily identified leaders excited about waste reduction. The principal and most teachers, parents, and administrators were already on board. The challenges were bureaucratic red tape, which affected the logistics of reducing waste, and getting students excited, and united, behind recycling and composting. At the apartment complex, it took more time to identify community leaders and build excitement. Many managers and tenants did not want to waste their time on recycling and composting because they didn't care about it, initially.

Daniel also encountered unintended consequences. Every action you take will have a consequence that you probably did not think about. He described an instance where they removed a garbage compactor from an apartment complex. This machine crushes garbage so that it takes up less space. After working with the community, they anticipated an increase in recycling and composting and, therefore, less garbage. However, this did not happen and trash overflowed from garbage bins. This upset the custodial staff because it created more work for them. There was also now more garbage contaminating the environment. This problem was resolved by scheduling more frequent garbage pickups.

Sometimes, unintended consequences are beneficial. Daniel and his team were not able to change tenants' behavior around recycling and composting, but they got the CEO and managers excited about it. When a new apartment complex was opened, the company implemented many changes that Daniel had hoped to see at the existing complex: incoming tenants signed agreements to commit to recycling, the name of the waste rooms was changed to "waste and recycling" rooms, and both waste and recycling bins were put in common areas. Daniel was thrilled: "All of a sudden, the three of us were working with the company. I go in-and-out of being inspired, and that reinvigorated me. Even if your work hasn't been necessarily successful, if people see that you are trying and people see that you care, all of a sudden they start caring and then you get inspiration from them, and they were inspired by you originally."

Working on sustainability problems is hard due to the inherent nature of these types of problems. The challenges faced by Daniel and his team arose as a result of the fact that "wicked" sustainability problems are embedded in systems that exhibit complexity. Additionally, solutions are very specific to each unique situation. This chapter describes the aspects of sustainability problems that make them so wicked, and offers an overview of the methods that can be used to develop solutions.

Core Questions and Key Concepts

Section 2.1: Understanding Sustainability Problems

Core Question: Why are sustainability problems so difficult to resolve?

Key Concept 2.1.1—Sustainability problems are wicked problems exhibiting six characteristics that make them difficult to resolve using traditional scientific approaches.

Key Concept 2.1.2—Wicked problems are difficult to resolve because they are embedded in socioecological systems characterized by complexity.

Section 2.2: Resolving Sustainability Problems

Core Question: How can sustainability problems be resolved?

Key Concept 2.2.1—Current state analysis, future scenarios, visioning, and transition strategies are emerging tools that can be used to resolve sustainability problems.

Key Concept 2.2.2—Weak and strong sustainability represent two different beliefs regarding the extent to which tradeoffs among environment, society, and economy may be made when resolving sustainability problems.

Key Concept 2.2.3—Including a diversity of stakeholders is key to resolving sustainability problems, as it ensures that both local and expert knowledge are incorporated into problem-solving processes and that solutions have "staying power."

Key Terms

- wicked problems
- tame problems
- complex systems
- norms
- system
- component
- interaction
- simple system
- boundary
- open systems
- closed systems
- reductionist thinking
- holistic thinking
- cascading effects
- intervention points (aka. leverage points)
- scales
- dynamics
- stock (aka. reservoir)
- flow
- feedbacks
- nonlinear
- shifting dominance
- threshold (aka. tipping point)
- regime (aka. alternative stable state)
- regime shift
- fluctuations
- state

Key Terms (*continued*)

- stabilizing feedback (aka. negative feedback)
- reinforcing feedback (aka. positive feedback or amplifying feedback)
- albedo
- inertia
- complex adaptive system (CAS)
- Transformational Sustainability Research (TSR)
- current state analysis
- indicators
- scenario
- vision
- visioning
- transition
- tradeoffs
- weak sustainability
- strong sustainability
- participatory approach
- coproduction
- transdisciplinary

> *"A wicked problem is a complex issue that defies complete definition, for which there can be no final solution, since any resolution generates further issues, and where solutions are not true or false or good or bad, but the best that can be done at the time. Such problems are not morally wicked, but diabolical in that they resist all the usual attempts to resolve them."*
>
> —*Tackling Wicked Problems: Through the Transdisciplinary Imagination*

Section 2.1: Understanding Sustainability Problems

Core Question: Why are sustainability problems so difficult to resolve?

wicked problems difficult problems that cannot be addressed using only traditional approaches, such as scientific and technological advances, and that require continuous attention because they can never be completely resolved.

tame problems relatively simple problems that can be solved using traditional approaches, such as scientific and technological advances.

Sustainability challenges are commonly referred to as **wicked problems**. As illustrated in the opening quote, this is not because they are morally bad or wicked as opposed to morally good. It is because they are very complicated and seem to withstand traditional approaches to solving them. Traditional approaches refer to the scientific and technological advances that have solved many of society's problems in the past. These traditional approaches can solve **tame problems**. For example, the development of vaccines to prevent diseases such as smallpox, polio, typhus, and tetanus have greatly improved human health. The solution to these human health issues was arrived at using the scientific method. In these cases, everyone agreed disease was a problem that should be solved.

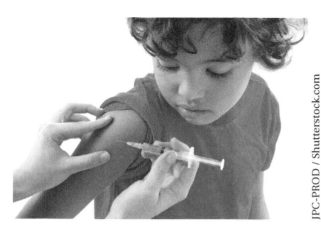 

Figure 2.1 The technical aspects of vaccine development and putting a man on the moon are tame problems, which can be solved using the relatively straightforward and linear type of thinking used by scientists and engineers.

The problem was thoroughly analyzed and adequately understood prior to developing solutions. The solution to the problem (a vaccine) was arrived at and mutually agreed upon through the traditional process of the science. Once the problem was solved, that was the end of the story.

Problems are tame problems when approaches to solving them involve moving from problem to solution in a step-by-step, linear fashion and relatively straightforward manner. It is clear when the problem is solved and there is nothing left to do. Other examples of tame problems include calculating the square root of 5,234, finding the shortest route from your home to your workplace using a map, repairing a car, or putting a man on the moon. An important point here is that a tame problem is not necessarily simple (Figure 2.1). In fact, especially in the case of putting a man on the moon, it can be incredibly technically complex. Thus, a major distinguishing feature between tame and wicked problems is not that tame problems are simple. Rather, it is that wicked problems are embedded in **complex systems**. Such systems have certain features such as high interconnectedness; cascading effects through multiple spatial and temporal scales resulting in unintended consequences; large uncertainties due to nonlinearities, path dependence, and inertia; and self-regulation due to feedbacks. This chapter will begin by describing six general characteristics that define wicked problems and will then explain how features of complex systems render wicked sustainability problems so difficult to resolve.

complex systems systems in which the many parts that compose the system are interconnected in an irreducible way, such that the whole system is greater than the sum of the parts.

Section 2.1.1—Introduction to Wicked Problems

Key Concept 2.1.1—Sustainability problems are wicked problems exhibiting six characteristics that make them difficult to solve using traditional scientific approaches.

The term *wicked problem* was first mentioned in the 1970s urban planning literature in a famous article written by University of California, Berkeley, professors Horst Rittel and Melvin Webber. They describe 10 characteristics of a wicked problem, but for the sake of simplicity, 6 general characteristics are given here:

1. Vague problem definition
2. Undefined solution
3. No endpoint
4. Irreversible
5. Unique
6. Urgent

This section will first describe these characteristics and then illustrate their application (**Box 2.1**). Then the section will conclude by briefly comparing tame problems with wicked problems using the six characteristics in order to highlight the difference between these two types of problems.

Characteristic 1: Vague Problem Definition. Wicked problems have a *vague problem definition*. This is because there are multiple and diverse stakeholders involved. Diversity among stakeholders is the result of many specific factors including living in different geographic locations, under different governance regimes with different capacities to deal with problems, and in societies with different cultures, values, beliefs, and informal **norms**. Diversity among stakeholders also means that not everyone will agree on the extent to which some issue is a problem or if it is even a problem at all.

Characteristic 2: Undefined Solution. The second characteristic of a wicked problem is that there is an *undefined solution*. In other words, there is no one definite solution to the problem; also, multiple and diverse stakeholders are involved. Even if there is consensus on the problem definition, not everyone will agree on when the problem has been resolved and how effectively.

Whether a solution to a wicked problem is "the right one" or "the wrong one" depends on who you ask. This can be contrasted with a tame problem, such as finding the square root of 256. No matter who you ask, the correct solution to this problem is 16. The solution is objective. Solutions to wicked problems are subjective. They are not black and white in an absolutely right or wrong sense like a math problem, but instead cover different shades of gray. Some are better or worse than others, or are good enough or not good enough, relative to a specific situation and the resources available to work on solving the problem. A solution that is good enough for one place and time might be not good enough for another.

---

norms define what is approved of or disapproved of within a society and, as a result, tend to shape or guide people's actions.

Characteristic 3: No End Point. The third general characteristic of a wicked problem is that there is *no end point*. When a solution is implemented, new problems arise because wicked problems are embedded in interconnected and complex systems, which makes them prone to cascading effects and unintended consequences (ideas presented in Section 2.1.2). Also, conditions change through time, and solutions must be continually adapted to meet new conditions. As a result, there is no obvious point at which we should cease trying to solve the problem. There are no final solutions to wicked problems. Rather, addressing wicked problems is an ongoing process. As a result, the word *resolving* rather than *solving* is used in this textbook to refer to attempts at addressing wicked problems. Although these terms are often used interchangeably, in this text, the former meaning is intended.

In reality, the implementation of new solutions often stops when the resources needed to solve a problem, such as money, time, and energy, become scarce. Even when the problem-solving process has stopped due to resource constraints, problems remain and new ones continue to arise. Thus, when devising and implementing solutions to wicked problems, it is important to try to anticipate and be prepared for new future problems as they arise.

Characteristic 4: Irreversible. The fourth characteristic of a wicked problem is that it is *irreversible*. This means that the effectiveness of a solution cannot be verified prior to implementation through low-stakes trial-and-error testing. Implementing a solution creates changes in the world that cannot be undone and will have real consequences.

Characteristic 5: Unique. The fifth characteristic of a wicked problem is that it is *unique*. This means that the same solution will not work effectively in all places. Every specific situation is distinct because the cultural, political, social, environmental, technological, economic, and other important aspects will be different in particular contexts.

Characteristic 6: Urgent. The sixth and final characteristic of wicked problems is that they are *urgent*. These problems are urgent because a failure to act will result in permanent harm to human and natural systems. The urgency of wicked problems presents challenges for resolving them. Because these problems are so urgent, solutions must often be pursued prior to fully understanding the problem. Given this, in addition to the fact that financial and other resources required for problem solving are limited, convincing key stakeholders to use their resources to take actions toward pursuing solutions without full information can be extremely difficult. It is because of these challenges that sustainability problem solvers warn of "paralysis by analysis," cautioning against too much problem analysis before action is taken. That said, there is a fine line between taking too long to act and acting too fast, as rushing a solution without collective decision-making can create more problems for people and ecosystems.

BOX 2.1   FREEING FEMALES FROM TWENTY-FIRST CENTURY OPPRESSION, A WICKED PROBLEM

An extremely serious, and tragically under-recognized, sustainability problem in the world today is oppression of women. In wealthy developed countries, oppression is associated with issues such as unequal salaries or career opportunities. However, in many developing countries around the world today, women are oppressed in much more harmful, and life-threatening, ways: baby girls are not given the same high quality medical care as boys; brides are burned to death for inadequate dowries or perceived disobedience; ultrasound technology results in abortion of female fetuses; young girls are kidnapped or sold into sex slavery, then locked in brothels against their will and beaten for noncompliance; mass rape of women is used by soldiers as a weapon to terrorize other tribes and ethnic groups; women are stoned to death if suspected of losing their virginity before marriage; and the unfortunate list goes on and on. These appalling human rights violations stem from the profound, deeply embedded inequalities between men and women in many countries around the world. It is estimated that 60 million to more than 100 million women are missing worldwide, and that more have died from oppression over the past 50 years than the number of men in all twentieth century wars combined. In the illustration of the characteristics of a wicked problem below, sex trafficking—one of the many forms of oppression that plague girls and women around the world—is used as an example.

Characteristic 1: Vague Problem Definition. Most readers of this book will agree that sex trafficking is a problem. However, deeply embedded social and cultural norms exist in the societies where oppression occurs, such as social stratification or a general attitude that women are insignificant and, therefore, disposable. The existence of these norms means that the people directly involved with the problem, who have the most power to stop it, are in disagreement about the severity, or even existence, of the problem. For example, many girls living in poor communities are sold into sex slavery by their families, who are desperate for money, because they are not valued as much as the money acquired in the sale. Local police have been known to instruct girls who escape from brothels, and turn up at police stations for help, to return to the brothel or, worse, sell them to another brothel for a profit. Further, police officers themselves often patronize these brothels. In India, girls forced into sex slavery are often low-caste peasants. The caste system is so imbedded into Indian society that many Indians don't protest this practice, or at least they look the other way.

Characteristic 2: Undefined Solution. Many proposed, and tested, solutions to the problem of sex trafficking exist. However, inaction comes when there is disagreement on the effectiveness of a given solution for a certain place and time. Three of the proposed solutions to the problem of sex trafficking are financial assistance, employment, and education (Figure 2.2). American Assistance for Cambodia, an aid group that assists females in starting a new life after escaping from forced sex slavery in a brothel, offers girls $400 to set up small businesses. Girls living in touristy areas become street vendors, selling clothing,

Figure 2.2 Women can be liberated from oppression by working as vendors in local markets or as assembly line workers in factories, or by attending elementary school.

jewelry, food, and drinks to travelers. In the short term, they are able to support their families. Over time, through saving and reinvestment, some expand their operations into permanent shops, becoming successful business women. Despite the success stories, people can be resistant to donating money for such loans because of the stereotype that poor people live in poverty because they are lazy.

In several East Asian countries, women were recently given the freedom to work in factories. Such work gives them a way to support themselves and their families, which can help prevent sex trafficking in the first place. Most of these assembly line factories produce the clothing, toys, and countless other products sold in stores of the wealthy Western world. In many cases, the vast majority of employees are women. Factory owners cite many advantages to hiring women instead of men, such as their smaller fingers for delicate detail work, their acceptance of lower pay, and their tendency to work harder. Allowing women into the labor force in these countries has resulted in many benefits, including a boost to East Asian economies, improved national savings rates, and increased education of women, as they could now pay for schooling. Despite the benefits to individual women, and their countries, not all would agree that this is a good solution. Some view these factories as "sweatshops," where overworked women earn low wages and are subjected to hazardous working conditions.

A success story in educating women involves American students at Overlake Elementary School, located in Redmond, WA. These students, led by their teacher Frank Grijalva, raised $13,000 through talent shows, bakes sales, and car washes, to donate to American Assistance for Cambodia (AAC), which focuses on preventing sex trafficking through education. With the $13,000, which was matched by both the Asian Development Bank and the World Bank, AAC built an elementary school in Pailin, a town on the Cambodian-Thai border known for its brothels. Almost 300 students attend the school. To keep girls in school, AAC started the Girls Be Ambitious program, which offers families $10 per month, in exchange for perfect

(continued)

attendance of their daughters in school. This program is a way for everyday people to help combat sex trafficking for $120 per year. However, some claim that school programs are not worth the investment because they can fall apart if funding is lost or an aid organization leaves an area.

Characteristic 3: No Endpoint. One reason that sustainability problems are so wicked is that they never end. When a solution is implemented, new problems arise. One example is when aid organizations work with families and national police officials to forcefully pull girls out of brothels. When girls are brought back to their homes, life can become dangerous and more challenging for them and their families. Brothel owners in their home community may have connections to the brothel from which the girl was rescued, creating tension within the community. Gang members affiliated with the brothels, who kidnap girls and sell them, seek revenge by threatening to kidnap other children in the family. Aid organizations are threatened with violence, and may be forced to move out of an area because of it.

Women who are given the freedom to work in factories take on the responsibility of paid employment, in addition to the unpaid household activities, such as cooking, cleaning, and childcare, that they are expected to continue to perform. In addition to the "double burden" of paid employment and continued duties in the household, under these working conditions some women are still oppressed and put in harm's way, as they are paid less than men, and are generally more willing to passively accept hazardous or unhealthy working conditions.

One extreme example of an unintended consequence of solution implementation is when a U.S. Senator, Tom Harkins, introduced legislation in 1990 to ban products manufactured by young workers in Bangladesh sweatshops. The mere introduction of this legislation caused factory owners to fire tens of thousands of young workers, many of them girls. Many of these girls, suddenly without employment, turned to brothels to earn a living, which resulted in many deaths due to AIDS.

Some unintended consequences do not result directly from an implemented solution, but instead are more of an extension of the problem that requires further solutions. If a young girl graduates from a newly built elementary school, continuing school through to college can be difficult. Sometimes, a middle school or high school might not exist locally. Daily travel to a school in another town is possible, but in violent regions where gangs roam the landscape, the journey is dangerous. Even if she graduates from high school, the cost of a college education can be prohibitive. The problem-solving process never ends.

Characteristic 4. Irreversible. The family of the young girl who was liberated from a brothel will forever live in danger and fear. Many women seeking work in factories leave their families behind to migrate to cities. A young girl graduates from elementary school, filled with hope and a new perspective on the world, only to have her hopes dashed when further education is not accessible. More Bangladeshi girls work in brothels because of the well-intentioned actions of one U.S. Senator, introducing legislation halfway around the world. All of these are changes in the world, in people's lives, that occur as a result of solutions that are tested through implementation. These changes cannot be undone.

Characteristic 5. Unique. If a solution worked to resolve the problem in one place, it may not work in other places. Financial assistance for girls to become street vendors may work in a large city frequented by tourists, but it would not work in a remote region of Afghanistan, for example, that is not frequented by tourists. Employment in factories cannot work for women living in areas where factories do not exist, or where women do not have the means to travel to areas where factories exist, regardless of whether they are free to work. In places where aid organizations exist, schools can be built and the education of girls sustained. However, some regions are extremely unsafe due to gang violence, such that foreign aid organizations hesitate to set up offices in these areas. Every situation is different due to the particular context in which it exists.

Characteristic 6. Urgent. Despite uncertainties, failure to act toward resolution of the problem results in permanent harm to girls and women, who endure unimaginable physical, mental, and emotional damage on a daily basis. Even future generations are affected, as children born in brothels to girls or women held as sex slaves grow up to be brothel servants (boys) or be forced into prostitution like their mothers (girls). With wicked problems, swift action is imperative, even though we might not fully understand, prior to action, the problem, the most effective solution, or the unintended consequences that might arise from a solution.

Wicked Problem—Tame Problem Comparison. In summary, wicked problems cannot be solved by the methods traditionally used to solve many problems for human societies. To demonstrate how wicked problems differ from tame problems, the characteristics of a wicked problem will be applied to the vaccine example (a tame problem) that opened this section. Wicked problems have *vague problem definitions* because there are multiple and diverse stakeholders. The smallpox disease is a tame problem because most stakeholders would agree that it is horrible disease that should be cured. Thus, smallpox is a clearly defined problem.

Wicked problems have *undefined solutions*. The technical solution to the smallpox problem was to develop a vaccine. Once a person acquired the smallpox virus, there was no cure so the only solution was prevention. In Medieval England, before the smallpox vaccine was developed, people used to think that hanging red curtains around the patient's bed would cure the disease. However, this did not work. Even if someone preferred this solution to a vaccine, it would be clear in the end that red curtains do not work. There is a defined solution to tame problems.

Wicked problems have *no end point*. Once the smallpox vaccine is developed, there is an obvious point where we cease problem solving in the medical research laboratory where the vaccine was developed. Although a small number of people may have complications from the vaccine, which would be considered an unintended consequence, overall it is a successful solution to the problem of smallpox and further efforts to produce solutions to the problem are not needed.

Wicked problems are *irreversible*. A scientist working in a laboratory to develop a smallpox vaccine learns by trial and error. This does not have real consequences for society as a whole. If one vaccine does not work, the scientist can try another one without much of a penalty for making a mistake.

Wicked problems are *unique* and require context-specific solutions. However, the smallpox vaccine is not a unique solution because the same solution to this disease works effectively in places around the world. Basic laws of physics, chemistry, and biology make this so. On the other hand, the socio-ecological systems in which wicked problems develop are not completely governed by such universal principles and local context is an important factor to consider when devising solutions.

Finally, wicked problems are *urgent*. They require that action be taken toward solutions prior to a thorough understanding of the problem. This is not to say that the smallpox epidemic was not urgent, but the problem could be sufficiently identified and analyzed before proposing solutions.

Section 2.1.2—Wicked Problems, Socioecological Systems, and Complexity

Key Concept 2.1.2—Wicked problems are difficult to resolve because they are embedded in socioecological systems characterized by complexity.

> *"Just how complex is a particular system? Is there any way to say precisely how much more complex one system is than another? . . . Is a human brain more complex than an ant brain? Is the human genome more complex than the genome for yeast? Did complexity in biological organisms increase over the last four billion years of evolution? . . . These are key questions, but they have not yet been answered to anyone's satisfaction and remain the source of many scientific arguments. [N]either a single science of complexity nor a single complexity theory exists yet . . . [R]ather [there are] several different sciences of complexity with different notions of what*

> *complexity means. . . . Intuitively, the answer to these questions would seem to be "of course." However, it has been surprisingly difficult to come up with a universally accepted definition of complexity that can help answer these kinds of questions."*
> —Melanie Mitchell, Complexity: A Guided Tour, 2009

The major reason that wicked problems are so difficult to resolve is that they are embedded in socioecological systems (SESs) that exhibit complexity. Thus, resolving them requires an understanding of their complexity. All systems, a car or an ecosystem or an economy, are made up of individual components and interactions among those components. Thus, this section will start by defining what is meant by a system, any system. However, the degree of complexity exhibited by different types of systems is not always the same, and defining how complexity differs among various systems is not straightforward. As implied in the opening quote to this section, precisely articulating just how complex some systems are relative to others is an active area of research. As such, this section will introduce a set of tools for thinking about complexity rather than a comprehensive set of hard-and-fast rules that define complexity because the latter simply does not yet exist.

The tools for thinking about complexity presented here, and in much more detail in later chapters (especially Chapters 5 and 6), can be considered as part of a "toolbox." When confronted with a repair, such as fixing a leaky kitchen sink, you might need a pipe wrench and slip joint pliers, but you will not likely need a screwdriver or a handheld circular saw. Similarly, you will not necessarily need all the tools for thinking about complexity presented in this text for understanding and resolving wicked problems in every single SES that you focus on. However, it is good to have all of these tools at your disposal in case you do need them in some cases. The tools are derived from ideas, theories, and frameworks for thinking about the world as used and developed in several fields that consider complexity as part of system analysis. These include dynamical systems theory, economics, thermodynamics, information theory, computation, evolutionary biology, ecology, and many others. For a more detailed look at how different fields of study have shaped thinking about complexity, see *Complexity: A Guided Tour,* an excellent book by Melanie Mitchell.

For the purposes of this introductory textbook, the tools for thinking about complexity will be focused on those that build an initial and general understanding of SES behavior. Because there is no universally accepted definition of complexity, terms are clearly defined as having a specific meaning in this text and may differ from how they are defined elsewhere. This is not to say that other definitions are "wrong" or inaccurate by any means, but the definitions provided here are meant to ensure a coherent presentation of the concepts introduced in this text. The systems outlook necessary for comprehending SES behavior must holistically incorporate the relatively well-characterized processes relevant to both human and natural systems, as well as accommodate historical legacies and a wide range of indeterminate human actions. To meet these conditions, with specific attention to sustainability, tools for thinking about SESs have been drawn from several sources:

- Basic systems thinking concepts derived from systems modeling (e.g., Meadows 2008)
- Dynamical systems theory as applied to natural and human systems to understand patterns of change (e.g., Scheffer 2009)
- Resilience approaches to understanding and managing complex adaptive systems (e.g., Walker and Salt 2006; Marten 2001; Chapin et al. 2009)
- Adaptive cycles framework for thinking about long-term change and resilience in evolving systems (Gunderson and Holling 2002; Marten 2001)

This text will only scratch the surface of the ideas presented in many of these books and readers are encouraged to consult them for more in-depth knowledge.

Socioecological Systems. The wicked problems described in the previous section are difficult to resolve because they are embedded in SESs that exhibit varying degrees of complexity. Before delving into tools for thinking about complexity, you will need a more precise understanding of what is meant by an SES and why SESs are necessary for understanding wicked problems. For a long time, humans studied natural systems as outside observers figuring out how to fix problems such as environment degradation. However, this removed humans from the system. In reality, humans and their actions are integral parts of the systems from which wicked problems arise. As described in Chapter 1, human societies have always influenced, and been influenced by, natural systems. Many of the problems facing natural systems today are rooted in human systems and the opposite can also be true (such as with natural disasters). The goal of sustainability is to simultaneously promote healthy ecosystems, human well-being, and viable economies. All of this means that human and natural systems must be studied together rather than separately, as socioecological systems (SESs, also known as coupled human-environmental systems).

The SES framework represents the dynamic, two-way interaction between human and natural systems (Figure 2.3). Humans interact

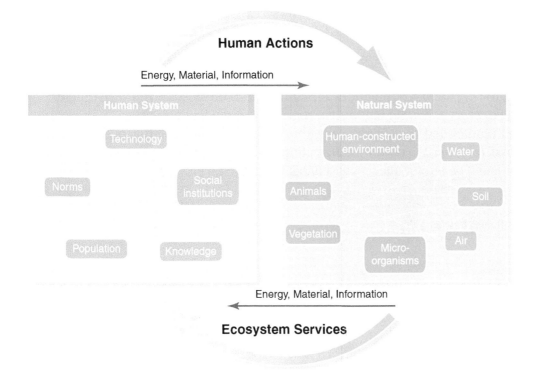

Figure 2.3 In socioecological systems, the interactions between human and natural systems are two-way.

with natural systems in order to gain some useful resource from them or take advantage of ecosystem services. As a result, they change the natural system in some way through human activities. If the change is substantial, it can reduce the capacity of natural systems to provide human systems with the resources and ecosystem services needed for human well-being and survival. When human systems become aware of such a problem, people and institutions may take action or make changes in response. The action can further affect natural systems, which can further affect human systems. The two-way interaction between human and natural systems continues in this way indefinitely. These ideas are illustrated with an example of changes in agricultural systems during the Industrial Revolution (**Box 2.2**).

Do SESs really exhibit that much complexity? After all, the example given in Box 2.2 seems relatively straightforward. The short answer is: yes. SESs can exhibit certain behaviors that make them incredibly difficult to understand. These features arise from the *dynamic network of interactions among system components* and the *ability of systems to continually adapt to changing external conditions*. There is a great deal of meaning behind many of the italicized words in the previous sentence. Precise definitions of some of these words are given in the remainder of this section, while others are merely introduced and

78 *Sustainable World: Approaches to Analyzing and Resolving Wicked Problems*

BOX 2.2 CONCEPT ILLUSTRATION

In an agricultural system (Figure 2.4), people grow crops such as corn in order to have food to eat. During the Industrial Revolution, in an effort to increase the amount of food produced for a growing human population, agriculture became mechanized such that fossil fuel–powered tractors and other equipment replaced human labor (Figure 2.5). However, there were unintended consequences to this action in terms of impacts on natural systems. The fossil fuels used to power tractors added excess CO_2 to the atmosphere, which affected the climate. The tractor allowed vast areas of land to be plowed, which were often

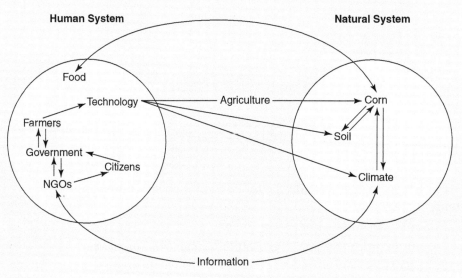

Figure 2.4

Figure 2.5 Before the Industrial Revolution, fields were plowed using human and animal power (*left*). Today, most agricultural fields are plowed by tractors, powered by fossil fuels (*right*).

left fallow after a crop was harvested. The bare soil on this land was more susceptible to erosion by wind and water than soil covered with a crop. The stability of the climate and the health of soils both impact the capacity of a plot of land to support crops. If this capacity was decreased, less food would be available for people to eat, and human well-being would be negatively impacted. This new information about the natural system was transmitted to the human system. People working in environmental nongovernmental organizations

Figure 2.6 Cover crops, such as the yellow mustard plant used in this vineyard, are planted to literally cover the soil to prevent soil erosion and bring nutrients back to the soil.

(NGOs) noticed the impacts of mechanized agriculture on soils and climate. The NGOs directly lobbied the government to put in place policies to protect soils and climate. They also raised awareness with citizens, who also lobbied the government. Governments worked with farmers to develop a plan to reduce soil erosion, such as the use of cover crops (Figure 2.6). Next, the use of synthetic nitrogen fertilizer arose as another technological advance for agriculture. However, fertilizer use resulted in excess N_2O gas, which affects climate. Climate change caused severe drought in some regions and flooding in others, both of which decreased food production. New information about the natural system again fed back into the human system and the process of response by the human system started all over again.

in-depth descriptions are left for later chapters of this text (especially Chapters 5 and 6). Three types of systems are defined in the remainder of this section: simple systems, complex systems, and complex adaptive systems. The different types of systems are characterized by different levels of complexity. All three types have implications for understanding SESs and resolving the wicked problems embedded in them.

Complexity Level 1: Simple Systems. Regardless of the level of complexity exhibited, a **system** consists of a set of interconnected components structured in such a way that the interactions among the components accomplish some purpose that cannot be achieved by any component alone. A **component** is a distinct part of a system, such as a steering wheel in a car system, trees in an ecosystem, the stomach in the digestive system, consumers in an economic system, and elected officials in a

system a set of components and the interactions among them that function together as a whole to accomplish or serve some purpose.

component a single element in a system that plays a specific role as part of the overall system.

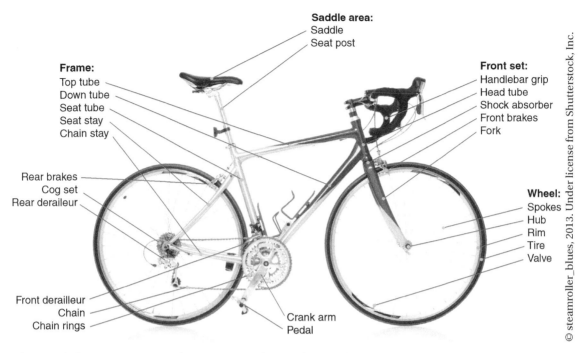

Figure 2.7 A bike is a simple system made up of many interacting components.

---

interaction a process through which a subset of system components relate to each other.

---

simple system a system in which components interact with each other to serve some purpose, but are not connected in an irreducible way as in a complex system.

---

political system. An **interaction** is a process through which different system components associate with each other, such as how a steering wheel interacts with a gear through physical forces to turn the car, how trees interact with CO_2 in the air through photosynthesis, or how a consumer interacts with a producer through purchasing in an economic system. In these systems, the steering wheel and gears, trees and CO_2, and consumers and producers are all components, whereas physical forces, photosynthesis, and purchasing are the interactions between components.

The purpose accomplished by each of these systems—steering a car, turning CO_2 into food using sunlight energy, exchanging goods and services—cannot be achieved by any of the individual components alone. This is an important point because not all collections of random things are systems, such as rocks lying on a sidewalk, because they are not interacting in a way that achieves some purpose that none can accomplish alone. If one or two rocks are taken away, the rest are still just a bunch of rocks lying on the sidewalk. However, if the gears were removed from a car, the trees removed from a forest, or the consumers removed from an economy, then these systems would no longer accomplish the same purpose.

A bicycle is an example of a **simple system** made up of many components and interactions among those components. Some components of the system that collectively serve the purpose of moving the bike forward include the pedals, crank arm, chain ring, chain, cog set, and front and rear wheels (**Figure 2.7**) Interactions among these components that move the bike forward are different physical forces exerted among the components. The person riding the bike uses force to push down on the pedals, and this causes the crank arm to rotate the chain rings. The chain rings move the chain, which is connected to the cog set on the back wheel. The movement of the chain causes the cog set to rotate, which results in the rear wheel turning and the bike moving forward.

In addition to components and interactions, all systems have a **boundary** that separates the internal components and interactions from the external environment. Boundaries can be real, such as a chicken wire fence surrounding a garden to keep out cats who like to use it as a litter box, or they can be imaginary. Imaginary boundaries are more important to understanding the ideas presented in this text. Imaginary boundaries are lines drawn around a system to mentally separate it from the external environment that influences a system's behavior to varying degrees. All systems considered in this text are **open systems** such that energy, materials, and information move across system boundaries from the external environment into the system, or vice versa. As a result, the processes occurring outside of a system's boundaries can affect the system, and cannot be ignored just because they are outside of the boundaries. This is opposed to **closed systems**, which do not interact with their external environments. In open systems, external processes are often crucial to fully understanding system behavior.

If the external environment can influence a system's behavior, then why bother separating a system from its external environment using boundaries in the first place? Boundaries are defined to simplify real-world systems, which contain seemingly limitless numbers of components and interactions among those components. As such, boundaries are delineated to include those components and interactions most important to understanding the problem being addressed. There is no one valid boundary for any given system, and boundaries vary not only as a result of the problem being addressed, but also for the same problem and system based on the perspective of the person defining the boundary. This is especially important for SESs, which are studied by a wide range of natural and social scientists and incorporate practical knowledge (see Section 2.2.3). Ecologists, economists, anthropologists, and practitioners might all draw different boundaries around the same system to address the same problem based on how their diverse perspectives shape their understanding about how the world works. They might also choose to include different components and interactions for the same reasons.

For now, let's use the bike example to illustrate how the problem being addressed determines how system boundaries are defined. Pretend that your bike will not move forward. What components and interactions will you need to include in your system to understand this problem? **Figure 2.8** shows three possible system boundaries: A, B, and C. The boundaries are portrayed with dotted lines, rather than solid

boundary an imaginary line drawn around a system to conceptually separate the internal components and interactions from the environment external to the system.

open systems systems in which energy, materials, and information can move freely across system boundaries from the external environment into the system and vice versa.

closed systems systems in which energy, materials, and information cannot move across system boundaries from the external environment into the system nor vice versa.

82 *Sustainable World: Approaches to Analyzing and Resolving Wicked Problems*

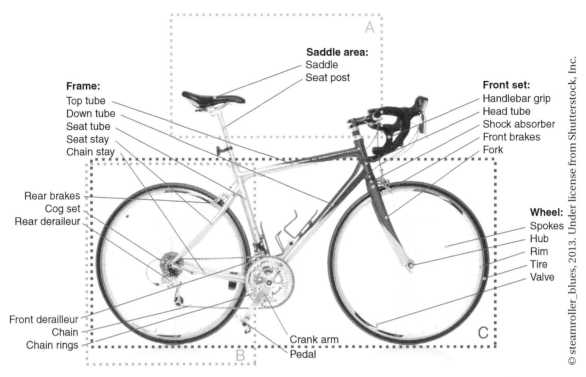

Figure 2.8 Defining system boundaries isolates the components and interactions that are most important for solving the problem.

lines, to indicate that they are open systems influenced by processes in the external environment. If you suspect that a flat rear tire is the reason your bike will not move forward, then you could define the system boundary to be Boundary B. If you suspect that a broken chain is the problem, then you would need to broaden your boundary to Boundary C. Defining your system boundary to be Boundary A is unlikely to help you understand why your bike will not move forward. Formally defining system components, interactions, and boundaries in this way in order to understand the bike, and solve the problem of your bike not moving forward, may seem unnecessary and excessive for such a simple system. Although simple systems can be complicated, such as a bike or even a spacecraft meant for travel to other planets, these systems are not characterized as complex. However, ecosystems, economies, and political systems are complex systems, and the tools for thinking about complexity presented thus far are important, but not sufficient, for understanding and resolving wicked problems in SESs, which exhibit much greater complexity. Before moving on to tools for thinking about complexity beyond simple systems, three important ideas relevant to understanding SESs that build on the tools presented in this section will be described: the importance of holistic thinking, the potential for cascading effects, and the multi-scalar nature of SESs. Before moving on to these concepts, **Box 2.3** provides a sustainability example that illustrates the systems concepts presented thus far.

BOX 2.3 MEXICAN FARMERS AND THE WORST GLOBAL COFFEE CRISIS IN HISTORY

Mexico is one of the leading coffee producing countries in the world. As the structure and interactions within the global coffee market changed throughout the mid-twentieth century, and into the twenty-first century, many small coffee farmers in Mexico experienced serious crisis. For a long time, growing coffee was a stabilizing aspect of rural economies in Mexico. Many people living in these regions are indigenous subsistence farmers who integrate coffee into their diverse agricultural systems, which includes growing and harvesting corn, fruit, medicine, fibers, firewood, and building materials, in addition to coffee. In these systems, growers were not dependent on coffee for survival. Instead, it was a means to earn cash, which supplemented their livelihoods and insulated them against encroaching global market forces. Global markets can cause unpredictable price fluctuations in commodities, such as coffee, due to factors such as changes in weather, supply and demand, and the actions of commodity traders. If left unregulated, global markets can wreak havoc on the lives of small rural coffee growers.

In the 1960s, the International Coffee Organization (ICO) passed the International Coffee Agreement (ICA) to protect small farmers in developing nations from global market forces. The agreement, which was among ICO member nations producing and consuming coffee, set quotas on exports and imposed prices on coffee. The export quotas prevented producing nations from flooding the world market with coffee, which would cause the price of coffee to decline globally. The imposed prices kept the price of coffee stable, and relatively high, so that small growers knew what to expect from their coffee sales. With stable global coffee markets in place, many small growers shifted from diverse agricultural systems to more of a focus on growing coffee. When this happened, food crops, and other crops that supported their livelihoods, were reduced or eliminated. This made farmers more dependent on selling coffee to earn a living. Under a global coffee market regulated by the ICA, this dependence was not a big issue. However, it left small farmers extremely exposed to global markets should the ICA be dissolved. In 1989, largely as a result of actions taken by the United States to sabotage the ICA, and to a lesser degree a growing surplus from non-ICO nations and the changing preferences of coffee consumers, the ICA did collapse.

Once the ICA collapsed, coffee prices crashed overnight. Small coffee growers in Mexico instantly lost 70% of their income. In many cases, the cost of growing coffee became higher than the price paid for it. A brief and temporary rebound in prices occurred in the mid- to late-1990s, but then prices plummeted again in 1999, hitting their lowest price ever in 2001 (less than 50 cents per pound). The second crash in 1999 had to do with a new neoliberal economic agenda, which led the World Bank, and other regional development banks, to implement policies aimed at reducing poverty in developing countries by increasing their exports. This encouraged even more coffee production by farmers, which caused prices to decline further due to oversupply.

(continued)

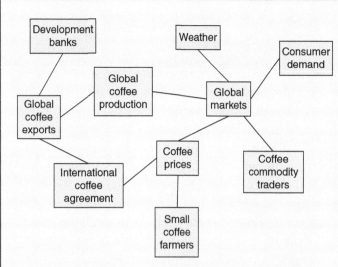

Figure 2.9 In this representation of the coffee crisis, components are represented by boxes and interactions are represented by the lines connecting the books.

The effects of global markets on small coffee farmers were devastating for many Mexican coffee farmers. Although some were able to revert back to their previous subsistence livelihoods, many could not. Some migrated to cities. However, due to land tenure policies in coffee-growing regions of Mexico, farmers who abandoned their land and migrated to cities would permanently lose their land and homes. Some went further than cities to the U.S. border, where they risked their lives in dangerous, and sometimes deadly, border crossings. Those who stayed lived in extreme poverty, being forced to cut back on expenditures, such as purchasing food and education for their children, and selling many of their belongings. Social upheavals and malnutrition became the norm in rural coffee producing regions.

Figure 2.9 shows one possible way that the system components and interactions for the coffee crisis could be represented. Defining the system helps in determining solutions that might be most effectively implemented. In some places, one successful solution to the problem is fair-trade coffee. Markets for fair trade coffee promise to buy coffee from small farmers at a fair and stable price, which is not subjected to the wild price fluctuations of global markets. Fair trade organizations do this by eliminating many of the "middlemen" or intermediaries, such as processors, exporters, brokers, importers, and distributors, who take a cut of the profit as coffee moves along the supply chain from field to café or retail shop. One strategy for shortening the supply chain is to have small farmers do some of the processing before selling their coffee beans, often by removing the parchment shells around the beans and then drying them. When small farmers do this, their product is referred to as "value-added," as it fetches them a higher profit. Finally, unlike in global markets, where coffee is sold to the highest bidder at any given moment, fair trade organizations establish long-standing relationships with farmers. As such, consumers can know who grew and processed the coffee they purchased. Fair trade organizations provide a more sustainable livelihood for many coffee farmers, who are no longer need to ride the violent wave of global market forces.

Holistic Thinking. The components of SESs are highly interconnected through their interactions to accomplish some purpose that cannot be achieved by any of the components alone. Therefore, it is best to look at the system as a whole when trying to resolve sustainability problems and not just a single component. Reductionist thinking has dominated knowledge production for centuries, as traditional disciplines were further divided into more and more specialized areas of study. For example, the discipline of biology was broken into different areas of study such as genetics, molecular biology, physiology, ecology, and many others. This specialization allowed for many discoveries that contribute to human well-being, and is still necessary for solving many problems, but a more integrated way of looking at the world through holistic thinking is necessary to ensure the sustainability of SESs into the future. This is analogous to the need to view a bike as a whole system, rather than its individual components, when figuring out why the system as a whole will not move forward. The importance of holistic thinking to resolving wicked problems can be illustrated with the story of the "The Blind Men and the Elephant":

> *Beyond Ghor, there was a city. All its inhabitants were blind. A king with his entourage arrived nearby; he brought his army and camped in the desert. He had a mighty elephant, which he used to increase the people's awe. The populace became anxious to see the elephant, and some sightless from among this blind community ran like fools to find it. As they did not even know the form or shape of the elephant, they groped sightlessly, gathering information by touching some part of it. Each thought that he knew something, because he could feel a part. . . . The man whose hand had reached an ear . . . said: "It is a large, rough thing, wide and broad, like a rug." And the one who had felt the trunk said: "I have the real facts about it. It is like a straight and hollow pipe, awful and destructive." The one who had felt its feet and legs said: "It is mighty and firm, like a pillar." Each had felt one part out of many. Each had perceived it wrongly. . . . This ancient Sufi story was told to teach a simple lesson but one that we often ignore: The behavior of a system cannot be known just by knowing the elements of which the system is made.*
>
> —Thinking in Systems *by Donella H. Meadows (p. 7)*

By viewing the world in a reductionist manner, we are like the blind men trying to resolve sustainability problems. We do not see the system holistically; we only see specific components of the system. As a result, conclusions about the behavior of a system can be wrong and unexpected behaviors (surprises) are more likely when system components are viewed in isolation. Another example can be seen in the

---

reductionist thinking a point of view that claims systems can be understood by studying each individual component and the interactions among those components as separate from the overall system.

holistic thinking a point of view that claims systems *cannot* be understood by studying each individual component and the interactions among those components in isolation from the overall system.

human body. High blood pressure causes the heart to work harder to pump blood; this can lead to a heart attack. High blood pressure can be caused by many things including diet, lifestyle choices such as smoking or lack of exercise, obesity, stress, genetics, age, thyroid disorders, or kidney disease. In order to resolve this problem, the circulatory system, which is what the disease affects, cannot be considered in isolation from other components. One must look outside the circulatory system to the urinary system (kidneys), immune system (thyroid), and beyond to a person's diet, overall lifestyle, and even genetic makeup.

Acknowledging the interconnectedness among SES components and the need to view them holistically is particularly important at this juncture in human history. For the first time ever, we live together in an intimately linked global society. Just like a problem in one part of the body can cause a heart attack in another part that can then lead to death, a problem in one part of the world can have effects on many other parts of the world that can then lead to suffering.

A final point to make about the interconnectedness of SESs is that there is no simple, singular solution to wicked problems. Instead, several solutions must be pursued simultaneously. This idea can be illustrated by drawing from an article on global food security written for the journal *Science* in 2010 by British biologist H. Charles Godfray and his colleagues. In this article, three challenges to the future of food security are presented: increasing wealth around the world for some will increase demand for meat and other energy-intensive foods (the A in the I = PAT model), the uncertain impacts of climate change on agriculture, and ensuring that increasing populations in the world's poorest regions have enough to eat. The authors propose several solutions to increase food production. These include technological innovation, reducing food waste, and changing diets so that people consume less meat and other energy-intensive foods. Recall from the very beginning of this chapter that creating a sustainable world requires altering individual behavior and consumption patterns, engaging values and changing social norms, promoting technological advances, and fostering the social, political, and economic institutions necessary to guide individual actions toward sustainable behavior. All of these factors are involved to some degree in the solutions proposed by Godfray and his colleagues.

Cascading Effects. When different components of SESs are very strongly interconnected, there is a potential for **cascading effects** that are often rapid and sometimes unpredictable. Cascading effects occur when the effect of a small action on the components, or the interactions among the components, in a system is amplified, such that the small action snowballs into a much larger change in the system. This is also referred to as the *butterfly effect*. It is important to note that the action can be natural or human-caused. For example, an equally destructive wildfire

cascading effects a chain of events set off by an action in one part of a system that results in relatively larger and typically unpredictable impacts on the rest of the system.

can be ignited by a natural lightning strike or by a cigarette thrown out of a car window. Either can be viewed as the "match that starts the fire" in a small square inch of ponderosa pine forest, which eventually cascades up to one pine tree, then to another tree, then to the entire forest ecosystem, and finally on to an entire landscape of forest, desert, scrub oak, and grassland ecosystems. This idea can also be applied to human systems. For example, the invention of Facebook occurred on a single college campus (Harvard) almost a decade ago and was initially used only by Harvard students. It was eventually expanded to other colleges in Boston and then to all Ivy League schools in the United States. It was later opened to high school students, then to anyone over age 13. Today, it is used by almost 1 billion people worldwide. This one innovation by a few college students has cascaded through sociotechnical systems to change how people interact and connect with one another socially.

Understanding cascading effects helps to explain problems in SESs, but it can also help find effective solution options by identifying **intervention points (aka. leverage points)**. These are locations in a system where a small change in a component or an interaction can lead to a large change in the overall system. If intervention points can be identified, more effective transitions to sustainability might be devised. A few small and effective actions aimed at problem solving, taken in one part of the system, can cascade through a system to have a large overall effect that can move the system toward sustainability. Like the lightning strike or cigarette that starts the fire, one small action at an intervention point can ignite major changes toward sustainability. Such interventions in SESs can have consequences that reach far beyond the initial effect to create changes in the systems at other places and time periods.

Multi-Scalar SESs. Sustainability challenges exist at multiple **scales**, and all scales must be considered when resolving these problems. This makes defining system boundaries, based on the internal system components and their interactions on the one hand and the external environment on the other, important but also seemingly arbitrary or fuzzy at times. SES scales can be both spatial (local, national, international) and temporal (now, next year, in a century). Thus, changes that affect SESs can occur on very small to very large spatial scales. For example, both a small pine bark beetle and human-caused global climate change affect forests. The 5 mm long mountain pine bark beetle is about the size of a grain of rice. Beetle populations play an important role in the natural life cycle of forest ecosystems in the western United States and Canada by infecting old, weak trees to make room for new ones. When trees are infected and die, their pine needles turn from green to red. The most severe destruction of forests by this beetle ever observed, from Alaska to southern California, began in the early 1990s and continues to the present day. Increased temperatures and drought in these regions resulting from global climate change are thought to be a major factor.

intervention points (aka. leverage points) efficient places to intervene in a system where relatively small changes or efforts cause a large overall shift in the behavior of an entire system.

scales a reference used to classify, arrange, and understand internal system components and their interactions in relation to their external environment.

Figure 2.10 Global climate change affects the tiny bark beetle (*left*), which exists at small spatial scales, such that entire pine forest landscapes (*right*), existing large spatial scales, become damaged.

Higher temperatures promote larger beetle populations by speeding up reproduction and decreasing mortality during previously cold, harsh winters. In addition, drought makes trees more vulnerable to beetle attack. Although the forest ecosystem may be the system of analysis around which boundaries are drawn in this case, both changes in components or interactions at small spatial scales within the system (e.g., a tiny beetle) and in the external environment outside of the system at large spatial scales (e.g., global climate change) both influence the system (Figure 2.10).

In addition to spatial scale, temporal scale is also a factor. Impacts on SESs occur on time scales ranging from seconds to millennia. For example, both asteroids and volcanoes have affected global biodiversity in the past. The Cretaceous-Tertiary mass extinction wiped out 70% of all biodiversity on Earth, including the dinosaurs. One proposed cause of this extinction is an asteroid impact that took seconds. However, it is the cascading effects of this impact that are thought to have caused the actual extinction. The impact blasted dust and aerosols into the atmosphere, which blocked sunlight from reaching the earth. Plants could not photosynthesize without sunlight and died. Because plants are the base of food chains, the effect of their death cascaded through food chains to affect all other organisms higher in the food chain. The other proposed cause of this extinction was the 800,000-year-long volcanic eruption known as the Deccan traps. This eruption could have contributed to mass extinction in a similar way as the meteor by adding dust and aerosols to the atmosphere, but over much longer time scales (Figure 2.11). When analyzing sustainability problems, spatial scales ranging from the tiny bark beetle to the entire global atmosphere, and time scales ranging from milliseconds to millions of years, must be considered. Solutions must also be implemented at multiple spatial scales to ensure sustainability at various locations around the world today, and at multiple temporal scales (now and in the future) to ensure intergenerational equity. **Box 2.4** applies these concepts, and illustrates the importance of holistic thinking and cascading effects, to a sustainability problem.

Figure 2.11 Both an asteroid impact (*left*), which took seconds, and volcanic eruptions (*right*), occurring over hundreds of thousands of year, may have contributed to the extinction of the dinosaurs and other life on Earth about 65 million years ago.

BOX 2.4   THE 2007–2008 FOOD CRISIS: HOLISTIC THINKING, CASCADING EFFECTS, AND MULTIPLE SCALES

The 2007–2008 Food Crisis is a good example of just how interconnected the world is today, and why this makes holistic thinking an essential tool for resolving sustainability problems. It also shows how small actions or changes to one part of the system lead to effects that cascade throughout the system, resulting in a large change in the system. Finally, the food crisis shows how interactions among system components, occurring at multiple spatial and temporal scales, come together to result in a food crisis. At end of 2007 almost 40 countries around the world were in a food crisis. Food became extremely expensive at this time. The countries in Africa were the worst hit, such as in Somalia where grain costs rose 350% and in Ethiopia by 500%. There was civil unrest as people protested high food prices in cities around the world. The cause of the food crisis is still up for debate, and most agree that there was no single cause, but multiple causes (Figure 2.12), an idea discussed in more detail in Chapter 3, Section 3.2 on causal chain analysis. The major proposed causes of the food crisis are *increased use of biofuels, rising oil prices, climate change, more demand for meat*, and *market speculation*.

When *biofuels* are made from food crops such as corn, there is less food available to eat. With a lower supply of food and an unchanged demand, the price of food rises. *Rising oil prices* contribute to the food crisis directly and indirectly. Both the use of oil-based synthetic

(continued)

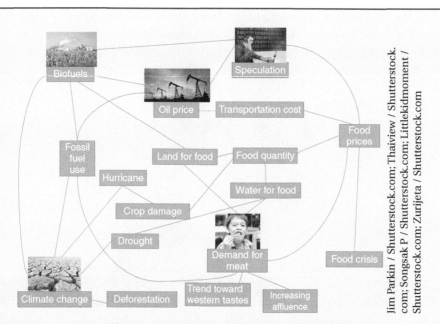

Figure 2.12 The food crisis had multiple causes, shown here with photos.

fertilizers, and the use of gasoline for transportation to market, directly affect food prices when oil prices rise by raising food production costs. Indirectly, high oil prices provide an incentive to seek out other fuel sources and contribute to the growth of biofuels. Severe droughts and flooding due to *climate change* can decrease food supplies, leading to an increase in food prices due to reduced harvest levels. Some countries were affected by this more severely than others. For example, wheat harvests in Australia were 50% to 60% below normal levels and 15% below normal in the United States. Overall, global grain production fell by 1.3% just before the food crisis. Rising populations and increased wealth in developing countries such as India and China led to *more demand for meat*. It takes about 7 kg of grain to produce 1 kg of beef. As with biofuels, diverting grain from the food supply for another use decreased the amount of grain available for humans to eat. Again, reduced supply caused food prices to increase. The role of *market speculation* is the least certain, but many suspect that this played a role as well. Basically, market speculation involves betting on the future price of some commodity. In this case, the speculators' bets were wrong, and this resulted in increased food prices. This example illustrates the vast interconnectedness of our world today and how problems in a few regions of the world can threaten the entire global system. The choice to use more biofuels in the United States can lead to people starving in Ethiopia.

Complexity Level 2: Complex Systems. Many of the the complex system concepts introduced in this section will be more thoroughly explained in Chapters 5 and 6. For now, only a few concepts are explained in some detail in order to illustrate the uncertainty inherent in complex systems, which is part of what makes sustainability problems so wicked. Complex systems exhibit certain **dynamics**, or patterns of

dynamics patterns of change exhibited by a system over time.

change over time, that distinguish them from the simple systems. One major distinguishing feature of complex systems is that system components can change over time as a result of the interactions among them. Cars and bicycles, on the other hand, are not complex systems because their components (e.g., gears, crankshaft, wheels) do not change as a result of interaction with other components through physical forces. In complex systems, the size of system components can change over time as a result of interactions, such as when trees grow in a forest due to photosynthesis or the number of producers in an economy grows due to increased purchasing by consumers. In systems-modeling jargon, each component is referred to as a **stock (aka. reservoir)** to reflect the amount of energy, materials, and information that it contains. Each system interaction is referred to as a **flow** to represent the amount of energy, material, and information added to or removed from components over a certain period of time.

Complex systems also exhibit **feedbacks** that can lead to **nonlinear** behavior. Feedbacks occur in complex systems when the components themselves regulate the processes defining the interactions among components. For example, two components of your personal finance system are the banking institution where you do business and your own personal bank account. One interaction between the banking institution and your bank account is the addition of money to your account based on interest rate, as defined by your banking institution. The amount of money being added to your account is regulated by the amount of money in your account at any given time. For example, if the interest rate is 1% and you have $1,000 in your bank account to start, then $10 is added when interest is compounded. The next time interest is compounded, the amount added is $10.10 because it is based on the new amount of money in your account which is $1,010. If you carry out these calculations farther into the future, you will see that your bank account grows in a nonlinear fashion as shown in Figure 2.13. This type of regulation of interactions among system components, by the size of a component, at any given time leads to nonlinear system behavior over time. These dynamics are not exhibited by simple systems, which tend to behave in linear ways. For example, if the chain ring (system component) on a bike were made to move faster by more intense pedaling, the physical forces (system interaction) that make it move would not change. This is in contrast to the way that the amount of money deposited in your bank account (system interaction) changes when there is more money existing in the account (system component).

One example of a sustainability problem as a complex system is the growth of technology, innovation, commerce, and human well-being. In their book *Abundance: The Future is Better Than You Think*, Peter Diamandis and Steven Kolter argue that technology has in the past, and will continue in the future, to increase the well-being of people

stock (aka. reservoir) a system component that reflects some quantity of energy, material, or information contained in a system.

flow a system interaction that represents the rate at which some quantity of energy, material, or information is transferred among system components or between the system and its external environment.

feedbacks specific types of interactions among system components, or between components and the external environment, that lead to the outcome of some process returning to affect the factor that originally initiated the process.

nonlinear an interaction defined by the fact that the magnitude of the factor causing some interaction in a system and the actual outcome of that interaction are not proportional.

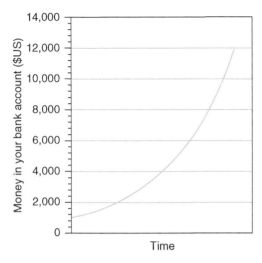

Figure 2.13 The nonlinear growth of your bank account over time is determined, in part, by the amount of money in your account.

Figure 2.14 In rural regions of Malawi, a country in southeastern Africa, women spend more than one-third of their time collecting and boiling water to obtain clean drinking water for their families.

and natural systems. To illustrate their argument, they give an example of how markets and technological innovation have increased human prosperity over time. Without technology and markets, they argue, each person or family would have to wash clothing by hand, acquire clean drinking water, collect firewood, and grow, harvest, and cook all their own food. These activities consume a large percentage of the total time available in a day, which reduces the amount of time available for education or other pursuits that contribute to human well-being. In contrast, if you focused your time on growing vegetables, and your neighbor focused on raising chickens, then you could exchange vegetables for eggs. In this case, both you and your neighbor still obtain a balanced diet, but now have more time for other activities. Focus on specific tasks is called specialization. Diamandis and Kolter argue that, throughout human history, specialization has promoted technological innovation, as each person has time to innovate and produce a product or service to trade in a market. More technologies, such as clean-running water to reduce the amount of time women spend obtaining clean drinking water, result in more time to pursue other things, such as an education (**Figure 2.14**). More time led to further specialization and innovation, and even more time saved. In this complex system, the amount of time available for pursuits that contribute to human well-being is analogous to the amount of money in your bank account: The more time there is for innovation, the more time-saving technologies are developed. Like the money in a bank account, the result is growth in free time.

Feedbacks not only cause systems to behave nonlinearly, but they can also result in unpredictable changes that contribute to uncertainty about how systems will behavior over time. Because of a **shifting dominance** between two major types of feedbacks, complex systems sometimes cross a **threshold (aka. tipping point)** into a new **regime (aka. alternative stable state)**. This **regime shift** results in a completely new system often characterized by a different set of interconnected components structured in new ways, such that the interactions among the components and feedbacks now support some new purpose. In short, regime shifts result in fundamental changes in system properties and behavior that may or may not be sustainable. In addition to regime shifts, feedbacks operating *within* a given regime lead to system **fluctuations** that vary the **state** of the system *within* that regime. Distinguishing normal systems fluctuations from complete regime shifts, and what system behavior looks like as it approaches a regime shift, is a difficult, but important, part of understanding SES complexity.

All of the bolded concepts in the previous paragraph will be thoroughly defined and discussed in much more detail in Chapter 5. For now, a few of these concepts are expounded on briefly as a general introduction to the factors contributing to uncertainty in complex systems. A simple way to understand how regime shifts are characterized by nonlinear and unpredictable system behavior is with a rubber band analogy. If you stretch a rubber band and then let it go, it will rebound to its original size. However, if you stretch it too far, it will break. Pick up a rubber band and begin to stretch the rubber band slowly using a constant pulling force. This is linear behavior because for each additional unit of force that you apply, the rubber band stretches the same additional distance. The nonlinear behavior starts as the rubber band approaches its breaking point. At this point, the rubber band feels more rigid and you have to apply much more force to move the rubber band the same distance until it finally snaps apart. When the rubber band breaks, you see a sudden change and the exact timing of that change was unpredictable. When the rubber band snapped, the rubber band crossed a threshold into a new regime that fundamentally serves a different purpose. The intact rubber band could hold a stack of cards together (Regime 1), but now it is a broken rubber band (Regime 2), which cannot be used to hold together a stack of cards in the same way

Sudden shifts in complex systems from one regime to another occur in both natural and human systems. Scientific analysis of these systems can often alert us to heightened risks of such sudden change, but it cannot yet forecast the exact point when change will occur. The importance of understanding these tools for thinking about complex systems behavior when it comes to sustainability of SESs is illustrated with an example of a regime shift in a coral reef ecosystem in **Box 2.5**. Coral reefs support high biodiversity, which contributes to ecosystem health, human well-being, and economic vitality.

---

shifting dominance a constant interplay between the two general types of feedback in a system that compete with each other for dominance and that determines the system's overall behavior at any given time.

threshold (aka. tipping point) the point beyond which a system shifts nonlinearly to a different regime.

regime (aka. alternative stable state) the dynamic and constantly changing yet characteristic pattern of conditions under which a system can exist and by which it supports certain functions or purposes.

regime shift a typically nonlinear change process by which a system transitions from an existing regime to a new regime under which the system supports different functions or purposes than before.

fluctuations periodic changes in the state of a system within a given regime that do not constitute a regime shift.

state the conditions under which a system exists at any given time.

BOX 2.5 SUDDEN CHANGES IN CORAL REEFS

Much of the time, change in ecosystems is linear. It is gradual and predictable. However, there are many examples of rapid nonlinear changes in ecosystems that surprise us. Coral reefs have experienced rapid shifts in species composition from being mostly corals to being dominated by algae (Figure 2.15). Two interacting factors often push a coral reef system toward this sudden change: overfishing and high nutrient inputs. In coral reefs, herbivorous fish graze on reef algae in the same way that cows graze on grass. Overfishing of herbivorous fish in reefs results in massive algae growth, just as grass would grow abundantly if cows stopped eating it. Algae also need nutrients to grow and more nutrients result in faster growth. Nutrient inputs to coral reefs include sewage inputs and runoff from agricultural fields. Overfishing and excessive nutrient inputs to reefs can occur over many decades and even centuries with no major change. Then suddenly, in a matter of weeks or months, the species composition can rapidly shift from coral-dominated to algae-dominated just as a rubber band suddenly breaks. Also like the rubber band breaking, these changes in coral reef systems can be irreversible and unpredictable. The two different reefs have different properties. Algae-dominated reefs harbor much less biodiversity than coral-dominated reefs; this can impact many human activities ranging from commercial fishing to tourism to subsistence reef harvesting

Figure 2.15 A healthy reef dominated by coral (*left*), which supports biodiversity and human livelihoods, can shift suddenly to an algae-dominated reef (*right*) that does not support either.

---

stabilizing feedback (aka. negative feedback) a general feedback type that keeps a system in its current regime.

It is the shifting dominance between two different types of feedbacks that result in complex systems shifting from one regime to another. A **stabilizing feedback (aka. negative feedback)** tends to keep a system in its current regime. They occur when two components of a system cause each other to change in opposite directions. For example, in hot climates, an air conditioner keeps a house cool. If the thermostat is programmed to keep the house at 75°F and the internal temperature of the house is currently 80°F, the thermostat will turn on the air conditioner. Once the temperature in the house reaches 75°F, the thermostat will signal the air conditioner to turn off. The two components of this system are temperature and the air conditioner. They cause each other to change in opposite directions: When the temperature goes above 75°F, the air conditioner causes the temperature to drop by turning on. When the temperature drops below 75°F, the air conditioner causes the temperature to rise by turning off. As long as this stabilizing feedback continues to operate, the internal temperature of the house will remain the same and will not be significantly affected by outside changes in temperature.

Feedbacks do not always create stability in complex systems and can instead result in change. A **reinforcing feedback (aka. positive feedback or amplifying feedback)** causes a system to change. Amplifying feedbacks cause systems to shift from one regime to another, such as in the examples of the rubber band and the coral reef. They occur when two components of a system cause each other to change in the same direction. For example, there are strong amplifying feedbacks operating when it comes to the melting of Arctic sea ice due to climate change. Since the 1970s when sea ice monitoring began, sea ice extent has dropped from an average of 7 million km^2 to 5 million km^2 in recent decades. This rapid melting is largely the result of a reinforcing feedback between sea ice extent and temperature. **Albedo** is a measure of the amount of solar radiation reflected from a surface rather than absorbed. When solar radiation is reflected due to a high albedo surface, it does not warm the surface. When it is absorbed due to a low albedo surface, it does. Higher temperatures in the Arctic have resulted in melting sea ice. Ice has a high albedo, so it reflects a lot of sunlight (about 30%–40% is reflected). As more and more ice melts, more water is exposed. Water has low albedo (it reflects < 10% of sunlight), so it absorbs more sunlight and heats up more than ice. This increases temperature in the water surrounding the ice, which causes the ice to melt even faster. The amplifying feedback operates between two components of this system: ice extent and temperature. Ice extent declines, temperatures increases (**Figure 2.16**). This causes ice to decline even more and temperature to increase even more. Eventually, when this amplifying feedback dominates over the stabilizing feedbacks keeping the system in its current regime (Regime 1: Ice-Covered Arctic), the Arctic system will shift to a new regime (Regime 2: No Ice Cover). Like the broken rubber band and the algae-dominated coral reef, an iceless Arctic will no longer support the same purposes as an ice-covered Arctic. For example, polar bears rely on ice as a platform for hunting seals, which are their main food source. An iceless Arctic will not be a system capable of supporting polar bears in the same way.

reinforcing feedback (aka. positive feedback or amplifying feedback) a general feedback type that causes a system to change and ultimately shift to a new regime.

albedo the fraction of solar radiation that is reflected from a surface rather than absorbed.

Figure 2.16 Light ice-covered surfaces (*left*) reflect sunlight and keep temperatures cool. As ice melts, darker water is exposed (*right*), causing more sunlight absorption, higher temperatures, and more rapid ice melting.

inertia the momentum of a system that keeps it moving in a certain direction and is difficult to resist when attempting to stop the system or steer it in a different direction.

In addition to playing a part in regime shifts, feedbacks also contribute to system **inertia**. This is another property of complex systems contributing to uncertainty of their behavior. Inertia has to do with an object's momentum, which keeps it moving in a certain direction. An object with a large momentum requires a lot of force to stop or steer in a different direction. Inertia is kind of like this, but it has to do with feedbacks. A system with strong inertia has a tendency to respond to external forces with reinforcing feedbacks, which send it strongly off in some direction. Once the system is moving off in that direction, it is difficult to stop it or steer it in another direction. Inertia results in a slow or delayed response of a system to additional forces that could eventually cause it to stop moving, or steer it in another direction.

Human societies regularly demonstrate inertia when confronted with the need for change. Individuals tend to align their own beliefs and attitudes with those people who make up the social group to which they are strongly tied, such as their family, friends, profession, or community. This tendency is so strong that even if individuals are presented with evidence that does not support their current beliefs, they will usually maintain them. This can make it incredibly difficulty for human societies to adapt to changing conditions, which can threaten their future survival. In his 2009 book *Collapse: How Societies Choose to Fail or Succeed*, scientist Jared Diamond argues that a society's response to a problem threatening its survival is one of the most important factors determining whether the society collapses or succeeds. One example is the Icelandic Vikings, who moved from Europe to colonize Greenland in the 10th century. In their homeland, they relied on cows as a major food source and brought them to Greenland (**Figure 2.17**). Even though cows were not as well adapted to the harsher, colder Greenland climate, they continued to depend on them for food and failed to adapt to their new environment by learning techniques for efficiently hunting

Figure 2.17 Cattle were an integral part of Viking society, as shown in this reconstructed Viking village in present-day Denmark (*left*), and with this drinking horn made from cattle (*right*), which was commonly used by Vikings.

local food sources, such as seal and caribou. In the early 15th century, Greenland plunged into an ice age, which created conditions under which the cattle could not survive. The Greenland Viking society experienced famine and eventually disappeared.

Inertia is also a property of natural systems. It can lead to continued change in a system even after the factor driving the change has stopped directly affecting the system. The climate system contains a lot of inertia. This is one reason the exact timing and extent of human-caused climate change is difficult to predict. For example, CO_2 emissions into the atmosphere are a major factor driving climate change. Even if humans stopped emitting CO_2 today, sea level rise due to thermal expansion of ocean water would continue for hundreds of years. Thermal expansion is the change in volume that a body of water experiences in response to a change in temperature. In the mid to late-20th century, sea level was rising at a rate of about 2 mm/year. However, the rate of thermal expansion has increased to *more than* 2 mm/year at present. This rate would continue to rise even if CO_2 emissions were stopped this instant. To understand why, just think of what happens when you heat water on the stove. When you place a pot of water on a burner, it does not boil instantly. It takes time. The oceans behave in the same way, but the response takes longer (centuries rather than minutes) because the global ocean is so immense. The melting of continental ice sheets due to climate change will continue for thousands of years, even if we stop CO_2 emissions today. Ice melting also does not occur instantaneously when exposed to heat. Put an ice cube out in the sun and you will see that it takes a few minutes to melt. Because the length of time between an actual impact on a system, and the expression of the full consequence of that impact in a system, is difficult to predict, system inertia adds an element of uncertainty.

complex adaptive system (CAS) a system capable of evolving over time in a manner that helps it adjust to changing conditions and typically in ways that promote its survival.

Complexity Level 3: Complex Adaptive Systems. In addition to the features of simple systems and complex systems already mentioned, a **complex adaptive system (CAS)** has several features that allow it to adapt to changing conditions, often in a way that promotes its survival. Thus, the distinguishing features of CASs are that adaptation to changing conditions plays a large role in their dynamic behavior and in defining their complexity. As laid out by Melanie Mitchell in her book *Complexity: A Guided Tour,* CASs have three major characteristics that define their complexity. First, they exhibit emergent properties that cannot be explained by the behavior of each individual system component alone. In other words, the whole is greater than the sum of its parts. This behavior emerges somewhat spontaneously, without a central controller organizing and directing individual component behavior, such that CASs seem to "take on a life of their own" (Figure 2.18). Second, CAS behavior is determined by a two-way exchange of matter, energy, and information among internal system components and processes on the one hand and the environment external to the system on the other. This shapes the behavior of CASs over time. Finally, CASs adapt their behavior to changing conditions over time by passing on and receiving information through learning and evolution. Following a brief introduction to adaptive cycles at the end of Chapter 5, which is a feature of CASs, other tools for thinking about CAS complexity are covered in great detail in Chapter 6.

All of the tools for thinking about SES complexity introduced here will be applied or expanded on in later chapters of this text. The tools are useful for understanding the SESs in which wicked problems are embedded. An overview of a problem-solving framework, called Transformational Sustainability Research (TSR), that can actually be used to go about resolving sustainability problems, based on an understanding of their complexity, is introduced in the final section of this chapter.

Figure 2.18 From Occupy Wall Street (*left*) to the Arab Spring (*right*), the recent use of social media, such as Facebook and Twitter, by individuals has resulted in large protests that "take on a life of their own," without a distinct leader or organizer.

Section 2.2: Resolving Sustainability Problems

Core Question: How can sustainability problems be resolved?

Three general topics will be presented in this section. First, a general problem-solving framework used by sustainability scientists and professionals to solve sustainability problems will be introduced. Only a brief overview of the problem-solving framework will be given here, but each component will be discussed in more detail in later chapters. Second, the importance of considering tradeoffs among environment, society, and economy, when resolving sustainability, problems, will be presented within the framework of weak versus strong sustainability. Finally, the importance of transdisciplinary and participatory approaches throughout the entire problem-solving process will be discussed.

Section 2.2.1—Overview of the Transformational Sustainability Research Framework

Key Concept 2.2.1—Current state analysis, future scenarios, visioning, and transition strategies are emerging tools that can be used to resolve sustainability problems.

Professor Arnim Wiek, a sustainability scientist at Arizona State University's School of Sustainability, developed a problem-solving framework for resolving sustainability problems. It was developed in his Sustainability Transition and Intervention Research lab at ASU in collaboration with many international colleagues. The framework, called **Transformational Sustainability Research (TSR)**, builds on various research, planning, governance, and decision-making frameworks, including ideas presented in *Great Transition: The Promise and Lure of the Times Ahead* by the Global Scenario Group, a work published in 2002 to assess the necessary conditions for a societal transition to sustainability. *Great Transition* asks four key questions: (1) Where are we? (2) Where are we headed? (3) Where do we want to go? (4) How do we get there? It should be noted before moving on that the reference to "we" in these questions is left purposefully broad and vague so that this framework can be adapted to specific contexts around the world where sustainability problems exist. Depending on the specific problem of interest, and the stakeholders involved, "we" could mean the members of a small village, residents of a city, all people in a nation, or our global society and humanity as a whole.

Transformational Sustainability Research (TSR) a general problem-solving framework that can be used to understand and resolve sustainability problems.

current state analysis a process used to evaluate and understand the present-day situation of a given socio-ecological system, and for a certain wicked problem facing that system, with the aim of producing information and insight for thinking about the future of that system and devising strategies for the transition of that system to a more sustainable future.

indicators quantitative or qualitative data used to determine the current state of a system, how far it is from some ultimate goal, and which way it is changing or whether it is even changing at all.

The four questions correspond to the four major components of the TSR framework presented in Figure 2.19. The first question is answered by analyzing the current state. **Current state analysis** (Chapter 3) involves assessing and understanding the current situation so that this information can be used to think about the future, and transitions to a more sustainable future. It involves classifying system drivers, identifying causal linkages among drivers, and a stakeholder analysis to examine the factors guiding human behavior toward actions that contribute to an unsustainable situation. This also means considering a number of factors that shape behavior, such as rules, regulations, resources, technologies, and the norms, beliefs, and values of stakeholders. The effects of actions on all three pillars of sustainability—environment, society, and economy—should be considered in this analysis, as well as the short-term and long-term consequences. After completing a current state analysis, a detailed answer can be given to the question: Where are we?

During the current state analysis, and also for looking toward the future (Chapter 7) and managing transitions (Chapter 8) to sustainability, **indicators** (Chapter 4) are used to track progress toward or away from sustainability. They are also used to understand SES complexity by distinguishing system fluctuations from more fundamental system changes, and identifying the *feedbacks* that stabilize systems or cause them to cross thresholds to become new systems (Chapter 5). Finally, they can be used to understand, and track over time, SESs that exhibit

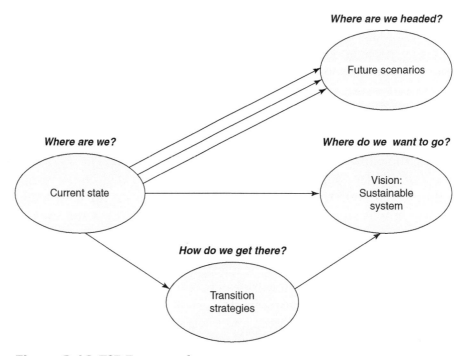

Figure 2.19 TSR Framework.

the levels of complexity characteristic of complex adaptive systems (CAS, Chapter 6).

To answer *where we are headed?*, and understand the impacts of our present actions on the future, *future scenarios* (Chapter 7) are developed. Sustainability is about long-term perspectives. As a result, the effects of our actions (or inaction) today, on the future, must be considered. A **scenario** is a carefully constructed story about alternative paths that a system might take as the current state moves into its future state. It is important to intentionally explore alternative, possible futures because, in an SES, many factors may combine in unpredictable ways that lead to a surprising future (recall the coral reef regime shift example in Box 2.5). Future scenario development is about trying to anticipate these alternative, plausible futures. Scenarios are developed to identify the events, actors, and processes that might lead to each alternative future. Future scenarios can be based on quantitative models that use mathematics, qualitative narratives that employ stories and pictures, or a mix of both. Engaging relevant stakeholders in the scenario development process is critical to promoting social learning, collective understanding, and action.

Future scenarios help us to figure out where we are headed, but the answer to *where do we want to go?* is our **vision** (Chapter 7). The **visioning** process brings together relevant and diverse stakeholders to attempt to reach consensus on what the SES should look like in the future. Some future scenarios, such as rapid sea level rise, followed by millions of climate refugees moving to other countries and destabilizing political systems, are undesirable for a sustainable future. Other scenarios, such as well-planned climate change adaptation measures, implemented in today's policies, to ensure human well-being in the future if sea levels were to rise, are more desirable. Visioning is used to collectively create a vision for a preferred future state and to set goals for achieving that preferred state. Recognizing that real, lasting change takes time, visions are focused on where we want to be in the long-term, typically 20 to 30 years into the future. Visions are developed through a series of meetings that take into account the perspectives and interests of all stakeholders.

The last part of the problem-solving framework is *transition strategies* (Chapter 8). During this phase, strategies are developed to move from the unsustainable current state to the desired sustainable future vision, such that the SES undergoes a **transition**. One major aspect of developing transition strategies is determining intervention points at a variety of spatial and temporal scales in the system of interest. Intervention points are places or times in an interconnected SES, in which a wicked problem is embedded, where the root of the difficulties really resides. If intervention points and required actions for sustainable change can be identified, then leverage can be gained and the

scenario a carefully constructed quantitative model or qualitative story about the many plausible alternative pathways a system might take into the future and what its future state might look like.

vision a desirable future ideal envisioned by a society or a subset of society.

visioning the process by which a vision is collectively created by a society or a subset of society.

transition a profound change in the way an existing socioecological system functions to meet the needs of society.

capacity to influence the future state of a system through small actions, that have relatively large effects, can be realized. As a result, intervention points are also called *leverage points*. Transitions strategies help us figure out how to steer or guide systems toward sustainability, and how to avoid undesirable future scenarios identified during the scenario development phase. A big part of designing transition strategies is weighing the tradeoffs among environmental, social, and economic priorities that will be necessary for a viable transition to sustainability.

Section 2.2.2—Making Tradeoffs for Sustainability

Key Concept 2.2.2—Weak and strong sustainability represent two different beliefs regarding the extent to which tradeoffs among environment, society, and economy may be made when resolving sustainability problems.

tradeoffs situations in which one or more aspects of a pillar of sustainability are lost in exchange for gaining one or more aspects of other sustainability pillars.

weak sustainability a belief that the resources and services provided by natural systems can be substituted for technologies developed by human systems and that extensive tradeoffs can be made among components of the three pillars of sustainability.

strong sustainability a belief that the substitutability of natural system components by technologies developed within human systems is limited and that tradeoffs among the three pillars of sustainability are also limited.

Tradeoffs help us reconcile the inherent conflicts among the environment, society, and economy pillars of sustainability. These conflicts are inherent because, as the integrity of one or more pillars is eroded, other pillars are added to or enhanced. For example, economic development unavoidably involves some degree of natural resource extraction and pollution. This adds to the economic pillar, in terms of GDP growth, but takes away from the environment pillar. Economic development to alleviate poverty by meeting people's basic needs requires use of natural resources and creates pollution that compromises the integrity of the environment. This adds to the society pillar, but takes away from the environment pillar. Under our current economic system, economic development takes away from society in some instances. This is evidenced by the rise of labor unions and child labor laws to address these problems, and the cases of sweatshop labor and worker exploitation that still occur around the world today. Solving almost all problems related to sustainability involves making tradeoffs. There are rarely perfect solutions with no costs to the environment, society, or economy, and there are often winners and losers. Striving for win-win-win situations in which some action enhances sustainability, by simultaneously adding to all three pillars, is the ideal goal. In reality, this is difficult to achieve.

There is disagreement between advocates of **weak sustainability** and **strong sustainability** regarding the extent of tradeoff that is possible or desirable. (**Box 2.6** provides a sustainability case study that illustrates these concepts.) Advocates of weak sustainability view the resources and services provided by natural systems as interchangeable with technologies developed by human systems. They are not concerned about environmental degradation because they believe technology alone can solve such

BOX 2.6 TRADEOFFS FOR RESILIENCE TO CLIMATE CHANGE IN SOUTHEAST ASIA, MANGROVES OR SEA WALLS?

Southeast Asia has over 108,000 km, or 67,000 miles, of coastline, much of which is covered in mangrove forests (Figure 2.20).Southeast Asian countries contain more than a third of coastal mangrove forests globally, and they are essential to the livelihoods of many rural coastal communities in these regions. Mangrove ecosystems provide a range of important ecosystem services, including fuel and timber, protection from coastal flooding, soil erosion prevention, nutrient cycling, carbon sequestration, oxygen production, and habitat for fish species important to local people. Under some conditions, it may be possible for mangroves to increase the elevation of coastal land areas, to counter sea level rise, as they trap and accumulate organic matter and sediments. Despite these benefits, Southeast Asian mangroves are threatened by aquaculture, and other types of coastal development, driven by global demands and market forces. Since these economic activities began in the 1980s, over a third of Southeast Asian coastal mangrove area has been lost. Protected areas and replanting projects are helping to conserve and restore these forests in parts of Southeast Asia (Figure 2.21a and 2.21b).

Climate change is causing sea levels to rise, and resulting in more frequent and intense storms. As a result, flooding is becoming more frequent and intense, having serious consequences for people living in these regions. In 2000, drought plus monsoon-flooding killed thousands of people in India and Bangladesh, while also displacing millions from their homes, causing food shortages due to failed crops, and damaging infrastructure. Natural disasters, such as earthquakes and accompanying tsunamis, cause even more extreme coastal flooding and devastation. During the 2004 Indian Ocean earthquake and tsunami, it is estimated that 280,000 lives were lost. In areas where mangrove forests do not exist, there is evidence that flooding and the subsequent effects of people's live have been worse.

Figure 2.20 Mangroves have complex root systems that can tolerate salt, wave action, and other harsh aspects of coastal environments, where tides cause water levels to rise (*left*) and fall (*right*) on a daily basis.

(continued)

Figure 2.21 (a) A fisherman tends to his net in Ream National Park, Cambodia, where mangrove habitat is protected and a third of people living in the park depend on fishing as their livelihood. (b) In Phetchaburi, Thailand, volunteers and students plant mangrove trees as part of a reforestation project.

Figure 2.22 Seawalls, such as this one built in Havanna, Cuba at the turn of the twentieth century, are not a new idea and have been used for quite some time to protect cities from stormy ocean waters.

In areas where mangrove ecosystems have been destroyed, some propose human-built sea walls as a way to mitigate flooding and adapt to rising sea levels (Figure 2.22). For example, in Jakarta, where only a tiny fraction of native mangrove forests still exist, the city is spending $40 billion over 20 years to build a 25-mile-long and 80-foot-tall sea wall. More than 25 feet of the wall will stand above water. Due to a combination of sea level rise and rapid local groundwater extraction, Jakarta is one of the fastest sinking cities in the world. Children walk to school and people ride motorcycles through flooded streets. Something needs to be done, and quickly, but not everyone agrees that a sea wall is the answer. In addition to the tremendous cost to build and maintain the sea wall, a recent study by the Indonesian government found that the sea wall could cause damage to surrounding ecosystems

> and local people. Storm waves bouncing off sea walls and back out into the ocean could damage coral reefs and erode surrounding islands out of existence. The wall would trap nutrients washing out of rivers, making eutrophication in Jakarta Bay even worse. Coral reef damage and eutrophication will reduce biodiversity and the abundance of important fish species. Thousands of people may be forced to relocate, especially those relying on fish as a source of food and income.
>
> Should sea walls be built? Can they protect people and ecosystems from sea level rise in Jakarta, and in Southeast Asia as a whole? Should economic activities that destroy mangrove ecosystems be allowed to continue, despite their value in protecting coastal areas and the other ecosystem services they provide? Putting the costs of building the sea wall aside, advocates of weak sustainability would argue that, with the right technologies in place, sea walls could substitute for mangrove forests when it comes to coastal protection. In addition, they might also argue that the ecosystem services lost by building the sea wall could be replaced by other technologies. Advocates of strong sustainability would argue that the tradeoffs are too great: on the one hand, people would be protected from flooding; on the other hand, people would be displaced and their livelihoods harmed, ecosystem services would be lost, and further environmental damage would ensue. In this case, the tradeoffs exist both within the society pillar, and between society and environment. When sustainability is the goal, the tradeoffs inherent in the sea wall solution are much greater than those in the situation where mangrove ecosystems are conserved or rehabilitated.

problems. In other words, nature has substitutes. In this view, it is possible to make tradeoffs between the environment and economy pillars without end. In the words of economist Robert Solow in 1974: "If it is very easy to substitute other factors for natural resources, then there is in principle no 'problem.' The world can, in effect, get along without natural resources, so exhaustion [of natural resources] is just an event, not a catastrophe." For example, the cellulose used to make paper pulp comes from trees. If all trees were cut down to make paper pulp and none remained, the cellulose from other natural resources, such as hemp or agricultural wastes from corn stalks could act as substitutes with the proper technologies. The weak sustainability view is understandable because technology has solved many human problems in the past, such as sanitation with sewage treatment and disease with medical advances. It is reasonable to think that it will continue to do so in the future.

Advocates of strong sustainability would disagree. They warn that we must be careful of the extent to which we make tradeoffs because there are limits to how much we can substitute the elements of one pillar for those of another. If we go beyond this limit, they argue, irreversible damage will occur as systems shift from one regime to another. Because human and natural systems are so interconnected, tradeoffs that are too large will leave one or more pillars too degraded, and this will harm the entire socioecological system. For example, as mentioned above, economic development takes away from the environment, and can decrease human well-being, and it adds to the economy. However, if too large a tradeoff occurs and too much is taken away from the environment, it will become so degraded that economic development will no longer be possible and the resources required for alleviating poverty will disappear. If too large a tradeoff occurs between society and economy, and people are plagued by hazardous or unjust working conditions, then economic development can be harmed by civil unrest and instability, or loss of consumer interest in a good or service produced under such inequitable conditions. Although some tradeoffs are usually inevitable, there is a limit. Strong sustainability arose from the fact that humans have traditionally given the economy, in terms of GDP growth, priority over the environment and society. Advocates of strong sustainability would also argue that technology cannot replace many resources and ecosystem services. For example, one could argue that a water treatment plant could replace the water purification services offered by ecosystems such as forests and wetlands. However, how can we replace clean air with technology? What about the photosynthetic processes that form the base of our food chains or the global water cycle? Advocates of strong sustainability would argue that we cannot replace them, and that sustainability requires behavioral change, such as reduced consumption patterns, not just green technology development.

Section 2.2.3—Participatory and Transdisciplinary Approaches

Key Concept 2.2.3—Including a diversity of stakeholders is key to resolving sustainability problems, as it ensures that both local and expert knowledge are incorporated into problem-solving processes and that solutions have "staying power."

participatory approach a democratic problem-solving process that incorporates both the specialized knowledge of experts and the practical knowledge of local communities as a basis for defining and understanding problems and devising solution options.

Sustainability means different things to different people. The weak sustainability versus strong sustainability debate is one of many examples of conflicting perspectives regarding what needs to be done to move our societies toward sustainability. It is important to incorporate these different perspectives into every stage of the problem-solving framework—current state analysis, indicator development, visioning, future scenario development, and transition strategies—by engaging all relevant stakeholders in the process. Traditionally, we have relied on mostly expert knowledge for problem solving. However, this knowledge alone has failed to resolve sustainability problems and we need a new approach. This new approach is generally referred to as a **participatory approach**. At its most basic level, this simply means that local communities are included in the problem-solving process rather than only involving scientists and government officials. Many sustainability scientists avoid the words *government* or *management*, which imply a *top-down* approach to problem-solving in which only a few people or organizations are involved. Instead, sustainability scientists use the word governance, which implies a more *bottom-up* approach of broad involvement by all people who affect, and are affected by, a situation and who work together toward a collective goal.

When communities work together to resolve problems, the approach is often referred to as *bottom-up*. A *top-down* approach is when primarily natural scientists, social scientists, government officials, academics, and other experts work on finding solutions. What is needed is a merger of the bottom-up approach with the traditional top-down method, which has failed to successfully resolve sustainability problems on its own.

There are several advantages to a bottom-up approach. Local knowledge about the current state of a situation may be more accurate than that of experts, who come into the community as outside observers of a problem. As conditions change through time, local communities will observe this and governance strategies can be adapted to changing conditions. When a solution is developed, communities may be the best judge of whether the proposed solution will be viable in a real-world local setting. Importantly, community participation provides communities with the capacity, in terms of education and skills, to continue a project after the volunteers leave or the money stops flowing. For example, in one instance, U.S. Peace Corps volunteers were working to help people in a remote village build a well for easy access to water. About midway through the project, a wealthy company swooped in to build wells immediately and at no cost to the villages. At first, the wells were fine. However, a year later, after the Peace Corps volunteers and the company had left, the wells broke. The villagers did not know how to fix them and they were back to square one. The ability of communities to address future problems is sometimes more significant than seeing immediate results. Finally, community buy-in or acceptance of a project is extremely important to the longevity and staying power of a plan for action toward sustainability. If the solution that is implemented is not the solution that people want, then time and resources have been wasted. Community engagement in problem solving can be a costly process in terms of time, effort, and money, but it is more likely to lead to a long-term, sustainable solution than a "quick-fix" expert solution.

Many types of knowledge, both expert and local, are needed to resolve sustainability problems. When a singular perspective on a problem is taken, as in "The Blind Men and the Elephant" parable, the complexity of the problem cannot be seen. When experts work together with local communities on sustainability challenges, it is often referred to as the **coproduction** of knowledge. This can be viewed as expert knowledge embedded into a specific local situation that has its own practical knowledge. The new knowledge produced from this process is **transdisciplinary** knowledge (Figure 2.23). The importance of

coproduction a process through which knowledge about a situation is generated by blending the specialized knowledge of experts with the practical knowledge of local communities.

transdisciplinary knowledge and understanding based on both the specialized expertise of academic experts and the practical real-world know-how of communities outside of a university.

Figure 2.23 The transdisciplinary knowledge needed to resolve sustainability problems comes from both the knowledge of experts, such as scientists and government officials, and the practical knowledge of people living in communities where problems exist.

interdisciplinary problem solving has been recognized for some time. This is when experts from many disciplines, such as biology, political science, and sociology, attempt to integrate concepts and ways of thinking from their individual disciplines to solve a problem in ways that would not have been possible using only a single discipline. However, all of this knowledge is still *expert knowledge.* Transdisciplinary knowledge goes beyond this by moving into the real world to include *practical knowledge* and ways of thinking in addition to expert knowledge. This book is focused on addressing sustainability challenges in the real world using the TSR framework (Section 2.2.1). In addition to the case study in **Box 2.7**, which ends this chapter, case studies are presented throughout the chapters of this book to give you examples of how practical knowledge can contribute to problem resolution. Each local problem is different, so possible ways to incorporate local knowledge into sustainability problem-solving vary. If you want to get real experience with this right away, then go out there and get involved in resolving sustainability problems, and find out for yourself!

BOX 2.7 PROTECTING NORTH AMERICAN CARIBOU: THE IMPORTANCE OF LOCAL INDIGENOUS KNOWLEDGE

By Joshua MacFadyen, Assistant Professor, Arizona State University

North American Caribou once occupied all of Canada and Alaska with the exception of the Northern Great Plains, Southern Ontario, and the Lower Mainland of British Columbia. The species was so ubiquitous that it was called the "bison of the north." It was as culturally and economically important to the Inuit, Dene, and Cree indigenous peoples, who lived in what is now northern Canada, as the bison had been to indigenous peoples who lived on the grasslands of the present-day United States and Canada (Figure 2.24). However, since Europeans settled in North America, caribou habitat has decreased by almost 40%. Today, caribou live on increasingly fragmented sections of land in Canadian tundra and taiga forests.

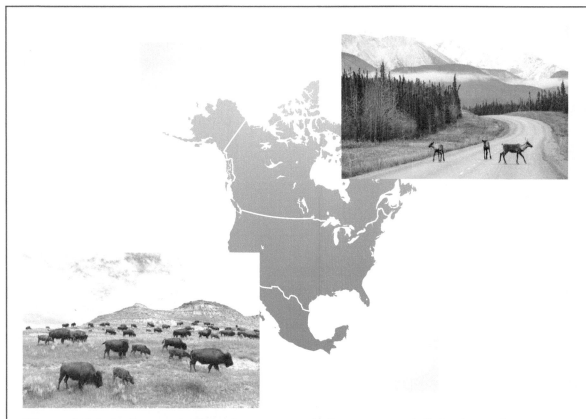

Figure 2.24 Caribou were just as important to indigenous people living in northern Canada as the herds of bison, which roamed the Great Plains, were to indigenous people who lived in territory that now belongs to the United States.

The decline was not for lack of expert scientific knowledge. In the postwar period, Canadian Wildlife Service (CWS) scientists used aerial observations, aerial photography, and ground observations to count the enormous herds. In 1948, CWS mammologist A.W.F. Banfield used the new wartime advances in aerial reconnaissance to conduct a survey of caribou across an incredibly vast section of the Arctic (Figure 2.25). These new observation techniques yielded groundbreaking scientific results, including the first evidence that caribou lived in distinct herds. When Banfield's team compiled the data, they found a shocking revelation. The North American herd, which explorers and naturalists from the early twentieth century had estimated at several million, had decreased dramatically to a population of only 670,000. The result was so stark that some questioned its accuracy, but it pointed to a looming crisis for this animal. Nobody wanted to see a repeat of what happened to the grassland bison in the nineteenth century.

Wildlife managers responded quickly. Waves of scientists and technicians trekked to Arctic communities on the ground or flew over to conduct targeted aerial studies. Teams of CWS scientists and other agencies debated the nature of the problem, and the best way to protect

(continued)

Figure 2.25 During WWII, fighter aircraft, such as the Spitfire PR Mk XI shown here, were adapted for reconnaissance using technological advances in optics and pilots trained for this purpose. After the war, these new technologies became available to scientists.

and restore the herd. Rather than question their baseline estimates, federal officials ignored local knowledge and treated the scientific studies as fact. They assumed that the problem was caused by wolf predation and overhunting by indigenous people. One technician studying caribou, Farley Mowat, took an opposing view. In his popular novels, including the fictionalized account *Never Cry Wolf*, he described the unique ways of knowing that he experienced by living with indigenous people. Mowat argued that wolves were not destroying the herd, and ultimately he was right. However, indigenous voices were denigrated, and restrictive wildlife management practices were enhanced to the point where, in 1961, all species of caribou were placed on the endangered species list. The result was an unenforceable hunting ban that only served to enhance anger and frustration among the indigenous people who relied on caribou for food and materials.

Like many sustainability challenges, this crisis was only partially understood because expert knowledge was given precedence over local knowledge. The physical world behaves according to the fixed laws of science, but scientific knowledge and ways of knowing about the physical world change over time. As such, what we can know about the physical world changes, depending on the knowledge and technologies available at the time. In this case, Banfield and his team were mesmerized by the advanced technologies newly available in the postwar period. These technologies provided a new way of knowing about caribou herds, but it ignored local indigenous knowledge, which was based on centuries of direct interaction with caribou herds. Despite this, the nature of the problem and decisions about how to protect herds were based largely on scientific information, and indigenous people were not included as stakeholders. The CWS did not adequately consider other sources of valuable information arising from local knowledge of caribou. In the 1960s and 1970s, it became clear that the CWS caribou counts were wildly inaccurate. As a result, in 1985, many species were removed from the endangered species list.

Today, northern conservation is a much more participatory effort. It includes post-normal science, which is scientific investigation that includes perspectives from stakeholders outside of the world of professional science. In caribou conservation today, this is evident in many ways. Inuit and First Nations groups play important roles on the boards of management agencies, such as the Beverly and Qamanirjuaq Caribou Management Board. Participatory mapping tools have been developed, such as by the Canadian Wildlife Conservation Society. Traditional ecological knowledge (TEK) has been employed in recent conservation efforts, such as the Chihchaga caribou protection plans to comply with Canada's Species at Risk Act.

Sustainability problems are, by definition, vague. Sometimes the problem is not really what a stakeholder thinks it is, and by including multiple voices and methods of identifying solutions through the co-production of knowledge, collaborators often redefine the problem. When we try to understand sustainability problems, we use a great deal of usually very reliable scientific knowledge. However, sometimes that knowledge is inaccurate. We need to recognize that this knowledge is as much a product of our time, based on the scientific knowledge and ways knowing that are presently available, as it is an objective observation. Participatory and transdisciplinary approaches are excellent ways to ensure that the problems, and the tradeoffs, are clear and transparent and that the solutions are effective, diverse, and equitable.

Copyright © Joshua MacFadyen. Reprinted by permission.

Bibliography

Beverly and Qamanirjuaq Caribou Management Board. 2016. *Safeguarding caribou for present and future generations.* Accessed December 23, 2016. http://arctic-caribou.com/.

Blakemore, E. 2015. Jakarta is building a gigantic bird-shaped seawall: but will the Great Garuda project be en*ough to save a sinking city? Smithsonian Magazine.* Accessed August 31, 2016. http://www.smithsonianmag.com/smart-*news/jakarta-building-gigantic-bird-shaped-seawall-180957536/*?no-ist.

Chapin, F. S., C. Folke, and G. P. Kofinas. 2009. "A Framework for Understanding Change." In *Principles of Ecosystem Stewardship: Resilience-Based Management in a Changing World,* edited by F. S. Chapin, G. P. Kofinas, and C. Folke, 3–28. New York: Springer.

David Suzuki Foundation. 2014. *Caribou TEK* and restoration study. Accessed December 23, 2016. http://davidsuzuki.org/public*ations/ reports/2016/caribou-tek-and-restoration-study/*.

Diamandis, P. H. and S. Kolter. 2012. *Abundance: The Future is Better than You Think.* NY, USA: Free Press, 386 pp.

Diamond, Jared M. 2005. *Collapse: How Societies Choose to Fail or Succeed.* New York: Viking, 575 p.

Elyda, C. 2015. S*ea wall an environmenta*l disaster. *The Jakarta Post*, October 7. Accessed August 31, 2016. http://www.*thejakarta*post.com/news/2015/10/07/sea-*wall-environmental-disaster-study.html*.

Ghosh, J. 2012. Women, labor, and capital accumulation in Asia, *Monthly Review: An Independent Socialist Magazine*, 63(8). Accessed Augu*st 18, 2016,* http://monthlyreview.org/2012/01/01/women-labor-and-capital-accumulation-in-asia/.

Giri, C., E. Ochieng, L. L. Tieszen, Z. Zhu, A. Singh, T. Loveland, J. Masek, and N. Duke. 2011. Status and distribution of mangrove forests of the world using earth observation satellite data. *Global Ecology and Biogeography* 20(1):154–59.

Global Scenario Group. 2002. *Great Transition: The Promise and Lure of the Times Ahead.* pp. AccessedJuly 17, 2013. http://www.tellus.org/documents/Great_Transition.pdf.

Godfray, H. Charles C et al. 2010. *Food security: the challenge of feeding 9 billion people,* Science, 327(5967): p. 812 – 818.

Gunderson, L. H., and C. S. Holling. 2002. *Panarchy: Understanding Transformations in Human and Natural Systems.* Washington, DC: Island Press.

Harris, J., Brown, V.A., and J. Russell. 2010. *Tackling Wicked Problems: Through the Transdisciplinary Imagination.* Taylor & Francis, 336 p.

Hurst, A. 2004. Barren Ground caribou co-management in the eastern Canadian Arctic: lessons for bushmeat management. *ODI Wildlife Policy briefing* (5).

Jaffee, D. 2007. *Brewing Justice: Fair Trade Coffee, Sustainability, and Survival.* Berkeley, CA: University of California Press, 331 pp.

Koch. W. 2015. *Could a titanic sea wall save this quickly sinking city? National Geographic.* Accessed August 31, 2016. http://news.nationalgeographic.com/energy/2015/12/151210-could-titanic-seawall-save-this-quickly-sinking-city/#/01jakartaseawall.ngsversion.1449846007915.jpg.

Krauss, K. W., K. L. McKee, C. E. Lovelock, D. R. Cahoon, N. Saintilan, R. Reef, and L. Chen. 2014. How mangrove forests adjust to rising sea level. *New Phytologist, 202(1): 19–34.* Accessed August 31, 2016. http://onlinelibrary.wiley.com/doi/10.1111/nph.12605/pdf.

Kristof, N. D. and S. WuDunn. 2009. *Half the Sky: Turning Oppression into Opportunity for Women Worldwide.* New York, USA: Alfred A. Knopf, 294 pp.

Lang, D. J., A. Wiek, M. Bergmann, M. Stauffacher, P. Martens, P. Moll, M. Swi*lling, and C. T*homas. 2012. "Transdisciplinary Research in Sustainability Science—Practice, Principles and Challenges. *Sustainability Science* 7 (Supplement 1): 25–43.

Loo, T. 2017. Political animals: barren ground caribou and the managers in a 'post-normal' age. *Environmental History.* 2017, 1–27, https://doi.org/10.1093/envhis/emx027.

Ludwig, D. 2001. "The Era of Management Is Over." *Ecosystems* 4: 758–64.

Marten, G. 2001. *Human Ecology: Basic Concepts for Sustainable Development.* London, UK: Earthscan Publications. http://gerrymarten.com/human-ecology/tableofcontents.html.

Meadows, D. H. 2008. *Thinking in Systems: A Primer.* White River Junction, VT: Chelsea Green Publishing.

Mitchell, M. 2009. *Complexity: A Guided Tour.* Cary, NC: Oxford University Press.

Ott, K. 2003. "The Case for Strong Sustainability." In *Greifswald's Environmental Ethics*, edited by K. Ott and P. Thapa, 59–64. Greifswald, Germany: Steinbecker Verlag Ulrich Rose. Accessed *May 25, 2012. www.scribd.com/doc/30376702/The-Case-for-Strong-Sustainability.*

Rittel, H. W. J., and M. M. Webber. 1973. "Dilemmas in a General Theory of Planning." *Policy Sciences* 4: 155–69.

Sandlos, J. 2011. *Hunters at the Margin: Native People and Wildlife Conservation in the Northwest Territories.* UBC Press.

Scheffer, M. 2009. *Critical Transitions in Nature and Society.* Princeton, NJ: Princeton University Press.

Skaburskis, A. 2008. "The Origin of 'Wicked Problems.'" *Planning Theory & Practice* 9 (2): 277–80.

Solow, R. 1974. "The Economics of Resources or the Resources of Economics." *American Economic Review* 64 (2): 1–14.

Valiela, I., J. L. Bowen, and J. K. York. 2001. Mangrove *forests: one of the world's th*reatened major tropical environments. *Bioscience* 51(10):807–815.

Walker, B., and D. Salt. 2006. *Resilience Thinking: Sustaining Ecosystems and People in a Changing World.* Washington, DC: Island Press.

WCS Canada. 2016. *Caribou and mining story map: how is mineral exploration affecting caribou in Ontario.* Accessed December 23, 2016.https://canada.wcs.org/Resources/Caribou-Story-Map.aspx.

Whyte, K. P., and P. B. Thompson. 2011. "Ideas for How to Take Wicked Problems Seriously." *Journal of Agriculture and Environmental Ethics* 25 (4): 441–45. doi:10.1007/s10806-011-9348-9.

Wiek, A., and D. J. Lang. 2013. *Transformational Sustainability Research—From Problems to Solutions.* School of Sustainability, Arizona State University, Tempe, AZ.

Wiek, A., B. Ness, F. S. Brand, P. Schweizer-Ries, and F. Farioli. 2012. "From Complex Systems Analysis to Transformational Change: A Comparative Appraisal of Sustainability Science Projects." *Sustainability Science* 7 (Supplement 1): 5–24.

Wiek, A., L. Withycombe, C. L. Redman, and S. Banas Mills. 2011. "Moving Forward on Competence in Sustainability Research and Problem Solving." *Environment: Science and Policy for Sustainable Development 53 (2): 3–12.*

*Wiek, A. 2009. "What Is a Sustainability Problem?" Working Paper, Sustaina*bility Transition and Intervention Research Lab, School of Sustainability, Arizona State University, Tempe, AZ.

*Wiek, A. 2010. "Analyzing Sustainability Problems and Developing Solution Options—A Pragmatic Approach." Working Paper, Sustainability Transition and Inter*vention Research Lab, School of Sustainability, Arizona State University, Tempe, AZ.

World Wildlife Fund and Equilibrium. 2008. *National security: protected areas and hazard mitigation*, 128 pp. Accessed August 31, 2016. file:///C:/Users/sonya.doucette/Downloads/natural_security_final.pdf.

End-of-Chapter Questions

General Questions

1. Fill in the table that follows to briefly and generally explain why each of the sustainability issues listed in the first column is a wicked problem. The six characteristics of a wicked problem discussed in this chapter are listed across the top of the table. Explain how each problem exhibits each of the six characteristics.

	Vague Problem Definition	Undefined Solution	No Endpoint	Irreversible	Unique	Urgent
Climate Change						
Biodiversity Loss						

	Vague Problem Definition	**Undefined Solution**	**No Endpoint**	**Irreversible**	**Unique**	**Urgent**
Population Growth						
Deforestation						
Overfishing of the Oceans						
Nonrenewable Resource Use						
Poverty						
Economic Growth and Overconsumption						

2. In order to resolve wicked problems, tradeoffs must be made among the three pillars of sustainability: environment, society, and economy. The extent to which these tradeoffs can be made is expressed by the two different belief systems of weak sustainability and strong sustainability. Determine whether each of the following statements expresses the weak sustainability or the strong sustainability viewpoint.

   a. Running out of nonrenewable, energy-dense fossil fuel energy resources is not a problem. When these resources do run out, new equivalent resources for energy production will be discovered and human society will be able to continue at its current levels of consumption and economic growth.

b. Indigenous societies of the Amazon basin have subsisted in the rainforests of this region for centuries. During this time, much of their well-being was derived from these traditional subsistence activities. While some commercial logging of tropical hardwoods in this region is okay, too much will be very harmful to the well-being of these indigenous societies.

c. Bees and other insects provide a valuable ecosystem service because they act as pollinators for crops that provide human societies with food. Without them, crops would not be pollinated and would not produce food. If bees and other pollinators become extinct, there is no technology presently available or that could be invented to replace this ecosystem service. Therefore, bees and other pollinators are essential for the survival of human societies and should be protected from extinction.

d. Indigenous societies of the Amazon basin have subsisted in the rainforests of this region for centuries. During this time, much of their well-being was derived from these traditional subsistence activities. Today, commercial logging of tropical hardwoods occurs in this region. This activity is beneficial to the well-being of indigenous societies because it leads to economic development. Any well-being experienced by indigenous societies in past centuries through traditional subsistence activities will be replaced by the well-being brought by economic development.

Project Questions

This is the first of many Project Question sections located at the end of each chapter. Beginning with this chapter, you will choose a wicked problem to focus on throughout this book. This might be a case study that you will work on virtually in the classroom or a problem that you will try to actually resolve in the real world. The problem you choose might be focused on many different issues, including food, water, energy, waste, transportation, human health, poverty, and population growth, among others. You will need to conduct research and gather additional information about your specific problem in order to fully answer the questions in the Project Question sections throughout this book, and your answers will require continual revisions as you incorporate new information. The Project Question section for this chapter is focused on beginning a current state analysis, which is the first step in the sustainability science problem-solving framework presented in this text (Figure 2.5).

1. Define Your Wicked Problem: Define your wicked sustainability problem using the six characteristics of a wicked problem, as you did for several wicked problems in Question 1 in the General Questions section. Your focus could be on one of many possible scales, such as a problem relevant to a small town, large city, county, state, country, region, or even our entire global society. You may want to skip back and forth between this question and the next, as it may help to answer both questions simultaneously.

2. **Define Your System:** As noted above, your wicked problem could be focused on a town, city, county, state, country, region, or our entire global society. Being specific about the scale of your problem helps you define your system. Define your system by detailing the system components and the interactions among those components, such as by using boxes to represent components and arrows to represent interactions as shown in Figures 2.9 and 2.12. Also define the boundaries of the system, such that the system components and the interactions among those components lie within the boundaries and the other aspects important to the problem lie outside of the system boundaries. Remember, wicked problems are embedded in open systems, which means that the components inside of the system boundaries interact with those outside of the boundaries. Also, recall that the purpose of defining your system is to help you simplify the complexity of the real world and that no one system definition is absolutely correct. The system that you define here is a first draft and will likely be revised many times as you work through the Project Questions sections at the end of each chapter and as you gather more information about the wicked problem that you have chosen to work on.

Chapter 3

CURRENT STATE ANALYSIS

Students Making a Difference

Photo courtesy of Stacey Freeman.

Creating a Common Ground for Sustainability

Molly Cashion is from the southeastern United States and first became interested in the social barriers to sustainability when she noticed how strongly people's stance on environmental issues was influenced by their Christian religious beliefs. Molly feels that beliefs, especially religious ones, are difficult to talk about, but they can't be ignored because they influence behavior: "Religion is one of these topics that we don't talk about. It makes me uncomfortable to talk about it. Maybe we should be talking about it more. [In sustainability] we are approaching hard questions. You can't leave out hard topics like beliefs because this is reality. It's okay and important to look at your belief system, and it's okay that it influences you, as long as you are aware of it." In sustainability challenges, we need to work with all sorts of people, regardless of our own beliefs. And things are usually more complex than we first assume.

One of the first complexities encountered by Molly was how the same religion—Christianity—could lead to very different actions around sustainability. While working at a law firm in the southeastern United States, she noticed that some people support the use of coal, in spite of scientific evidence to the contrary, because they believe God gave us coal and He wants us to use it. While working on her Master's degree project at a secular environmental education center (EEC), founded by an Episcopalian summer camp director, she observed how religion can compel people to action on environmental issues out of obligation and a sense of responsibility for the well-being of people and the planet. Molly's curiosity, and her courage to delve into belief systems, allowed her to explore these contradictions. As she delved deeper, things got more complicated. She drew concept maps to keep track of details and make connections, in order to gain a better understanding of a situation: "I had these big pieces of paper hanging on my walls."

The people at the EEC were *mostly* children being raised in religious families and *mostly* educators who were not religious. As such, the groups seemed to be "religious kids" and "secular educators." However, we cannot classify people so simply. In one instance, Molly observed educators planning a lesson on the dangers of CO_2 emissions from coal. To explain this to kids, it was necessary to talk about when coal formed. Science says it formed millions of years ago. Christianity says the Earth is 6,000 years old. Some environmental educators were religious and believed in a 6,000-year-old Earth. The educators were in a bind: they could not present the scientific evidence because it would conflict with religious beliefs, and they could not present religious beliefs because it would conflict with state laws prohibiting religious instruction in schools. Beliefs and laws were influencing how the educators could behave, and what they could say. Educators found common ground, among themselves and with the children, and in a way that complied with the law, by saying that coal formed "a really long time ago" and leaving out the details.

From her work, Molly has a renewed sense of hope, as she has started to see the potential for finding common ground among people with different ideas, beliefs, and perspectives: "I feel more hopeful that people can work together [on sustainability challenges]." In this chapter, you will learn about tools for dealing with the complexity of sustainability challenges and for working with the different stakeholder groups involved. Like Molly's big paper concept maps, you will learn how to draw your own maps to help you better understand the current state of the system in which your sustainability problem is embedded. You will also explore the idea of stakeholder groups and learn strategies for finding common ground.

Core Questions and Key Concepts

Section 3.1: Defining the System and Classifying Drivers

Core Question: What is regulating the system in which a sustainability problem is embedded?

Key Concept 3.1.1—Defining a system means identifying the boundaries, components, processes, and drivers relevant to a problem that is embedded in a specific place, time, and context.

Key Concept 3.1.2—Natural and human drivers regulate change in socioecological systems over many spatial and temporal scales.

Key Concept 3.1.3—Natural and human drivers can influence socioecological systems either directly and unequivocally, or indirectly and more diffusely.

Section 3.2: Causal Chain Analysis

Core Question: How can system regulation be understood in a way that helps to resolve sustainability problems?

Key Concept 3.2.1—Causal chain analysis promotes deeper understanding of sustainability problems by connecting drivers to the problem.

Key Concept 3.2.2—Unraveling complex causal chains requires an understanding of the relative extent of influence of each driver on the system, current driver trends, and information about the specific cultural, political, social, technological, economic, and environmental context in which the problem is embedded.

Key Concept 3.2.3—A western Indian Ocean case study illustrates how information gleaned through driver classification and causal chain analysis is specific to the problem of solid waste pollution and the local contexts for four different island states.

Section 3.3: Stakeholder Analysis

Core Question: How can stakeholder analysis contribute to an assessment of the current state of a sustainability problem?

Key Concept 3.3.1—Stakeholders can be broadly defined as people who have an interest or stake in some policy, conflict, or organizational goal, but identifying stakeholders is not a straightforward process.

Key Concept 3.3.2—Factors influencing stakeholder behavior result in actions that affect the sustainability of a system, and awareness of stakeholder interest in and influence on a sustainability problem reveals power dynamics, beneficial alliances, and significant risks.

Key Concept 3.3.3—Investigating stakeholder relationships provides insight into how stakeholder relationships should be managed and to what extent, and how, different stakeholders groups might influence each other.

Key Terms

direct driver	infiltration	secondary stakeholders
indirect driver	percolation	key stakeholders
causal chain analysis	stakeholders	active stakeholders
indicators	critical transition	passive stakeholders
cultural eutrophication	order of magnitude	value
Q_{10} factor	primary stakeholders	beliefs (aka. worldviews)

This section of the textbook will introduce in detail the first component of the TSR framework: current state (Figure 3.1). Recall that this framework is used to guide problem analysis and solution development. The methods used in this framework come from several different disciplines ranging from business to ecology to urban planning. Because sustainability is a process, there is no final solution to sustainability problems. As a result, word the *resolving* is often used rather than *solving*. These two words are used interchangeably in this text, but the meaning is always *resolving*. This chapter is focused on analyzing present-day system conditions to give a detailed answer to the question: *Where are we?* Recall from Chapter 2 that the "we" in this question is purposefully broad and vague so that this framework can be adapted to specific contexts around the world. Understanding the current state of a system lays the groundwork upon which the other components of the TSR framework, future scenarios and visions (Chapter 7), and transition strategies for movement toward more sustainable systems (Chapter 8), are founded.

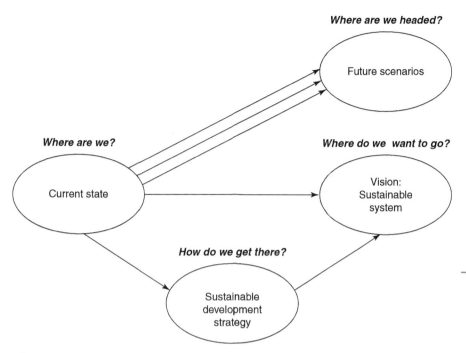

Figure 3.1 TSR Framework.

direct drivers drivers that clearly and unequivocally influence the behavior of a system

indirect drivers drivers that influence the behavior of a system in a more diffuse way by altering one or more direct drivers.

causal chain analysis a tool for analyzing socioecological systems by classifying key drivers, establishing the relationships among drivers, and determining their relative influences on a system.

indicators quantitative or qualitative data used to determine the current state of a system, how far it is from some ultimate goal, and which direction it is changing or whether it is even changing at all.

Determining present-day system conditions requires an understanding of the problem through analysis so that strategic intervention points in the present system may be identified, and leveraged, as a means for resolving sustainability problems. Specifically, analysis of the current state involves defining the system in which the problem is embedded, classifying the **direct drivers** and **indirect drivers** of the system that are relevant to the problem, conducting **causal chain analysis** that links the problem back to the underlying drivers producing the problem, assessing the extent of influence and the trends of relevant drivers, pinpointing factors influencing human activities that lie at the root of these problems, and identifying **indicators** useful for tracking progress toward or away from sustainability. All of these topics are covered in this chapter, and the next (Chapter 4), albeit in a rather static way. Chapters 5 and 6 build on these topics by portraying them more dynamically using a complex systems perspective, including discussion of feedbacks, thresholds, stability landscapes, and resilience. Chapters 7 and 8 will move away from the current state to thinking about plausible futures (scenarios), desirable and sustainable futures (vision), and how to get to the future vision from the current state (transition strategies).

Section 3.1: Defining the System and Classifying Drivers

Core Question: What is regulating the system in which a sustainability problem is embedded?

Drivers are the governing forces that act on a system, either causing it to change or to remain in its current state. They are analogous to the driver of a car that can move the car forward, steer it in a certain direction, or keep it in the same place (**Figure 3.2**). Drivers can push systems toward sustainability, pull them away from it, or prevent systems from changing at all. Identifying and understanding important system drivers in the current state informs efforts aimed at exploring plausible and desirable futures (Chapter 7) and devising transition strategies to move current systems toward sustainability (Chapter 8). The definition of drivers varies across academic disciplines and system analysis frameworks. This chapter uses the definitions of drivers from the Millennium Ecosystem Assessment. Drivers are variously referred to as *stresses*, *disturbances*, or *pressures*, which often denote the influence of human actions on socioecological systems. Indeed, human actions have become central drivers of change in these systems on a global scale. However, drivers are not restricted to human actions.

In the socioecological systems of concern to sustainability, drivers can derive from natural processes or human activities, such as a wildfire caused by a lightning strike or by a campfire left unattended. They can also cause both natural and human systems to change, such as an ecosystem changing from grassland to shrubland, or a political system from an authoritarian government to a democracy. Other examples of drivers in different systems might include cultural norms, religious beliefs, population growth and migration, urbanization and sprawl, technological innovation, production and consumption of goods, globalization, climate change and variability, fertilizer and pesticide application, deforestation, invasive species, diseases, drought, war, hurricanes, earthquakes, economic crisis, and the list goes on and on. Whether an example from this list is a driver will depend on the specific system. Thus, drivers cannot be classified until the specific system is defined.

Figure 3.2 Like a person at the steering wheel driving a car, drivers determine where a system goes.

Stock-Asso / Shutterstock.com

Drivers can be natural or human-caused, can act across multiple spatial and temporal scales, and can be direct or indirect. Changes in complex systems are almost always brought about by multiple interactions among many drivers, rather than by a single driver. Drivers can cause change in a system, but can also change themselves as the system changes. Causal chain analysis is described in Section 3.2 as a tool for analyzing this complexity by classifying key drivers, establishing the relationship among drivers, and determining their relative influences on a system. Drivers can also provide clues for identifying useful indicators to measure change, which will be discussed in Section 3.4 of this chapter. In Section 3.1, the focus is on understanding how to classify drivers based on the specific system of interest, how they operate across spatial and temporal scales, and their direct and indirect nature.

Section 3.1.1—Defining the System

Key Concept 3.1.1—Defining a system means identifying the boundaries, components, processes, and drivers relevant to a problem that is embedded in a specific place, time, and context.

When classifying drivers for a system, the system boundaries, components, and interactions must also be defined. System boundaries, components, and interactions were described in detail in Section 2.1.2 using the bike example, which is a simple system, and the fair-trade coffee example, which was a complex system. The details of this process will not be repeated here. Instead, this section will focus on two case studies of **cultural eutrophication** in two of the biggest coastal dead zones on Earth located in the Black Sea and in the East China Sea (**Figure 3.3**). These case studies will illustrate how systems must be defined differently when the same type of sustainability problem exists in two different places, times, or contexts. When systems are defined in a way that is specific to a local context, then local drivers can be accurately identified and appropriate solutions to the problem more effectively developed.

cultural eutrophication response of aquatic ecosystems to the addition of excess nutrients that are a by-product of human activities.

Cultural eutrophication is not new. It occurs naturally over geological time periods, and has been present in major human civilizations throughout history, such as in ancient Mesopotamia and Medieval Europe. In the past, cultural eutrophication was localized. Today it has become a global issue, and it poses one of the most serious threats to rivers, lakes, and coastal waters worldwide, affecting ecosystem health, human well-being, and economic vitality. Eutrophication of coastal waters creates dead zones at the mouths of large, or multiple, rivers that flow into the sea with excess nutrients. The excess nutrients can result in toxic algae blooms that contaminate water, and cause low oxygen conditions that kill fish and other freshwater organisms. This harms coastal livelihoods and economies, including subsistence and commercial fishing, and tourism (**Figure 3.4**).

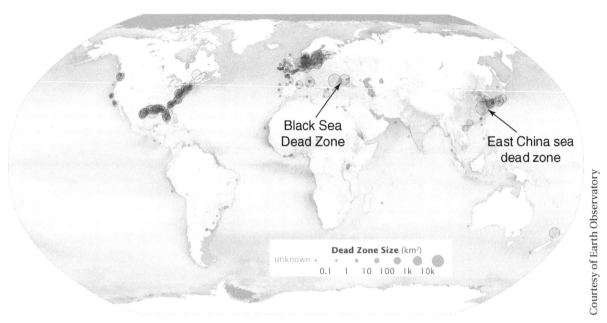

Figure 3.3 Coastal Dead zones, which lack oxygen and make life difficult or impossible for many organisms, are located where rivers carrying nutrient pollutants meet the sea.

Figure 3.4 Cultural eutrophication can kill fish and other respiring organisms, and make water toxic for human consumption and recreation.

The dead zones in the Black Sea and the East China Sea are two of the largest in the world (Figure 3.3). When defining systems for the Black Sea and the East China Sea, there are some similarities: the direct drivers of eutrophication and the system boundaries. In both cases, eutrophication is caused by excess nutrient inputs to coastal areas from rivers, where extra nutrients allow algae and other coastal vegetation to photosynthesize more. More photosynthesis means more biomass, which sinks down to the bottom sediments when the algae and vegetation die. Bacteria eat the dead biomass, using oxygen in the process. A lot of biomass means a lot of oxygen is used up. In dead zones, so much oxygen is used that organisms, such as fish, clams, crabs, and other respiring creatures, need to move out of the affected area (e.g., fish) or, if they cannot move (e.g., clams), live under stressed and often fatal conditions. In both cases, the severity of eutrophication is directly affected by the presence or absence of wetlands, the flushing rate of coastal waters, and the sediment load in rivers. Also, in both cases, system boundaries are defined as the physical boundaries of the watershed. This is because the entire area containing all of the land that drains into these coastal regions must be considered when determining the sources of excess nutrients (Figure 3.5). Only water falling within the watershed boundary will carry excess nutrients into these coastal regions. The rain that falls outside of the boundary will drain into other regions.

Although eutrophication is caused by the same basic process, and system boundaries are defined in a similar way, the sources of excess nutrients and other factors that affect eutrophication vary (Table 3.1). For the East China Sea, 90%–95% of excess nutrients come from a single river: the Yangtze River (Figure 3.6). The Yangtze is the longest river in China, with a total watershed area

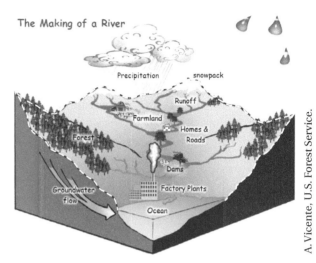

Figure 3.5 Watershed boundaries (yellow dotted line) define the system boundaries of interest for understanding cultural eutrophication.

Table 3.1 Factors Affecting Eutrophication

Contributing Factors	Explanation
Watershed area	Larger basin areas have a larger land areas draining into the river
Population density	Higher population densities mean that more people occupy a given area of land
Human activities on land	Synthetic fertilizers, raw sewage, municipal waste water, animal husbandry, and deforestation all add nutrients to rivers
Natural sediment load	Higher sediment loads mean that coastal areas have higher sediment concentrations, which can result in turbid waters that limit light penetration and therefore photosynthesis by plankton
Presence of dams	Dams trap sediment behind their walls, decreasing the natural sediment load of a river and sediment concentrations in coastal areas where rivers flow into the ocean
Flushing rate	Slower replacement of coastal waters by oxygenated water results in lower oxygen concentrations in coastal waters
Coastal wetland area	Wetlands acts as filters that remove nutrients from waters that flow through them
Government policies	Government regulations can help reduce nutrient inputs
Cultural norms	Informal rules about behavior influence how people act

of 1.8 million km² and a discharge of 944 km³ of water each year. Four hundred million people live in the Yangtze River basin, giving the basin an average population density of 200 people/km². Population growth is rapid. Eighty percent of cities in the basin have no wastewater treatment, which means that raw human sewage, which contains nutrients, flows into the East China Sea. Since the 1990s, use of inorganic fertilizers for agriculture has rapidly increased and China is now the world's largest consumer of these fertilizers. Fertilizers, containing nitrogen and phosphorous nutrients, are added to crops to promote growth, but excess nutrients wash into rivers. In addition to agriculture, aquaculture composes 95% of all freshwater fisheries in the country. In aquaculture, farmed fish are raised in pens at high populations densities, adding lots of excess nutrients from their waste to freshwaters and, ultimately, coastal

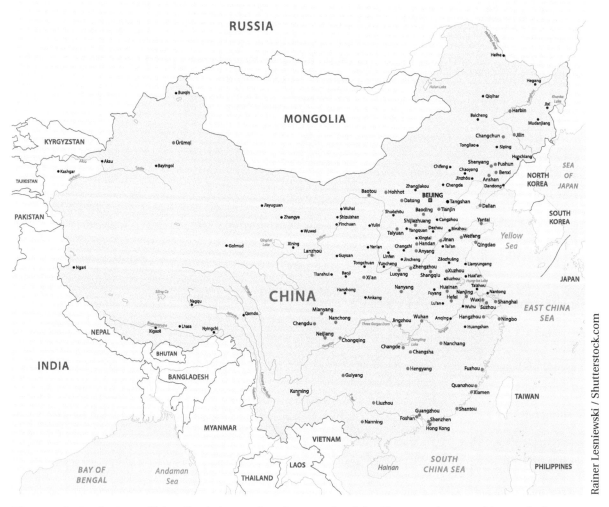

Figure 3.6 The East China Sea is located at the mouth of the Yangtze river, and is nestled among the countries of China, South Korea, Japan, and Taiwan.

regions. Owing to its source in the mountainous Himalaya, the Yangtze River has high quantities of sediment present in its waters naturally. Dams built on the Yangtze retain a lot of sediment behind their walls. With less sediment blocking light from entering the water, there is more photosynthesis and more eutrophication can occur. Active deforestation and intensive agriculture in the basin result in soil erosion, which causes sediment to increase in the river. Forest and agricultural soils both contain nutrients, so erosion adds nutrients. Coastal wetlands filter nutrients from waters as rivers flow through them and into the sea, but in China, wetlands are being destroyed by shrimp aquaculture and land reclamation.

The Black Sea is largely a land-locked sea (Figure 3.7). Its only outlet to a major world ocean is the narrow Bosphorous Strait. Unlike the East China Sea, more than one river contributes excess nutrients to the Black Sea. The three major rivers are the Danube flowing through central and eastern Europe, the Dnipro flowing through the Ukraine, and the Don flowing through Russia. Many other smaller rivers, in other

Figure 3.7 The Black Sea is surrounded by many different countries from both Europe and Asia, and it flows between these two continents through the Bosphorous Strait in the southwest and into the Mediterranean Sea.

countries surrounding the Black Sea, also contribute. The basin area that flows into the Black Sea touches 23 different countries and is 2 million km². About 162 million people live within the basin area, giving the basin a population density of about 80 people/km². Population growth is relatively stable. Many cities within the basin have wastewater treatment, although some do not, especially coastal cities where sewers discharge directly into the sea. The natural sediment quantities present in the rivers flowing into the Black Sea come from lowland regions, with the exception of a small part of the Alps, and are much lower than in the Yangtze. Dams exist along these rivers, but soil erosion from deforestation and agriculture is much less intense than in the Yangtze. Protected coastland wetland areas exist in many different countries with coasts along the Black Sea. Although some governments have taken action to reduce nutrient inputs, coordinated action across 23 different countries is challenging and many people still do not view eutrophication as a pressing problem.

Figures 3.8a and 3.8b shows possible ways that the two systems described above could be represented. The systems are similar, in that fertilizers and raw human sewage are indirect drivers of eutrophication. River sediment load, lack of coastal wetlands, and aquaculture are additional indirect drivers of eutrophication in the East China Sea, but they are not as important, or not present at all, in the Black Sea region. The Yangtze River enters the East China Sea as an estuary, which is flushed with oxygen-rich ocean waters more frequently than the enclosed basin of the Black Sea. In addition to place, time and context are also important when defining a system and its drivers. In this example, drivers vary with time. For example, prior to the rapid increase in fertilizer use in China in the 1990s, inorganic fertilizer use contributed much less to eutrophication than at present. The system would be defined differently in the past than it is today. Finally, the institutional context for watershed management in the Black Sea is more complicated, due to the many different countries involved,

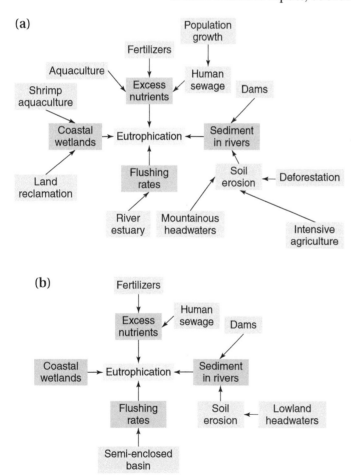

Figure 3.8 System representations for (a) the East China Sea and (b) the Black Sea, where the *boxes* are system components and the *arrows* are interactions between the components. *Blue boxes* are direct drivers and *orange boxes* are indirect drivers.

compared to management in the Yangtze River basin, which is contained entirely within the boundaries of one country.

More comparisons could be made between these two specific cases, but the point has been made that systems must be defined for a specific place, time, and context. Figures 3.8a and 3.8b show that nutrient sources and other factors that affect eutrophication are different between the two cases. When a system is defined for one place, time, or context, the system definition cannot be transferred to and used in another place, time, and context. If a system is not defined in a way that is specific to a local context, then appropriate solutions to a problem within that specific context may not be effective.

Section 3.1.2—Classifying Driver Scale

Key Concept 3.1.2—Natural and human drivers regulate change in socioecological systems over many spatial and temporal scales.

Drivers Span Many Spatial Scales. Once a system and it's drivers are clearly defined, it is important to look beyond and within a system's boundaries to larger and smaller spatial scales. This is necessary for a thorough understanding of the drivers regulating the system because socio-ecological systems are open systems, meaning that energy, matter, and information flow across system boundaries. For example, phosphorus fertilizer (a type of matter) might be brought into the Yangtze river basin from a large mining operation located in Morocco, which is a country located outside of the Yangtze watershed boundary. Morocco holds about 75 % of the world's phosphate reserves, and is the world's third largest exporter of phosphorus. If Morocco were to experience a disruption in phosphorus exports, such as due to a war affecting the country, then this could affect the amount of phosphorus fertilizer used in China. Thus, to identify all potential drivers regulating a system, factors at many different spatial scales must be carefully considered. These scales range from local to regional to national and finally to global, and include drivers from both social and natural systems (**Figure 3.9**).

In the coastal eutrophication example, natural system factors important to understanding the problem operate at different spatial scales. At the global scale, climate change generally results in warmer temperatures, which might influence how nutrient pollution affects coastal regions. Warmer temperatures increase photosynthesis rates, causing more nutrients to be taken up. More nutrient uptake means more biomass produced by plankton and aquatic plants than at lower temperatures (Q_{10} **factor**). More production of biomass means that there is more biomass to be degraded by respiration, and this will use up more oxygen, which is less soluble in warmer water. The overall

Q_{10} factor
a dimensionless quantity used in both biology and chemistry to define how reaction rates change with every 10°C change in temperature, which is typically by a factor of about 2 to 3 times.

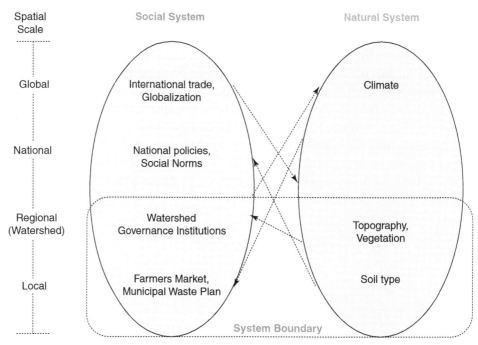

Figure 3.9 The direct and indirect drivers that influence the behavior of systems operate at many different spatial scales.

result could be lower oxygen conditions than at cooler temperatures, which would make eutrophication worse.

At the regional watershed scale, topography and vegetation cover are important (Figure 3.9). If the watershed is composed of land with a generally low elevation and gradual topography, the watershed may contain rivers lined with wetlands and riparian zones. One service provided by these ecosystems is to clean nutrients out of the water (Figure 3.10). In contrast, with steeper mountainous topography, water flows more quickly into rivers and then into coastal regions without much nutrient removal by wetlands or riparian areas. This problem could be amplified if vegetation cover is altered. Forests tend to hold onto water and soils. With less forest cover, more nutrients could reach coastal regions.

At the local scale, soil type can be important. Some soils, such as those containing sand, are more permeable than others. The more permeable the soil, the more water will flow down into the soil, through **infiltration** and **percolation** processes, rather than across the land surface and directly into rivers. Microorganisms living in soils take up nutrients, which in a permeable soil might prevent nutrients from flowing into rivers. In the coastal eutrophication example, climate, topography, vegetation, and soil type are possible natural system drivers operating at several different spatial scales (Figure 3.9).

---

infiltration the downward passage of water through the soil.

percolation the general movement of any liquid, including water, through porous material such as sediment, gravel, or soil.

There are a whole additional suite of possible drivers in the social system (Figure 3.9). At the global scale, international trade and globalization mean that more food is grown for export. This often results in large-scale production in an industrial-style food system focused on maximizing efficiency through application of large quantities of inorganic phosphorus fertilizer mined from rocks. This trend has been influential in the Yangtze River basin, where a lot of rice is grown for export to global markets. Activities at the national level can also influence nutrient inputs to coastal regions. In the 1960s, inorganic fertilizer use intensified in river basins flowing into the Black Sea. However, two changes at the national scale affected nutrient inputs to the Black Sea. In the 1990s, nutrient pollution from the Danube decreased slightly due to progressive agricultural policies implemented by the European Union. During this same time period, reduced agricultural activities in the Dnipro and Don basins, following the collapse of the Soviet Union, also caused fertilizer inputs to decrease. Social norms can also play a role. In China, the government has not done much to reduce nutrient inputs, or protect wetlands, due to a perception that land and ocean resources are inexhaustible. In contrast, institutions in the Black Sea region have demonstrated a greater concern for sustainability, based on their policies and conservation efforts. Finally, a national demand for freshwater fish in China drives the abundance of aquaculture operations.

Figure 3.10 Riparian zones occur where land meets water, such as the forested areas shown in this picture, and are so important to water quality that they are retained even when land around them is developed.

At regional watershed scales, the types of institutions available for management of nutrient inputs may influence pollution levels. The Yangtze River is located entirely within one country. In theory, this should make management of nutrients within the river basin relatively straightforward, as all activities adding nutrients to the river are subject to the laws and regulations of only one governing body. In contrast, rivers flowing into the Black Sea touch 23 different countries, which means that nutrient management requires international collaboration. The formation of the European Union has helped the situation, as countries hoping to join have been required to adopt EU management policies prior to accession (Figure 3.11). If conservation groups, for example, have more power over decision making in this type of organization than profit-maximizing business interests do, then this would be a driving force toward sustainability. In addition to governments, international agreements have played a role in the Black Sea. For example, the RAMSAR Convention, an intergovernmental treaty established in the 1970s, has played a role in the conservation of coastal wetlands in the Black Sea (Figure 3.12).

At the local level, the presence or absence of institutions demanding sustainably produced food, such as farmers' markets, could affect

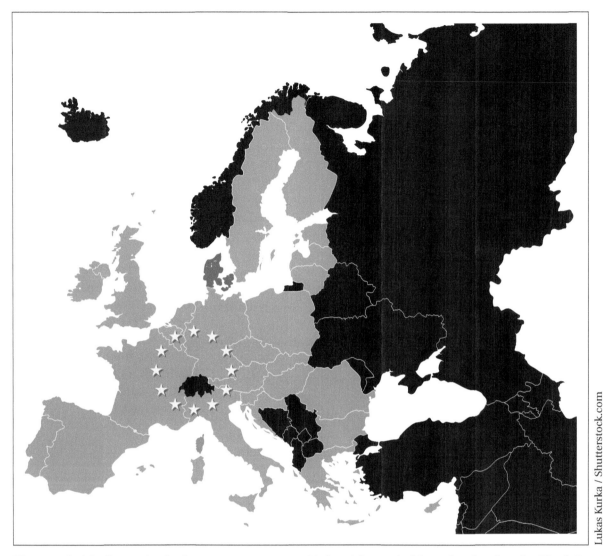

Figure 3.11 Countries in the eastern European Union (shown in blue) that border the Black Sea, namely Bulgaria and Romania, adopted EU agricultural policies when they joined.

the type of food grown by individual farms. In some towns, people might demand farmers' markets with sustainably produced food, whereas citizens of other towns might not prefer this type of food. Local municipalities have different methods for handling human waste from households. For example, underground septic systems that leak sewage (and nutrients) into rivers might be legal in some towns. In the absence of national policies, some towns might completely lack sewage treatment, while others require it prior to release of water back into rivers.

Drivers Operate across Many Temporal Scales. As you can see from the coastal eutrophication example thus far, systems are regulated by interactions among multiple drivers at several spatial scales. These drivers are a mix of ecosystem processes, government policies, economic activity, and cultural factors. In addition to operating across many spatial scales, drivers also span several time scales. They can work slowly over long time scales, such as with population growth and the accumulation of anthropogenic greenhouse gases in the atmosphere. They can also impact systems intermittently and suddenly, such as with droughts, wildfires, wars, hurricanes, and economic crises. These ideas concerning the time scales over which drivers operate will be covered in more detail in Chapter 6 by exploring the dynamics of fast and slow drivers. For now, it is enough to recognize that various drivers regulating a system operate over different time scales and that classifying the time scale for each driver depends on the specific sustainability problem of concern and the local context within which that problem is embedded.

Figure 3.12 In addition to cleaning water of nutrients, wetlands provide habitat for wildlife, such as the great white pelican shown in this photo from the Danube Delta Wildlife Reserve.

In the coastal eutrophication example, the processes of climate change and globalization may occur on long time scales of tens to hundreds of years or more. Changes in vegetation cover, and in national or international policies, can occur over years to decades. Decisions by vendors at a single town's farmers' market to transition to sustainable agriculture can happen within a few years or less. If you compare these times scales to the spatial scales shown in Figure 3.9, you can see that they are roughly correlated: Processes operating at large spatial scales tend influence systems over long time scales, while those operating at small spatial scales often influence systems over shorter time scales. Time and space scales do not always correlate exactly like this, but it often works this way. There are many examples of this phenomenon. Before describing two examples below, an important caveat to note is that it is possible for change at larger scales to happen very rapidly. This is often referred to by sustainability scientists as a catastrophic change or a **critical transition**. For this type of change to happen, certain conditions must be met, as described in Chapter 5.

critical transition a nonlinear threshold-based regime shift that is hard to reverse and often difficult to anticipate in socioecological systems.

Spatial and Temporal Scale Correlation, and Orders of Magnitude. This section offers two examples, one from a natural system (Figure 3.13a) and one from a social system (Figure 3.13b), of how drivers at similar spatial and temporal scales often align. The first example of correlation between temporal and spatial scales is based on drivers regulating boreal forest ecosystems (Figure 3.13a). In boreal forests, space scales

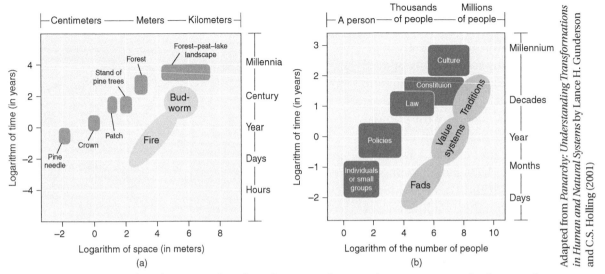

Figure 3.13 Spatial and temporal scales, shown as the x- and y-axes, respectively, are often directly correlated.

range from a centimeter-long pine needle, to a one hundred square meter stand of trees, to a forest–peatland–lake landscape covering thousands of square kilometers (**Figure 3.14**). In terms of time scale, the centimeter-long pine needle takes months to grow, a stand of pine trees in this forest takes almost a century to mature, and landscapes in this system evolve over 10,000-year time scales. Two important drivers of change in boreal forest systems are fire and budworms (shown as light green circles in Figure 3.13a). Natural fires in these systems affect many spatial scales ranging from the pine needle up to the forest. Fire regulates these systems over time scales ranging from days to years. The other driver of change is the spruce budworm, which plays an important role in the natural life cycle of forest ecosystems. The budworm affects the boreal forest at the spatial scale of the stand and the time scale of years.

Important note on logarithms: Notice in Figure 3.13 that the axes are shown as logarithmic scales. Logarithms allow you to express any number as a power of 10. For example, the log of 100 is 2 ($10^2 = 10 \times 10 = 100$) and the log of 10,000 is 4 ($10^4 = 10 \times 10 \times 10 \times 10 = 10,000$). This basic mathematical concept is important for understanding what is meant by scale in socioecological systems because as you move from lower to higher scales, and vice versa, you jump from one power of 10 to another. Each new power of 10 is called an **order of magnitude**. This term is used to describe differences among the various spatial and temporal scales relevant to understanding sustainability problems embedded in socioecological systems. In this boreal forest example, the landscape (10^6 meters) is four orders of magnitude larger than a forest stand (10^2 meters), which is in turn four orders

order of magnitude an expression of the approximate magnitude of something based on a power of 10 difference between that quantity relative to another.

of magnitude larger than a pine needle (10^{-2} meters). A pine needle (10^{-2} meters) is eight orders of magnitude smaller than a boreal forest landscape (10^6 meters). The landscape, forest stand, and pine needle represent three different spatial scales. As you move from one order of magnitude to another, you move from one spatial or temporal scale to another. The order of magnitude concept describes precisely what is meant by different space and time scales.

The second example of how time and space scales are correlated applies to institutional change (Figure 3.13b). The spatial scale in this system is the number of people involved (the x-axis on Figure 3.13b) Institutional change can happen on a small spatial scale as a result of actions taken by individuals or groups of tens to hundreds of people. New policies are made by larger groups of 100 to 10,000, which may be composed of legislative bodies and their staff (Figure 3.15). At the largest spatial scale, institutional change occurs within an entire culture made up of millions to hundreds of millions of people. In general, actions taken by individuals or small groups are fast and take anywhere from hours to months. Policies take longer, and change at this level occurs on the order of years. Cultural change is often the slowest type of institutional change

Figure 3.14 One-third of all forests worldwide are boreal forests, whose landscape of wetlands, rivers and lakes, peatland, and forest (as shown in this Canadian boreal forest) provide important ecosystem services, including carbon sequestration and water filtration.

Figure 3.15 The United Nations general assembly, which is composed of delegates from 193 member nations, is viewed as an important institution for meeting global sustainability challenges.

and can take centuries to millennia to occur. The drivers that regulate institutional changes made by small groups, with policies, or for entire cultures could be factors such as fads, values, and traditions (shown as light blue circles in Figure 3.13b). Like fire and the budworm in boreal forest ecosystems, these drivers regulate institutional change over a range of spatial and temporal scales in human systems.

Section 3.1.3—Classifying Driver Influence

Key Concept 3.1.3—Natural and human drivers can influence socioecological systems either directly and unequivocally, or indirectly and more diffusely.

The final aspect of drivers explored in this chapter is whether they directly or indirectly regulate a system. Direct drivers are drivers that clearly and unequivocally influence the behavior of a system. Indirect drivers are drivers that influence the behavior of a system in a more diffuse way, by altering one or more direct drivers. Direct and indirect drivers were defined in this way by the 2005 Millennium Ecosystem assessment (MEA). The MEA used drivers to understand biodiversity loss. First, they identified five direct drivers of biodiversity loss: habitat change, climate change, invasive species, overexploitation of species, and nutrient pollution. Five general classes of indirect drivers of biodiversity loss were also identified: demographic, economic, sociopolitical, cultural and religious, and scientific and technological.

Demographic drivers refer to characteristics of a population, such as growth rate. Economic drivers include factors such as supply and demand, global market forces, and consumption patterns. Sociopolitical drivers refer to factors that influence decision making, such as public participation. Cultural and religious drivers include informal social norms, beliefs, and values that influence human behavior. Scientific and technological drivers include physical objects or knowledge used by humans to control or adapt to their environment.

All of the MEA drivers are shown in **Figure 3.16**, which illustrates one way to explicitly represent direct and indirect drivers in a system. The arrows drawn from the five direct drivers to the problem, which is biodiversity loss, are larger and thicker than other arrows in the diagram to illustrate their unequivocal influence on the system. *Unequivocal* generally means that something is very clear or beyond any doubt. Based on the assessment of thousands of studies in ecosystems around the world, it was clear to the MEA authors that these five drivers are the direct cause of biodiversity loss. The arrows drawn from the indirect drivers to the direct drivers are dashed and faint to illustrate that they indirectly affect biodiversity by influencing the direct drivers. Thus, indirect drivers regulate systems more diffusely than direct drivers. Normally, one direct driver is simultaneously influenced by several indirect drivers, making causal chain analysis (Section 3.2) complicated. **Figure 3.17**, which defines direct and indirect drivers of eutrophication in the East China Sea using the same framework as in Figure 3.16, gives an example of this. As with system definition (Section 3.1.1) and scale

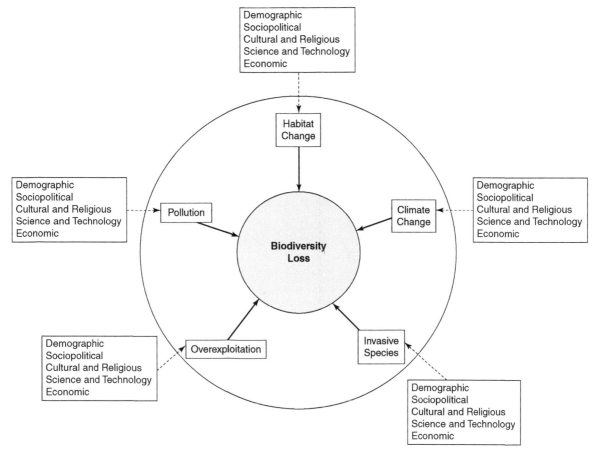

Figure 3.16 The Millenium Ecosystem Assessment (MEA) used five direct drivers (green boxes) and five indirect drivers (shown in each yellow box) to study biodiversity loss.

(Section 3.1.2), the direct or indirect influence of drivers on a system depends on the specific problem of concern and the time, place, and context in which that problem is embedded.

The drivers used by the MEA are focused on the natural systems component of socioecological systems. Other possible drivers, such as education, housing, employment, access to health care, population growth, war and conflict, and drought, may be focused on changes in the human component of socioecological systems (see Box 3.1 for a human systems example). It is important to emphasize that not all drivers push natural and human systems toward an unsustainable condition. They can also push systems toward sustainability or keep systems from moving at all. It is crucial to be aware of how drivers might influence a system when devising solution options. The feedbacks and other dynamics underlying drivers of system change are described in more detail in Chapters 5 and 6. The next section is focused on how the systems and drivers defined in this section can be used to figure out the root causes of a sustainability problem.

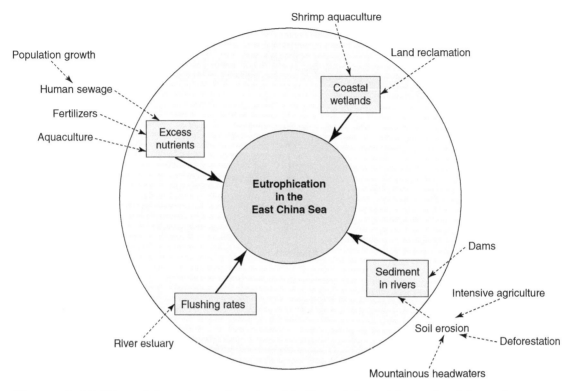

Figure 3.17 Direct drivers (*green boxes*) and indirect drivers (*shown in white region outside of the blue circles*) of eutrophication in the East China Sea.

BOX 3.1 THE SOCIAL AND ENVIRONMENTAL COSTS OF CHEAP CLOTHING

The Problem. As revealed by the 2015 documentary *True Cost*, most clothing sold in the United States, and other wealthy nations, is produced by the fast fashion industry (Figure 3.18). Fast fashion produces cheap clothing quickly and inexpensively, such that new styles are constantly available to consumers at low prices. The number of fashion seasons, when people buy the latest clothing styles newly available in stores, has increased from two seasons per year to weekly. Online ordering has made purchasing more convenient, as it can now happen at work, in a car, and outside of regular business hours. Today's clothing is inexpensive, such that the average consumer does not think twice about discarding damaged or out-of-date clothing. Advertisements, and other forms of marketing, convince us that buying the latest fashions will make us beautiful, accepted, healthy, and happy (Figure 3.19). (This is contrary to happiness research, suggesting that a focus on material wealth causes depression, anxiety, and other problems leading to unhappiness.) Overall, consumers buy 800% more clothing today, compared to only two decades ago, and clothing companies earn more profit, despite the rising costs of clothing production.

Figure 3.18 H&M, the second biggest clothing company in the world, is notorious for its fast fashion model.

Figure 3.19 The well-advertised Black Friday shopping tradition leads to mass consumption where people wait in long lines, and even camp overnight in tents, to be the first in line to, in the words of Steven Colbert, spend "money we don't have on things we don't need to give to people we don't like."

There are hidden costs, to people and the environment, embedded in cheap-and-fast clothing. The costs have to do with the way clothes are made. For example, only about 3% of the clothing worn by people in the United States today is actually made in the United States. The other 97% of clothing production is outsourced to developing countries, where costs are lower. (This is in contrast to the 1960s, when 95% of clothing worn by people in the United States was made in-country.) From growing cotton to making clothing and accessories, workers in developing countries are exploited by the clothing industry in unimaginable ways, all in the name of achieving the lowest possible production cost and the highest possible profit.

Exploitation begins with the cotton farmer, who produces the crop used to make the majority of clothing today. India is the top cotton producing country in the world (Figure 3.20).

(continued)

Figure 3.20 Indian farmers, such as this man from Maharashtra state, produce more than a quarter of the world's cotton.

Small cotton farmers in India are not paid a fair price for their cotton, which is one way that clothing companies cut production costs. However, this is only a part of the problem. When farmers buy seeds and pesticides, from companies such as Monsanto, they go into debt. In the past, farmers saved their own seeds to use in the next season's planting. When a crop failed, it costs them nothing other than the loss of their crop, and they had other crops grown alongside cotton as a safety net. Enticed by claims that Monsanto's cotton seeds would produce high yields, Indian farmers started to buy them, often on credit, and plant cotton exclusively. When seeds were bought, failed crops forced farmers into debt, as they could not sell their cotton to pay for seeds they had already used, and their safety net of other crops was gone. Monsanto seeds often produce lower yields than native cotton plants. When more pesticides were offered as a solution to low yields, farmers bought these too, incurring further debt. Suicide, often by drinking a bottle of pesticide, is common practice by Indian farmers who find themselves deeply in debt. In the first decade of the twenty-first century, about 100,000 Indian farmers committed suicide. Pesticide use also causes health problems. In Punjab, a major cotton growing state, hundreds of people in every village suffer from birth defects, cancers, and mental illness.

Further along the fashion supply chain, factory workers are exploited to cut costs. With no other options for work, many people in developing countries are forced into "sweatshops," where they labor for long hours, often under dangerous working conditions, for extremely low wages. In Bangladesh, 85% of workers are women. Owing to long hours, and lack of childcare, many women are forced to leave their children with family and friends in their home villages, away from cities. In rural villages, children have access to education and an overall better life, but they only see their mothers a few times per year. Factory workers face constant job insecurity, as clothing companies will move to new locations if costs are lower elsewhere. Workers who form unions or protest working conditions are often beaten by factory managers or violently attacked by riot police, such as in the 2013–2014 riots in Cambodia. These uprisings concern governments, such as Bangladesh where clothing export is 70% of their GDP. To prevent factories from relocating to places where wages are lower, and there are less disruptions from workers, governments enforce low wages, ignore labor laws, and support police brutality.

The most tragic way that companies keep costs low is to disregard safety issues in their factories (Figure 3.21). This has led to deadly tragedies for workers in developing countries. Factory fires are common, but the single worst disaster in the clothing industry's history occurred in 2013 in Rana Plaza, located in Dhaka, the capital of Bangladesh. For weeks, factory workers, who earned $2 per day, had been complaining about dangerous looking cracks in the walls and ceilings of the eight story building, but their concerns were ignored. On April 24, 2013, the build collapsed, killing 1,129 people and seriously injuring hundreds of others. It is telling that the year of this tragedy was also the fashion industry's most profitable year ever.

In addition to catastrophes such as the sudden collapse of a building, pollution from factories harms people in more subtle, less obvious ways. The Ganges, which is considered a sacred river by Hindi culture, is polluted with chromium VI and other toxic chemicals from leather tanneries in Kanpur, India (Figure 3.22). These factories manufacture cheap leather used by the clothing industry to make many products, including shoes, purses, and belts. In addition to contaminating their spiritual life, the chemicals are present in drinking water, and in the soils and water used for local agriculture. People living in this area are plagued with skin rashes, stomach ailments, jaundice, and liver cancer.

The clothing on our backs is filled with the blood, misery, and heartbreak of millions of people. According to *True Cost*, familiar brands contributing to the fast fashion model include H&M, Zara, Primark, Forever 21, Top Shop, GAP, Mango, and Joe Fresh. The workers trapped in this system are some of the most vulnerable and powerless people in the world. They form the essential base of the fast fashion clothing supply chain, yet they have no voice in it. There are environmental costs too, including soil and habitat degradation, biodiversity loss, pollution of water and soils, and growing landfills where clothing sits for more than 200 years before biodegrading. Some clothing is donated to charity and reused, but 90% of it is shipped to developing countries, such as Haiti, where local people buy in bulk at reduced cost and local clothing economies vanish. Wealthy nations are not immune to harm. In Texas, the largest cotton producing state in the United States, brain surgeons remove brain tumors most often in men aged 45–65 that work in agriculture. The clothing industry is part of a larger

Figure 3.21 Garment factory fires, such as this one, have injured hundreds of workers, often as a result of sparse exits or fire escapes, and otherwise unsafe working conditions.

(continued)

Figure 3.22 The Ganges, a holy river to Hindi people, has been polluted with garbage, industrial waste, and other forms of pollution.

capitalist system focused on ever-increasing profits, in which human rights and ecosystems are traded for economic benefits.

Solutions. Figure 3.23 presents a tangled web of drivers, which are part of a global supply chain, that lead to the costs of fast fashion. Where can we intervene in this system to make it more sustainable? What drivers should we focus on? The Decent Working Conditions and Fair Competition Act, which would have prevented the sale of products produced in sweatshops in the United States, was introduced to Congress, but failed due to objections that it would impede free trade. Free trade advocates favor voluntary codes of conduct, such as the Ethical Trading Initiative, which are almost always unsuccessful. Some argue that sweatshops raise the quality of life for workers, who have no other alternatives for work. Owing to close relationships between governments and corporate interests today, policy changes in this arena are difficult. However, individuals and companies can and do make a difference.

Fair trade clothing companies are taking shape and breaking down the powerful drivers of the social and environmental costs of fast fashion (Figure 3.23). People Tree, the first company certified by the World Fair Trade Organization, is a great example. People Tree pulls entire families out of Dhaka slums, away from sweatshop work, and back to their home villages to work in producer groups. They teach workers garment-making skills and tailor fashions to local resources. Clothing is often made by hand, rather than machines, which cuts CO_2 emissions (Figure 3.24). In some cases, clothing is made by workers in their own homes. In other cases, People Tree sets up rural production centers, and uses company profits to fund schools and daycare centers. People Tree also works with fair trade cotton farmers, such

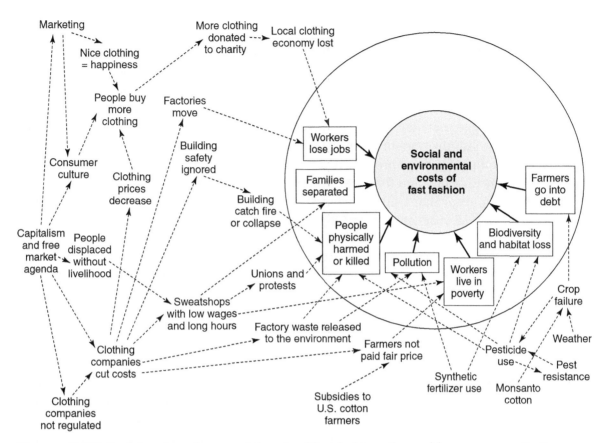

Figure 3.23 Social and environmental costs of fast fashion, the problem, are a result of many direct drivers (*solid arrows*) and indirect drivers (*dashed arrows*).

as Agrocel in India and Sekem in Egypt, to incorporate organic cotton into their products. Safia Minney, Founder and CEO of People Tree, has worked over the past 20 years to create a powerful network of individuals to promote this model of production in countries around the world. In her book, called *Naked Fashion*, she highlights the sustainability work of many others in the fashion industry, including photographers, designers, buyers, models, and other members of the fashion supply chain.

Figure 3.24 Traditional handlooms, which make beautiful fabrics, are part of south Asia's traditional culture.

(continued)

Starting a company is one powerful way to combat the fast fashion model to create a more sustainable system. The everyday actions of individuals is also powerful. Protests after the Rana Plaza disaster garnered support from the Prime Minister of Bangladesh, who was facing re-election at the time, to increase wages. Although the minimum wage in Bangladesh factories remains the lowest in the world, it was their first raise in 10 years and resulted in a 77% pay increase. A shift in behaviors and attitudes of consumers—that's YOU!!—is powerful. One could argue that, like garment workers and farmers, clothing producers are also stuck in the fast fashion system: when consumers seek out the cheapest products possible, clothing companies must provide cheap clothing to stay in business. As a consumer, you can stop creating demand for cheap clothing by consuming less and buying fair trade brands. In the words of Sharti, a Bangledeshi woman who was laid off from her factory due to an injury: "After 16 years of working 12 hours a day, I have nothing. I'm like an old dairy cow. You might as well slit my throat. … If you pay a little more, we can live a little better." To raise the wage of a Bangladeshi factory worker to a living wage, the price of a pair of jeans would have to increase from $32 to $33.30. Wear clothing longer, so that you buy less. The cost to your wallet could end up being about the same, or less.

Section 3.2: Causal Chain Analysis

Core Question: How can system regulation be understood in a way that helps resolve sustainability problems?

"[T]here are no ends to the causal chains that link interacting open systems . . . The planner who works with open systems is caught up in the ambiguity of their causal webs."

—H. Rittel and M. Weber, 1973,
Dilemma in a General Theory
of Planning

Cause is complicated, but establishing cause is necessary for finding targeted solutions to sustainability problems. Drivers regulate systems across multiple spatial and temporal scales, directly and indirectly, and can be from natural or social systems. When dealing with sustainability problems, there are always multiple and interacting drivers. This makes it difficult to figure out simple cause–effect relationships between drivers and the problem. For sustainability problems, clear and linear cause-and-effect linkages between a single driver and the problem rarely exist. As a result, the identification and understanding of drivers can be incredibly complicated.

Causal chain analysis (CCA), as defined at the beginning of Section 3.1, helps trace cause–effect pathways from the problem back to the drivers regulating the socioecological system, and considers the relationship among interacting drivers. CCA does not identify one single, most important driver, but rather accounts for a wide range of interrelated factors that led to the problem. Analyzing causal chains helps determine which aspects of a system to focus on when devising solutions by identifying the most important drivers, their relative influences, and the current trend for each driver in terms of how their influence on the system is changing through time (increasing, decreasing, or constant). This type of information can help in devising appropriate solutions, as problem solvers are better positioned to support sustainable and cost-effective interventions. CCA improves understanding of complex problems by demonstrating why they exist, who or what is responsible, and how to solve them. Without this understanding, solution options for addressing a problem are difficult to devise. Several aspects of CCA are presented here: assessing causality, determining extent of influence and current trends, and identifying the specific conditions (context) under which certain drivers become important.

Section 3.2.1—Assessing Causality

Key Concept 3.2.1—Causal chain analysis promotes deeper understanding of sustainability problems by connecting drivers to the problem.

Establishing causality in a system is a basic activity that occurs both informally in your daily life and formally during a rigorous formal assessment of a sustainability problem. When you establish causality, you make clear the relationship between a set of drivers (causes) and the influence (effect) on a system. In short, you establish a cause–effect relationship. For example, if you have a stomachache after eating cereal with milk that smelled a bit sour, then you can establish a cause–effect relationship between the sour milk (cause) and your stomachache (effect). Causal chains are *chains* because there is usually an effect that can be traced back to one cause (Direct Driver), which is also affected by some other cause (Indirect Driver). In the milk example, the reason that you have sour milk in your fridge might be because you were so busy this week that you didn't have time to go to the grocery store to get fresh milk. The causal chain then becomes: hectic week (Indirect Driver) → sour milk (Direct Driver) → stomachache (Effect). Thus, a causal chain looks something like this: Indirect Driver → Direct Driver → Effect. Causal chains work backwards from the observed effect, which can be desirable or undesirable, to the causes of that effect. When written in words, as opposed to illustrated in diagrams, effective causal chains should be expressed as brief, powerful explanations of causation so that the connection among the factors is clear. Box 3.2 provides examples of causal chains from the clothing supply chain problem presented in Box 3.1 by presenting excerpts from the writing first and then rewritten as brief, powerful causal chain explanations that make clear connections among drivers, and between drivers and the problem.

Causal chains work to clearly connect the dots between drivers and the problem. The milk–stomachache example is simple. The clothing supply chain example in Boxes 3.1 and 3.2 illustrates how causal chains are helpful when conducting an in-depth analysis of sustainability problems, which are much more complicated. Figure 3.23 (Box 3.1), where the system and drivers are first diagrammed, appears as a tangled web. Causal chain analysis helps the problem analyst to look at cause-effect relationships within this tangled web, while still viewing the problem holistically. This avoids the tendency for reductionist approaches that simplify complex problems to one or

BOX 3.2 CAUSES OF PHYSICAL HARM TO GARMENT FACTORY WORKERS

One of the major social costs of fast fashion is that people are physically harmed or killed. The original writing, from Box 3.1, that describes this problem is: "The most tragic way that companies keep costs low is to disregard safety issues in their factories. This has led to deadly tragedies for workers in developing countries. Factory fires are common, but the single worst disaster in the clothing industry's history occurred in 2013 in Rana Plaza, located in Dhaka, the capital of Bangladesh. ... Workers who form unions or protest working conditions are often beaten by factory managers or violently attacked by riot police, such as in the 2013–2014 riots in Cambodia. These uprisings concern governments, such as Bangladesh where clothing export is 70% of their GDP. To prevent factories from relocating to places where wages are lower, and there are less disruptions from workers, governments enforce low wages, ignore labor laws, and support police brutality."

This statement mentions physical harms (injury and death) and a list of possible explanations for these harms (low wages, unsafe buildings, policy brutality, government corruption, cutting costs, fires, and building collapse). What is missing are brief powerful statements of causation that clearly make the connection between the effect (problem) and the causes (drivers). To figure out the root causes of physical harm to garment factory workers, one causal chain from this communication can be written as: "Workers who unionize or protest are injured or killed (Effect). They form unions and protest (Direct Driver) because of low wages and dangerous working conditions (Indirect Driver), which are both a result of cost cutting efforts of clothing companies (Indirect Driver 2)" (Figure 3.25, top). Another causal chain that

(continued)

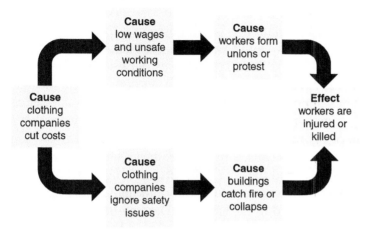

Figure 3.25 Two causal chains, which both start with the same cause (or driver), show two ways that garment workers are harmed.

reveals root causes of physical harm to garment factory workers is: "Factory workers are injured or killed (Effect) because building catch fire or collapse (Direct Driver) due to clothing companies ignoring safety issues (Indirect Driver) to cut costs (Indirect Driver 2)" (Figure 3.25, bottom). Using these brief, powerful statements, causal chains can be constructed using diagrams, as shown in Figure 3.25.

The root cause of the problem, in both cases, is that clothing companies try to cut costs. In both cases, attempts to cut costs lead to a chain of events that result in injury or death for garment factory workers. To put a stop to these injuries and deaths, one could try to stop clothing companies from cost cutting. This would be an effective focus of a problem-solving effort for this sustainability problem. Possible solutions could include passing policies to ban import of clothing from companies who do not offer living wages and provide safe working conditions, or urging consumers to avoid purchasing clothing from such companies. As was mentioned in Box 3.1, these are exactly some of the solution options being pursued.

a few factors, which must be resisted when dealing with sustainability problems. Such simplistic approaches can lead to policies or institutional arrangements that cause additional harm. A holistic approach using a thorough CCA is needed to gain a wider, more intricate, and deeper understanding that can then be used to pursue multiple solution options simultaneously.

Causal chains are intertwined in causal webs, with many interacting drivers leading to the problem. With all this complexity, how can the most important drivers be assessed so that focused and targeted solutions to sustainability problems may be devised? The answer lies in assessing the *relative influences* and *current trends* for drivers, and also paying attention to the *specific contexts* in which problems are embedded because there are no universal solutions. Before delving into these aspects of causal chain analysis, one more note on causality must be made: Correlation ≠ Causation.

Correlation ≠ Causation. The phrase "Correlation ≠ Causation" means that if two variables are correlated, one does not necessarily cause the other. If correlation is always assumed to mean causation, then some causal chains would be inaccurate, leading to ineffective solutions. This idea is best illustrated with examples. In the late 1940s, a nationwide study found a strong correlation between the incidence rate of new cases of polio among children in a community and per capita ice cream consumption in that community. In other words, as ice cream consumption increased, polio incidence increased. This led some to believe that polio was caused by ice cream consumption. However, other researchers realized that the data sets showing a strong correlation between high ice cream consumption and high incidence of new polio cases were gathered in the summer. Data gathered in the winter showed a lower incidence of new polio cases and a lower rate of ice cream consumption. We now know that polio is a viral infection that spreads more easily when children gather together to play in unsanitary conditions, such as during summer vacation, than when children play together under more sanitary school conditions during the winter. Imagine if policymakers had listened to researcher's first explanation of causation: "Polio is caused by ice cream consumption." This could have led to an unnecessary ice cream ban! It also would have delayed the development of a vaccine for the disease (**Figure 3.26**).

Another example is the bubonic plague. Prior to the onset of the plague, cats were associated with witchcraft and were systematically eliminated. When the plague took hold and spread in the mid-1300s to the 1700s, cats were blamed for the disease, which resulted in even greater elimination of cats. This elimination of cats allowed the rat population to multiply exponentially. It became illegal in some cities to have a cat, but some cat lovers secretly kept their cats despite the prohibition. Those people who had cats in their homes were less susceptible to getting the plague. It was eventually discovered that the plague was actually transmitted by fleas carried on rats. This is another example of getting the causal chain wrong. In this case, when action was taken based on the perceived causal relationship, it created more of a problem as a result: People killed the cats that preyed on the rats, which caused further increases in rat populations, which resulted in further spread of the bubonic plague (**Figure 3.27**).

Figure 3.26 Polio, which can result in paralysis of the legs, is still a problem in some countries where the polio vaccine is not readily available, such as for this child living on the street in India.

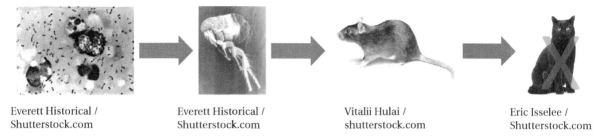

| Everett Historical / Shutterstock.com | Everett Historical / Shutterstock.com | Vitalii Hulai / shutterstock.com | Eric Isselee / Shutterstock.com |

Figure 3.27 The bubonic plague was caused by *Yersinia pestis* bacteria, which were carried by fleas, which were carried by rats. When cats, which killed rats, were targeted as the cause and eliminated, this made the bubonic plague epidemic worse.

The strong correlation between ice cream consumption and polio incidence did not prove that ice cream consumption caused polio. The causal relationship assumed between cats and the bubonic plague was also incorrect, and actually resulted in more problems. The lesson from these examples is that correlation does not necessarily equal causation. This is something you should always keep in mind when conducting a CCA by constantly asking yourself: Does this correlation really mean that the driver is a cause of the problem? Many types of information can help answer this general question, including quantitative and qualitative data, and expert and practical knowledge. The information being used to determine causality should not be restricted to expert, quantitative information, which has tended to be privileged in the recent past. Although this type of information is important, if attention is paid only to this type, then many pieces of the puzzle are missed. Finally, whatever type of information is considered, remember that the information needed to establish causality simply may not be available. Lack of evidence of a cause (driver), however, does not mean that the cause does not exist! The failure to acknowledge this in both the polio and bubonic plague cases, in which the nascent field of disease epidemiology had not yet made the link between microorganisms and disease, led to inaccurate CCA and faulty solution options.

Section 3.2.2—Extent of Influence, Current Trends, and the Importance of Context

Key Concept 3.2.2—Unraveling complex causal chains requires an understanding of the relative extent of influence of each driver on the system, current driver trends, and information about the specific cultural, political, social, technological, economic, and environmental context in which the problem is embedded.

Extent of Influence and Current Trends. To unravel the complex causal relationships underlying sustainability problems, which can lead to useful information for devising effective solution options, CCA must account for the broad range of drivers, the relative influence of each, current trends, and the specific context. To illustrate these ideas initially, we will revisit the MEA assessment of biodiversity loss from ecosystems at a global scale. Then, in the next section (Section 3.2.3), these ideas will be explored through a more extensive case study describing a sustainability assessment conducted in the western Indian Ocean Islands by the UNEP Global International Waters Assessment project aimed at determining the status of international waters, and possible solutions, in selected regions around the world.

Recall that the MEA identified five direct drivers of biodiversity loss worldwide (Figure 3.16). MEA scientists used the *extent of influence* of each of these different drivers of biodiversity loss, and the *current trends* for each driver, to determine the key factors driving global biodiversity loss is. *Extent of influence* is a measure of how much a given driver contributes to the problem, relative to other drivers. *Current trend* is a measure of how a driver's influence is changing over time: increasing, decreasing, or staying the same. **Figure 3.28** shows the results of their analysis. The direct drivers of biodiversity loss are listed across the top of the figure and the different ecosystems that were assessed are listed down the left-hand side. The extent of influence of each driver was assessed over the past 50 to 100 years. Each driver's extent of influence on biodiversity loss in each ecosystem is rated from low (light yellow) to very high (red). The low rating means that the *extent of influence* of a driver on biodiversity loss is minimal, while a very high rating means that a driver has significantly altered biodiversity in that ecosystem over the past 50 to 100 years. The other dimension in this analysis is the *current trend* of a driver. If a driver's influence on biodiversity loss in a given ecosystem is declining, then an arrow sloping downward and to the right is shown. A horizontal arrow pointing to the right indicates that the current level of influence is expected to continue. Arrows that point up, *increasing impact* and *very rapid increase of the impact*, mean that the driver's influence on biodiversity loss is increasing.

This information—*extent of influence* and *current trends*—identifies the drivers that should be the focus when devising solution options. For example, in boreal forests of the northern hemisphere, pollution and climate change are the two most important drivers of biodiversity loss (Figure 3.28). Thus, if biodiversity loss is to be reduced, these two direct drivers, and their associated indirect drivers (which are not shown in Figure 3.28), should be the focus. Pollution is the most critical driver,

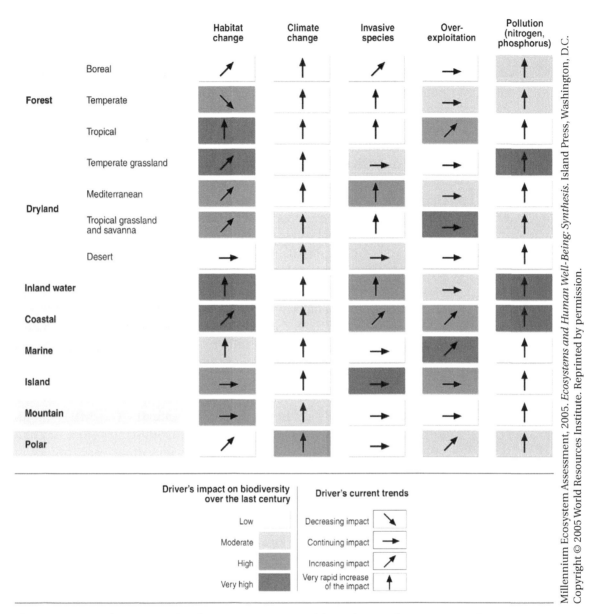

Figure 3.28 The extent of influence (represented by color) and current trend (represented by an arrow) of the five direct drivers of biodiversity loss used by the MEA.

since the extent of influence is moderate and the trend is very rapidly increasing. The current trend of the climate change driver is very rapidly increasing, and the extent of influence is low, so this driver may become more important in the future. A focus on drivers with a large influence on the problem, and with an influence that is increasing, does not mean that other drivers should be ignored. However, given limited resources in terms of time, money, and effort necessary for devising and implementing effective solution options, this type of analysis can help narrow down which drivers to target.

It is also important to recognize the changing relationships between drivers, as they do not regulate systems independently of each other and do not remain constant through time. For example, northern latitudes, where boreal forests are located, are expected to experience more rapid warming than tropical regions. This could exacerbate eutrophication problems caused by nutrient pollution, due to increased algal productivity at higher temperatures. Drivers are interacting and dynamic, Also, the relative importance of different drivers could change over time, sometimes rapidly. For example, as climate change becomes more severe, it could exert a much larger influence on biodiversity loss. In boreal forests, climate change will affect carbon storage (Figure 3.29), potentially making climate change worse for the entire global biosphere. The MEA assessment was conducted in 2005 and things may be different in 2020. It is important to be aware of driver interactions, and changes in driver influence through time, when devising solutions for systems that are constantly interacting and changing. Such dynamic systems are the rule, rather than the exception, when it comes to solving sustainability problems. Adaptive governance and long-term thinking are required.

Importance of Context. The MEA biodiversity assessment illustrates that certain drivers are more important than others in different ecosystems, at distinct times, and across various spatial scales. In other words, the most relevant drivers regulating ecosystems depend on the *specific context*. Today, in boreal forests, pollution and climate change are key drivers of biodiversity loss (Figure 3.28). In contrast, habitat change and overexploitation of species (by harvesting vegetation and hunting animals) are stronger drivers of biodiversity loss in tropical forests today. However, climate change, invasive species, and pollution drivers for tropical forests cannot be ignored. The current trends for all three are rapidly increasing and these drivers may become more important than others in the future. These expected increases must be anticipated and built into solution options. The spatial scale at which these different drivers influence ecosystems also provides clues for the most effective means for resolving

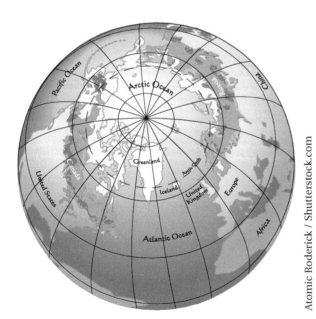

Figure 3.29 As climate changes, scientists predict that boreal forest ecosystems around the world, shown in green and brown, will shift northward and be replaced by grassland ecosystems, which store less carbon and allow more CO_2 to remain in the atmosphere.

biodiversity loss. Climate change occurs at a global scale, so solutions aimed at climate mitigation must engage the global community. Solutions focused on climate change adaptation can occur at the national, regional, or local levels. Addressing nutrient pollution of waterways should target institutions at regional, local, and possibly national scales, whereas engaging global institutions in a worldwide effort to stop eutrophication might be less efficient. However, solution options at this level cannot be completely ignored due to the aspects of globalization and international trade that affect local and regional watersheds.

In addition to ecosystem type, and temporal and spatial scale, drivers of biodiversity loss will be specific to different regions around the world due to variations in cultural, political, social, technological, economic, and environmental systems. Figure 3.28 presents driver influences and trends on a large spatial scale. In addition to region-specific variation in ecosystems, the human systems interacting with these ecosystems in different parts of the world will vary. They will be composed of different stakeholders whose actions are driven by a variety of values and beliefs, informal social norms, formal rules and regulations, and available resources and technologies. Stakeholder actions based on all of these factors will be explored in more detail in Section 3.3. As a result of differences in human systems, people in different places around the world will have varying capacities to deal with biodiversity loss and other sustainability problems. In the next section, a case study of pollution and global change in the western Indian Ocean illustrates the varying capacities of different island nations in this region to deal with their problems.

Section 3.2.3—Drivers and Causal Chains: A Western Indian Ocean Case Study

Key Concept 3.2.3—A western Indian Ocean case study illustrates how information gleaned through driver classification and causal chain analysis is specific to the problem of solid waste pollution and the local contexts for four different island states.

The islands of concern in this case study are located off the east coast of Africa in the western Indian Ocean near the equator, between 5°N and 30°S latitudes (Figure 3.30). These island states are Comoros, Madagascar, Mauritius, and the Seychelles. Solid waste pollution is the primary challenge to sustainable development in this island region. Solid waste includes any nonorganic waste, such as objects made from plastics, metals, or glass, and any liquid or gaseous chemicals contained

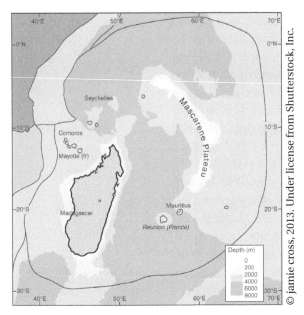

Figure 3.30 The four island states shown here — Comoros, Madagascar, Mauritius, and the Seychelles — must handle their solid waste problems differently due to varied local contexts.

within these materials (Figure 3.31). This type of pollution affects the environment, human well-being, and economic development on the islands. Solid waste can travel thousands of kilometers after entering the ocean. Lost fishing gill nets, six-pack rings, and other plastics threaten marine life by entanglement and ingestion, which makes eating, breathing, and swimming difficult and sometimes fatal. Toxic waters from unsanitary open waste dumps can leach into coastal waters and groundwater supplies, causing accumulation of trace metals and PCBs (polychlorinated biphenyls) in the tissues of animals and humans living close to dumps. School children, scavengers, and waste workers who spend time near these dumps can develop health problems, ranging from asthma to infectious diseases, such as malaria. Burning waste contributes to toxic air pollution. Solid waste washed up on beaches and into reefs harms the tourism industry. Other economic impacts include environmental cleanup costs, damage of fishing gear by entanglement with garbage, and flood damage caused by drains blocked by solid waste debris. Dealing with the solid waste problem involves managing the entire progression from waste generation, to collection, through to proper long-term disposal. Juggling all of these aspects of solid waste management requires an integrated and adaptive approach, which can be informed by knowledge gleaned from a CCA.

Direct and Indirect Drivers. The direct and indirect drivers of solid waste pollution relevant to a CCA for this case study are shown in Figure 3.32. Population growth is an indirect driver that influences all of the drivers discussed below. Thus, the specific influences of population on each direct driver will not be described explicitly. One direct driver is increased consumption on the islands. More solid waste is produced when more goods and services are consumed. This driver has a very strong influence on solid waste pollution and is expected to increase. The four major factors that influence consumption, and thus indirectly drive pollution, are shown in Figure 3.32. Globalization has resulted in exposure to, and a subsequent desire for, the high standards of living enjoyed by people of industrialized nations. The high consumption lifestyles associated with these living standards have been brought to the islands through technologies such as television and the Internet. Increased purchasing power by locals due to industrialization and international trade has also resulted in higher consumption patterns. Tourism increases consumption not only indirectly, by exposure to Western lifestyles, but also by directly increasing the demand for, and import of, material goods onto the islands to supply tourist facilities. These recently available products are purchased by both tourists and locals. Finally, international trade is an indirect driver affecting consumption patterns by making available a greater variety of products for purchase and is related to the other indirect drivers.

The presence of solid waste left unmanaged in the environment, such as in unsanitary open waste dumps and littering along roadways, is the second direct driver of pollution. Most of this waste is dumped on land, but a large portion ends up clogging ravines and drains, damaging coastal wetlands and mangroves downstream, or floating off into the open ocean. Four indirect drivers lead to the prevalence of unmanaged waste across many landscapes on these islands. First, technological resources may not be available to construct properly engineered sanitary landfills. Even if these resources are available, the financial resources may not be available to implement and operate landfill projects. Regulations and funding must be put in place to ensure coordinated waste collection services and other waste reduction activities such as recycling and composting programs. Another indirect driver is public education about the environmental, social, and economic costs of solid waste. Regulations, such as fines for littering and illegal waste dumping, can also control unmanaged solid waste in the environment.

Figure 3.31 Solid waste found in open dump sites (*top*) can be washed into the ocean (*middle*), where it harms marine life, and is often burned to reduce volume (*bottom*), which releases toxic chemicals into the air.

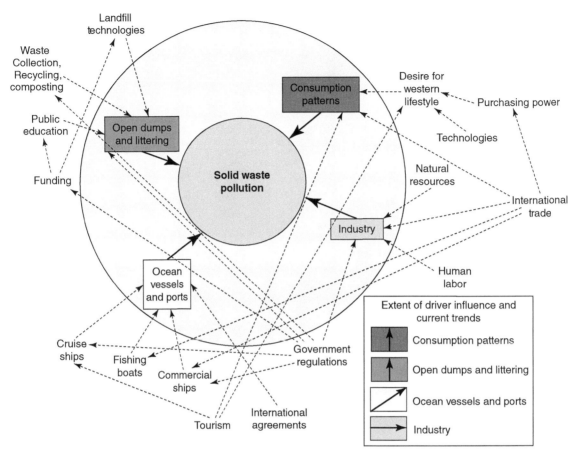

Figure 3.32 The four direct drivers of solid waste pollution are shown within the blue circle. Indirect drivers are shown in the white region outside of the circle.

The third direct driver of pollution in this region is the dumping of solid waste directly into ocean waters by commercial transport ships, fishing boats, and cruise ships. With increases in international trade, an indirect driver here again, more commercial ship and fishing boat traffic has led to more solid waste dumping. The typical cruise ship carrying 3,000 to 5,000 passengers generates about 50 tons of solid waste per week. More tourism, another indirect driver, will only augment the amount of solid waste offloaded into ocean waters. Elimination of solid waste from ocean waters not contained within the Exclusive Economic Zone (EEZ) of a country, and therefore regulated by the laws of that country, requires cooperation among nations. International agreements about solid waste dumping in unregulated open ocean waters can affect the amounts dumped by vessels. Finally, national policies concerning waste management requirements for ships docking in a country's ports can indirectly influence solid waste pollution by regulating commercial vessels and cruise ships.

The final direct driver of importance here is industry, which manufactures goods for resident consumption or export using local resources. Agriculture, fisheries, and mining are three prominent industries in this region. Due to emerging opportunities for international trade, an indirect driver here, industry is not expected to grow significantly. Products brought in from overseas will preclude the need to manufacture those same products locally. The second indirect driver for industry includes both natural and human resources. Solid waste generation will depend on the type of industry, which will depend on the local natural resources available as inputs to the production process. The availability and cost of human labor will also influence whether industries are established in an area. The final indirect driver of solid waste production that influences industry is government regulation. Policies that require industries to take responsibility for the wastes they produce can result in pollution reduction.

Extent of Influence, Current Trends, and Specific Context. In addition to classifying all of these drivers as direct and indirect, Figure 3.32 shows the *extent of influence* of each direct driver and the *current trend*, using the same color and arrow scheme as the MEA biodiversity loss assessment (Figure 3.28). This CCA reveals the major drivers of solid waste pollution, the relationships among the drivers, the relative extent of their influence, and current trends into the future. All of this information can help devise solution options for the solid waste problem in this region. However, to devise effective solutions, information is needed about the differences among the islands, which will affect their varying capacities to deal with the pollution problem. As mentioned above, each *specific context* in which a sustainability problem is embedded will have its own unique cultural, political, social, technological, economic, and environmental conditions. This is because stakeholder actions are driven by different values and beliefs, informal social norms, formal rules and regulations, and available resources and technologies in different types of systems. Stakeholder analysis is a tool for further defining the specific context and will be presented in detail in Section 3.3. A comparison of the contextual differences among the four island nations in this case study will be given here, before moving on to stakeholder analysis. Examples include differences in environmental, cultural, technological, and economic systems among the islands.

Freshwater resource systems and coastal ecosystem types are two environmental factors that vary across the four islands. Solid waste dumped on land can contaminate surface and ground waters, which are a major source of drinking water for local populations. Impacts of pollution on freshwater resources vary from island to island. With the lowest annual rainfall and no perennial streams (e.g., streams that flow year-round), Comoros has the scarcest water resources and cannot afford to contaminate them (Figure 3.33). In contrast, Mauritius has the highest rainfall of all the islands and has many perennial streams. More abundant water resources mean that Mauritius may be better able to cope with polluted water than Comoros. In fact, over 99% of the human population has access to potable water in Mauritius, while less than 50% of people in Comoros do. In addition to drinking water, solid waste pollution affects fishermens' livelihoods and local economies, which are dependent on fish species living in reefs. All four islands are surrounded by coral reefs to some extent (Figure 3.34). However, only Madagascar and the Seychelles also have highly productive sea grass beds that harbor fish in their coastal waters. This makes them better able to cope with pollution affecting coral reefs because, if coral reefs were to become less productive due to pollution, they would still have sea grass beds in which to fish. Both freshwater resource abundance and coastal habitats that support livelihoods are examples of environmental resources that vary among the four island nations.

Social and technological factors also affect solid waste management. In Madagascar and Comoros, it is normal for local residents to dump solid waste on beaches and in the ocean and rivers. This is common practice throughout the region, but no longer occurs in the Seychelles and Mauritius because regulations have constrained the behavior of stakeholders who dump waste. For example, the Seychelle government invested over US$8 million into a national solid waste management program over a 10-year period. Mauritius has a similar program. In contrast, the government of Madagascar has invested only a small amount of public funds into improving waste management relative to other expenditures. In both Comoros and Madagascar, a policy process for solid waste management is nonexistent (Figure 3.35). Technical knowledge of waste management systems in the more service-oriented economies of Mauritius and the Seychelles has led to the opening of a sanitary landfill in each country. In the more manufacturing-based economies of Comoros and Madagascar, there are no sanitary landfills for solid waste due to lack of technical expertise and institutional capacity to support such services.

There are many other factors that define the varying capacities of these four island nations to deal with solid waste pollution, but the discussion of

Figure 3.33 Compared to other western Indian Ocean islands, Comoros has a very dry landscape, as shown here behind their national flag.

Figure 3.34 All four islands are surrounded by coral reefs with high biodiversity, as shown here for Coco island in the Seychelles, which provide each country with valuable opportunities for fishing, tourism, and other activities.

this specific case study and CCA will stop here. Once the system is defined and the drivers are classified, CCA can help establish causal relationships by considering the relationships among many interconnected drivers, the extent of influence of each driver, and each driver's current trend. In order to devise sustainable and effective solution options, all of this information must be considered for the specific context in which the problem is embedded. Stakeholder analysis, described in the next section, will help you further refine the details of what is meant by *specific context* by focusing explicitly on drivers of human actions.

Figure 3.35 Tropical beaches littered with garbage on the islands of Comoros and Madagascar result from a lack of waste management policies and technical expertise.

Section 3.3: Stakeholder Analysis

Core Question: How can stakeholder analysis contribute to an assessment of the current state of a sustainability problem?

"Stakeholder analysis means many things to different people. Various methods and approaches have been developed in different fields for different purposes, leading to confusion over the concept and practice of stakeholder analysis."

—Mark Reed, *Who's in and why?*, 2009

Up to this point, this chapter has focused on defining the current state of a sustainability problem by identifying and classifying drivers, and using these drivers to conduct a causal chain analysis. However, the current state analysis of a system is not complete until the people and organizations with a stake in the sustainability problem are considered. This is where stakeholder analysis comes into the picture. As conveyed by the opening quote to this section, there are a wide variety of methods available for stakeholder analysis and there is no one universal method appropriate for all contexts. In business management, for example, stakeholder analysis is concerned with business success and focused on identifying allies and threats so that allies can be mobilized and threats defeated. In sustainability, stakeholder analysis is more concerned with empowering traditionally marginalized groups so that they can contribute their knowledge to, and be part of, the problem-solving process. Also important to sustainability is the representation of diverse perspectives and agendas so that proposed solution options have "buy in" from the many groups affected, leading to more effective and lasting solutions.

In any analysis of stakeholders for a given sustainability problem, stakeholders first need to be identified. After identifying stakeholders, uncovering the factors affecting their behavior can help understand why they take certain actions and make specific decisions, which both affect the sustainability of a given situation. Classifying the degree of interest and influence that each stakeholder has over a situation can reveal power dynamics, alliances, and risky situations. Finally, mapping relationships between stakeholders provides insight into the type and strength of relationship between each group and, thus, to what extent, and how, different stakeholders might influence each other.

Section 3.3.1—Identifying Stakeholders

Key Concept 3.3.1—Stakeholders can be broadly defined as people who have an interest or stake in some policy, conflict, or organizational goal, but identifying stakeholders is not a straightforward process.

There is no one correct method for selecting stakeholders. However, three general types of methods can be employed: focus groups, interviews, and snow-ball sampling. When implementing these methods, it is important to keep in mind two general guidelines to ensure equitable and accurate representation of different stakeholders. First, identifying stakeholders is an ethical issue. Those who are in charge of the process decide, consciously or unconsciously, based on their own biases and perceptions, who to include and who to exclude. This can lead to unidentified groups and, more seriously, exclusion of marginalized groups, rendering these groups even more vulnerable and powerless. As a result, those in charge of the process must be sensitive to and remain aware of these issues throughout the identification process, using participatory processes (Section 2.2.3) whenever possible. Second, stakeholder analyses are iterative, meaning that who is, and who is not, a stakeholder can change over time, as conditions and other circumstances changes. As such, conveners of a stakeholder analysis should periodically check in with previously identified stakeholder groups in order to gauge their present level of involvement.

The following methods can be used, often in sequence, to identify stakeholders: focus groups, interviews, and snow-ball sampling interviews (**Figure 3.36**). In focus groups, small work groups are convened to identify stakeholders by categorizing different stakeholder interests

Figure 3.36 Focus groups use flip-charts, post-it notes, and other tool to brainstorm about stakeholder categories (*left*), whereas interviews provide more in-depth insight into stakeholders through one-on-one conversations (*right*).

and other important attributes. The results of a causal chain analysis can be a starting point for focus groups, as it offers a clear understanding of the issues at hand and the boundaries of the problem. Focus groups require skilled facilitators to obtain quality results, but are generally cost-effective to implement and produce a list of stakeholders rather quickly. Using the results of the focus group, interviews can be conducted with representatives from each stakeholder group. The purpose of these interviews is twofold. First, interviews with actual stakeholders, who have been perceived in a certain way by the focus group, can verify the accuracy, or inaccuracy, of the stakeholder categories created by the focus group. Second, documenting the stories and experiences of individual stakeholders adds a layer of richness and complexity, which may be more representative of reality. These benefits come with costs, however. The interview process is costly, mostly in terms of time. Also, interviews sometimes adds so much complexity, and highlight the "gray areas" so well, that it becomes difficult to reach consensus on different stakeholder categories, defeating the whole purpose of the process. For example, if the issue is climate change, two identified stakeholder categories might be "climate activists" and "CO_2 emitters." However, some climate activists are also CO_2 emitters, such as when driving a car, heating their home, or using public transit, so things are not as cut-and-dry as they might first seem.

Snow-ball sampling takes a different direction than the interview process just described. Instead of using interviews as a means to verify and supplement the stakeholder categories defined by the focus group, snow-ball sampling uses interviews to identify new stakeholder categories and obtain contact information for additional interviewees from these categories. In this case, interviews can be easier to obtain, such that interview requests are declined less, but the information used to define stakeholder groups can become biased by the social networks and connections of the first stakeholders interviewed.

The previous paragraphs describe guidelines and methods for identifying stakeholders, but what are stakeholders in the first place? There is no one agreed upon definition. In this book, **stakeholders** are defined as individual persons, organizations, or other entities that are positively or negatively affected by other's actions or carry out the actions causing the positive and negative effects in the first place. Simply stated, they have a stake or interest in a certain issue. Thus, those causing an unsustainable situation, and those affected by that situation, are both considered stakeholders. The issue of concern can be a variety of things, including a conflict, policy, organizational goal, or a governance process.

Several different types of stakeholders can be identified during a stakeholder analysis, and there is more than one way to classify them. One way is to break them into primary, secondary, and key stakeholders. **Primary stakeholders** are those that strongly affect other stakeholders by their actions or are profoundly affected, either positively or negatively, by other stakeholders' actions. **Secondary stakeholders** are intermediaries who indirectly affect other stakeholders, or are only slightly affected by other stakeholders' actions. **Key stakeholders** are especially important to identify because they have the power to significantly influence or change a given situation. They can be primary or secondary stakeholders themselves, but not always. The same sort of vagueness and ambiguity applies to these three stakeholder designations as the "climate activist" and "CO_2 emitter" categorizations: a person can be more than one type of stakeholder if he or she has more than one type of stake or interest in the issue. In reality, stakeholder categories and classifications are generalizations, not completely black and white, and not mutually exclusive (**Figure 3.37**).

Important stakeholder groups not directly represented in sustainability are other species and future generations. Other species may be indirectly represented by government organizations, such as the Environmental Protection Agency, or nongovernmental advocacy groups, such as the Environmental Defense Fund. Representing future generations can be trickier. Sustainability is concerned with intergenerational equity, but it is challenging to represent future generations in today's stakeholder engagement processes for reasons noted in the 1987 Brundtland Report: "We borrow environmental capital from future generations with no intention or prospect of repaying. . . . We act as we do because we can get away with it: future generations do not vote; they have no political or financial power; they cannot challenge our decisions." Although representation of future generations as stakeholders is not yet widespread,

---

stakeholders
individuals, organizations, or other entities that benefit or are harmed by other's actions, or who carry out the actions causing the benefits or harms, and as a result have an interest in some policy, conflict, organizational goal, or other issue that will influence their future actions.

primary stakeholders
a class of stakeholders who strongly affect other stakeholders by their actions or are profoundly affected by the actions of other stakeholders.

secondary stakeholders
a class of intermediary stakeholders who indirectly affect other stakeholders or who are only slightly affected by the actions of other stakeholders.

key stakeholders
a critical class of stakeholders, who can be primary or secondary stakeholders or neither, and who have the power to significantly influence or change a given situation.

efforts have begun. For example, Israel has a Commissioner for Future Generations and Hungary has an Ombudsperson for Future Generations. These individuals or groups act as representatives within the governing bodies of these nations. Individuals and informal groups are also taking on this challenge. Twelve-year-old Canadian Severn Suzuki gave a speech to the council at the 1992 Rio Earth Summit as a representative of future generations. Twenty years later, at the 2012 Rio+20 Summit, Severn Cullis-Suzuki returned as a young mother to advocate not only for her generation but also for that of her child.

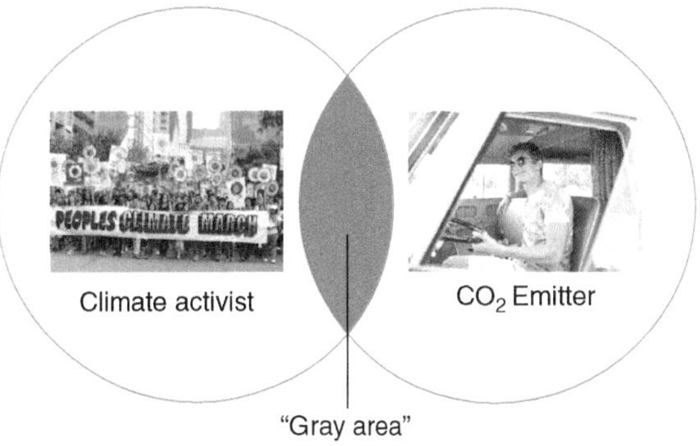

Climate activist

CO_2 Emitter

"Gray area"

a katz / Shutterstock.com

Syda Productions / Shutterstock.com

Figure 3.37 Stakeholder categories are never cut-and-dry and there are always "gray areas" where categories overlap, such as with this climate activist who participates in climate marches to bring awareness to climate change, but also drives a vehicle powered by fossil fuels.

Section 3.3.2—Analyzing Stakeholder Behavior, Interest, and Influence

Key Concept 3.3.2—Factors influencing stakeholder behavior result in actions that affect the sustainability of a system, and awareness of stakeholder interest in and influence on a sustainability problem reveals power dynamics, beneficial alliances, and significant risks.

Once stakeholders are identified, it is useful to understand the factors shaping their behavior, and their degree of interest in or influence on a situation. A framework developed by sustainability scientist, Professor Arnim Wiek, can be used for the former. An interest–influence matrix can be used for the latter. Box 3.3 applies this framework, and an interest–influence matrix, to a sustainability problem focused on water supply in rural African villages. These are only two of many existing methods for analyzing stakeholders and the interested reader should consult the sources provided in the bibliography at the end of this chapter, as a start. Further, Section 3.3.3 presents another method for analyzing stakeholders by looking at relationships between different stakeholder groups or individuals.

A framework developed by sustainability scientist Arnim Wiek at Arizona State University is a tool that can be used to understand how actions taken by stakeholders affect a sustainability issue, and how stakeholders actions are shaped by values and beliefs, informal social norms, formal rules and regulations, and available resources and technologies (Figure 3.38). This analysis of stakeholder actions is focused on what are sometimes referred to as **active stakeholders**, or those

active stakeholders a class of stakeholders who have direct or indirect influence over an issue.

passive stakeholders a class of stakeholders who are affected by the actions of active stakeholders, but who do not necessarily influence the problem.

who influence the issue to some degree, whether directly or indirectly (this is opposed to **passive stakeholders**, who are affected by the actions of active stakeholders, but do not necessarily affect the issue). An assessment of the factors that guide stakeholder behavior toward actions that result in an unsustainable situation can help in further understanding a sustainability problem and, subsequently, can point to potential solution options. The framework presented in **Figure 3.38** is a simplification of a framework developed by one sustainability scientist. There are many ways to portray the factors that shape human behavior and different social scientists, such as economists, sociologists, anthropologists, or psychologists, might frame or think about these factors differently. Thus, the framework presented in this section is only one of many possible ways of conceptualizing why stakeholder behavior results in certain actions.

Wiek's framework (Figure 3.38) considers a number of factors that shape stakeholder behavior, such as values and beliefs, informal social norms, formal rules and regulations, and available resources and technologies. Each is introduced below in separate subsections. Information about these aspects of each stakeholder, whether an individual or a group, can be obtained using similar methods as were used to identify stakeholders: focus groups and different interviewing methods.

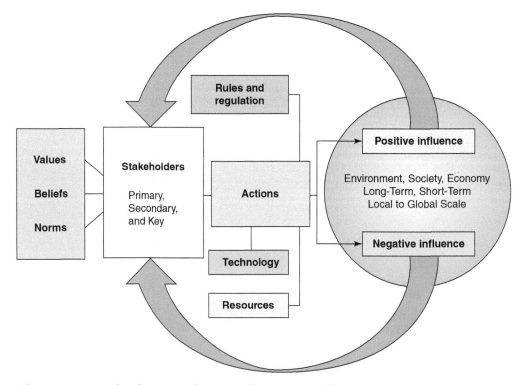

Figure 3.38 This framework, created by sustainability scientist Professor Arnim Wiek, helps understand how various factors guide stakeholders to take actions that contribute to, and are affected by, a sustainability problem.

Whatever the method used, participatory processes usually provide better information when it comes to sustainability problems.

Behavior results in actions, which influence the sustainability, or unsustainability, of a given situation. (Actions, in this case, refer to both actual actions and also decisions.) For each stakeholder category identified, the positive and negative influences of their actions on sustainability should be noted. In Figure 3.38, the positive and negative influences of stakeholder actions are represented by different colored boxes. The arrows point from those boxes back to the stakeholder box to illustrate how positive and negative influences of their actions might shape future actions. This is known as a feedback, a concept discussed in more detail in Chapter 5. The three pillars of sustainability (environment, society, and economy) should be considered in this analysis, ranging from local to global spatial scales, as well as the short- and long-term consequences. Short- and long-term environmental health, social equity, and economic viability must all be considered simultaneously, across multiple spatial scales, to effectively advance sustainability.

Values, Beliefs, and Norms. Values, beliefs, and norms are shown as one box in Figure 3.38. **Values** are general principles and standards for defining what is worthwhile or important. They provide criteria for defining action based on a fundamental sense of what is valuable and what is not. For example, people value things such as freedom, independence, equity, or a structured hierarchy. Values are not right or wrong, or good or bad, and they are generally difficult to change in the short term. Thus, when working on sustainability problems, the focus should be to engage stakeholder values rather try to change them.

> **value** a general principle or standard that defines what is important and provides criteria for action based on a fundamental sense of what is worthwhile and what is not; cannot be deemed correct or incorrect.

Beliefs and norms, on the other hand, are not as deeply entrenched as values and are specific rather than general. Although they can also be difficult to change in the short term, they are relatively less so. Thus, beliefs and norms can be challenged and, sometimes, changed. **Beliefs (aka world views)** are a sense of how the world works and can be fundamentally "right" or "wrong." A person may believe that the world works in a certain way when, in reality, it definitely does not. As a result, beliefs are susceptible to knowledge and can be shaped by learning. Norms were defined in Section 2.1.1 and have to do with what is approved or disapproved of within a group, or larger society, and they guide specific actions.

> **beliefs (aka. worldviews)** a sense of how the world works that is susceptible to knowledge and, therefore, can be fundamentally correct or incorrect.

Formal Rules and Regulations. Formal rules and regulations can take many forms, such as laws prescribed by a government, internal self-regulation by an industry, or a legal contract between two or more parties. Whatever the format, formal rules and regulations are intended to guide human behavior to some end. They are represented together in a box in Figure 3.38.

Resources. The resources available to stakeholders in a given place, or at a certain time, also influence their actions. These are represented by their own box in Figure 3.38. Broadly, resources are anything of limited availability that is consumed to obtain a benefit. Resources can be environmental, social, or economic in nature. Because resources are consumed by stakeholders to obtain a benefit, if they are too intensively used, then they may eventually become exhausted and unavailable for future use. As a result of the possibility for resource depletion, resources often require allocation and governance through formal or informal institutions. Of course, this applies only to renewable resources. Nonrenewable resources are depleted independently of any governance system attempting their allocation, simply because they are used at much faster rates than they can be regenerated (e.g., fossil fuels such as coal, oil, and natural gas).

Technology. Technology significantly affects the ability of human societies to control and adapt to their natural environments. It involves the use of physical objects, such as tools or crafts, and also knowledge, such as techniques, methods, or systems of organization. Technology has benefited society and allowed it to thrive in many ways throughout human history (e.g., the wheel, the plow, the Internet), but some technologies have also been harmful due to unintended consequences (e.g., pesticide impacts on people and ecosystems). Thus, it is important to think through all of the consequences of adopting a new technology to prevent harmful unintended consequences. For example, nuclear power was intended as a "clean" (non-fossil fuel) energy source, but it has great destructive potential as weapons, is a non-renewable resource, and generates a legacy of toxic waste that must be carefully managed. Technology is represented by a separate box in Figure 3.38.

In addition to understanding how certain factors shape stakeholder behavior, another way to analyze individual stakeholders, or stakeholder groups, is to construct an interest–influence matrix (**Figure 3.39**). Such an analysis is useful because it can identify beneficial alliances and significant risks, and also explicitly illuminate the power dynamics of a situation, which increases the transparency and equity of any future decision-making process involving the stakeholders. To construct the matrix, each stakeholder

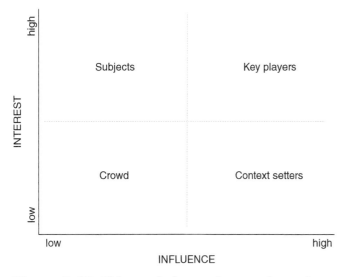

Figure 3.39 This graph shows where to plot various stakeholders, based on their degree of interest in and influence on a sustainability problem, and can provide information about powerful actors, potential alliances, and important risks.

BOX 3.3 3 PLAYPUMPS AND WATER IN SMALL RURAL AFRICAN VILLAGES: GOOD INTENTIONS, BAD OUTCOMES

By 2009, Trevor Field's charity organization, PlayPump International, had installed 1,800 PlayPumps in the African countries of Mozambique, South Africa, Swaziland, and Zambia (Figure 3.40). Using skills acquired during a career in advertising, and an entrepreneurial spirit, Field quit his job in 1995 to dedicate himself to his charity full time. Just 5 years prior, he had bought the patent for the PlayPump, which was a new type of water pump invented by water engineer Ronnie Stuvier. The pump took the form of a merry-go-round which, when spun, would pump groundwater to the surface. In his eyes, this invention could help make daily water retrieval easier for women living in small African villages by harnessing the natural drive of their children to play. Instead of pumping water by hand, sometimes after walking miles to a water source, children playing on the merry-go-round would pump the water for them (Figure 3.41).

Field needed to raise money for his project and his first sponsor was Colgate Palmolive. He also used his own money, at first, to pay for the pumps and their installation. By seeking connections with government entities and businesses in the country of South Africa, he was able to acquire even more funding. By 2000, he had installed enough pumps to gain recognition for his achievements by receiving the World Bank's Development Marketplace Award. This award brought even more funding from more sponsors, including the Case Foundation, a British fund-raising charity, a multi-million dollar grant from First Lady Laura Bush, and even money raised from singer Jay-Z's tour called "The Diary of Jay-Z: Water for Life." The media and celebrities brought further attention to his efforts. Witty headlines, such as *The Magic Roundabout* and *Pumping Water is Child's Play*, appeared in magazines and newspapers. It seemed that Field had made the most amazing breakthrough in water pumping technology for developing nations in Africa and was on a roll.

About 20 years after his PlayPump adventure began, reports released by two different development

Figure 3.40 By 2009, PlayPump International had installed about 1,800 PlayPumps in the four African countries indicated by red stars on this map.

(continued)

Courtesy of U.S. Department of State   John Wollwerth / Shutterstock.com

Figure 3.41 In 2011, Peace Corps volunteers living in a South African village installed a PlayPump (*left*) to replace a hand pump, similar to the one used in the South Sudan village (*left*), and returned later that year to find the pump broken, some livestock dead, and villagers having to travel to nearby communities for water.

agencies, UNICEF and the Swiss Resource Centre and Consultancies for Development (SKAT), caused things to turn sour. The reports cited cases where children were paid to push the pumps because, unlike normal merry-go-rounds that were fun to play on, the PlayPumps did not pick up momentum as force was applied and, instead, required constant pushing. In some villages, children were harmed while pushing the pumps, including broken limbs and vomiting due to excessive spinning. As a result, women had to push the merry-go-round. They found this exhausting and humiliating, and reported that much more effort was required to extract the same amount of water using a PlayPump than the hand-pumps that the PlayPumps had replaced. In some villages, locals liked the PlayPumps but became frustrated when they broke, which frequently happened a few months after installation. Unlike the hand pumps, they were not able to make repairs on their own and, instead, needed to obtain outside help by calling a phone number provided along with their pump. In most cases, this outside help never arrived.

Who are the stakeholders in this case study? What factors caused them to take the actions that they did? How did stakeholder influence and interest result in this unfortunate situation?

Based on the information given above, the stakeholders in this case study can be identified as the people living in African villages who need to pump their own water, Trevor Field, Ronnie Stuvier, financial sponsors and donors, the media and celebrities, and development agencies (both the World Bank, who promoted the project, and UNICEF and SKAT, who ultimately denounced it).

Values, Beliefs, and Norms. Local people living in African villages value a dependable drinking water supply. Based on the information given, it is hard to know whether they believed that the PlayPumps would contribute to this end. Social norms dictated that women would obtain water for the family each day, with children in tow. In many societies, women are not as valued and, therefore, do not have a voice. As a result, they may have been less likely to speak up about problems with the pumps. Trevor Field values making a difference in the lives of people he perceives as less fortunate. He believed that his effort would help people living in Africa villages. The "top-down" approach, an enduring norm in the international development community, led him to forge ahead with his new technology without consulting the local people who would be using it. Ronnie Stuvier, as a water engineer, likely values technological advancements. He likely believed that the PlayPump would be a useful invention. Again, the norm within the science and technology community is often to create a new technology without consulting those who would use it. That said, Trevor Field bought Stuvier's patent and may have changed the intended direction and use of the PlayPump, so it is hard to surmise Stuvier's actual beliefs and intentions. Like Field, the financial sponsors and donors of the project value helping people who they perceive as less fortunate. They believed that funding the PlayPump project would meet this end. They also followed the pervasive norm of bypassing local communities, as they provided funds to a charity to install a specific type of pump in communities rather than directly to communities to purchase what they actually needed. The media values stories that will maintain or increase readership, so they latched onto the PlayPump. However, they renounced the PlayPump venture when the UNICEF and SKAT reports came out, using similarly witty quips such as "money down the drain" and "troubled waters." Like the other stakeholders, they may value helping people who are less fortunate, but this is not their main focus or agenda. The World Bank, as a development agency disbursing the Marketplace Award, values innovative development projects with high impact potential. They believe that offering this award will lead to the success of projects that will improve conditions for people living in developing nations. Like other stakeholders, they did not consult local people to ask if the PlayPumps should receive the award. UNICEF and SKAT, as development agencies investigating the success or failure of the PlayPump project "on the ground," value the collection of empirical data to support conclusions about project success. They believe that this approach provides the most accurate data and, in this case, bucked the trend (norm) of bypassing the input of local people.

Formal Rules and Regulations. In this case, a lack of formal rules and regulations may have contributed to the problem. In the African countries where the PlayPumps were installed, villages did not have access to any sort of water infrastructure set up by their government. As a result, they had to figure out how to obtain water on their own. This meant that the means for acquiring water was sort of a "free for all," open to any invention or idea that came along. Rules and regulations within the PlayPump International charity, and some donor organizations and development agencies, that effectively ensured representation of local people were lacking. Even when there were rules and regulations, such as a promise by PlayPump International to fix and maintain broken pumps by supplying a contact number, they were not followed in many cases.

Resources. The two major resources guiding stakeholder behavior and actions in this case study are water and financial resources. Water is a renewable resource essential to survival and access to clean water is a major problem in the developing world. In this case, depletion or scarcity of water resources was not a central factor, as it often is for sustainability problem. Instead, the issue driving stakeholder behavior was efficient and better access to water resources. The PlayPump was supposed to make the daily task of retrieving water easier for women living in African villages. The other major resource guiding stakeholder behavior was financial resources. It could be argued that the need to obtain financial resources from donors early on may have led Field to showcase and promote his PlayPump, using the advertising skills acquired from his former career, before first assessing the true success of the technology. On the surface, it seemed like a win–win situation that should be promoted and funded—children gained playground equipment and the women's task of pumping water became easier—but the realities turned out to be different. A lack of financial resources to support maintenance of the PlayPumps, once installed, may have also contributed to the problem.

Technology. The ability to pump water from the ground using technology, regardless of the type of pump used, allows people to subsist in small African villages, away from an established water infrastructure. At first, the PlayPump seemed like a better technology for pumping water than the hand pumps normally used by women. However, like many technologies, there were harmful unintended consequences, such as injured children and women that were more tired from pushing the merry-go-round on their own.

In addition to studying the factors that affect stakeholder behavior, constructing an interest–influence matrix can be useful (Figure 3.42). If you were a sustainability scientist trying to remedy the PlayPump problem, and prevent other similar problems from happening in the future, you would need to actively engage the key players in the problem-solving process to make this happen. It should not be difficult to engage them, at least in theory, as they have a high interest in the problem. Context setters do not have a strong interest, necessarily, in fixing the PlayPump problem, or preventing similar technology snafus in the future. As a result, it may be more difficult to engage them. However, they have a good deal of influence on the situation, and pose a significant risk to the success of your effort, so thoughtful strategies for

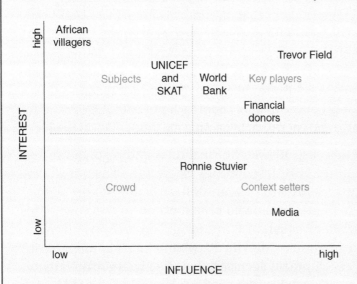

Figure 3.42 This interest–influence matrix, specific to the PlayPump debacle, can help sustainability scientists interact with stakeholders in appropriate and effective ways.

> engagement may need to be devised. The media is definitely a Context setter in this situation and developers of technologies, such as water engineer Ronnie Stuvier, may be as well. However, Ronnie Stuvier could also be viewed as part of the Crowd, which are stakeholders that require little engagement. In this view, technology developers are not liable for the technologies they develop and, instead, it is the people that implement them that need to be made aware of the detrimental effects. This is debatable, and gets into ethical issues surrounding technological development, which is why Ronnie Stuvier is placed on the line between the Crowd and the Context setters. Subjects have high interest in the situation, but not a lot of influence. As a sustainability scientist engaging stakeholders, you will want to empower these groups by bringing them to the table, on a level playing field, with other more powerful stakeholders, when solutions are being devised and decisions are being made.

is classified as a key player, context setter, subject, or crowd. The actions of key players have a relatively strong influence on an issue and they also have a strong interest in the issue. Context setters have a strong influence, but little interest. Subjects have high interest, but little influence. As such, subjects are often the marginalized stakeholders that stakeholder analysis, when conducted for sustainability purposes, aims to empower. Crowd stakeholders have both weak influence and little interest in the issue.

Section 3.3.3—Investigating Stakeholder Relationships

Key Concept 3.3.3—Investigating stakeholder relationships provides insight into how stakeholder relationships should be managed and to what extent, and how, different stakeholders groups might influence each other.

Social network analysis (SNA) seeks to understand the relationships between different stakeholders, and the strength of the relationships. Relationships can be positive or negative, and they can be strong or weak. Positive relationships are characterized by things such as trust, mutual support, and friendship, whereas negative relationships can involve distrust, fear, and animosity. This information can be useful for determining which stakeholders will work well together and for identifying which relationships will require more delicate management. It is important to recognize strong relationships because these are the stakeholders that are most likely to influence one another, such as through communication, sharing resources, and advice. Strong relationships are also hard to break. Weak relationships occur between stakeholders that are dissimilar to, and unfamiliar with, one another. As such, stakeholders with weak relationships are not likely to influence one another and the relationships are easily broken. However, weak relationships are important because they can bring diverse information and resources to the table, which would not otherwise be present. SNA was introduced in order to emphasize the importance of understanding stakeholder relationships, but the details of this analysis will not be discussed further. The interested reader should consult this chapter's bibliography. Additionally, there are other means of investigating stakeholder relationships, such as actor-linkage matrices and knowledge mapping, which can also be accessed by consulting the bibliography.

Once drivers have been classified, causal chains analyzed, and a stakeholder analysis conducted for the specific context in which a sustainability problem is embedded, it is useful to determine indicators of sustainability. Indicators are the topic of the next chapter. They will help you to understand exactly where a system lies with respect to sustainability, and which way it is currently moving (toward or away from sustainability).

Bibliography

Ascher, W. 2007. "Policy Sciences Contributions to Analysis to Promote Sustainability." *Sustainability Science* 2: 141–49.

Belausteguigoitia, J. C. 2004. "Causal Chain Analysis and Root Causes: The GIWA Approach." *Ambio* 33: 7–12.

Bianchi, T.S. and M.A. Allison. 2009. "Large-river delta-front estuaries as natural "recorders" of global environmental change." *Proceedings of the National Academy of Sciences* 106(20): 8085–8092.

Borysova, O., A. Kondakov, S. Paleari, E. Rautalahti-Miettinen, F. Stolberg, and D. Daler. 2005. "Eutrophication in the Black Sea region: impact assessment and causal chain analysis." Global International Waters Assessment. Accessed September 1, 2016. http://www.unep.org/dewa/giwa/areas/reports/r22/giwa_eutrophication_in_blacksea.pdf

Chapin, F. S., C. Folke, and G. P. Kofinas. 2009. "A Framework for Understanding Change." In *Principles of Ecosystem Stewardship: Resilience-Based Management in a Changing World,* edited by F. S. Chapin, G. P. Kofinas, and C. Folke, 3–28. New York: Springer.

Collins, S. L., et al. 2010. "An Integrated Conceptual Framework for Long-Term Social-Ecological Eesearch." *Frontiers in Ecology and Environment* 9(6): 351–57. doi:10.1890/100068.

DeFries, R., et al. 2005. "Analytical Approaches for Assessing Ecosystem Condition and Human Well-Being." *Millennium Ecosystem Assessment,* 37–71. Accessed April 21, 2016. www.millenniumassessment.org/documents/document.271.aspx.pdf.

Eisenburg, J. N. S., et al. 2006. "Environmental Change and Infectious Disease: How New Roads Affect the Transmission of Diarrheal Pathogens in Ecuador." *PNAS* 103 (51): 19460–65.

Fioriani, D. C. 2007. "Application of the Causal Chain Analysis to Protected Areas, a Case Study: Environmental Protected Areas of Anhatomirim (SC–Brazil)." *Natureza & Conservação* 5(1): 132–40.

Gunderson, L. H., and C. S. Holling, eds. 2001. *Panarchy: Understanding Transformations in Human and Natural Systems.* London: Island Press.

Horrigan, Leo, Robert S. Lawrence, and Polly Walker. 2002. "How Sustainable Agriculture Can Address the Environmental and Human Health Harms of Industrial Agriculture." *Environmental Health Perspectives 110(5): 445–56.*

Howarth, R., F. Chan, D.J. Conley, J. Garnier, S.C. Doney, R.Marino, and G. Billen. 2011. "Coupled biogeochemical cycles: eutrophication and hypoxia in temperate estuaries and coastal marine ecosystems."*Frontiers in Ecology and Environment* 9(1):18–26.

Kajikawa, Y. 2008. "Research *Core* and Framework of Sustainability Science." *Sustainability Science* 3: 215–39.

Kates, R. W., and P. Dasgupta. 2007. "African Poverty: A Grand Challenge for Sustainability Science." *PNAS*, 104 (43): 16747–50.

Koven, C.D. 2013. "Bor*eal carbon loss* due to poleward shift in low-carbon ecosystems." Nature Geoscience 6: 452–456.

MacAskill, W. 2015. Doing Good Better: How Effective Altruism Can Help You Make a Difference. New York, NY: Gotham Books, 258 pp.

Minney, S. 2011. Naked Fashion: The New Sustainable Fashion Revolution. Oxford, UK: New Internationalist Publications Ltd, 176 pp.

Nelson, G. C., et al. 2005. "Drivers of Change in Ecosystem Condition and Services." *Millennium Ecosystem Assessment*, 173–222. Accessed April 21, 2016. www.millenniumassessment.org/documents/document.331.aspx.pdf.

Odada, E. O., D. O. Olago, K. Kulindwa, M. Ntiba, and S. Wandiga. 2004. "Mitigation of Environmental Problems in Lake Victoria, East Africa: Causal Chain and Policy Options Analyses." *Ambio* 33: 13–23.

Parris, T. M., and R. W. Kates. 2003. "Characterizing a Sustainability Transition: Goals, Targets, Trends, and Driving Forces." *PNAS* 100 (14): 8068–73.

Payet, R. A., et al. 2004. *Indian Ocean Islands, GIWA Regional Assessment 45b*. Accessed June 30, 2012. www.unep.org/dewa/giwa/areas/reports/r45b/giwa_regional_assessment_45b.pdf.

Qu, J., Z. Xu, Q. Long, L. Wang, X. Shen, J. Zhang, and Y. Cai. "East China Sea: GIWA regional assessment 36." Global International Waters Assessment. Accessed September 1, 2016. http://www.unep.org/dewa/giwa/areas/reports/r36/giwa_regional_assessment_36.pdf.

Reed, M.S., A. Graves, N. Dandy, H. Posthumus, K. Hubacek, J. Morris, C. Prell, C.H. Quinn, and L.C. Stringer. 2009. "Who's in and why? A typology of stakeholder analysis methods for natural resource management." *Journal of Environmental Management* 90: 1933–1949.

True Costs. 2015. Directed by Andrew Morgan, produced by Michael Ross, executive producers Lucy Siegle, Livia Firth, Vincent Vittorio, and Christopher L. Harvey, associate producer Laura Piety, and narrated by Andrew Morgan.

United State Embassy. 2012. "Peace Corps installs "merry-do-round" water pump in South Africa, 12 August 2012."Accessed December 16, 2016. http://iipdigital.usembassy.gov/st/english/article/2012/08/20120810134505.html#axzz4T1mcqO00.

Wiek, A. 2010a. "Analyzing Sustainability Problems and D*eveloping Solution Options—A Pragmati*c Approach." Wor*king Paper*, Sustainability Transition and Intervention Research Lab, School of Sustainability, Arizona State University, Tempe, AZ.

Wiek, A. 2010b. "Analyzing Sustainability Problems from a Pragmatic and Systemic Perspective." Working Paper, Sustainability Transition and Intervention Research Lab, School of Sustainability, Arizona State University, Tempe, AZ.

Zhang, J., et al. 2010. "Natural and human-induced hypoxia and consequences for coastal areas: synthesis and future development." *Biogoescience* 7: 1443–1467.

Webber, M. W. 1973. "Dilemmas in a General Theory of Planning" Poli*cy Sciences 4*(2): 155–169.

End-of-Chapter Questions

General Questions

1. Natural and human drivers influence socioecological systems either directly and unequivocally, as direct drivers, or indirectly and more diffusely, as indirect drivers. In each of the following scenarios, identify the direct and indirect drivers. Use the diagram type shown in Figures 3.17 and 3.23 to relate direct and indirect drivers to each other and to the socioecological system problem.

a. Diabetes is one of the leading causes of death in the United States, and Type 2 diabetes is the most common form of the disease. Type 2 diabetes is thought to be caused by a variety of factors including genetic factors and obesity. People with variants of the TCF7L2 gene seem to be more susceptible to the disease, with an 80% chance of developing Type 2 diabetes, than those who do not carry this gene. However, obesity is another factor contributing to the development of this disease. Obesity occurs when more calories are consumed than are lost due to physical activity. Many highly processed foods produced by today's agricultural systems are more calorie dense than less-processed whole foods. Many people are less physically active today for an array of reasons, including sitting at desk jobs all day and living in places with transportation infrastructures that promote driving rather than walking or biking.

b. Human-induced climate change can occur by two general processes: altering the chemistry of the Earth's atmosphere through increased greenhouse gas concentrations and by changing the reflectivity of the Earth's surface to sunlight (for example, a portion of the Earth's surface covered by ice and snow reflects more sunlight than a portion of the Earth's surface covered by water, which absorbs sunlight and retains more heat). Greenhouse gases are added to the Earth's atmosphere by way of several human activities, including fossil fuel burning and deforestation (CO_2), raising cattle (CH_4), and use of excessive synthetic nitrogen fertilizer (N_2O). The reflectivity of the Earth's surface decreases when forest is converted to asphalt roadways, which reflect less sunlight than forests and absorb more heat.

c. According to the Bureau of Labor Statistics, the unemployment rate in 2012 was about 8%. This relatively high unemployment rate is largely thought to be due to the 2008 financial crisis, but in addition to this, there are many other reasons for unemployment. When the tasks of workers are done with machines or advanced technologies, such as robots or computers, unemployment can occur. Because technological developments can lead to economic growth in general, advancement of emerging technologies is often promoted by governments through funding initiatives. Outsourcing of jobs to other countries with lower labor costs and less stringent environmental regulations also cause workers to lose their jobs. During recessions, which are a part of the boom-and-bust of business cycles, consumer demand can decline to low enough levels that employers need to lay off workers to maintain profit.

d. Eutrophication is a form of pollution in water bodies that can harm people, the economy, and the environment. Eutrophication is caused when excessive nutrient inputs, in the form of nitrogen and phosphorus, result in low oxygen conditions in water bodies. When farmers apply nitrogen fertilizers to their crops, some is used by plants for growth while up to 40%–60% is left in soils to run off into water bodies when it rains or when a field is irrigated. There is so much nitrogen in the Mississippi River that a New Jersey–sized coastal "dead zone" forms in the Gulf of Mexico at the river's mouth each summer, when the photosynthetic activity of plankton increases due to more sunlight. Recent technological improvements have increased fertilizer application efficiency in many countries, such that plants take up a larger percentage of the applied fertilizer with less left over to run off into water bodies.

2. In addition to identifying direct and indirect drivers, it is important to pay attention to the extent of influence and current trends of each driver. This helps in figuring out the most important drivers for a specific situation and, therefore, which ones to focus resources on when attempting to resolve an unsustainable situation. For each scenario below, first draw a driver identification diagram as you did for the last question and as shown in Figures 3.17 and 3.23. Then, use the four-option color scheme to indicate extent of influence and the four different arrows to indicate current trends, as was done for global biodiversity loss (Figure 3.28) and western Indian Ocean solid waste pollution (Figure 3.32). Based on your analysis, determine which direct drivers should be the focus of a sustainability problem-solving effort for each situation by listing the direct drivers in order of most important to least important.

   a. Drivers of Rainforest Loss. In the state of Rondonia, located in the Brazilian Amazon rainforest, the loss of rainforest has resulted from many factors over the past several decades, including road building, agriculture, and climate change. Based on satellite images, 4,200 km^2 of rainforest were cleared in Rondonia in 1978, 30,000 km^2 in 1988, 53,300 km^2 in 1998, and 67,764 km^2 in 2003. In the 1970s and 1980s, the main causes of deforestation were highway and road expansion supported by Brazilian government programs. The government supported these programs to provide easier access to the interior of the Amazon basin for a few groups, such as mining operations, logging companies, and struggling farmers who might move west from overcrowded and overburdened eastern regions to colonize new areas of the Amazon basin to help alleviate poverty. Many of the main highways and roads that are part of this government expansion program have already been built. In the 1990s and 2000s, industrial agricultural operations that clear forest for the purposes of large-scale soybean farming and cattle ranching are increasingly large contributors to deforestation. With projected future increases in population growth and consumer demand for soy-based products and beef, in developed and developing countries respectively, such trends are expected to continue as long as there is more Amazonian forest land available for growing food and raising cattle. Drought due to human-induced climate change is another increasingly important contributor to rainforest loss and, with the 50- to 100-year time lags associated with human-caused climate change, this will continue to increase into the distant future.

b. **Drivers of Poverty.** According to the International Monetary Fund, the poorest country in the world in 2012 was the Democratic Republic of Congo (DRC) with a per capita income of US$364.48. As a result, it has extremely high poverty rates, especially in rural areas (75%). Causes of rural poverty include ethnic conflict, disease, and lack of basic social services such as education. The DRC is one of the most ethnically diverse nations in the world and this, combined with economic hardship following independence from its Belgian colonizer in 1960, has resulted in intense ethnic conflict, especially over the past 20 years. These conflicts appear to be continuing unabated and are punctuated by only short periods of tenuous and readily violated peace agreements. The conflict is fueled by many factors, including access to basic resources such as water and valuable mineral resources. High incidences of disease, such as HIV/AIDS, also contribute to poverty. In the early 1980s, the DRC was one of the first countries in Africa to diagnose this disease. In 2003, 4.2% of adults were living with HIV/AIDS. By 2011, the rate of adult HIV/AIDS prevalence had dropped to 3.3%. Many factors are thought to be contributing to the spread of HIV/AIDS in the DRC, including large migrations of soldiers and refugees, difficulty of obtaining safe transfusions of blood, and lack of education, HIV/AIDS testing sites, and access to condoms in rural areas far from cities. Lack of basic social services, such as education, is another contributor to poverty. Many studies show that education, especially of women, can help alleviate poverty in many instances. Since the 1960s, the public education budget in the DRC has declined from 25% of the nation's total budget to 5% in the early 2000s. In 2013, however, the national government reversed this trend and increased the education budget to almost 14% of national income. In addition to the lack of resources for the national DRC government needed for school programs, another factor that prevents children from attending school is the fear that they will be abducted by rebel armies involved in ethnic conflict to serve as child soldiers while they are away at school.

3. Apply Wiek's framework, as presented in Figure 3.38, to the following case study. Determine who the stakeholders are, why they took the actions that they did, and the positive and negative influences of their actions. Determine the primary, secondary, and key stakeholder groups. Identify the values, beliefs, and norms of each stakeholder group that you think may have shaped their actions and identify technologies, resources, and rules and regulations. You may need to do research beyond the content presented here in order to fully identify all of the relevant factors, as they are not all included in the brief case study presented below. Finally, construct an interest-influence matrix, as presented in Figures 3.39 and 3.42. This case study was shortened and rewritten from Case Study 2 in *Resilience Thinking* by Brian Walker and David Salt. After you have applied Wiek's stakeholder analysis framework to this case study, think about how stakeholder actions might be different in a different context.

Case Study: The Goulburn-Broken catchment (GBC) is located in southeastern Australia. The region is a naturally fertile and productive watershed. Rainfall in the region follows a "feast-or-famine" pattern, with years of drought followed by very wet years. The GBC regional economy is supported largely by dairy farming and horticulture. Although both types of agricultural activities require water, dairy farming requires more. The region is experiencing a record drought and is in need of rain, but only a certain amount of rain. Too much will infiltrate soils and raise groundwater table levels. This is a problem because groundwater in the region is naturally salty and pasture grasses and horticulture crops do not tolerate salt well. Thus, if the groundwater levels come within 2 meters of the land's surface, the land's productivity will be harmed. Irrigation systems exist and can be used to alleviate the drought conditions, especially due to a dam and the resulting water supply from the reservoir Lake Eildon. However, farmers cannot irrigate too much, as this will have the same effect on the water table as too much rain.

One reason that the groundwater table issue is such a concern is that it has risen over the past several decades. Over the past century or so, since Europeans settled in the region and began large-scale agriculture, over 98% of native vegetation has been removed to make room for annual crops and grazing cattle. Much of the native vegetation was perennial and had very deep roots capable of extracting groundwater from deeper in the soil than the present-day annual crops, which have shallow roots. Thus, native vegetation kept groundwater levels low, up to 20 to 50 meters below the land surface, and prevented the threat of salty water harming the land's agricultural productivity. As crops and pasture grasses with shallow roots increasingly dominated the landscape, groundwater levels rose.

Dairy farmers are concerned with the productivity of the land to ensure that their large-scale agricultural operations continue to turn a profit. They are interested in expansion, reinvestment, and economic growth. Two stints of high rainfall throughout the 1950s and then again in the 1970s, in addition to their irrigation system, left farmers with the notion that they would have the water they needed to continue expanding their agricultural operations. The status quo underlying their agricultural production systems has been efficiency of agricultural production, advanced technologies, and hard work to solve any problems that arise. To solve the problem of salty groundwater when water table levels have approached the land surface too closely, pumping groundwater back into the river to lower water table levels has become common practice. The other farmers in the area are horticulturalists growing stone fruit crops, such as apples and peaches, and they generally have the same concerns and ideas as dairy farmers.

Other stakeholder groups include local communities, central government agencies, and environmental organizations. Local and regional communities depend on farmers' economic prosperity and success. The Murray-Darling Basin Commission is a centralized, joint state and federal government agency charged with maintaining the health of rivers in the GBC in terms of salinity. When pumping of salty groundwater to lower water table levels occurred, it increased the salinity and affected the use of river water by other stakeholder groups. Thus, part of their job was to regulate this pumping to ensure acceptable water quality in GBC rivers. Their standard mode of operation for years had been large-scale centralized regulation without much input from local or regional communities. However, they eventually realized the benefits of local and regional community input for watershed governance. As a result, some watershed governance was handed over to Catchment Management Authorities (CMAs) composed of community members from a given region. Although problems still exist, this institutional reform contributed to better watershed governance. Finally, "landcare" coalitions with a stewardship ethic were established to address the broad, system-wide problems that farmers could not resolve alone.

Project Questions

1. **Identifying Direct and Indirect Drivers:** Use the diagram type shown in Figures 3.17 and 3.23 to relate direct and indirect drivers to each other and to the sustainability problem that you started to explore in the Chapter 2 Project Questions section.
2. **Using Causal Chain Analysis to Identify Key Drivers:** Build on the figure that you drew from Question 1 to determine the extent of influence and current trends of the direct drivers that you identified. Use color and arrow schemes as was done in Figures 3.28 and 3.32. Use this information to rank the direct drivers from most to least important, as you did for Question 2 in the General Questions section of this chapter.
3. **Stakeholder Analysis in Your Specific Context:** Identify the stakeholders, use Wiek's stakeholder analysis framework (Figure 3.38) and construct an interest-influence matrix (Figures 3.39 and 3.42) to help you answer the following questions regarding the sustainability problem that you are investigating.

   a. Identify at least four stakeholders in your system and describe their values, beliefs, and norms. Also, indicate whether each identified stakeholder is a primary, secondary, or key stakeholder and explain why.

	Values	**Beliefs**	**Norms**
Stakeholder 1:			
Stakeholder 2:			
Stakeholder 3:			
Stakeholder 4:			

   b. Determine the primary actions leading to your sustainability problem, which you defined in the Project Question section of Chapter 2. Describe the technologies, resources, and rules and regulations governing or influencing these actions. Add more rows to the table below if you need to define more actions.

	Technologies	**Resources**	**Rules & Regulations**
Action 1:			

Action 2:			
Action 3:			
Action 4:			

    c. Describe the positive and negative influences of the actions defined in Question 3b above. Make sure to focus on influences on all three pillars of sustainability (environment, society, economy) at all relevant temporal (long and short term) and spatial (local to global) scales.

    d. Which stakeholders will need to be part of the process, from start to finish, for your efforts to be successful? Of these, which are marginalized groups that will be empowered when you make them a part of decision-making processes and the overall problem-solving effort? Which stakeholders need to be engaged occasionally, and why? Are their any stakeholders that do not need to be actively engaged?

Chapter 4
INDICATORS OF SUSTAINABILITY

Students Making a Difference

Studying Locust Swarms to Promote Food Security in Senegal

In rural southeastern Senegal, locust swarms destroy food crops, leading to food insecurity for already vulnerable communities. Climate change and land degradation are likely making the swarms worse. Degraded soils produce grass with a low nitrogen content that is a treat for locusts and lead to outbreaks. The Senegalese locust is often considered the main pest of the African Sahel. Outbreaks are costly and can have lasting impacts on livelihoods. Mira Word works with farmers in this region as part of the Living with Locusts project, which seeks to understand the factors that cause locust outbreaks and develop alternatives to pesticide-based management.

During her fieldwork, uncertainty runs high as she ventures through mud hut villages that lack running water, bridges that wash out at a moment's notice, and a culture very different from her own. Despite its challenges, working in this place prepares her for sustainability work in general: "Being flexible is so important. I try to be a sponge and be really open. Your brain is going to get scrambled in ways that you didn't even realize. Go in with a really open mind and try not to rationalize things that don't make sense. Just roll with it. The more you do really challenging things that are outside of your comfort zone, the more you are able to live within that space of flexibility where you don't need to define every single thing right now."

Mira is an ecologist and is most interested in soil, but she's working with a team of scientists, farmers, and other partners, including the Senegalese plant protection agency and the University of Dakar, to develop a broad set of indicators to help predict what kinds of agricultural practices might keep locusts at bay. She runs experiments to determine which crops locusts like to eat, testing if they prefer carbohydrate-rich plants such as millet over protein-rich crops such as groundnuts (peanuts). She observes where locusts like to hang out, testing if they prefer degraded habitats with low biodiversity, such as overgrazed rangelands. For her own project, she is studying characteristics of soil, such as nitrogen, texture, and organic matter content. All of these variables—locust diet, preferred habitat, soil characteristics—are indicators that might help empower farmers to better manage locusts and increase their livelihoods.

Mira uses her knowledge of applied ecology, but notes the importance of local knowledge: "We couldn't do this work without the different partnerships." She knows that any plan to help farmers to manage crops and soils needs to mesh with the value systems and worldviews of people living in the region, and needs to consider the resource constraints of rural Senegal. Because of this, she spends time walking with farmers in their fields, learning about when and where it would be best to collect her samples. She learns about the history of why farmers choose to plant certain crops at certain times, and how and where they graze their animals, as there is a whole cultural system underlying these things. It is information that she cannot find in a scientific journal article. She knows that the most reliable and effective indicators are developed in collaboration with local communities and within a local context. In this chapter, you will learn about sustainability indicators and how they are best developed within a local context through participatory processes.

Core Questions and Key Concepts

Section 4.1: Introduction to Sustainability Indicators

Core Question: What types of indicators are useful for resolving sustainability problems?

Key Concept 4.1.1—New indicators that holistically convey information about the environment, society, and economy across many spatial and temporal scales are needed to understand a system's sustainability.

Key Concept 4.1.2—Six general characteristics define useful indicators of sustainability and these characteristics should be used to guide indicator selection for sustainability problems.

Section 4.2: Sustainability Indicators, a Case Study from the Primorska Region of Slovenia

Core Question: How are indicators developed through participatory processes involving relevant stakeholders?

Key Concept 4.2.1—Sustainability indicators, and their boundaries, are developed for a specific context through a participatory process, involving several stakeholders who collectively develop indicators by sharing knowledge and assessing contentiousness and feasibility.

Key Terms

indicator
state indicators
pressure indicators (aka. control indicators)
response indicators
contentiousness
feasibility

> *"[W]e have been chasing a moving shadow called sustainability.... [T]here is no single meaning and there is no agreement on how it is measured and recognized in an objective sense.... Scientists and professionals took (or were given) the impossible task of achieving definitive measurement of this word. The impossible task was to measure what was never potentially measurable: the immeasurable 'sustainability'.... Why have many tried to show that sustainability = 42?"*
>
> —Simon Bell and Stephen Morse, 2008, *Sustainability Indicators: Measuring the Immeasurable?*

In the last chapter, methods for analyzing the current state of a system were presented. These methods can help us resolve sustainability problems by answering the following questions: What is regulating a system (driver classification)? What causes a system to be sustainable or unsustainable (causal chain analysis)? What factors guide stakeholder actions within specific contexts (stakeholder analysis)? How all of these things are explicitly known is another matter. Specifically, how do we know about the current system, and that there are undesirable influences on the system leading it to an unsustainable state? How do we know what drivers are regulating the system? How do we know about the causal relationships among these drivers, their relative influences on the system, and the current trends of these drivers? How do we know which factors are shaping stakeholder behavior?

The short answer to all of these questions is: we use data to indicate what is going on. These data come in the form of what are known as **indicators**, which is the topic of this chapter. Indicators are useful not only for understanding the current state, but also for constructing future scenarios and transforming desirable and sustainable future visions (Chapter 7) into tangible goals. They are also used when devising strategies for sustainability transitions (Chapter 8). Indicators generally help in understanding where the system currently stands with respect to sustainability, how far the system must go to reach sustainability, and which way the system is headed (toward or away from sustainability), or whether the system is changing at all. In essence, indicators are about measuring sustainability. However, as will become clear by the end of this chapter, measuring sustainability using indicators is easier said than done.

---

indicator quantitative or qualitative data used to determine the current state of a system, how far it is from some ultimate goal, and which way it is changing or whether it is even changing at all.

Section 4.1: Introduction to Sustainability Indicators

Core Question: What types of indicators are useful for resolving sustainability problems?

Although indicators can be quantitative and technical, they can also be qualitative and anecdotal. You informally use indicators every day to make decisions based on the current state, what might change that state, and why. For example, a blue sky and 85°F temperatures on a July morning *indicate* that you should wear a T-shirt, shorts, or other light clothing that day (current state of the weather). If you are familiar with the climate in which you live, then you can also anticipate what might cause the weather to change that day and why. If you live in the southwestern, and much of the western United States, you know that typical weather patterns in July and August cause the weather to change from clear blue skies in the morning to rainy thunderheads by the afternoon (**Figure 4.1**). Based on this information, you might choose to bring along an umbrella because climate records *indicate* what might cause the current state to change (the *drivers* regulating climate).

Delving more deeply into the climate science behind these patterns can explain why monsoon season weather causes the system to change throughout the day. One example is the relationship between heat and moisture. There is moisture in the air during the monsoon. The air temperature rises from 85°F in the morning to 110°F in the afternoon. The moist, hot air rises and, as it rises, it cools. Cold air holds less moisture than hot air, so, at some high altitude in the atmosphere, the maximum capacity for the cooler air to hold moisture will eventually be exceeded, and it will rain. An understanding of these causal relationships through a *causal chain analysis indicates* why thunderstorms occur in the afternoon

Figure 4.1 Information about typical weather patterns in the desert indicate that a relatively cloudless quiet summer morning (*left*) can turn into a violent thunderstorm in the late afternoon (*right*).

and early evening. *Indicators* used as data to come to these conclusions include air temperature, humidity, clouds or lack of clouds in the atmosphere, latitude, and longitude. Regardless of whether you realize it, you used indicators to determine what clothing to wear today. Indicators are everywhere and offer information about the world around us.

Section 4.1.1—Sustainability Indicators versus Traditional Indicators

Key Concept 4.1.1—New indicators that holistically convey information about the environment, society, and economy across many spatial and temporal scales are needed to understand a system's sustainability.

In sustainability, indicators are intended to help us understand where a system is, which way it is going, and how far it is from sustainability. A good indicator helps us recognize an unsustainable situation before it gets too bad, and offers clues about what needs to be done to resolve the problem. Indicators are most effective when they are part of a larger suite of indicators, created to holistically assess sustainability at many different spatial and temporal scales, from local to global and from the present into the future. They allow one to see where the problem areas are in order to devise solutions and to aid in the tracking of system changes into the future. (Indicators can change over time, however, so it is important to regularly reassess indicators and what they are really telling us.) Indicators relevant to sustainability must convey information about the environment (e.g., water quality, air quality, and natural resource availability), society (e.g., education, health, poverty, and crime rates), and economy (e.g., stockholder profits, materials for production, and job availability). Also, socioecological systems are characterized by dynamic interactions among the three pillars of sustainability. Thus, indicators must capture links among the three pillars and gauge how the linkages are leading the system toward or away from sustainability, or regulating it in such a way that it is not moving at all.

An example that illustrates these ideas is provided in **Figure 4.2**. The natural resource base provides the materials for production on which jobs and stockholder profits depend. Jobs affect the poverty rate, and the poverty rate is related to crime. Materials for production can affect air quality and water quality, which can both have an effect on health. Water quality can have an effect on stockholder profits: if a process requires clean water as an input, then cleaning up poor-quality water prior to processing is an extra expense that reduces stockholder profits. Likewise, health problems, whether due to general air quality problems or exposure to toxic materials from the materials used for production, have an effect on worker productivity and contribute to the rising costs

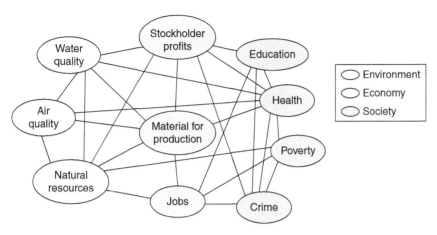

Figure 4.2 Sustainability indicators should provide information about all three pillars of sustainability and capture the relationships among them.

of health insurance. There are many linkages among the environmental, social, and economic aspects of sustainability. Sustainability requires this type of holistic, integrated approach to problem-solving. Effective sustainability indicators must be multidimensional.

Traditional indicators, such as the gross domestic product (GDP), are not appropriate (on their own) to measure sustainability. GDP, also referred to as the gross national product (GNP), is a well-publicized traditional indicator that measures the amount of money being spent in a country. It is generally used to indicate the economic well-being of a country. The assumption behind GDP as an economic indicator is: "The more money being spent, the better the overall economic well-being." However, GDP is not a good sustainability indicator on its own. GDP only reflects the amount of economic activity, regardless of the effect of that activity on environmental health and human well-being. GDP can be trending upwards, while environmental health and human well-being are declining. For example, when there is a ten-car pileup on the highway, the GDP goes up because of the money spent on medical fees, repair costs, and cleanup and disposal fees (Figure 4.3). However, human well-being declines because people are injured, and environmental health is negatively affected as ten cars are added to a landfill. On the other hand, if those same ten people did not buy cars, then environmental health and human well-being may increase. People may be healthier walking or biking to work rather than driving. This would also be better for the environment, as it reduces CO_2 emissions contributing to climate change and other tailpipe emissions that cause acid rain. However, when people do not buy cars, GDP does not increase. Using traditional indicators on their own as indicators of sustainability is not effective. According to Hazel Henderson in her book Paradigms of Progress: "Trying to run a complex society on a single indicator like

Figure 4.3 A rise in GDP following a car crash reflects only the amount of money spent (*left*), and not harms to people and the environment (*right*).

the Gross National Product is like trying to fly a 747 with only one gauge on the instrument panel . . . Imagine if your doctor, when giving you a checkup, did no more than check your blood pressure."

An example of a better indicator for sustainability that might replace GDP is the Index of Sustainable Economic Welfare (ISEW), which was developed by ecological economist Herman Daly in the late 1980s. It is not yet in wide usage, mostly because more data are required for calculation, but it is a better measure of sustainability than GDP (**Figure 4.4**). It accounts for environmental and social dimensions, as well as economic development, by subtracting the harmful environmental and social consequences of economic activity from the GDP. For instance, the ISEW accounts for air pollution by estimating the cost of damage in terms of negative human health impacts per ton of five key air pollutants emitted from vehicles. Another example is that it accounts for nonrenewable resource depletion by estimating the cost to replace a barrel of oil with the same amount of energy from a renewable source. It estimates the cost of climate change due to greenhouse gases per ton of emissions. The cost of ozone depletion is also calculated per ton of ozone depleting substance produced. All of these costs are subtracted from the GDP. In addition to subtracting costs from the GDP, ISEW also corrects the GDP to include significant beneficial economic activities that are not incorporated into the GDP because they do not have a dollar value. Thus, activities such as unpaid domestic household labor, or community childcare among friends and family, are added to the GDP. These unpaid services contribute to human well-being, but are not included in traditional calculations of GDP because they do not involve spending money.

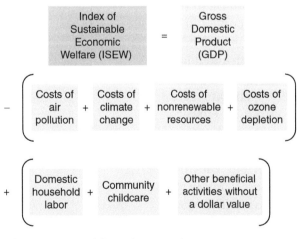

Figure 4.4 Although it requires more information to calculate, the Index of Sustainable Economic Welfare (ISEW) is a better sustainability indicator than GDP because it accounts for all three pillars of sustainability.

A final general note about sustainability indicators is that they can change over time. Therefore, it is important to regularly reassess indicators and what they are really telling us. **Box 4.1** provides a recent example of this.

BOX 4.1 CO_2 EMISSIONS AND GDP: INDICATORS THAT HAVE CHANGED OVER TIME

CO_2 emissions and GDP are two commonly used indicators that are relevant to sustainability. CO_2 emissions tell us how climate change has, or could potentially, affect the earth system. GDP tells us about the amount of money being spent in a country. Historically, these two indicators have been directly correlated: as CO_2 emissions went up, GDP increased. Owing to this correlation, people thought that economies could not continue to grow if CO_2 emissions were reduced. This has proven to be a big barrier to sustainability and a shift to clean energy. However, according to a 2016 study by the Brookings Institute, this may not be the case for much longer, as the relationship between these two indicators appears to be changing. In at least 35 countries around the world, including more than 30 states in the United States, trends over the past 10–20 years show that the amount of fossil fuels that we burn no longer correlates with economic growth (**Figure 4.5**).

This apparent trend in the "decoupling" of economic growth from CO_2 emissions means that we may no longer be able to use one as an indicator of the other. In other words, more CO_2 emissions do not necessarily mean better economic growth, at least according to trends over the past two decades. This opens the future to more sustainable energy sources, including renewables and other clean energy technologies, as we let go of the anxiety surrounding CO_2 emission reductions and economic growth.

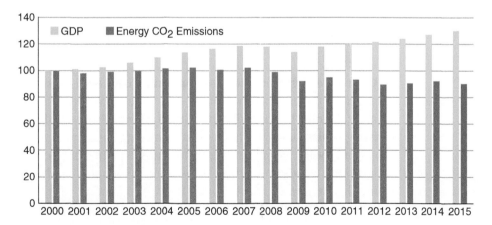

Figure 4.5 Since 2000, in the United States, CO_2 emissions have declined overall while the economy has continued to grow (y-axis values are indexed to 100 in 2000).

Section 4.1.2—Characteristics of Effective Sustainability Indicators

Key Concept 4.1.2—Six general characteristics define useful indicators of sustainability and should be used to guide their development for sustainability problems.

Section 4.1.1 hinted at some characteristics of effective sustainability indicators. In this section, six characteristics are discussed in detail. There is no broad consensus on the best sustainability indicators and, as with everything related to sustainability, the specific context within which a problem is embedded will ultimately determine the most appropriate indicators. The characteristics of effective indicators of sustainability presented in this chapter are based largely on those described in a book published in 2008 by indicator experts Simon Bell and Stephen Morse titled *Sustainability Indicators: Measuring the Immeasurable?* Much of the information about indicators presented in the book is based on their experiences in collaboratively developing sustainability indicators with stakeholder communities. In Section 4.2 of this chapter one of these efforts, facilitated by Simon Bell, is described in order to illustrate how sustainability indicators are actually developed and used in a real-world situation. In this section, the following general characteristics of effective sustainability indicators are presented with many examples to promote understanding:

1. Use appropriate time and space scales
2. Data are realistically available
3. Easy to understand, even by non-experts
4. Focused on current state, drivers, and responses
5. Collectively determined by stakeholders
6. Representative of real-world complexity

Characteristic 1: Use Appropriate Time and Space Scales. As is always the case when dealing with the complex systems of sustainability, scale must be carefully considered. This is also true when selecting and using indicators. Whether considering temporal or spatial scales, some starting point and reference frame must be chosen in order to evaluate indicator trends toward or away from sustainability. This must be done intelligently, however, because the starting point and reference frame choice will influence conclusions about a system, as revealed by the indicators. The following climate change example illustrates these ideas.

Two major indicators of human-caused global climate change are atmospheric CO_2 concentration and global temperature. The graphs in **Figure 4.6** show that both atmospheric CO_2 concentration and global

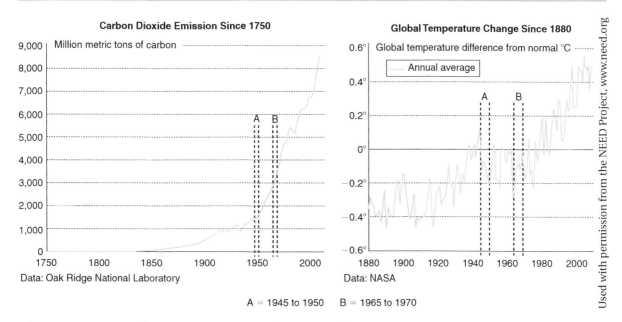

Figure 4.6 Depending on the time period referenced, two major indicators of global climate change—CO_2 emissions and temperature—can lead to different conclusions about human-caused climate change.

temperature have increased over the past few hundred years. However, CO_2 concentration rises more smoothly than global temperature, which increases overall, but tends to fluctuate up and down. When choosing an indicator for detecting trends in human-caused climate change, the starting point and time frame are important because they will influence conclusions. Conclusions can also depend on which indicator is used. For example, if the CO_2 concentration is used over time period A, from 1945 to 1950, the conclusion is that concentrations are increasing. If this indicator is considered over time period B, from 1965 to 1970, the conclusion is the same. If the same two time periods are considered for the global temperature, then the conclusion about climate change is different. During time period A, global temperature shows an overall decrease. However, during time period B it shows an overall increase. Thus, if the global temperature indicator is used to gauge how human activities influence climate, then different conclusions will be made depending on the time period used. For both of these indicators, long-term data sets on the order of hundreds of years are needed to make accurate conclusions about human effects on climate because climate changes over long time scales.

The same ideas apply to spatial scale. For example, Farmer Bill is concerned with the health of the soil on a piece of land that he just purchased from another farmer. He knows that the other farmer worked the soils intensely through multi-cropping and deep plowing. He chooses soil phosphorus concentration as the indicator of soil health. He knows that soil can take years to recover and reacquire the nutrient contents that are

196 *Sustainable World: Approaches to Analyzing and Resolving Wicked Problems*

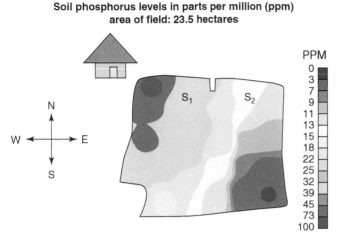

Figure 4.7 Red and orange regions have the lowest phosphorus concentrations, and green and blue areas have higher soil concentrations, such that concentrations very spatially across the field.

high enough for growing crops. He decides to use a sampling location next to his house (S1 in Figure 4.7). He applies soil amendments to his entire field each month and monitors soil concentrations by collecting samples at S1. After a year, he sees that the soil concentrations have recovered to about 40 ppm. (Parts per million, ppm, is a unit of measurement used to measure nutrient concentrations in soil.) The recommended minimum soil phosphorus concentration for the crop he wants to grow is 25 ppm, so he concludes that there is more than enough!

Based on the information obtained from his indicator, he plants his entire field and does not bother applying fertilizer, because the concentration is 15 ppm above what is needed. Within two months, about half of his crop has failed, with the exception of a small area near his house. To learn why, he brings in a soil consultant who surveys 25 different points across the entire field and creates the map in Figure 4.7. Farmer Bill learns from this that soil phosphorus concentrations vary spatially across his field and, in order to use soil phosphorus concentration as an accurate indicator of soil health, he needed to sample at finer spatial scales.

Characteristic 2: Data Are Realistically Available. The best indicators give an accurate picture of the current state of a system relative to sustainability over time, and help answer the question: is the system approaching sustainability, moving away from it, or staying the same? Unfortunately, although one can dream up the perfect indicators to answer this question, data are often not available, or cannot be made available. One common issue related to availability is that indicators must provide information while there is still time to act and correct the problem. If they do not provide information early and often enough, then they are not good indicators. For example, imagine a gas gauge that only shows the amount of gasoline in the tank when the engine starts. After driving for several hours without stopping, the reading would no longer be useful. Knowledge about how much gasoline is in the tank at each moment is critical to prevent the problem of running out of gas. Similarly, a report card distributed a week before graduation, rather than twice a year during each year of high school, would arrive too late for failing students. Students who need remedial help would not know that they need it while there is still time to act, and they would not graduate from high school as a result.

One reason for the challenges surrounding data availability for sustainability indicators has to do with our limited understanding about complex system behavior. When systems are changing, they do not always move linearly and predictably toward, or away, from sustainability. Sometimes, socioecological systems experience unexpected and rapid changes. Such nonlinear, often unpredictable, behavior is a defining characteristic of complex systems (recall the rubber band and coral reef examples in Section 2.1.2) and is discussed in more detail in Chapter 5. For now, it is enough to note the current need for identifying indicators that can warn of sudden and rapid transitions in socioecological systems. This has been a major research area for many types of ecosystems, such as grasslands and shallow lakes. For example, decades of overgrazing in a grassland can suddenly cause a shift, in a matter of years or less, from a healthy grassland ecosystem that supports cattle to a desert scrub landscape that cannot support cattle, nor the human systems dependent on those cattle (Figure 4.8). Finding holistic sustainability indicators that warn of such rapid shifts in social, environmental, and economic systems, while there is still time to act and prevent them from happening, is an important and active research area in sustainability science.

A second reason for the challenges surrounding data availability for sustainability indicators has to do with availability of resources for broad data collection across spatial and temporal scales. Even when the best sustainability indicators for a certain situation are known, they may simply not be available. Frequently, the best indicators have no data, while the indicators with abundant data are the least representative of sustainability. This leads many to choose traditional indicators, such as GDP, that have readily available and abundant data that are comparable across many different countries. However, such traditional indicators are not holistic representations of sustainability. In order to create sustainability indicators from existing and readily available indicators of environmental health, societal well-being, and economic activity, separate indicators are often combined into indices (e.g., ISEW, Human

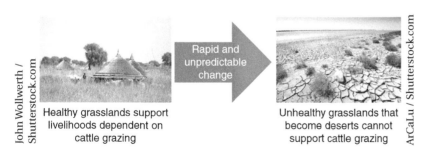

Figure 4.8 **Indicators that forecast rapid and unpredictable shifts in systems, such as those that occur when grasslands are overgrazed by cattle, as shown here, are presently not available.**

Development Index, Bhutan's Gross National Happiness Index, and Environmental Sustainability Index). The intent of these indices is to capture sustainability in a single number that does not require much additional analysis, which is appealing to policymakers and other practitioners. However, single number indices oversimplify complex systems and important information is lost. These issues are discussed in more detail in their respective sections in the following pages.

The third reason for the challenges surrounding data availability for sustainability indicators has to do with indicator availability over long periods of time that extend far enough into the past as well as the future. If there are not enough data from the past, often referred to as "baseline data," then changes leading to unsustainable trends cannot be detected. For example, scientists have suspected for a long time that extra atmospheric CO_2 leads to a greenhouse effect that causes the Earth's surface to warm, and the climate to change. Swedish chemist Svante Arrhenius is credited with first pointing out the heat-trapping properties of CO_2 in 1896. However, routine measurements of CO_2 concentration did not begin until the 1950s. Today, there are hundreds of measurement stations around the world. CO_2 concentrations prior to the 1950s are determined using measurements from tree rings, lake sediments, ice cores, and other methods. If these measurements had not been made, then CO_2 concentration would not be a good indicator to detect current trends because there would not be enough baseline data from the past (Figure 4.9). Data collection must also continue into the future. Atmospheric CO_2 concentration is a relatively easy measurement to make, but sometimes limited resources prohibit indicator data collection. For example, sampling soil frequently enough, and at small enough spatial scales to provide accurate information, can be costly in terms of time, money, and effort. As a result, soil indicators typically do not extend far back into

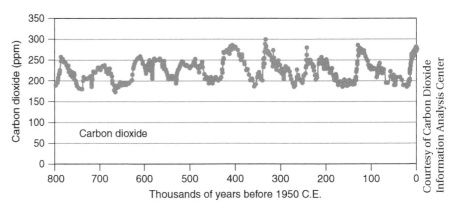

Figure 4.9 Without the baseline data from ice cores shown here, we would not know that present CO_2 concentrations, which exceed 400 ppm, are "off the charts" and have not existed for at least the past 800,000 years.

the past, and it is unlikely that intensive data collection can continue year after year into the future.

Characteristic 3. Easy to Understand, even by Non-experts. Often times, experts design indicators based on their research, or other highly specialized or technical activities. However, indicators must be understandable by all stakeholders involved with sustainability problems, including policymakers, practitioners, and other non-experts, who will use them to make decisions. If stakeholders do not know what indicators tell them about a system, then they will not know how to act. Even worse, they may take inappropriate actions that cause harm. For example, high schools have different ways of indicating academic progress. Some schools have letter grades A through F. Other schools use percentages from 100 to 0 or a 4.0-point scale. Some schools do not give grades and instead provide written comments. These different indicators all express a student's progress, or lack of progress, in a way that is understandable to the person reading the report card, such as members of a college admissions committee. However, if a report card gave grades in ancient Greek script, the meaning of the grades would be a mystery to most people (Figure 4.10).

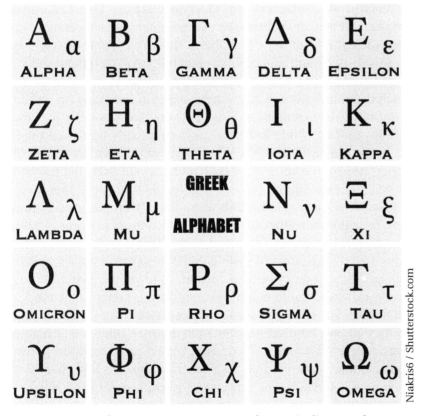

Figure 4.10 If your school reports grades, an indicator of academic progress, using the ancient Greek alphabet (shown here), then it is likely that college admissions committees would not have a clear indication of how well you did in school.

If an admissions committee attempted to understand the ancient Greek script, but misinterpreted the information, then students deserving admission might be overlooked and unqualified students might gain admission. Indicators that cannot be understood by decision makers are less likely to be used to support decision making, or may lead to inappropriate decisions and actions. This would make it hard for an admissions committee to decide whether a student should be accepted at a college.

In order to assess a situation, and know what type of action is needed, and when, indicators must be understandable. Understanding indicators has a lot to do with understanding the assumptions upon which indicator calculation is based and how indicators are qualified. Without this information, it can be difficult to determine whether a certain indicator is appropriate for a specific context. Unknowingly applying indicators to situations where they are not fully appropriate can lead to decisions based on incomplete or faulty information. Box 4.3 presents a sustainability example to illustrate the importance of having a clear understanding of underlying assumptions when making decisions based on indicators; and how unclear indicators can lead to misunderstandings, and possibly additional harm.

Characteristic 4: Focused on Current State, Drivers, and Responses. It is overwhelming to choose among a large number of indicators, with abundant data availability, that are capable of tracking environmental, social, and economic sustainability. An overabundance of indicator choices arises when a wide range of stakeholders are involved in their selection. In Section 4.2, which describes a case study on collaborative indicator development in Slovenia, stakeholders initially came up with a list of 75 possible indicators, and eventually narrowed them down to 20, for practical purposes. When narrowing down a large list of indicators, trying to maximize distinctive and relevant data that provide a broad systems view should be a central focus. To achieve this, three general groups of sustainability indicators have been suggested: state indicators, pressure indicators, and response indicators (Figure 4.11). These groups were originally suggested at the 1992 UN Rio Earth Summit, where a list of indicators of sustainable development was compiled.

state indicators
a class of indicators focused on problem outcomes by tracking undesirable conditions within a system, or other indirect factors that shape these conditions, and render the system unsustainable.

State indicators describe the current state of the system. They are focused on undesirable conditions within the system that render it unsustainable. For example, soils low in nutrients are unsustainable because they cannot be used to grow food. Farmer Bill's soils became phosphorus depleted because of extensive plowing and multi-cropping. The unsustainable condition that resulted from this was low quantities of phosphorus in the soil. The amount of phosphorus in the soil is a state indicator in this case. If phosphorus measurements are not available, other more commonly available measurements can be used to indirectly indicate low soil phosphorus in a field that has been extensively plowed and cropped. For example, sandy soils are less likely to hold

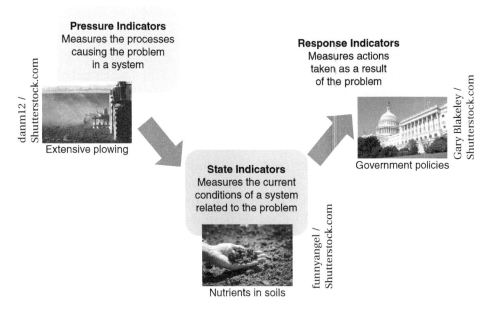

Figure 4.11 Choosing indicators from these three general categories (examples of each shown) when selecting indicators ensures a broad systems view of a problem and the possible solutions.

onto phosphorus than other soil types. Soils on steep slopes are more likely to wash away during rain storms, taking phosphorus with them, than soils on flat areas. Thus, in this example, soil texture and topography could also serve as state indicators of unhealthy soils, although they are less direct than phosphorus measurements. Examples of state indicators for human well-being include income, life expectancy, crime rates, and infant mortality. For economic conditions, GDP or household income could be used.

Pressure indicators (aka. **control indicators**) track processes regulating a system that lead to certain, usually unsustainable, conditions. They are also referred to as control indicators and, within the context of this textbook, are focused on drivers (Chapter 3). An example of a pressure indicator from the Farmer Bill scenario is cropping rate. Each time a new crop is planted, more phosphorus is removed from the soil by the plants. The more frequently this happens, and the less time the soil has to recover between plantings, the more phosphorus will be removed from soils. Over time, this will result in unhealthy soils that cannot be used to grow crops. By tracking cropping rate over time, the causes of unhealthy soils can be determined. For example, if cropping rate is increased from two to six crops per year, and soil phosphorus declines within a year, this is a good indication that cropping rate is causing unhealthy soils. Examples of pressure indicators used to understand the factors that influence human well-being include the availability of healthcare or funding for

pressure indicators (aka. control indicators) a class of indicators focused on the cause of a problem by tracking the processes regulating a system's behavior that lead to certain conditions within a system.

education. Pressure indicators in economic systems include export or manufacturing rates.

Response indicators track the actions that are taken in response to a problem. For example, farmer education could foster understanding that extensive plowing and multi-cropping depletes the soil of phosphorus. If, after 10 years of tracking response indicators such as the availability of educational programs and changes in farmers' actions, there is no change in soil health, or its causes, based on state or pressure indicator data, then it may be necessary to change the solution strategy. Another possible avenue could be to lobby for government policies that encourage healthy soils through regulations or incentive structures. In reality, to solve a sustainability problem such as this, several solution strategies would need to be pursued simultaneously. Whatever the responses, they should be tracked using response indicators so that solution strategy effectiveness can be assessed, and the strategy changed if necessary.

These three groups of indicators are not mutually exclusive, as the categories overlap in some cases. However, paying attention to these indicator groups when narrowing down the list of possible indicators will lead to more effective indicators for problem solving. A broad selection is necessary in order to gain a holistic picture of a sustainability problem and the system in which it is embedded. By ensuring that indicators of all three types are included, we can determine the directional movement of a system toward or away from sustainability, or its failure to move at all, and assess which responses are effective and which must be changed.

Characteristic 5: Collectively Determined by Stakeholders. Indicators tell us where a system currently stands with respect to sustainability, how far it must move to reach sustainability, and which direction it is headed (toward or away from sustainability) or whether it is even moving at all. However, an important question has not yet been explored: where, exactly, is sustainability located? If the location of sustainability is not defined in terms of indicators, then it will not be possible to know when sustainability is reached. As discussed in Chapter 1, sustainability means different things to different people and there is no one clear definition. Therefore, stakeholders must be involved in indicator development, in order to define sustainability indicators that are appropriate for a specific context, and should also be involved in collecting data for indicators, so that accurate data is obtained. Including a broad range of stakeholders is extremely important if indicators are to be effectively used for solving sustainability problems embedded within specific local contexts.

Before using indicators to track the state of a system, it is important to first define which indicators will be used. During this indicator

---

response indicators
a class of indicators focused on evaluating the effectiveness of solutions by assessing the actions that were taken in response to a problem and its cause.

selection process, it is important to choose indicators that will be useful for the specific local context in which the sustainability problem is embedded. This means stakeholders who are familiar with, or have direct experience with, a specific local context should be deeply involved, or at least consulted, when deciding on indicators of sustainability to be used within that context. Input about indicators from knowledgeable stakeholders will help problem solvers to most accurately and effectively track the sustainability of the system within a specific local context. This should lead to more effective problem solving efforts. The case study in **Box 4.2** illustrates the importance of stakeholder input for understanding the problem within a local context.

> ### BOX 4.2 DEWORMING: AN UNEXPECTED INDICATOR OF STUDENT SUCCESS
>
> In the 1990s, a Dutch charity called Investing in Children and Their Societies (ICS) wanted to use money from their child sponsorship program to improve attendance and test scores of Kenyan students. Their plan was to improve these things by hiring and paying more teachers, buying new textbooks and other classroom materials, and providing free uniforms to school children. They choose to take these actions because, in their experience, the availability of these resources to students is often an indicator of student success. Before implementing their plan on a large scale, two researchers convinced ICS to study the effectiveness of their planned improvements before rolling out the program at large scales. ICS agree to allow the researchers to collect data as part of their pilot program.
>
> The researchers, educated at Oxford and Harvard universities, implemented one improvement at a time in order to test the effectiveness of each. They collected data in a total of 14 different schools in Kenya, with seven of the schools implementing the improvements and the other seven making no changes or improvements. First, they bought new textbooks for seven of the schools, and not the others, but saw no difference between the two groups of schools in attendance or test scores. Next, they hired more teachers, but they again observed no difference between the two groups of schools. They continued to try other improvements, such as different classroom materials and free school uniforms, but still saw no change in student outcomes. The availability of these resources was often an indicator of student success, in their experience, so they could not understand why they did not work in Kenya. Undeterred, they continued to look for ways to improve educational outcomes for Kenyan school children, but did not find an answer until they talked to someone knowledgeable about local conditions: a friend who worked at the World Bank.
>
> Their World Bank friend was more familiar with issues affecting children in developing countries and suggested that they test deworming. Parasitic intestinal worms are an issue that affects Kenyan children, along with more than 1 billion people in developing countries
>
> *(Continued)*

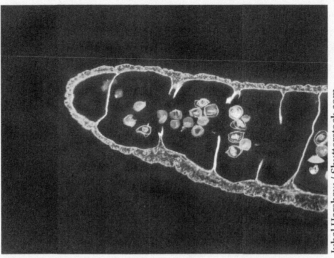

Figure 4.12 Infections by parasitic worms that live in human intestines, such as the tapeworm shown here (*top*), makes it difficult for Kenyan students, such as these children at a school near Tsavo National Park (*bottom*), to maintain attendance and succeed.

around the world. The worms cause problems such as anemia and immune system suppression, which make people more vulnerable to other diseases, including malaria. The physical ailments that result from an infection with these worms make it difficult for children to attend and excel in school (Figure 4.12). Most people living in developed countries are not aware of this issue. The problem can be solved for children with an inexpensive medication, available for only five cents per day, which can be easily administered to students by their teachers. The researchers conducted more trials in the schools but this time, instead of new classroom materials, clothing, and more teachers, they provided deworming medication.

The results of the deworming trials were astounding. Children treated with the medication spent an extra 2 weeks in school each year, on average, compared to their school attendance in the past. Even more astounding, when the researchers surveyed the children 10 years later, they were working more hours each week and earning 20% more income than those who had not been offered the medication in school as children. Based on their knowledge and perspectives prior to their 1990 trials in Kenyan schools, the researchers would never have guessed that the availability of deworming medication would be such an effective indicator of the success of children in school and later in life! This highlights the importance of involving people familiar with a local context, in this case, the friend from the World Bank, when developing indicators of sustainability for a specific place and time.

Considering local knowledge when collecting indicator data to determine where a system is with respect to sustainability, at any given time, is also important. Ever since the scientific revolution during the 18th century Enlightenment, Western society has privileged quantitative information, derived using the scientific method, and discounted other forms of qualitative information as less useful. This thinking still dominates, and decision makers are often obsessed with quantification. Quantification definitely has its benefits, which is one reason why it has become so privileged, but a singular focus on the types of knowledge that can be quantified often excludes a lot of important information.

An example of a situation where quantification is beneficial is for indicators used to track changes in the global climate system. One important indicator used in this case is CO_2 concentration. Based on climate models, scientists have determined that the upper safe limit of CO_2 in the atmosphere is 450 ppm (some say it is even lower at 350 ppm). The CO_2 molecules accumulating in the air cannot be seen with the naked eye (Figure 4.13). Thus, quantitative information that associates an increase in atmospheric CO_2 concentrations with global temperature increases, and with other phenomena related to climate change, is very important here. Also, climate change happens slowly over decades, or even over hundred-year time scales, such

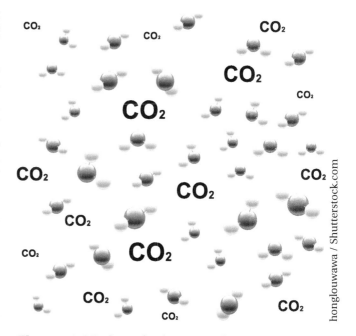

Figure 4.13 Quantitative scientific knowledge about the CO_2 molecule, which cannot be seen with the naked eye, has been important for understanding the cause of climate change.

that warming of the global environment is not likely to be decisively perceived over the lifetime of an individual. In this case, one could argue that quantitative indicators are necessary for tracking changes in a system.

On the other hand, sometimes qualitative knowledge is a more appropriate and accurate indicator than quantitative information. The first reason is simply that not all human experience can be quantified. When trying to gauge how human well-being has changed over the past 10 years, information for indicators may come from stories of people living in an area, rather than from measurements. For example, when trying to gauge human happiness and life satisfaction, survey responses might indicate that people seem "worse," "better," or "much better", rather than 10% worse, 50% better, or 80% better. In these cases, quantitative information may be unavailable or simply unnecessary. This is not only true of human systems, but natural systems as well. The second reason is that, in some cases, qualitative knowledge can be a more accurate indicator than quantitative knowledge. For example, the indigenous Cree people of the eastern Canadian subarctic zone use many of the same indicators as scientific wildlife experts to monitor caribou populations: spatial distribution, individual behavior, migratory patterns, age structure, fat content, and community interactions, such as predator abundance. Out of all of these indicators, the Crees' focus on fat content more than any other, and they are more accurately able to gauge caribou population health than scientists. Cree assessment of fat content is based on qualitative observations made over years of hunting the animal, whereas scientific analysis involves quantitative physical and chemical measurements made in a laboratory. Thus, the Crees' practical knowledge of caribou populations is arguably a better indicator than quantitative data collected by scientists, and should therefore be considered as valid information by decision makers who are trying to prevent caribou population collapse (Figure 4.14). In reality,

Figure 4.14 The Cree, shown in this illustration from 1860 (*left*), have lived in northern Canada for hundreds of years and their knowledge about caribou (*right*) is based on extensive direct practical experience rather than scientific knowledge.

quantitative and qualitative information should be viewed as complementary rather than in opposition, when determining and collecting data for sustainability indicators because different types of indicators can be more accurate under certain conditions. In the caribou population example, qualitative information related to fat content is useful when population growth is limited by the amount of food available within the caribou's home range, but not when growth is limited by an overabundance of predators. In the latter case, quantitative scientific information has proven to be a more useful indicator.

Characteristic 6: Representative of Real-world Complexity.
All indicators, by necessity, simplify reality to some degree by reducing complexity. That is, after all, the point of using indicators. The question then becomes: how much simplification is safe so that policymakers and practitioners, who use indicators to make decisions, have sufficiently accurate information about a system to make good decisions? As illustrated by the marine fisheries example in **Box 4.3**, misunderstandings can lead to bad decisions that move systems away from sustainability. This is because, when any situation is simplified, there is an unavoidable loss of information. A major criticism of aggregate sustainability indicators, such as the Environmental Sustainability Index (ESI), Human Development Index (HDI), and the Human Sustainable Development Index (HDSI), is their intent to summarize complexity into a single measure. These three aggregate sustainability indicators are described next to illustrate the problem with too much simplification. Following this, alternatives to too much simplification, that are still useful for decision makers, are described.

BOX 4.3   MANAGEMENT OF THE PERUVIAN ANCHOVY FISHERY IN THE 1970S

The Peruvian fishmeal industry was established in 1959 to produce feed for land animals, such as pigs and chickens, and for fish farmed in aquaculture. The feed is composed of ground up anchovy, which is an ocean fish that, prior to overfishing, was abundant in the eastern Pacific Ocean located along the coast of Peru (Figure 4.15). The industry is very important to Peru's economy, as it is the country's second largest export, following copper, and provides jobs for thousands of people. By the mid-1960s, the anchovy fishery was in trouble and a commission, the Instituto del Mar del Peru (IMARPE), was established by the Peruvian government, with funding from the United Nations, to obtain management advice from scientists. Scientists used Maximum Sustainable Yield (MSY) as an indicator to guide management.

(Continued)

Fish are a renewable resource: as long as they are fished at a rate that allows fish populations to be replenished fast enough, to keep up with losses from the population due to fishing, then fishing can continue sustainably into the future (Figure 4.16). The problem is that, if left unmanaged, most ocean fisheries are fished at a rate that is faster than the rate of replenishment. MSY has been used by fisheries managers to guide decisions about the amount of fish that can be caught in a given year, without harming the entire population of fish, so that fishing can continue the next year. MSY is a concept developed by natural scientists for the purpose of understanding population ecology, not for managing fisheries. When MSY has been used to manage fisheries, without considering the assumptions underlying its calculation, it has led to mismanagement of the fishery resource. This happened during management of the Peruvian anchovy fishery in the 1970s.

Figure 4.15 Anchovies are a salt-water forage fish found in many of the world's oceans, and those caught off the coast of Peru have provided the export-based fishmeal industry with raw material since the late 1950s.

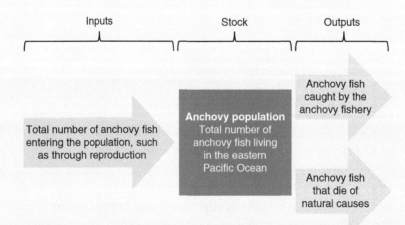

Figure 4.16 A renewable resource is used sustainably when the inputs into the resource's stock equal the outputs from the stock over time, such as the number of anchovy added to and lost from an anchovy population from one fishing season to the next.

There are several simplifying assumptions built into the MSY calculation. These assumptions are appropriate when MSY is used by population ecologists to understand natural system dynamics, but not when used by fisheries managers to determine the absolute number of fish that can be sustainably caught. One issue with MSY is that it is calculated using a logistic growth curve, a concept from population ecology, which assumes that the growth rate for a fish population depends only on time. In essence, this is saying that the number of fish in a population depends on time only. In reality, the population size at Time 2 is also dependent on initial population size at Time 1. For example, if there are two mating pairs of fish at Time 1, and each mating pair produces four offspring that survive, then the population size at Time 2 will be 12 fish (Figure 4.17a). However, if there are three mating pairs of fish at Time 1, and each mating pair produces four offspring that survive, then the population size at Time 2 will be 18 fish (Figure 4.17b). Thus, population size at Time 2 is dependent not only on time, but also on the initial population size at Time 1. Not accurately accounting for initial population size in MSY calculations often leads to the conclusion that there are more fish in the ocean than actually exist. When decisions are based on this indicator, it has led to the extraction of too many fish from the ocean.

Another assumption underlying the calculation of MSY is that the carrying capacity of an ocean environment for a fish population does not change. Carrying capacity is a concept from ecology that determines the total number of individuals, such as fish, that a given environment can support. In reality, carrying capacity for fish populations is not constant, but changes from year to year based on food availability, predator abundance, and many other factors. Assuming that it is constant from year to year can result in overfishing. For example, if a carrying capacity of 200 individuals is assumed, and the carrying capacity in reality is only 50, then MSY calculations are based on an overestimate of the number of fish that an ocean environment can support in a given year. In addition, carrying capacity is difficult to accurately determine for a fish population in the ocean. This is especially true for the

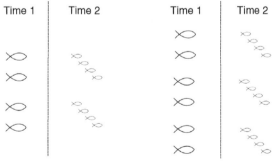

Figure 4.17 The growth rate of a fish population depends not only on time, as assumed in MSY calculations, but also on the number of individuals in the population initially.

ocean waters located along the coast of Peru, which is an upwelling zone experiencing wide fluctuations in carrying capacity due to variations in nutrient availability from year to year.

There are still more assumptions that make MSY use for fisheries management a bad idea, but they will not be discussed further in this book. The point is that MSY often overestimates the number of fishes that can be sustainably caught from an ocean environment. This is what happened in Peru, and it contributed to the eventual collapse of the anchovy fishery in the 1970s. At the time, Peruvian scientists determined the MSY to be 10 million tonnes of anchovy. Leading up to the 1970s collapse, anchovy populations were fished within this predicted MSY, but still became heavily exploited. Then, around the time of the fishery's collapse, an El Nino event created unfavorable climate conditions for the anchovy. The fishmeal industry also lobbied the Peruvian government to raise the catch limits beyond the MSY, and the government often complied. This made things even worse, since the MSY used as a starting point was already an overestimate of the number of fish that could be sustainably caught. Together, these environmental and social factors led to the eventual collapse of the fishery. It was not until the late 1980s and early 1990s that signs of recovery began to appear.

If policymakers and practitioners use MSY to make decisions, but do not understand or account for its assumptions, then bad decisions will be (and have been) made. The use of MSY to manage fisheries was not a problem of the Peruvian anchovy fishery alone. Its use is thought to have contributed to the demise of many fish populations around the world. MSY is an example of an indicator that reduces the complexity of an ocean fishery to a single number, which is appealing to decision makers and practitioners. However, there is often a tradeoff between offering a simple overview to decision makers and practitioners through a single number, and offering real-world complexity, which might result in indicators that are too extensive and cumbersome for decision makers to use. The need for indicators to be representative of real world complexity is discussed in further detail in the following pages.

The problem with too much simplification. Reducing complexity is not a problem unique to sustainability indicators, and is confronted any time people try to make sense of the real world. To understand real-world complexity, the world is often broken down into manageable pieces, with the goal of first understanding how each piece works on its own, and then how each piece functions as part of a larger system. This approach is known as *reductionism* and it is criticized as an insufficient means for understanding the complex systems of sustainability. There are so many pieces, and interactions among the pieces, that is it not possible, or productive, to study every single one. You may have heard this phrase in biology class: "The whole is greater than the sum of its parts." According to this statement, by reducing the complexity of a system when studying its pieces individually, a part of the picture is lost. Although it is important to understand different components or pieces of a system to make sense of its complexity, systems also need to be viewed holistically.

The world does need to be simplified to some degree when developing indicators, but how and to what degree is not straightforward. If it is not done right, it can result in inaccurate indicators that are then used by policymakers and practitioners to make decisions. One example of a recently developed aggregate indicator that is thought to have influenced national policies, but has also received criticism for being too simplistic, is the Environmental Sustainability Index (ESI). This index was developed jointly by researchers at Columbia and Yale. It rates each country's sustainability on a scale of 0 (least sustainable) to 100 (most sustainable). Figure 4.18 shows the ranking of the top 25 countries based on the 2005 ESI. The single number for each country shown in this graph is calculated using mathematical equations, that will not be explained here, by researchers at Yale University. Seventy-six indicators are used to calculate one number for the ESI, an aggregate indicator, for each country. The categories for these variables are listed in Table 4.1. Category I is composed of mostly state indicators, Category II of mostly pressure indicators, and Categories III through IV are largely response indicators. The advantage of compressing all of these individual indicators into one number is that it provides decision makers with an at-a-glance, comparative snapshot of environmental sustainability in countries around the world. The disadvantage is that it reduces complexity and creates misrepresentations in some cases.

Many of the misrepresentations paint developing nations in an unfairly poor light. For example, companies in many developed nations, with strict environmental regulations and labor laws, move factories overseas, where

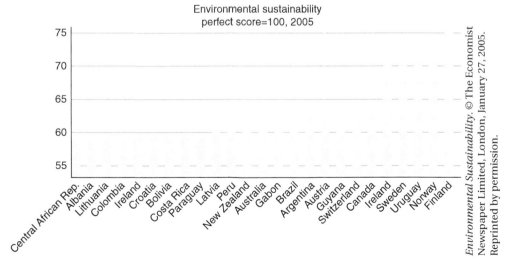

Figure 4.18 Ranking of top 25 countries according to the Environmental Sustainability Index.

Table 4.1 Indicators Used to Calculate the Environmental Sustainability Index (ESI)

Category I. Environmental Systems (17)
Air Quality (4): NO_2 concentration, SO_2 concentration, TSP concentration, indoor air pollution from solid fuel use
Biodiversity (5): % territory threatened, % threatened bird species, % threatened mammal species, % threatened amphibian species, National Biodiversity Index
Land (2): % total land area with low human impact, % total land area with high human impact
Water Quality (4): dissolved oxygen concentration, conductivity, phosphorus concentration, suspended solids
Water Quantity (2): freshwater availability per capita, internal groundwater availability per capita

Categoty II. Reducing Environmental Stresses (21)
Reducing Air Pollution (5): coal consumption, human NO_2 emissions, human SO_2 emissions, human VOC emissions, vehicle usage (all of these variables are per populated land area)
Reducing Ecosystem Stress (2): forest cover change rate, acidification
Reducing Population Pressure (2): change in project population, total fertility rate
Reducing Waste and Consumption Pressures (3): ecological footprint, recycling rates, hazardous waste generation
Reducing Water Stress (4): industrial organic pollutants, fertilizer use, pesticide use, % under severe water stress
Natural Resource Management (5): overfishing, salinized farmland, sustainably managed forest area, World Economic Forum (WEF) subsidy survey, agricultural subsidies

Category III. Reducing Human Vulnerability (7)
Environmental Health (3): intestinal infectious disease death, child death from respiratory illness, infant mortality
Basic Human Sustenance (2): proportion undernourished, % with access to an improved drinking water source
Reducing Environmental Disasters (2): deaths from floods and tropical cyclones, Natural Disaster Exposure Index

Category IV. Social and Institutional Capacity (24)
Environmental Governance (12): protected land, gasoline price, Rio to Johannesburg Dashboard, knowledge creation, International Union for the Conservation of Nature (IUCN) member organizations, local Agenda 21 initiatives, corruption, rule of law, civil and political liberties, WEF Environmental Governance, government effectiveness, democracy measure
Eco-efficiency (2): energy efficiency, hydropower and renewable energy production
Private Sector Responsiveness (5): Dow Jones Sustainability, Innovest EcoValue rating, ISO 14001 certifications, WEF private sector environmental innovation, participation in the Responsible Care Program of the Chemical Manufacturers Association
Science and Technology (5): Innovation Index, Digital Access Index, female education, school enrollment rate, # of researchers

Category V. Global Stewardship (7)
International Collaborations (3): environmental NGO member, funding of environmental projects/development, international agreements
Greenhouse Gas Emissions (2): carbon emissions per GDP, carbon emissions per capita
Reducing Transboundary Environmental Pressures (2): SO_2 exports, import of polluting goods and raw materials

they can pollute and take advantage of cheap labor (Figure 4.19). When this happens, the resulting environmental and social ills lower ESI ratings for developing nations and artificially inflate it for the developed nations that have moved their unsustainable practices to other countries. Another problem, not necessarily related to developing versus developed nations, is that many of the indicators used to calculate ESI are response indicators, such as protected land area, local Agenda 21 initiatives, and government effectiveness. Although such response indicators are useful for resolving sustainability problems, especially when used alongside state and pressure indicators, in the ESI these response indicators boost the final score. This is not a problem if the human actions measured by these indicators are actually advancing sustainability, but it becomes an

Figure 4.19 Foreign countries who outsource manufacturing to developing nations have artificially high ESI rankings when their factories pollute the environment, such as this factory on the Yellow River in China (*top*), and take advantage of cheap labor, such as by this child in a garment factory in Old Delhi, India (*bottom*).

issue when they are not, because they give credit where credit is not due. In this case, more information is needed to determine the actual effectiveness of these response indicators in advancing sustainability.

The ESI focuses on the environmental aspects of sustainability, but there are other indices that consider the social and economic dimensions: the UN's Human Development Index (HDI) and an alternative to the HDI, developed by Mongolian researcher Chuluun Togtokh, called the Human Sustainable Development Index (HSDI). The HDI considers economic and social indicators, such as income, education, social security, employment, literacy rates, life expectancy, technological innovation, and economic mobility. Thus, a HDI increase for a country indicates that the factors that improve human well-being have increased. However, as the economies of developing countries grow, fossil fuel-based energy use, and the accompanying CO_2 emissions, also increase. HDI does not account for the environmental aspects of development, such as CO_2 emissions. The HSDI adds to the HDI by accounting for CO_2 emissions, such that the more CO_2 emitted by a country, the lower the HSDI. The hope is that the HSDI ratings will encourage developing nations to grow sustainably. When country rankings for these two indices are compared, many less developed nations with low CO_2 emissions move up in rank or appear for the first time under the HSDI rankings (**Table 4.2**). These include Hong Kong, Spain, Singapore, Greece, and Andorra. Also, notice that many developed nations with high HDI, but poor environmental records, such as Australia, Canada, and the United States, drop to a lower rank or disappear from the list completely within the HSDI ranking. Finally, other countries performing equally well across the environmental, social, and economic domains of sustainability remain at about the same position. These include Norway, New Zealand, and Germany.

When all three indices are compared (Table 4.2), many countries appear in the ESI ranking that do not appear in either the HDI or the HSDI. These include countries in South and Latin America, Eastern Europe, and even one African country. These disparities raise questions: Which of these indices is the best measure of sustainability? Which will allow us to gauge where a nation currently stands with respect to sustainability, how far it has to go to reach sustainability, and which way it is headed (toward or away from sustainability), if the system is in fact shifting? It is hard to tell without knowing the details behind each index, which brings us back to square one. The purpose of these indices is to simplify complexity in order to aid in decision making. But can sustainability really equal one number, such as 42? Although these indices do provide societal motivation for countries to move toward sustainability by providing international rankings that shame some countries, and showcase others, they do not provide a holistic picture of sustainability nor consistently indicate where a country is with respect

Table 4.2 Comparison of the Top 25 Countries According to the aggregate indicators ESI, HDI, and HDSI

Ranking	ESI	HDI	HDSI
1	Finland	Norway	Norway
2	Norway	Australia	Sweden
3	Uruguay	Netherlands	Switzerland
4	Sweden	United States	Hong Kong, China (SAR)
5	Iceland	New Zealand	New Zealand
6	Canada	Canada	Israel
7	Switzerland	Ireland	Iceland
8	Guyana	Liechtenstein	France
9	Austria	Germany	Ireland
10	Argentina	Sweden	Germany
11	Brazil	Switzerland	Netherlands
12	Gabon	Japan	Denmark
13	Australia	Hong Kong, China (SAR)	Japan
14	New Zealand	Iceland	Austria
15	Peru	Korea	Spain
16	Latvia	Denmark	Slovenia
17	Paraguay	Israel	Korea
18	Costa Rica	Belgium	Italy
19	Bolivia	Austria	Belgium
20	Croatia	France	Singapore
21	Ireland	Slovenia	Finland
22	Columbia	Finland	United Kingdom
23	Lithuania	Spain	Greece
24	Albania	Italy	Canada
25	Central African Republic	Luxembourg	Andorra

to it. Is there a way to present indicators in a relatively simple way, that is easy for decision makers to understand, but that is also more representative of system complexity?

Alternatives to too much simplification. The best indicators find an appropriate balance between too much simplification and enough simplification, such that indicators are useful for policymakers and decision makers, but still adequately capture system complexity and provide holistic information. For sustainability, holistic information includes indicators about environment, society, and economy. A nonprofit organization called the Sustainable Society Foundation, which was founded in The Netherlands in 2006, developed the Sustainable Society Index (SSI). This index represents the three pillars of sustainability using 21 indicators, each representative of one of the three pillars of sustainability (**Table 4.3**). The SSI does not reduce sustainability to a single number by combining individual indicators into an aggregate indicator, as do the other indices already discussed in this chapter (ESI, HDI, and HSDI). Instead, two different strategies are taken. First, a type of graphic called a radar chart, also referred to as an AMEOBA diagram or spider chart, is used to shown all 21 indicators together (Figures 4.20a and 4.20b). Second, a single number

Table 4.3 Twenty-one indicators used for the Sustainable Society Index (SSI)

Human Well-being	Basic Needs	1. Sufficient food 2. Sufficient water to drink 3. Safe sanitation
	Health	4. Education 5. Healthy life 6. Gender equality
	Personal and Social Development	7. Income distribution 8. Population growth 9. Good governance
Environmental Well-being	Natural Resources	10. Biodiversity 11. Renewable water resources 12. Consumption
	Climate and Energy	13. Energy use 14. Energy savings 15. Greenhouse gases 16. Renewable energy
Economic Well-being	Transition	17. Organic farming 18. Genuine savings
	Economy	19. Gross domestic product 20. Employment 21. Public debt

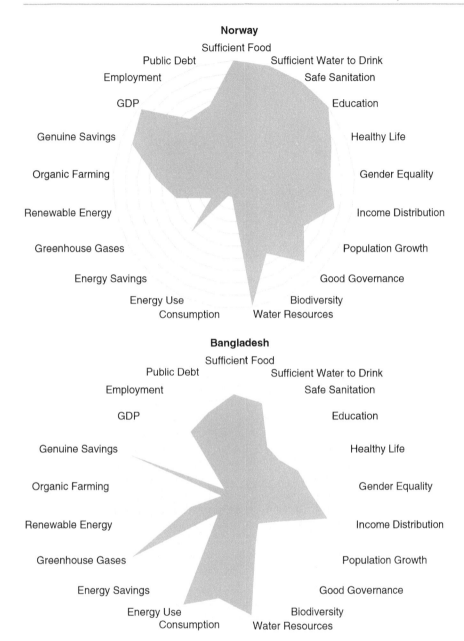

Figure 4.20 These radar diagrams for Norway (a) and Bangladesh (b), based on 2014 indicators from the Sustainable Society Index (SSI), provide a holistic "snapshot" of sustainability for each country.

is calculated for each of the three sustainability pillars (**Table 4.4**), rather than combining all aspects of sustainability into a single number. As such, the SSI generally does a better job at capturing complexity in a holistic way than do the ESI, HDI, and HDSI indicators.

Table 4.4 2014 SSI rankings for each of the three pillars of sustainability

Rank	Human Well-Being	Environmental Well-Being	Economic Well-Being
1	Finland	Guinea-Bissau	Norway
2	Iceland	Malawi	Switzerland
3	Germany	Nepal	Sweden
4	Japan	Mozambique	Denmark
5	Sweden	Central African Republic	Estonia
6	Denmark	Zambia	Luxembourg
7	Norway	Rwanda	Australia
8	Austria	Congo. Dem. Rep.	Czech Republic
9	Hungary	Burkina Faso	Finland
10	Ireland	Burundi	Slovenia
11	Czech Republic	Tanzania	South Korea
12	France	Sierra Leone	Latvia
13	Slovak Republic	Uganda	Austria
14	Netherlands	Kenya	Lithuania
15	Slovenia	Cameroon	Mexico
16	Malta	Ethiopia	Argentina
17	Lithuania	Zimbabwe	Slovak Republic
18	Spain	Liberia	United Arab Emirates
19	Switzerland	Madagascar	Poland
20	Estonia	Guinea	Uruguay
21	South Korea	Tajikistan	Romania
22	Belgium	Benin	Qatar
23	Canada	Gambia	Kuwait
24	Taiwan	Haiti	Turkey
25	Greece	Chad	New Zealand

Another way to capture complexity, but still provide easy-to-access information for policymakers and decision makers, it to create a user-friendly internet-based database called a dashboard. As implied by its name, a dashboard allows any person with an internet connection to be the "driver" or "pilot" when accessing information about indicators. An example of a dashboard is one created by The World Bank that can be used to monitor, for a world region or country, the 2016 Sustainable Development Goals (SDGs) using information from the World Development Indicators (WDI) database (http://datatopics.worldbank.org/sdgs/). For example, a policymaker or decision maker interested in knowing about the current state of SDG #1 (No Poverty) in Bangladesh, could go to the dashboard and view four different indicators of poverty using WDI data available for this country: (1) percentage of the population living on less than US $1.90 per day, (2) percentage of the population below the national poverty line, (3) percentage of the rural population below the national poverty line, or (4) percentage of the urban population below the national poverty line. The data are automatically displayed in a user-friendly graphic format. Dashboards are different from radar diagrams or indices because, rather than viewing data preselected and precompiled by an analyst, a policymaker or decision maker has more control over which data to view at any moment. Although this is not the case for the present format of the WDI SDG dashboard, a dashboard could be set up as a radar diagram for a more holistic assessment.

Section 4.2: Sustainability Indicators, a Case Study from the Primorska Region of Slovenia

Core Question: How are indicators developed through participatory processes involving relevant stakeholders?

To end this chapter, this section gives an example of successful sustainability indicator development, selection, and presentation by a group of stakeholders in Slovenia. It will encompass many of the characteristics of effective sustainability indicators discussed in Section 4.1.2. The idea of establishing boundaries for each indicator, instead of a single number, is a new concept that will be introduced in this section. Indicator boundaries are tools that can be used when multiple stakeholders collectively determine indicators through participatory processes. Boundaries are useful in this case because the different beliefs, values, and preferences of stakeholders will influence their definitions of sustainability. They provide a compromise about where, exactly, sustainability is located with respect to each indicator.

Section 4.2.1—The Participatory Process of Indicator Development

Key Concept 4.2.1—Sustainability indicators, and their boundaries, are developed for a specific context through a participatory process, involving several stakeholders who collectively develop indicators by sharing knowledge and assessing contentiousness and feasibility.

The story begins in the Primorska region of the country of Slovenia (**Figure 4.21**). This region is located in western Slovenia and includes a coastal region and inland mountainous peaks up to about 3,000 feet in altitude (**Figure 4.22**). When it comes to sustainable development,

Figure 4.21 The Primorska region of Slovenia is indicated here by a red star.

Figure 4.22 The Primorska region of Slovenia has coastal towns, such as the Port of Koper located on the Adriatic Sea (*left*), and mountainous regions containing lakes, castles, and islands, such as Lake Bled located in the Julian Alps of northwestern Slovenia (*right*).

the region faces many issues, such as development pressures from tourism and the private sector, urbanization, solid waste disposal, drinking water quality, sewage drainage and treatment, and public transportation. In order to address these problems, the *Imagine* project was carried out in 2005 as a series of five workshops over a six-month period that involved the participation of over 50 stakeholders. Stakeholders included representatives from different groups, including nongovernmental organizations (NGOs), private sector interests, the general public, and local, regional, and national governments. A major focus of these workshops was to engage stakeholders in defining indicators to track sustainable development of the coastal Primorska region.

Stakeholders began indicator development by breaking into groups representative of common interests. Each group was told to brainstorm a list of sustainability indicators, no matter how outrageous or controversial they may appear to other interest groups. Expert stakeholders, such as scientists and policymakers, were more likely to come up with quantitative indicators similar to the MSY, ESI, or HDI. Local people devised indicators focused on their quality of life and general human welfare, many of which were of a more qualitative nature. Stakeholders with a financial interest in sustainable development, such as touristic business interests, banks providing loans, and some sectors of the government, were more likely to formulate indictors of profit and economic activity. After this initial brainstorming stage, the groups had collectively developed 76 possible indicators of sustainability. Each group's indicators covered different aspects of sustainability, based mostly on their own interests. Including diverse stakeholders in the indicator development process ensured representation of many interests. It also provided a working list of indicators that could be used to develop a holistic view of the systems of concern by including environmental, social, and economic aspects. The next steps were to assess indicator contentiousness and feasibility.

Contentiousness and **feasibility** are assessed by presenting all indicators to all stakeholders. Some will be controversial and there can be much debate at this stage. This is useful, however, because stakeholders will learn about each other's interests and viewpoints during this process. Indicators devised by expert stakeholders, such as MSY or ESI, may mean nothing to non-expert stakeholders. Thus, this is a time for clarification and understanding of certain indicators by various stakeholder groups, followed by acceptance or rejection. It is advantageous to have all stakeholders involved when checking the feasibility of proposed indicators because some stakeholders will know what information is important to have, while others will know which information is actually possible to obtain. For example, an ecologist concerned with water quality degradation due to heavy metals leaching out of solid waste might want to know the total amount of soluble arsenic and lead in garbage dumped into landfills each month. Public waste officials know that this

---

contentiousness the degree of controversy or disagreement surrounding an indicator, in terms of its accurate representation of sustainability, resulting from the different beliefs, values, and preferences of diverse stakeholder groups

---

feasibility the degree to which an indicator can be successfully used to track sustainability based on issues such as data availability or acquisition

information is not available, but they can provide scientists with information that they do keep track of, such as the number of kilograms of garbage collected per person over a time period. Scientists can use this information, plus their own measurements of the average soluble arsenic and lead content of a typical kilogram of garbage, in order to get the data they need. Other feasibility measures, such as past and future data availability, resources required to obtain data, time and space scales, and understandability of data by all stakeholders are assessed at this point. The goal is to narrow the list of indicators down to about 20 to 30.

It is important to note here that, in deciding which indicators to use, stakeholder interests are taken into account. However, the alignment of these collective interests with sustainability must occur, as sustainability is not a "free for all." Something is sustainable when it does not diminish the well-being of people living today, or in the future, and when it preserves the Earth's natural systems. Thus, during stakeholder negotiations, indicators must be selected with these guidelines in mind.

Indicators were narrowed down by stakeholders to a list of 20 that broadly covered the environmental, social, and economic pillars of sustainability (see the domain column in **Table 4.5**). "Tourism" is included as a separate domain because the influence of tourism is very large. There are two lists of indicators for two different regions of Primorska, Carst Region and Coast Region, because it was found during the workshops that the two regions face very different problems, and that the drivers regulating the two regions were not the same.

The next step for stakeholders was to collectively define the boundaries for each sustainability indicator. Defining boundaries can be particularly difficult because, as has been already mentioned several times in this textbook, there is no one definition of sustainability for everyone. At the end of this process, there must be agreement among stakeholders about the upper and lower limits of sustainability for each indicator. Those that were agreed upon by Primorska stakeholders are shown in Table 4.5. Because of the differences between the regions, when common indicators were used for both, there were often different boundaries. For example, "Number of beds per 100 inhabitants" is an indicator of the ability of a region to host tourists, and was used to track the sustainability of tourism in both regions. However, the boundaries were different for the two regions, with the less populated Carst region's boundary ranging from 5 to 8 beds, and the more populated Coast region's from 30 to 35. This is because the two regions have different capacities for hosting tourists. Using the same boundary for both regions would not make sense within the specific contexts of each. This is just one of the many possible examples illustrating why sustainability must be defined within a local context.

Finally, all of this information was brought together as a holistic presentation of the state of each region, relative to sustainability, by

Table 4.5 Sustainability Indicators Developed by Primorska, Slovenia stakeholders.

Carst Region

Indicator	Domain	Boundaries Lower	Boundaries Upper	Indicator Unit
1. Public waste removal	Environment	12	20	Kg per population
2. % households connected to sewer system	Environment	80	90	%
3. % of population working	Social	40	70	%
4. Daily commute for active working force	Social	1,500	2,500	Rate
5. Aging Index	Social	35	50	Rate
6. % of individuals with higher eduation	Social	20	30	%
7. Number nights of tourists per inhabitants	Tourism	250	350	Nights/100
8. Number of beds per 100 inhabitants	Tourism	5	8	Beds/100
9. Gross income tax base per capita	Economy	105	130	Index
10. Business net profit per loss per employee	Economy	300	600	Ln SIT '000

Coast Region

Indicator	Domain	Boundaries Lower	Boundaries Upper	Indicator Unit
1. Urbanization rate	Social	60	70	%
2. % households connected to sewer system	Environment	75	90	%
3. Drinking water quality (% unsuitable)	Environment	0	2	%
4. Sea water quality in public baths	Env/Tourism	90	100	%
5. % coastal land with regulation	Tourism	30	50	% of land
6. Investment in coastal nature protection	Economy	50	100	MIO SIT
7. Employment structure	Economy	2	3	Nights/100
8. Number of beds per 100 inhabitants	Tourism	30	35	Beds/100
9. Number nights of tourists per inhabitants	Tourism	3,000	4,000	Nights/100
10. Educational structure of inhabitants	Social	20	30	%

displaying indicators in a radar diagram (Figure 4.23), as described under Characteristic 6 of Section 4.1.2. This type of diagram provides a richer view of the system's sustainability because it allows you to see relatively quickly what, exactly, is unsustainable about a system. For example, in the Carst region, tourism indicators show that tourism is below the sustainable level, economy is at the lower limit, social indicators are within or beyond the boundaries, and one aspect of the environment exceeds the upper boundary while the other is below it (Figure 4.23). If two environmental indicators were averaged into one number, a number between the two that indicates environmental sustainability might be obtained. However, in reality, the environmental situation in the Carst region is not sustainable (according to these indicators) because one is too high and the other is too low.

At this point, you may be wondering about the meaning of the upper boundary for sustainability. After all, an upper boundary on sustainability represents a condition beyond which a system is too sustainable for a given indicator. This may seem silly because you may think: How can too much sustainability be undesirable? One answer to this question is a practical matter related to resource distribution and focus. For example, in the Carst region (Figure 4.23), the daily commute for the active working force (Indicator 4) exceeds sustainability, but the percentage of households connected to a sewer system (Indicator 2) has not yet reached sustainability. This analysis, if used by municipalities for decision-making, suggests that limited resources (money, time) should be focused on projects related to improving sewer systems instead of improving the transportation infrastructure for commuting. Another reason that an indicator falling above an

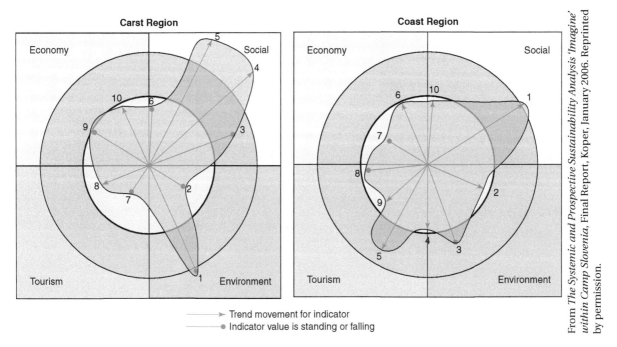

Figure 4.23 Radar diagrams for each region of Primorska, with sustainability boundaries shown in brown, for the four sustainability domains selected by stakeholders. The numbers shown in the diagrams correspond with the indicators, defined for each region, described in Table 4.5.

upper boundary can be problematic is that, sometimes, "too much of a good thing" can be harmful. For people, too much food, coffee, or wealth can be harmful. Too much economic growth can be harmful to other aspects of sustainability, such as people and the environment. An indicator that falls above an upper boundary can alert us to these possibilities.

These diagrams can also be used to track the state of sustainability over time. Three AMOEBA diagrams are shown in Figure 4.24 for three different years. They combine all 20 indicators (10 from each region) into one diagram to provide a snapshot of the region's overall sustainability. From these diagrams, overall trends over time can be observed to gauge whether the system is headed toward or away from sustainability. For example, since 1992 the economic sector (as represented by the four indicators used here) has been slowly approaching the lower sustainability boundary. The social sector has continued to move further beyond the upper boundary. For the tourism domains, sustainability has increased according to some indicators, but according to others has remained about the same. Environmental sustainability increased from 1992 to 1996 and then decreased from 1996 to 2003. It is important to note here that none of these indicators is necessarily an accurate representation of reality in terms of absolute values of measurement. Rather, each indicator should be viewed as a tool for making *relative* observations of trends in each indicator over time.

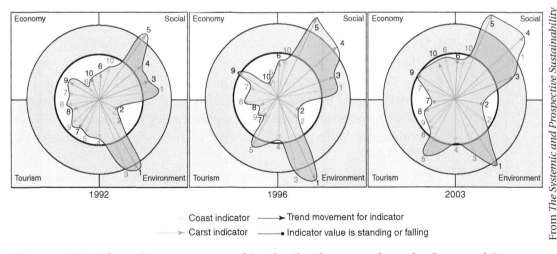

Figure 4.24 These diagrams are combined radar diagrams, from the Coast and Carst regions, intended to provide a picture of the entire Primorska region. Sustainability boundaries are shown in brown and numbers within the diagrams correspond with the indicators, defined for each region, described in Table 4.5.

This chapter explained how indicators can be used to analyze the current state of a system by providing information about the drivers regulating a system, causal relationships, and the factors shaping stakeholder action. They are also used to help track whether systems are changing and, if they are, whether they are moving toward or away from sustainability. Finally, indicators are useful for gaining insight into whether responses to problems are improving a system's sustainability. Indicators are useful not only for the current state, but also for thinking about the future through future scenarios and visioning (Chapter 7) and for understanding transitions (Chapter 8) toward sustainability. Before moving on to thinking about the future in Chapter 7, and how current systems might be guided toward a more sustainable future in Chapter 8, it is important to understand more about systems in general. The next two chapters are focused on building your understanding of the dynamic behavior of complex systems. Awareness of how complex systems behave will help in your overall understanding of current systems, future systems, and how to guide systems toward a more sustainable future.

Bibliography

Bell, S., and S. Morse. 2008. *Sustainability Indicators: Measuring the Immeasurable?* London: Earthscan.

Binder, C. R., et al. 2010. "Considering the Normative, Systemic, and Procedural Dimensions in Indicator-Based Sustainability Assessments in Agriculture." *Environmental Impact Assessment Review* 30: 71–81.

ESI Methodology. Accessed July 1, 2012. www.yale.edu/esi/a_methodology.pdf

Fraser, E. D. G., et al. 2006. "Bottom Up and Top Down: Analysis of Participatory Processes for Sustainability Indicator Identification as a Pathway to Community Empowerment and Sustainable Environmental Management." *Journal of Environmental Management* 78: 114–27.

Henderson, H. 1991. *Paradigms of Progress.* San Francisco: Berrett-Koehler.

Henderson, Hazel. 1991. *Paradigms of progress: life beyond economics*. Indianapolis, IN, USA : Knowledge Systems, 293 pp.

Laws, E. A. 2006. El Nino and the *Peruvian Anchovy Fishery, University Corporation for Atmospheric Research*, 58 pp. Accessed September 30, 2016. https://www.ucar.edu/communications/gcip/m12anchovy/m12pdf.pdf.

MacAskill, W. 2015. *Doing Good Better: How Effective Altruism Can Help You Make a Difference.* Gotham Books: New York, NY, 258 pp.

Maher, I. 2006. *The Systemic and Prospective Sustainability Analysis 'Imagine' within CAMP Slovenia.* Final Report, Regional Development Agency, South Primorska. Accessed June 30, 2012. www.papthecoastcentre.org/pdfs/pac_slovenije_final.pdf.

Saha, D and M. Muro. 2016. *Growth, carbon, and Trump: State progress and drift on economic growth and emissions 'decoupling,' Report of the Brookings Institute.* Accessed May 9, 2017. https://www.brookings.edu/research/growth-carbon-and-trump-state-progress-and-drift-on-economic-growth-and-emissions-decoupling/

Wiek, A. 2010. "Analyzing Sustainability Problems and Developing Solution Options—A Pragmatic Approach." Working Paper, Sustainability Transition and Intervention Research Lab, School of Sustainability, Arizona State University, Tempe, AZ.

Wiek, A., and C. Binder, C. 2005. "Solution Spaces for Decision-Making—A Sustainability Assessment Tool for City-Regions." *Environmental Impact Assessment Review,* 25 (6): 589–608.

End-of-Chapter Questions

General Questions

1. In the following hypothetical scenarios, pick the set of indicators (Option 1 or 2) that are a better measure of sustainability and explain why.

    a. For low-income city dwellers, safe and affordable housing can be hard to find in a large metropolitan area. Which is a better set of sustainability indicators that can be used to monitor this situation?

    Option 1: Percent of new housing units within a ½ mile of public transit, number of housing units located on sites with industrial contamination, and the neighborhood crime rate to average monthly rent ratio.

    Option 2: Vacancy rates of city housing units, annual rate of rent increase, and government housing subsidy increase relative to inflation.

    b. Although it is difficult to establish causation between global climate changes and an individual local weather event, the two are associated to some degree. Specifically, human-caused climate change will likely bring higher intensity hurricanes, with consequences for human societies. Which is a better set of sustainability indicators that can be used to monitor this situation?

Option 1: Insurance claims paid out each year by the federal government, changes in unemployment due to natural disasters, and costs of repair to public infrastructure.

Option 2: Changes in unemployment due to natural disasters, durations of refugee situations, and rate of fossil fuel emissions.

c. Hydraulic fracking is a procedure used to recover natural gas by injecting a pressurized mixture of water, sand, and chemicals deep into the ground. Which is a better set of sustainability indicators that can be used to monitor this situation?

Option 1: Groundwater contamination rates, depletion of freshwater resources, and air quality deterioration as a result of fracking operations.

Option 2: Groundwater contamination rates, number of jobs created per fracking site, and the number of people with immune system disorders developed in a fracking region.

2. Use the six characteristics of effective indictors found in Section 4.1.2 to explain why or why not each of the following hypothetical scenarios is a good indicator.

a. As of late May 2013, news stories have claimed that the Food and Drug Administration (FDA) in the United States was on the verge of approving genetically modified salmon. GMOs, or Genetically Modified Organisms, in general are very controversial. Thus, in order to add to the social and economic indicators already developed for this situation, the FDA convened a panel of scientists to develop indicators that could be used to monitor the environmental impacts of genetically modified salmon (GMS). One concern with GMS is that they will harm native, non-genetically modified fish populations. As a result, scientists have decided to monitor salmon populations along the western coast of the lower 48 states. This includes the states of Washington, Oregon, and California, where about 5% of chinook, coho, and sockeye salmon populations thrive. Most native salmon populations survive in Alaska, but these data are more difficult and more expensive to obtain due to the remote location. They plan to collect the following data on native fish populations twice per year: genetic makeup and population dynamics. The population dynamics data are relatively easy to obtain, whereas the genetic data are more difficult to obtain due to lengthy and costly laboratory analyses. The population dynamics data providing overall population density, natality rates, and survivorship will be reported to policy makers. In addition to genetic and population data, it would be helpful to know about any changes in the ways that native salmon populations interact with populations of other species. However, a community ecologist who studies such interactions will not be available for data collection.

b. About one-third of people living in urban areas worldwide live in slums. Young children are among the most vulnerable members of these populations, experiencing hardships such as malnutrition, disease, and lack of basic social services, such as clean water, sanitation, and education. The United Nations Human Settlements Programme (UN-Habitat) is studying the sustainability of urban slums. In order to add to a list of environmental and economic indicators, UN-Habitat is developing a suite of social indicators for young children living under slum conditions. It plans to assess the state of children living under slum conditions worldwide by collecting data for several social indicators in urban areas for the following nine countries: Cambodia, Nepal, India, Senegal, Rwanda, Uganda, Haiti, Honduras, and Brazil. They have obtained funding to collect these data over a three-year period to start, with a possibility to continue funding. In order to determine what types of social indicators to collect data for, UN-Habitat visited one city in each of the nine countries and talked to local people living in slum conditions. To track malnutrition issues, nutritional status will be tracked using the following indicators: vitamin A, iron, zinc, and protein deficiencies. Disease and medical issues are other threats to children living in slums, especially diseases such as diarrhea and pneumonia, and complications during birth. These are among the leading causes of death. The second leading cause of death to young children is traffic accidents. Thus, UN-Habitat decided that mortality rates for children aged 0–14 would be an effective indicator, along with the associated causes of death when available. One reason that the death of young children goes unreported in urban slums is that children are not registered at birth and, therefore, their existence not known. One solution to this problem has been registration programs. Thus, UN-Habitat will use registration rates (normalized to population growth) as another indicator. Lack of access to clean water is another major issue. One indicator of access to water is proximity to government water infrastructure. If this is not available, which it often is not in urban slums, families must wait in line to pay for clean water. One solution to waterborne diseases has been efforts to vaccinate young children, which can help prevent vaccine-preventable diseases. Another basic social service lacking in urban slums is education, so UN-Habitat will use total years of schooling as another indicator. Together, the indicators described here will provide a way to monitor the social domain of children in urban slums.

Project Questions

1. Create a set of nine indicators (three environmental, three social, and three economic) that could be used to gather information about your sustainability problem. Note whether each indicator is a state, pressure, or response indicator by circling one of the three in the first column of the table. Be sure to give a detailed explanation of each indicator, note the measurement units for each indictor if it is quantitative, note the present-day value of each indicator, and determine the lower and upper bounds that define the range within which each indicator must fall for a given domain to be considered sustainable.

Indicator	Domain	Explanation	Units	Present Value	Lower Bound	Upper Bound
1. state pressure response	Env					
2. state pressure response	Env					
3. state pressure response	Env					
4. state pressure response	Soc					
5. state pressure response	Soc					
6. state pressure response	Soc					
7. state pressure response	Econ					
8. state pressure response	Econ					
9. state pressure response	Econ					

2. Plot the indicators you selected in Question 1 of this section on the radar diagram below to get an overall snapshot of the degree of sustainability of the current state of your system.

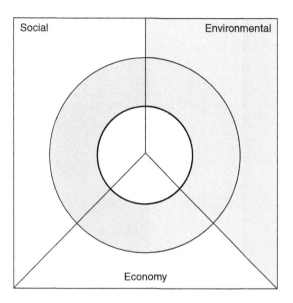

Chapter 5
RESILIENCE AND PATTERNS OF CHANGE

Students Making a Difference

Climate Change and the U.S. Military

Wes Herche is a research scientist and geospatial analyst with an MBA who studies sustainability science. He has collaborated on many projects with the Global Security Initiative (GSI) at Arizona State University, including a major effort focused on climate security in the Niger River Basin (NRB) in Africa. With the creation of U.S. AFRICOM in 2007, this region is an area of growing strategic importance for the U.S. military. The military has to decide where to make investments in operational capabilities, and determine what areas are more likely to experience conflict or struggles. However, in today's world, there is so much information at our finger tips that it is challenging to put it all together in a way that supports good decisions. With the help of many physical and social scientists, the NRB project created decision support tools, in the form of interactive web-based platforms, to help military and intelligence analysts visualize patterns of change in this region, related to climate change, over long- and short-time periods. According to Wes, the overwhelming amount of information available is one reason that tipping points are hard to predict: "You don't know when these tipping points—like a regime change in an African nation—will be reached. Political shifts, or food and water shortages, could lead to strife or full scale riots, and then there is no turning back. Maybe the U.S. military is being deployed somewhere or maybe we are in an all-out war. You can't tell when those exact points are, but you're trying to understand the long-term factors and detect the little bubbles before the pot boils because you can't always see what's going on at the surface." The NRB project set out to build the capacity to understand and detect "little bubbles", so that the military could anticipate crises and respond effectively.

The Department of Defense views climate change as a "force multiplier," and considers its impacts is critical to their mission. The interdisciplinary GSI team did not have fully comprehensive data sets to work with, this is almost never the case, but they did have a lot of data on water. They knew about historical, and more recent, rainfall patterns, and fluctuations in river discharge. Historically, in other regions, water has been tied to food and economics. Changes in rainfall could lead to changes in crop yields, which could lead to spikes in food prices, or severe and persistent food shortages. The GSI team pinpointed these areas and where they overlapped with areas of political instability. Although they could not predict the future, they could identify regions where problems are more likely to arise. Communication scientists worked with computer scientists to target areas of concern by trawling massive data sets for keywords, such as "climate change," "food shortage," or "water," across sources of publically available information, including internet forums, the local press, and even social media such as Twitter. They used this information to anticipate the "little bubbles" before a "boiling pot" of conflict and unrest erupted in a society.

In this chapter, you will learn about patterns of change that occur in the systems in which sustainability problems are commonly embedded. These systems are highly interconnected and, as a result, behave in ways that make them difficult to understand. As he described how changes in a country's water availability could affect everything from food availability and prices, to energy production, to international trade, Wes makes the important point that you can't fully understand such complex systems, but that you still have to be aware of them: "Things are never isolated. For every lever you pull, another lever moves somewhere else, or multiple ones. You can't understand it. You can't hold it all in your head all at one time. But you have to at least be cognizant of the fact that that is the type of system you are operating in."

Core Questions and Key Concepts

Section 5.1: Regime Shifts, Thresholds, and Resilience

Core Question: How do systems change from one regime to another?

Key Concept 5.1.1—There is an important distinction between regular fluctuations in a system and the type of transformation that occurs when a system shifts to a new regime.

Key Concept 5.1.2—Whether a system changes linearly or nonlinearly has consequences for detecting change, and also for influencing the system's movement toward or away from sustainability.

Key Concept 5.1.3—Resilience is a concept used for understanding how susceptible a system is to a regime shift. It can be assessed based on the proximity of a system to a threshold, as a result of driver interactions, and is often higher when systems contain diversity.

Section 5.2: Feedbacks and Resilience

Core Question: Why do systems cross thresholds into new regimes and experience fluctuations within a given regime?

Key Concept 5.2.1—It is the shifting dominance among stabilizing and reinforcing feedbacks that determines the state of a system at any given time and whether or not it will shift to a new regime.

Key Concept 5.2.2—One major reason for system fluctuations is the continual interaction among system components through feedbacks over time.

Section 5.3: Adaptive Cycles and Resilience

Core Question: How is the resilience of SESs affected over long time scales?

Key Concept 5.3.1—Complex adaptive systems pass through four general phases that affect their resilience, which influences their response to disturbances affecting the system.

Key Concept 5.3.2—Adaptive cycles have been applied to natural systems to understand ecological succession, and to economic systems to understand boom-and-bust cycles.

Key Terms

regime (aka. alternative stable state)
threshold (aka. tipping point)
feedback
resilience
adaptive cycle
regime shift
alternative stable state
fluctuation
state
nonlinear change
critical transition
desertification
functional diversity
response diversity
adaptive capacity
albedo
shifting dominance
reinforcing feedback (aka. positive feedback)
stabilizing feedback (aka. negative feedback)
niche

> *"Suppose you are in a canoe and gradually lean farther and farther over to one side to look at something interesting underwater. Leaning over too far may cause you to capsize and end up in an alternative stable state upside down.... [R]eturning from the capsized state requires more than just leaning a bit less to the side. It is difficult to see the tipping point coming, as the position of the boat may change relatively little up until the critical point. Also, close to the tipping point, resilience of the upright position is small, and minor disturbances such as a small wave can tip the balance.... Could coral reefs, the climate, or public attitude really tip over like a canoe?"*
>
> —Marten Scheffer, Critical Transitions in Nature and Society, p. 13

If sustainability is fundamentally about preserving the conditions necessary for the survival of prosperous human societies, as stated in the opening sentence to Chapter 1, then resilience is about finding ways to maintain the conditions needed for prosperity in the face of unanticipated disturbances and ongoing change. In this sense, sustainability and resilience are complimentary. In another sense, they are at odds with each other. Sustainability is not something that can be pinned down exactly: the definition of "prosperous," and the "conditions" needed to achieve this state of being, are far from straightforward and will depend on who you ask. Sustainability is inherently a normative concept: it is a statement of how things *ought* to be. (But sustainability is not completely open to interpretation and follows this general guideline: Something is sustainable if it preserves Earth's natural capital and does not diminish the well-being of people living today or in the future.) In contrast to sustainability, resilience is something that, at least in theory, can be specified.

This chapter describes the ways that resilience can be specified by explaining how the natural and human systems of concern to sustainability change, sometimes rapidly like the canoe in the opening passage to this chapter, from one **regime (aka. alternative stable state)** to another. This understanding involves insight about how **thresholds (aka. tipping points)** are ultimately crossed (sometimes unpredictably) as systems shift from one regime to another, and also why it can sometimes be hard to go back. These insights are gained by learning how **feedbacks** among system components regulate a system by determining its ability to absorb disturbances, by either changing the system or keeping it the same. The ability of a system to absorb disturbances, but still maintain the same structure, function, and feedbacks, is referred to as **resilience**. In other words, it is the ability of a system to remain in its current regime in the face of disturbances. The ability of a system to do this, or not, depends in part on changes that occur in a system as it passes through different phases of cyclical long-term change called **adaptive cycles**.

Resilience is a concept that accepts change in socioecological systems (SESs) as the rule, rather than the exception, and as a fundamental characteristic of how the world functions and evolves over time. A resilience view of the world has implications for how the sustainability of SESs may actually be brought about, which involves a change from traditional management styles that strive to avert change and diminish variability in systems. Instead, resilience-based approaches anticipate and work with change in attempts to guide SESs toward sustainability. In order to understand how this may be achieved, one must first understand how socioecological systems behave. This chapter, and the next, provide an in-depth look at tools for thinking about complexity in SESs, as outlined in Section 2.1.2 of Chapter 2, which can be used to increase understanding about these systems and guide them toward sustainability. Both chapters build on concepts introduced in Section 2.1.2, such as what is meant by a system, its boundaries, its components, the interactions among components, and feedbacks. It may be helpful to review that section before proceeding.

An exploration of the tools for thinking about complexity, presented in this chapter, will improve your understanding of how systems arrived in their current state and how they might develop into the future. Understanding the dynamic interactions among internal system components, and how they are influenced by external disturbances, will help you become a better analyst of sustainability problems, as you work to overcome them. The examples used in this chapter are mostly drawn from natural systems because regime shifts, thresholds,

---

regime (aka. alternative stable state) the dynamic and constantly changing, yet characteristic, pattern of conditions under which a system can exist and by which it supports certain functions or purposes.

threshold (aka. tipping point) the point beyond which a system shifts nonlinearly to a different regime.

feedback a specific type of interaction among system components, or with the external environment, such that the outcome of some process returns to affect the factor that originally initiated the process.

resilience the ability of a system to absorb disturbances and remain in its current regime with the same conditions, structure, function, and identity.

adaptive cycles four different phases of cyclical long-term change characteristic of many complex systems.

feedbacks, adaptive cycles, and resilience have been well-studied in these systems. However, keep in mind that these concepts can also apply to human systems and some examples are given. Ultimately, the concepts presented in this chapter, and the next, are a set of tools for understanding the complexity of SESs, and resolving the sustainability problems embedded in them.

Section 5.1: Regime Shifts, Thresholds, and Resilience

Core Question: How do systems change from one regime to another?

This section begins by defining what is meant by a regime and distinguishes change in systems that occur during **regime shifts** from fluctuations and changes happening within one type of regime. This is followed by a description of the different types of change—linear and nonlinear—that can occur in SESs, and the implications of this for recovery from an unsustainable regime or movement toward one that is more sustainable. Finally, the concept of resilience is used to explain why some SESs may be more susceptible to regime shifts than others.

Section 5.1.1—Regimes

Key Concept 5.1.1—There is an important distinction between regular fluctuations in a system and the type of transformation that occurs when a system shifts to a new regime.

In the last two chapters, the concepts important to understanding the current state of a system were explored. However, systems can exist in more than one state and the concepts from the last two chapters also apply to system states other than the current state. The different possible states are often referred to as **alternative stable states**. However, in this book, they are called regimes. Different possible regimes for several specific natural and human systems are given in **Table 5.1**. The term regime is used rather than alternative stable state because the words *stable* and *state* imply that a system is static, stagnant, or unchanging. In reality, socioecological systems are dynamic, fluctuating, and constantly changing within a given regime. The trick is to sort out the regular patterns, or **fluctuations**, and characteristic behavior of a system *within* a given regime, which determine its current **state**, from those patterns or behaviors that cause systems to change to different states or regimes, which may or may not be sustainable. This is, in large part, the focus of this chapter.

---

regime shift a typically nonlinear change process, by which a system transitions from an existing regime to a new regime, under which the system supports different functions or purposes than before.

alternative stable state a term often used in place of regime.

fluctuation periodic changes in the state of a system within a given regime that do not constitute a regime shift.

state the conditions under which a system exists at any given time.

Table 5.1 Possible regimes for a few natural and human systems

System Type	Regime 1	Regime 2
Coral reefs	Coral-dominated	Algae-dominated
Shallow lakes	Clear, vegetated	Turbid, non-vegetated
Governance systems	Democracy	Authoritarian
Economic well-being	Wealth	Poverty
Earth system	Glaciated	Non-glaciated
Drylands	Grassland with cattle	Shrubland without cattle
Coastal kelp forests	Dense kelp forests	Sparse kelp forests
Savanna	Woodland	Grassland

When resolving sustainability problems, preventing regime shifts from sustainable to unsustainable regimes (e.g., see the lake eutrophication example, **Box 5.1**) is not the only concern. Oftentimes, promoting change from undesirable regimes to desirable ones (e.g. see the governance example, Box 5.1) is the focus. Thus, the concepts of regimes and regimes shifts are useful for understanding how systems arrived in their unsustainable state, but also for figuring out how they might be returned to a sustainable state.

BOX 5.1   REGIME SHIFTS IN LAKES AND GOVERNANCE SYSTEMS

Lake Eutrophication: Clear, vegetated regime to turbid, unvegetated regime. Shallow lakes generally have two different regimes (Table 5.1). A healthy lake has clear water and contains abundant aquatic vegetation, which provides a habitat for many organisms such as zooplankton, fish, and birds (Figure 5.1a). When eutrophication occurs, a lake shifts into an alternate regime dominated by microscopic plankton and turbid water (Figure 5.1b). The turbid water is caused by abundant plankton that does not allow light to enter the water, which inhibits photosynthesis by aquatic plants. This results in a lake with minimal aquatic vegetation and very little habitat for other organisms. This regime is not desirable for sustainability because it has low biodiversity. Although the two regimes of *clear, vegetated*

(Continued)

Regime 1: A clear vegetated lake supports biodiversity

Regime 2: A turbid non-vegetated lacks biodiversity

Ricardo de Paula Ferreira / Shutterstock.com; Ricardo de Paula Ferreira / Shutterstock.com; Ricardo de Paula Ferreira / Shutterstock.com; Ihor Bondarenko / Shutterstock.com

Figure 5.1 Healthy lakes can support high levels of biodiversity and habitat for many animals (*left*), but unhealthy lakes that have undergone eutrophication (*right*) exist in regimes that support low levels of biodiversity.

(Regime 1) and *turbid, non-vegetated* (Regime 2) represent distinct system conditions, such that one supports biodiversity and the other does not, each regime experiences changes that do not cause a regime shift but do alter its *state*. For example, both lakes experience seasonal changes in temperature, and this is a pattern that repeats itself every year: in the summer, the temperatures are warmer and in the winter they are colder. These changes are *fluctuations* that alter the state of the system, but do not alter the overall condition of each lake so much that there is a shift toward a different regime: Regime 1 still supports biodiversity and Regime 2 does not, regardless of seasonal temperature fluctuations.

Governance Systems: Authoritarian government to democracy. Two types of governance systems for a given nation might include a democracy and an authoritarian government (Table 5.1). Democracy is a type of government where all citizens of a nation have the opportunity to collectively decide on laws, policies, and other regulations (Figure 5.2a). In contrast, in an authoritarian government, citizens do not have a voice because laws are dictated by a small group of powerful ruling elite without public input (Figure 5.2b). A democracy and an authoritarian government represent different governance regimes. In each of these regimes, the process by which laws are made is distinct. In the democracy (Regime 1) laws are shaped by citizens, whereas in an authoritarian government (Regime 2) laws are dictated by a small group. Laws and government policies may change through time within each regime, and these cause *fluctuations* within each system that change the system's state, but the processes by which laws and policies change remains the same within each, as long as the system remains within that regime. It is, however, possible for a system to experience a more pervasive type of change, beyond normal fluctuations, that causes the system's regime and the underlying processes by which laws and policies are made, to also change. Drivers often cause this type of change. For example, a driver such as civic engagement through protest could, over time, cause a governance system to shift from an authoritarian to democratic regime.

Regime 1: In democratic governments, citizen vote for elected officials to represent their voice in decisions related to laws and policymaking.

Regime 2: In authoritarian governments, a small group makes decisions related to laws and policies, and the citizen's voice is not heard, and sometimes silenced.

Roman Yanushevsky / Shutterstock.com; Nazar Gonchar / Shutterstock.com; Joseph Sohm / Shutterstock.com; Joseph Sohm / Shutterstock.com;

Figure 5.2 In democracies, citizens have a voice, usually by voting for elected officials to represent them in government (*left*), but citizens do not have a voice in authoritarian governments, of which Mao's 1970s Cultural Revolution in China is one of the most extreme examples (*right*).

Section 5.1.2—Thresholds and Regime Shifts

Key Concept 5.1.2—Whether a system changes linearly or nonlinearly has consequences for detecting change, and also for influencing the system's movement toward or away from sustainability.

In the last chapter, the use of indicators for tracking changes in a system over time was explored. You learned that they are used to indicate whether a system is moving toward or away from sustainability, or even changing at all. In the graphs shown in Figure 5.3, the y-axis represents an indicator being used to track change in the system (a state indicator, Section 4.1.2). The x-axis represents a key variable—the driver—that is thought to be regulating and contributing to change in the system (a pressure or response indicator, Section 4.1.2, both of which are drivers). Indicator trends show that some systems change linearly and predictably, such as is shown in Figure 5.3a. Systems can also change in nonlinear and unpredictable ways, such as in Figures 5.3b

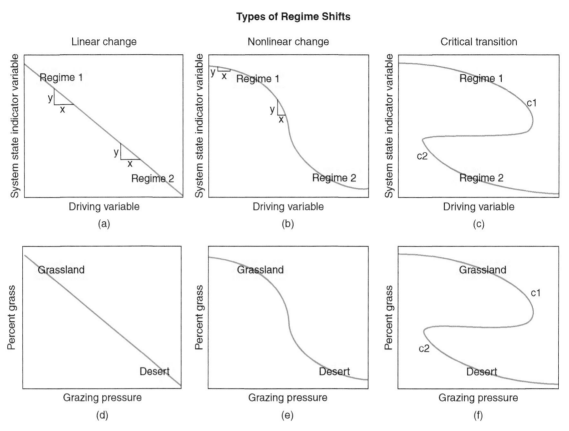

Figure 5.3 The graphs depict three patterns of change that can occur during a regime shift.

nonlinear change a type of change defined by the fact that the magnitude of the factor causing the change in a system, and the actual change that occurs in the system, are not proportional.

and 5.3c. These nonlinear changes are often surprising because large changes in the system, as revealed by the system state indicator variable (y-axes in Figure 5.3), can occur with only small changes in the drivers regulating change (x-axes in Figure 5.3). This idea of surprising, nonlinear change was illustrated with the example of the canoe tipping over in the opening passage of this chapter. Recall also the rubber band and coral reef examples in Section 2.1.2. These are all examples of **nonlinear change**. In the rubber band example, the driver causing change in the system is the pulling force of your hands (driving variable, Figure 5.3). The two regimes in this case are *intact rubber band* and *broken rubber band*. Although the pulling force of your hands remains constant, the rubber band "system" changes suddenly. At the tipping point, or threshold, an additional small pull will result in a large change in the system when the rubber band breaks.

Before delving into a more detailed description of the three types of change depicted in Figure 5.3, it is important to note that these are not the only patterns of change observed in socioecological systems. The graphs in Figure 5.3 should be viewed as more of a continuum ranging from linear change at one end of the continuum (Figure 5.3a) to nonlinear, hard-to-reverse **critical transitions** at the other (Figure 5.3c). Many other patterns of change exist, which fall between or are a mixture of, the patterns of change that are depicted in these graphs. Nonetheless, the following discussion of patterns of change will give you an idea of the different ways that regime shifts can occur in natural and human systems.

In general, the difference between linear and nonlinear change can be understood by looking at how the *system state indicator variable, on the y-axis,* changes relative to the *driving variable, on the x-axis,* in Figures 5.3a and 5.3b. In Figure 5.3a, the *system state indicator variable* (y) changes proportionally to the *driving variable* (x). For example, if x changes by 5 units, then y changes by 5 units. If x changes by 10 units, then y changes by 10 units. The change is proportional, in this case one-to-one (i.e., 5/5 = 1/1 and 10/10 = 1/1). Change that is proportional over the full range of values for both variables is called a linear change.

Nonlinear change, on the other hand is not proportional, as shown in Figure 5.3b. In this case, when x is a small number and y is a large number, (e.g., some point on the curve that falls in the upper left-hand portion of the graph), a large change in x results in only a small change in y. As you move from left to right along the x-axis, x becomes larger, and a smaller change in x results in a larger change in y. In other words, the change is not proportional for all values of x and y. For example, when x is small and y is large, a 10 unit change in x might correspond with a 1 unit change in y (x:y = 10:1). As you move from left to right along the x-axis, and x becomes larger, a 10 unit change in x might cause a 5 unit change in y (x:y = 10:5). Such nonlinear change can be surprising because, like the force tipping the canoe, a small disturbance or pressure on the system by the *driving variable* can cause the system to change drastically from Regime 1 (upright canoe) to Regime 2 (upside down canoe) when close to a threshold. In Figure 5.3b, the more vertical and steeper part of the curve in the middle of the graph is the threshold. When this type of nonlinear change occurs, the system has crossed a threshold into a different regime. In contrast, linear change (Figure 5.3a) does not involve crossing a threshold because change happens in a more steady and predictable way.

The exact point at which a system crosses a threshold to a new regime during nonlinear change is difficult to predict, especially because each system is slightly different, and there are many interacting drivers involved in system regulation. Nonlinear change is also difficult to predict because a system may seem undisturbed by some driver (e.g., grazing pressure in a grassland or phosphorus addition to a lake) until it suddenly shifts to another regime (e.g., grassland to desert shrubland, clear to turbid lake, see Table 5.1). At that point, it is too late to reduce the driver causing the disturbance. In addition, once a regime

> **critical transition** a nonlinear threshold-based regime shift that is hard to reverse and often difficult to anticipate in socioecological systems.

shift occurs, it can be extremely costly, if not impossible, to reverse. A nonlinear and hard-to-reverse regime shift is known as a critical transition (Figure 5.3c).

Because of the need to predict or anticipate nonlinear regime shifts, an active research area in sustainability is aimed at determining when natural and human systems are approaching thresholds. This began, at large, in 2009 when a research team led by sustainability scientist Johan Rockström of Stockholm University first attempted to characterize "A Safe Operating Space for Humanity" by defining thresholds (called boundaries in his research study) beyond which the earth system would shift, or has already shifted, into an unsustainable regime (Figure 5.4). In 2015, Will Steffen from the Stockholm Resilience Centre, along with Johan Rockstrom and many others, revised the original indicators and boundaries. (New indicators are shown along the outer edge of the "donut" in Figure 5.5.) Instead

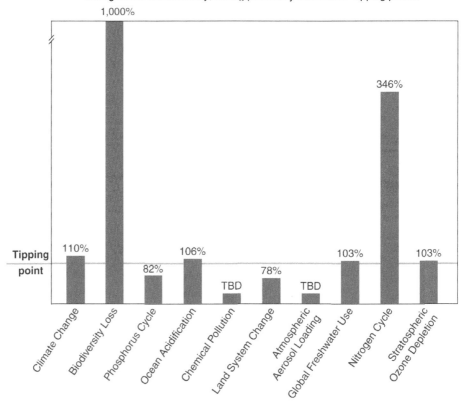

Figure 5.4 According to Rockstrom's 2009 study, which was refined in 2015, the green line represents a safe upper limit for Earth's life sustaining biophysical systems, beyond which humanity has moved into a risky area where the potential exists for crossing irreversibly into new regimes.

of one boundary, the potential for regime shift was characterized as safe, increasing risk, or high risk. According to this new study, on a global scale, only biogeochemical flows and the genetic diversity component of biodiversity are at high risk. Climate change and land system change fall into the increasing risk category. The other indicators, as shown in Figure 5.5, are either still in the safe zone (stratospheric ozone depletion, ocean acidification, and freshwater use) or are unknown (novel entities, atmospheric aerosol loading, and the genetic diversity component of biodiversity).

In 2012, Kate Raworth of Oxfam International, a global movement focused on eradicating poverty, brought social boundaries into the mix in her paper "A Safe and Just Space for Humanity: Can We Live Within the Donut?". She merged planetary and social boundaries together into one framework (Figure 5.5). In her framework, the outside edge of the donut represents the planetary boundaries explored by Rockstrom and Steffen, above which the environment becomes degraded in a way that it cannot support human systems as it has for the past 10,000 years. The inner edge of the doughnut is the social boundary, below which human deprivation exists in social systems.

A distinction can be made between the two types of change represented in Figure 5.3b and Figure 5.3c. Both types of change are nonlinear and involve crossing thresholds, but the critical transition depicted in Figure 5.3c is different because it is extremely difficult to reverse. For the linear and nonlinear changes depicted in Figures 5.3a and 5.3b, reversing the change in the system involves reducing the disturbance intensity to previous levels. Although the system may take time to recover, system recovery will eventually occur when the disturbance intensity is reversed back to its previous level, before the system crossed a threshold (Figure 5.3b) or before it changed linearly to a point beyond a sustainable state (Figure 5.3a). However, with critical transitions (Figure 5.3c), the driver intensity must be reduced to levels *far below* the level it was at when the regime shift occurred, if the system is to be returned back to the desirable regime. This means that the driving variable (e.g., grazing pressure on a grassland or phosphorus inputs in to a lake) would need to be reduced to very low levels—at or below c_2. C_2 is much lower than the driving variable at c_1, which marks the point just before the system crossed a threshold into a new regime. Although this may seem simple to do according to the graph, it is difficult to accomplish in reality. This idea, along with the concepts just described in the last several paragraphs, are illustrated in **Box 5.2**, which provides an example of patterns of change in dryland socioecological systems, specifically the desertification of grasslands.

Critical transitions have been observed in several natural and human systems (e.g., shallow lakes, grasslands, coral reefs, and public attitudes). These transitions may be the most important types of regime shifts to study right now because they are the least understood, the

desertification a phenomenon induced by human activity that occurs in dryland systems and results in the loss or decline of economic or biological yield as a dryland system shifts to a less productive system

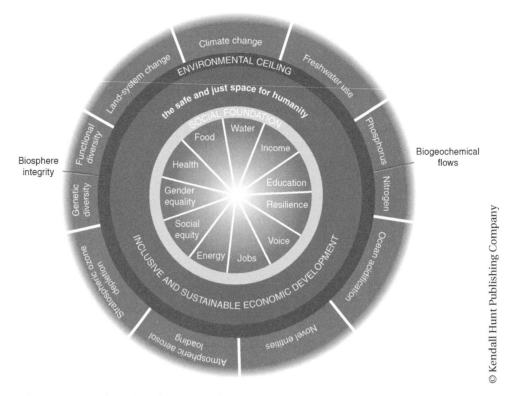

Figure 5.5 The "doughnut" model portrays a safe and just operating space for humanity by acknowledging boundaries for environmental systems as well as for social systems.

least predictable, and most the difficult to reverse. Researchers are attempting to understand critical transitions in systems ranging from small shallow lakes, to ocean circulation, to the entire planet. Recently, a University of California biologist, Anthony Barnosky, led a research team that attempted to define a critical transition to an unsustainable state for the entire earth system (**Figure 5.6**). According to this research, the driving variable (called global forcing on the x-axis in Figure 5.6) for this transition is human population size, which is the indirect driver underlying many direct drivers, such as climate change and habitat degradation. The results shown in Figure 5.6 suggest that our global society is very close to a critical transition. The regime shift that would occur at this critical transition would be from the present-day climatic regime of the past 11,000 years, since the end of the last ice age, to which *human societies have adapted* (Regime 1), to *different conditions* (Regime 2). The *different conditions* of Regime 2 may or may not support the survival of human societies as we know them today. According to this study, if our global society crossed a threshold into Regime 2, return to Regime 1 would require reducing human population to 2 billion people or less (Figure 5.6). Today, there are over 7 billion people living on earth.

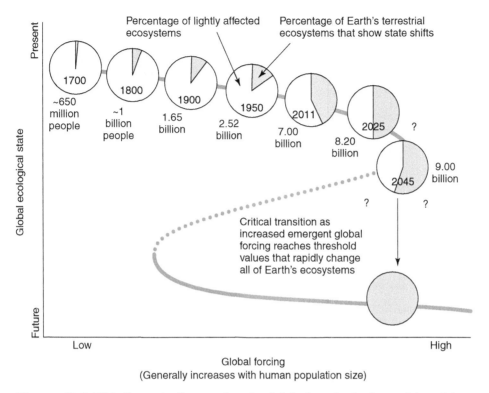

Figure 5.6 This figure indicates that the "global ecological state" (y-axis) is a function of population size (x-axis), but recall from the I = PAT model (Chapter 1) that the human impact (I) on ecosystems is a function not only of population size (P), but also affluence, lifestyle, and resource consumption (A) and technology (T).

BOX 5.2 PATTERNS OF CHANGE IN DRYLANDS: GRASSLAND DESERTIFICATION

Drylands are broadly defined by the Millennium Ecosystem Assessment as "all terrestrial regions where the production of crops, forage, wood, and other ecosystem services is limited by water." These regions make up about 40% of all terrestrial land area on earth (Figure 5.7). At least two billion people live in drylands worldwide, with 90% in developing countries. According to indicators of human well-being, such as GNP and infant mortality rate, these people also experience some of the worst living conditions on earth (Figure 5.8). The vegetative productivity of drylands, upon which all life inhabiting these systems depends, is limited by low soil moisture due to sparse rainfall coupled with high evaporation rates due to dry climates. Deserts are the most arid type of dryland, followed by grasslands, and then woodlands, which are the least arid.

(Continued)

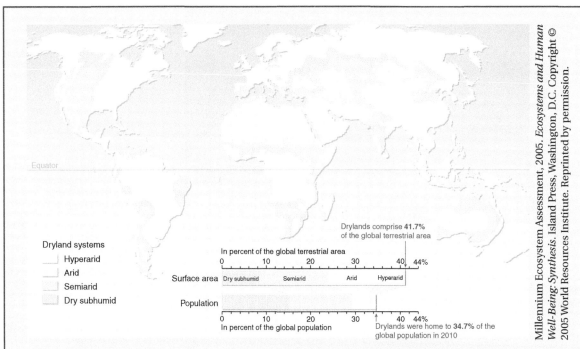

Figure 5.7 Colored regions show the total area of Earth's land surfaces covered by grasslands, which are divided into four types: hyperarid, arid, semiarid, and dry subhumid.

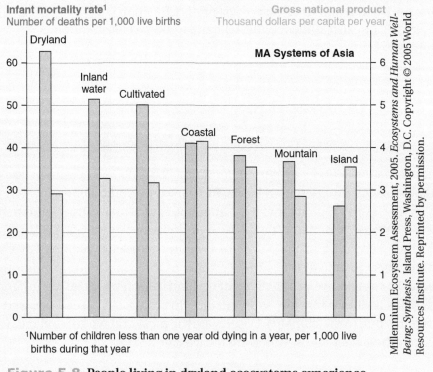

[1]Number of children less than one year old dying in a year, per 1,000 live births during that year

Figure 5.8 People living in dryland ecosystems experience some of the worst living conditions on earth.

A common problem in dryland systems is a phenomenon known as desertification, which is land degradation resulting in the "reduction of or loss in biological or economic productivity" (from Millennium Ecosystem Assessment). Thus, it affects both ecosystems and the human systems dependent on them. Desertification is caused by many drivers, including climate change and human activities, and is one of the greatest sustainability challenges of our time. In the early 1990s, the United Nations Convention to Combat Desertification (UNCCD) was created to tackle this problem. Worldwide, about 10–20% of drylands have already undergone desertification, or experienced a high degree of degradation approaching desertification. Dryland systems dominated by grasslands can have at least two possible regimes: *grassland with cattle* and *desert shrubland without cattle* (Table 5.1). Grasslands that can support cattle populations are beneficial to people who raise cattle as their livelihood. However, grazing too many cattle at once on a piece of land can lead to desertification and the loss of livelihoods when cattle can no longer graze.

Figures 5.3d through 5.3f show the different patterns of change by which grasslands might change to deserts by overgrazing. Figure 5.3d depicts linear change: for each additional unit of grazing pressure (driving variable), the percentage of grass remaining on an area of land (system state indicator variable) will decrease proportionally. In this case, the percent of grass on the y-axis is an indicator of the state of the system. For example, if it is known, such as from the practical experience of a cattle rancher or from scientific studies, that 40% or more grass cover defines a healthy grassland that can support cattle for a particular system, then managing grazing pressure so that cattle will not eat more than 60% of the grass will promote sustainability. Thus, if the relationship between grazing pressure and percent grass cover is linear, then it is easy to figure out how many cattle can be grazed on a piece of land at any given time. More often than not, the relationship between grazing pressure and percent grass cover is nonlinear (Figures 5.3e and 5.3f). In this case, it is difficult to predict how many cattle can be safely grazed without destroying the grassland because the point at which one more head of cattle will cause a precipitous drop in the percent grass cover below 40% must be known. When too many cattle are grazed, a threshold is crossed from grassland that can support cattle (Regime 1) to desert shrubland that cannot support as many or any cattle (Regime 2).

If the regime shift were a critical transition, human systems would have to be scaled back extensively to push the system back over the threshold to healthy grasslands. In the process, the livelihoods of many people would be destroyed. For example, if a critical transition has occurred, then grazing pressure must be reduced to c2 on Figure 5.3f. Before the critical transition, at c1 in Figure 5.3f, it was possible for 100 families to graze 50 head of cattle. At c2, only 10 families could graze 25 head of cattle. Reducing pressure below c2 means that many fewer cattle than before the transition can be grazed. The dilemma becomes: which families get to stay and graze their cattle, and which do not? Furthermore, what if returning to the sustainable *grassland with cattle* regime requires grazing pressures that are too low to be easily controlled by human societies? For example, wildlife, such as wildebeest and water buffalo, in African grasslands graze in the same areas as cattle. What if restoring

(Continued)

Figure 5.9 (a) Healthy Grassland (b) Desert Shrubland.

the system to a state that can support cattle requires eliminating all cattle, as well as some wildlife grazers, in order to reach the c2 level of grazing pressure? Potential issues such as these clearly present large dilemmas for human societies, which is why many argue that it is better to avoid critical transitions in the first place.

Critical transitions in grasslands are difficult to reverse because, once a land area has undergone desertification, conditions become too harsh for recolonization of the land by grass (Figure 5.9). Even if all herbivores, such as cattle and grazing wildlife, were completely removed, the barren desert soil makes it difficult for grass seeds to take root. When most vegetation cover, aside from scattered shrubs, is absent, then water and soil erosion become impediments to grass reestablishment. When grass is present, precipitation is held in the soil surrounding the grass creating a microclimate that is nourishing for grass seeds to start growth. When mature grass is not present, water quickly infiltrates into deep soil layers where neither grass seeds nor seedlings can reach it. Also, without grass to hold the soil in place, the nutrient-rich topsoil needed for grass seeds to establish is washed away by wind and water. The grass seeds themselves can also be washed away. Soil compaction by the cattle that previously occupied the land prevents water from entering soils and makes erosion worse. Extreme floods or droughts caused by climate change can also create conditions with too much water and erosion, or too little water, respectively.

Section 5.1.3—Resilience

Key Concept 5.1.3—Resilience is a concept used for understanding how susceptible a system is to a regime shift. It can be assessed based on the proximity of a system to a threshold, as a result of driver interactions, and is often higher when systems contain diversity.

Resilience is the ability of a system to absorb disturbances, but still maintain the same conditions, function, or identity. In other words, it is the ability of a system to remain in its current regime in the face of disturbances. If a system is resilient, it has the capacity to avoid crossing a threshold into a different regime. Resilience can be desirable or undesirable, depending on the goal that has been set for the system. If sustainability is the goal, and the system is currently sustainable, then resilience is desirable because it will keep the system in the current state. If a system is unsustainable, then it is desirable to decrease the resilience and encourage the system to change toward a more sustainable regime. Thus, when resolving sustainability problems, enhancing or preserving the resilience of sustainable systems, and eroding the resilience of unsustainable ones, is required.

Remember that sustainability must always be defined for each specific context. Similarly, resilience must be defined relative to the desired purpose of the system, and what is being sustained. For example, a piece of land in a grassland ecosystem can serve many purposes. It can support diverse species, such as wildebeest, water buffalo, lions, cheetahs, and elephants, for the purpose of wildlife conservation and ecotourism. If the land is to serve this purpose, then the resilience of the system must be preserved in terms of supporting biodiversity. Grasslands can also support cattle for the purpose of providing local families with livelihoods and incomes. If the land is to serve this purpose, then the resilience of the system must be preserved in terms of supporting cattle. To achieve sustainability, system resilience must be defined in terms of both biodiversity (environment pillar), and livelihoods and incomes (society and economy pillars).

Understanding Resilience Using the Ball-and-Basin Model. The ball-and-basin model is often used to explore system state, regime shifts, thresholds, and system resilience (Figure 5.10). The ball represents the system. The ball's position within the basin represents the system's state within a given regime, which is different in Figures 5.10a and 5.10b. The two basins represent two possible regimes. One is sustainable and the other is unsustainable. The black dashed line that separates the two regimes is the threshold. Once the system (the ball) crosses the threshold, it undergoes a regime shift and "rolls down" into the other regime basin. Resilience is represented by the horizontal distance between the position of the ball in the basin and the threshold. Thus, resilience is determined by the system's state within the regime.

Figure 5.10b shows how a driver causing change in a system can influence the resilience of the system. The driver pushes the system closer to a threshold, which reduces the system's resilience to additional disturbances by other drivers. In other words, it reduces the ability of a system to remain in its current regime in the face of more extensive or additional disturbances. In complex systems, there is always more than one driver causing the system to change (recall causal chain analysis from Section 3.2) and it is important to keep this in mind when gauging system resilience. There are also drivers that keep systems in their regime, rather

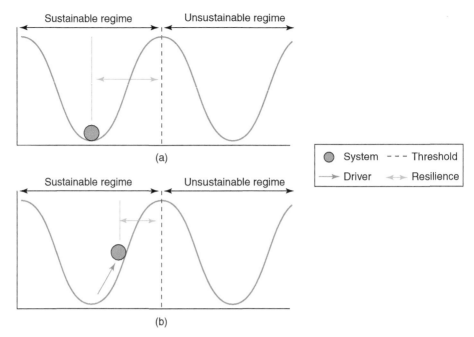

Figure 5.10 The ball-and-basin model is a simple way to view resilience.

than changing them, but this will not be discussed until Section 5.2 on feedbacks. It is necessary to consider both types of drivers, in addition to relevant spatial and temporal scales, in order to assess overall system resilience. The ball-and-basin model is a simplification of reality, and will be expanded upon in later sections of this chapter. For now, it is a good start. The usefulness of the ball-and-basin framework for thinking about system resilience is illustrated by the desertification example in **Box 5.3** and the economic regime change example in **Box 5.4**.

Diversity and Resilience. Diversity is an important component of resilience for both natural and human systems, and can be explained using concepts from both the natural and social sciences. **Functional diversity** is a concept from ecology that refers to the function, or "profession", that each species performs in an ecosystem. All healthy ecosystems require different species to perform certain functions. For example, peanut plants, cloves, and alder trees perform a nitrogen fixing function that removes N_2 gas from the atmosphere and converts it to a form that can be used by other organisms in ecosystems. Predators (such as foxes, bears, and bobcats) keep their prey populations (such as deer and elk) at low levels that are not destructive to vegetation. Wildebeest and water buffalo perform a grazer function in grasslands that return nutrients stored in grasses back to soils so that new grass can grow. When thinking about resilience of natural

---

functional diversity a component of biodiversity concerned with the function that each species performs in an ecosystem.

BOX 5.3   REGIME SHIFTS AND RESILIENCE: GRASSLAND TO DESERT SHRUBLAND

Figure 5.11 shows how the ball-and-basin model applies to grasslands undergoing grazing pressure. In Figure 5.11a, grazing pressure reduces the resilience of grasslands by changing the system's state and pushing it closer to the threshold, beyond which it would shift from grassland to desert shrubland. It is important to note that, even when human activities influence them, grasslands can still be sustainable. A common misconception is that ecosystems must be preserved in their natural, "pristine" state in order to be sustainable. However, it can be seen in Figure 5.11a that grasslands can be sustainably used in many different states inside the grassland regime, as long as grazing pressure does not become so great that the system undergoes a regime shift to a desert shrubland. The latter is not sustainable if the purpose of the system is to support a human society dependent on cattle ranching as a livelihood. For wildlife to be part of the system, so that livelihoods can involve ecotourism, cattle grazing pressures must be low enough so that sufficient forage is left for the wildlife species to survive.

There are other drivers of change in grassland systems, and Figure 5.11b shows how climate change is a second driver that can further change the system's state, decreasing the grassland's resilience even more. In dryland systems, climate change can cause extreme drought. As discussed in Box 5.2, low moisture conditions in dryland soils make it difficult for grass

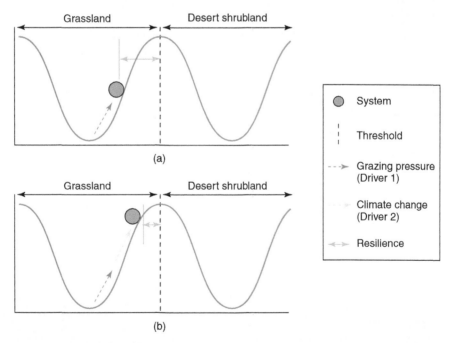

Figure 5.11 Grazing pressure and climate change are two drivers that reduce the resilience of grasslands.

(Continued)

seeds to grow. Thus, removal of grass by grazing cattle (and wildlife), combined with a decreased ability for grasses to regrow after they are eaten, can result in a system with very low resilience. A system such as this has very little ability to remain in its current regime in the face of additional disturbances. For example, if a wildfire sparked by a natural lightning strike burned all of the grass in a given area, this additional driver could be enough to push the system over the threshold into desert shrubland. If the system resilience were higher, such as in Figure 5.11a, with only grazing pressure affecting resilience, then the grassland may have withstood the impact of the wildfire.

Grazing, drought induced by climate change, and wildfires are three drivers identified here that cause the grassland system to change. However, other drivers regulate the system in a way that keeps it from changing. For example, seed dispersal is a process that occurs in dryland systems, mostly by wind or animals. If a small patch of grass is disturbed within an area of grassland, then seed dispersal will bring new seeds into the disturbed area. Thus, seed dispersal is a driver that keeps grasslands in their current state. Thus, regime shifts occur when the drivers of change "outcompete" the drivers that are keeping the system the same. This idea will be returned to in Section 5.2 on feedbacks.

In addition to identifying all of the drivers that change grassland systems (grazing, drought, and wildfire) and those that keep the grassland in the current state (seed dispersal), it is also important to consider different spatial and temporal scales when assessing system resilience. For example, when considering the resilience of an area of dryland ecosystem, a large enough area must be considered in order to get an accurate picture of its resilience. Figure 5.12 is a picture of a dryland ecosystem landscape in North America. If only the bare patch of land that composes the entire area, A, (bounded by the dotted line) is considered, then the grassland does not appear to be resilient to disturbance and, in fact, appears to have already shifted to desert. However, if the larger area, B, (which includes both a bare patch and a patch containing grass) is considered, then the system appears to be resilient because there is a "seed bank" available in the grass patch adjacent to the bare patch, which could restore the bare area within area B to grass.

Time scale must also be defined appropriately when defining system resilience with respect to the survival of human societies. In this example, loss of present-day

Figure 5.12 Specifying spatial scale is important for determining system resilience.

grasslands, resulting from overgrazing and climate change, may result in the loss of resilience for human societies and economies dependent on cattle grazing supported by the grasslands. However, if human activities stop contributing to the erosion of the resilience of the grasslands, then the grasslands may have a chance to recover, after thousands or millions of years. Thus, the grasslands are resilient with respect to their ability to support ecosystems over very long (geologic) time scales. However, this is not relevant to sustainability and the survival of prosperous human societies, which depend on resilient grasslands for ecosystem services that support societies today.

BOX 5.4 FOSTERING RESILIENCE THROUGH ECONOMIC REGIME CHANGE: A REGIME SHIFT IN THE MAKING

By Antonia Castro-Graham, City of Huntington Beach, Assistant to the City Manager/Energy & Sustainability Projects Manager

When it comes to cities, local governments must transform them, with the help of responsible citizens, businesses, and communities, into sustainable and resilient systems so that they have the capacity to endure. Creating economic opportunities, which also benefit people and the environment, helps to build these sustainable resilient communities. One example of a resilient economic system that is taking shape is a program in the State of California called the Orange County Recycling Market Development Zone (OC RMDZ, Figure 5.13). The following is the story of a regime shift in the making through the creation and implementation of the OC RMDZ.

The OC RMDZ was approved by the State of California in March 2016. However, 6 years of hard work were required to achieve this approval. The problem with gaining approval was that, for years, local municipalities tried to sell the idea of creating a robust, environmentally responsible manufacturing community within very conservative Orange County. In order to appeal to conservatives, a group of municipalities shifted the focus from environment to economy by focusing on the benefits to businesses, such as reduced disposal costs, reuse of materials, and reduced transportation costs (Driver 1 in Figure 5.14). This strategy was successful and, in 2015, the city of Huntington Beach worked with the cities of Garden Grove, Orange, Santa Ana, and Stanton, along with Orange County Waste and Recycling, to apply to become an RMDZ. This was to be the first multi-jurisdiction zone in Orange County. Today, the RMDZ is an economic development program that utilizes the growing supply of recyclable material to fuel new businesses, expand existing ones, create jobs, and divert waste from landfills.

The RMDZ program enables local governments to grow businesses within their communities that utilize recycled materials (secondary feedstock) in their manufacturing process or

Contributed by Antonia Graham. Copyright © Kendall Hunt Publishing Company. *(Continued)*

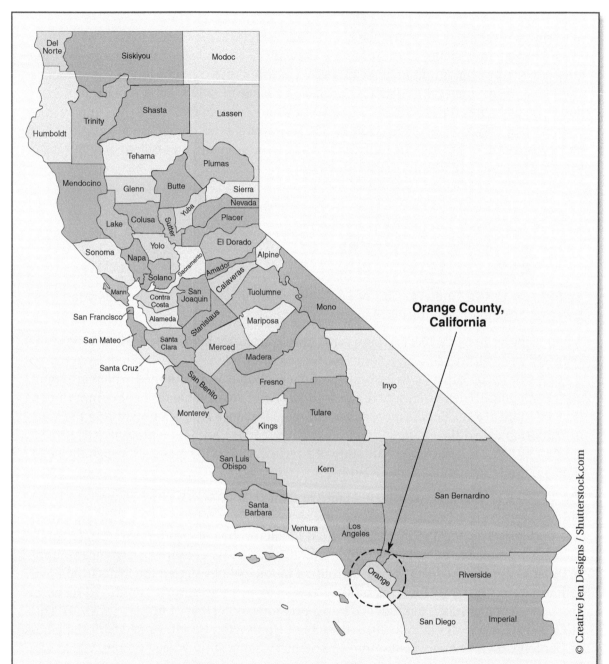

Figure 5.13 The OC RMDZ is located in Orange Country, small coastal region found in the state of California in the United States.

that seek to partner with companies who may want to use their discards (Figure 5.15). This type of program treats trash as a resource instead of a discard. The program offers financing for manufacturers to expand their business, develop new manufacturing technologies, hire new employees, and receive free technical assistance that would otherwise

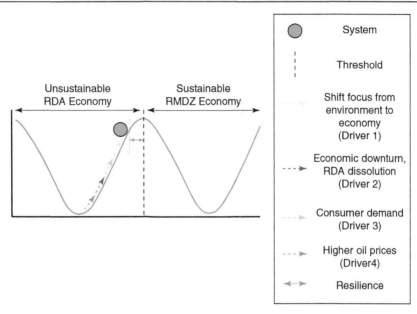

Figure 5.14 Several drivers are moving the old RDA economy toward a more sustainable RMDZ economy, by eroding the resilience of the old unsustainable system and causing a regime shift.

Figure 5.15 The RMDZ program stops recycled material from being sent overseas, such as the secondary feedstock in this recycling plant bound for Asia (*left*), and instead feeds these materials, such as discarded scrap metal (*right*), back into local manufacturing processes.

be cost prohibitive to a small manufacturer. In Fiscal Year 2016–2017, there were $8,000,000 available to manufacturers seeking to take advantage of the RMDZ. For Orange County as a whole, the RMDZ program creates a circular economy, in terms of material use and reuse, such that the program is ecofriendly. It also enhances the County's overall economic development potential and creates jobs.

(Continued)

The situation was different before the approval of the OC RMDZ in 2016. In response to the 2008 economic downturn, the State of California ceased operating local Redevelopment Agencies (RDAs) in 2012. RDAs had been in existence since the end of World War II. Their intended purpose was to rectify urban blight and create economic development opportunities in once blighted areas. Redevelopment efforts focused on creating economic opportunities in areas that were once plagued by issues such as crime, unsafe buildings, and crumbling infrastructure. Cities used RDAs to attract and encourage private sector investment that otherwise wouldn't occur because of blight.

With the dissolution of Redevelopment Programs in California (Driver 2 in Figure 5.14), many cities are struggling to find ways to help their business communities. Prior to 2012, municipalities could use RDA funds to incentivize and assist businesses. After 2012, municipalities had to get creative and find new ways to assist businesses. The RMDZ is a tool that cities can use to foster innovation, retain businesses, and attract business to Orange County, all of which diversify and strengthen local economies and create jobs. The RMDZ is a more sustainable form of economic development because it is not reliant on tax revenue. It simply seeks to utilize trash as a resource by reimagining it into new products. This reduces reliance on harvesting new natural resources.

In the short term, the RMDZ will provide existing manufacturers with financial assistance and technical expertise. The long-term vision for the RMDZ is to create a robust, resilient, and innovative manufacturing community that utilizes materials generated locally within the County and creates local jobs: in other words, it hopes to promote a resilient system focused on sustainability. The RMDZ also offers its members a commodity marketplace in which waste haulers, businesses located within the RMDZ, RMDZ cities, and the County will be able to buy, sell, and trade secondary feedstock and recycled materials. This will further increase the circular economy, reduce disposal costs for businesses, and provide haulers with local markets for their recyclables. This is in contrast to the RDA regime, in which recyclables were shipped overseas or to a landfill. By developing local markets for these secondary materials, the RMDZ reduces economic vulnerability caused by the pricing of recyclables, which are a considered a commodity, by external markets. Local markets also encourage greater levels of recycling and reuse in the region, and make low value recyclables more valuable to collect.

The biggest challenge to the OC RMDZ is the price of oil. When oil is cheap, it is cheaper to utilize virgin materials than it is to process recycled content. Using virgin natural resources is a less expensive alternative, but a more harmful environmental choice. It also hurts the recyclable commodity market, as these secondary materials become worthless and waste haulers find it cheaper to send them to a landfill. As consumers continue to become more "eco-conscious," they have the ability to influence the marketplace and demand products made from recycled content, regardless of the price. It is up to consumers to drive the current economic system to a more sustainable system (Driver 3 in Figure 5.14). If oil prices were to increase, they would be an additional driver that could help erode the resilience of

> the current and unsustainable RDA regime, and move the system toward a more sustainable RMDZ regime (Driver 4 in Figure 5.14).
>
> As local governments strive to retain and attract businesses, the RMDZ represents a unique solution: it is a tool for municipalities, a tool for haulers, an opportunity for local manufacturers, and illustrates the commitment to the environmental stewardship by reducing waste sent to our landfills. It helps the economy, creates jobs, and protects the environment. A true win–win situation for everyone involved. The RMDZ program is a perfect example of what sustainability really is: an economic system that works for both the environment and the business community. The RMDZ represents diversity and resilience. For an economy to be resilient, there must be a diversification of industries. The RMDZ provides the local economy with the ability to enhance resilience by investing in the local manufacturing economy.

systems, **response diversity** is very important (Figure 5.16). This type of diversity arises from the fact that different species have varied responses to different disturbances. For example, cloves, alder trees, and peanut plants are nitrogen fixers within ecosystems. If climate change causes drought conditions that make it impossible for peanut plants to grow, and grazing pressure eliminates cloves, then the ecosystem still has alder trees to perform the important nitrogen fixing function. This is because different species have varying responses to specific disturbances. The more species from the same functional group an ecosystem has, the higher the response diversity (and resilience) of that ecosystem. High response diversity gives an ecosystem a greater chance of remaining in its current regime in the face of disturbance.

Resilience in human systems also depends on diversity. For example, diversity of knowledge types can more effectively help traditional managers resolve natural resource challenges. As was seen in the example of the Cree's indigenous knowledge about caribou populations in Section 4.1.2, both scientific and traditional knowledge are valuable in predicting caribou populations from year to year. Scientific knowledge tends to be privileged above all others, but many different types of knowledge are important in order for human societies to have the ability to adapt to changing conditions. The resilience of a society can be enhanced by preserving cultural diversity, in addition to ecological diversity. Diversity builds **adaptive capacity**, which is important for the resilience of human systems. Adaptive capacity is the ability to innovate, reorganize, and learn when faced with change. A good example is the adaptive capacity of businesses that establish research and development (R&D) teams to foster innovation. In any business, there is a tension between production efficiency and creative innovation. Investing resources in

response diversity
the number of different species in the same functional group in an ecosystem that have varied responses to different disturbances.

adaptive capacity
the ability of a system to innovate, reorganize, and learn when faced with change.

Figure 5.16 The more organisms there are in an ecosystem that perform each function needed to keep ecosystems healthy, the higher the functional diversity and possibly response diversity in the face of disturbances such as climate, habitat degradation, and other problems facing ecosystems.

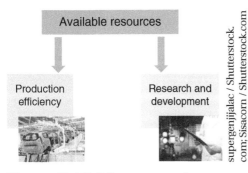

Figure 5.17 When companies invest resources in research and development, in addition to production efficiency, they forego some short-term profits but they will increase their long-term resilience in the face of changing consumer preferences and other market disturbances.

both can decrease overall profits, so companies may choose to specialize in production efficiency to maximize profits. However, if customer preference shifts, then companies that do not have a diversity of new products to draw from to meet these new preferences will not survive (**Figure 5.17**). For example, customer preferences might shift from gasoline-powered cars to hybrid cars. Companies with R&D teams already working on hybrid models will be more likely to survive this change, while others will not. Successful companies have learned this, and have responded by reorganizing their businesses into separate R&D and production units. In general, successful human systems have the capacity to innovate, reorganize, and learn in the face of change.

A Note on Thresholds and Multiple Regimes. Before leaving this section on regime shifts, thresholds, and resilience, it is important to reemphasize that not all systems experience threshold-based regime shifts, and that some systems have more than two possible regimes. For linear systems, as in Figure 5.3a, regime shifts do not involve crossing thresholds where a system rapidly shifts to a different condition. Recall that in these systems, change in response to the driver is proportional. As a result, rapid, surprising, and hard-to-reverse regime shifts do not occur. Rather, they can usually be anticipated and reversed. Although, recovery of linear systems from disturbance may be so slow that, from the user's perspective (e.g., the cattle rancher), the system might as well have crossed a threshold into a different regime because it takes such a long time to recover. In some cases, it may not recover within the lifetime of that user.

For many systems, there may be more than two alternative possible regimes. The simplified ball-and-basin model leaves the impression that systems have only two regimes, but many systems have more. For example, coral reefs have up to five possible regimes: coral-dominated, seaweed-dominated, algae-dominated, sea-urchin dominated with bare rock, and bare rock. What regime the coral reef is in depends on two different drivers: amount of fishing pressure and quantity of nutrient inputs into the system. If both fishing pressure and nutrient inputs are low, then the reef will be in the healthy coral-dominated regime. If both fishing pressure and nutrient inputs are very high, then the reef will be in an unhealthy algae-dominated regime. An algae-dominated regime does not have the capacity to support healthy levels of biodiversity. If fishing pressure is very high, and nutrient inputs are kept low, the reef shifts into a bare rock regime. In this regime, the reef lacks the capacity to support biodiversity, and also other important ecosystem services. At moderate fishing pressures and nutrient inputs, the reef will shift to a seaweed-dominated regime, in which some (but not all) ecosystem services will be supported. The shifts among all of these different regimes depend on the intensity level of each driver relative to the other and are often nonlinear. In fact, many complex systems of concern to sustainability experience nonlinear change due to the many dynamic interacting factors inherent in these systems, including feedbacks, which are the topic of the next section.

Section 5.2: Feedbacks and Resilience

Core Question: Why do systems cross thresholds into new regimes and experience fluctuations within a given regime?

When systems cross thresholds, a small disturbance by some driver of change causes a large change in the state of the system (Figures 5.3b and 5.3c). This type of change is often referred to as a runaway feedback, which snowballs the system into a new state, similar to a snowball that picks up momentum, growing bigger and bigger, as it rolls down a hill. **Box 5.5** provides an example of a runaway feedback that could occur as societies transition to more renewable sources of energy. A common

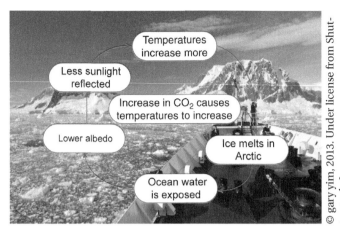

Figure 5.18 Melting of Arctic sea ice has been hypothesized to be caused by a "runaway" feedback.

runaway feedback heard in the news today is the melting of Arctic sea ice, which is happening much faster than scientists anticipated. The runaway feedback in this case works as follows (Figure 5.18): humans are adding CO_2 to the atmosphere and this increases global temperatures. Higher temperatures cause ice to melt in the Arctic sea and more water is exposed. Water has a lower *albedo*, which is a measure of the reflectivity of a surface. The lower the albedo, the less sunlight is reflected and the more it is absorbed. More sunlight energy is absorbed by water than ice, and this heats the system even more. The result is even higher temperatures, even more melted ice, even lower albedo, and more temperature increases. In this example, a small change in the driver (CO_2 emitted) results in a large change in the system (Arctic sea ice melting). Beyond this initial small change in the driver, it is not an additional increase in driver intensity that results in more ice melting. Instead, it is a *runaway* feedback causing the system to change. If this feedback is strong enough, the system will eventually shift from the current regime of *sea ice cover* (Regime 1) to a new regime of *no sea ice cover* (Regime 2). How strong does the feedback have to be for this to happen, and stronger than *what*?

albedo the fraction of solar radiation that is reflected from a surface rather than absorbed.

BOX 5.5 THE UTILITY DEATH SPIRAL, A "RUNAWAY" FEEDBACK

By Wes Herche, Research Scientist at the Arizona State University, Global Security Initiative

In the electric utility industry, there is fear among utility providers of the so-called utility death spiral brought on by distributed generation of solar energy. The price of rooftop photovoltaic panels for residential applications has dropped exponentially over the past 40 years (Figure

5.19). The cost of home battery storage is dropping at similar rates. This means generating electricity from your home's rooftop is becoming more cost-competitive with traditional sources of electricity, such as coal, natural gas, and nuclear, and thus more enticing for home owners.

As a backup, most households with rooftop solar want to remain connected to the grid that utility providers use to supply electricity (Figure 5.20). However, adoption of rooftop solar by households is not uniform: you might have solar panels on your roof, but others in your neighborhood do not. This means that utility companies have to maintain roughly the same amount of grid infrastructure, even though households with rooftop solar do not purchase their electricity most of the time. The problem for electric utility companies is that, even though less consumers purchase their electricity, their costs stay largely the same. The cost of transmission, distribution, and grid maintenance is a sizeable portion of a utility company's cost structure. Further, the facilities used to generate their electricity, such as coal, gas, and nuclear plants, are often financed by these companies over long time periods of 30 years or more. Thus, utility companies have a substantial financial commitment to their existing generation platforms. As consumers start to generate their own electricity, utility companies lose revenues, but their costs remain largely the same. As a result, costs have to be distributed across the remaining set of utility customers, which means higher electricity rates for these customers.

Figure 5.19 Many households are starting to install photovoltaic cells on their roofs, as shown here, as the costs of generating electricity using sunlight decrease.

Figure 5.20 Today, the vast majority of electricity is delivered through a grid network, which includes generating stations to produce electricity, and high-voltage power lines (shown here) and other distribution lines that carry electricity to consumers.

Contributed by Wesley Herche. Copyright © Kendall Hunt Publishing Company.

(Continued)

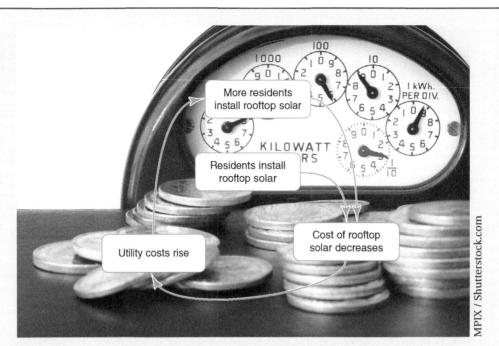

Figure 5.21 A "runaway" feedback between rising utility costs and decreasing costs of rooftop solar could cause a big regime shift toward more renewable energy sources.

The higher rates for electricity make rooftop solar, which is steadily declining in price, all the more attractive to homeowners, creating a "runaway" feedback (Figure 5.21). In Section 5.1, you learned about thresholds or "tipping points." There is a fear among utility companies that a threshold will be reached for distributed solar, and perhaps other renewables, with the result being the rapid unraveling of the present electric utility regime, which has operated in roughly the same capacity for over 100 years. While some may welcome the idea of disruptive change in the electric utility industry, and movement of energy production systems into a new regime based on renewables, our banking, communications, and other critical infrastructure are highly dependent on the present electric utility regime. For example, many essential services, such as traffic lights, the servers that hold our bank balances, and medical facilities, have contingency plans for short-term outages, but ultimately depend on a steady flow of electricity from the grid. If the present electric utility regime were to suddenly unravel in a "utility death spiral," how will these services be maintained? What are other potential negative or unforeseen consequences that we might need to think about? Does the complexity of the electric utility system, and its deep embeddedness into all facets of our society, make a massive transition in this space a "wicked problem," as you learned about in Chapter 2?

Section 5.2.1—Shifting Dominance of Stabilizing and Reinforcing Feedbacks

Key Concept 5.2.1—It is the shifting dominance among stabilizing and reinforcing feedbacks that determines the state of a system at any given time and whether or not it will shift to a new regime.

It is the **shifting dominance**, or the constant interplay between two types of feedbacks competing with each other for dominance, that ultimately determines a system's behavior and the regime in which that system is situated at any given time. Thus, feedbacks determine both the current state and any past or future conditions of a system. A feedback is a specific type of interaction among components of a system, as introduced in Section 2.1.2. They are given their name because they literally feed back to the beginning of whatever caused some process to start in the first place. As a result, the outcome of some process (i.e., melted Arctic sea ice) returns to affect the factor that originally set the process in motion (i.e., increased temperatures). *How* that outcome affects the original factor depends on the type of feedback. **Reinforcing feedbacks,** also known as positive, amplifying, or runaway feedbacks, change the current state of a system by *increasing* the magnitude of the original factor. For the Arctic sea ice example (Figure 5.18), the outcome of melted sea ice (i.e., lower albedo) increases the magnitude of the temperature, which is the original factor that caused the sea ice to melt in the first place. **Stabilizing feedbacks,** also known as negative or balancing feedbacks, keep a system in its current state because the outcome *decreases* the magnitude of the original factor. In the case of the Arctic sea ice example, if the temperature did not increase to melt the ice, the albedo would stay the same. This means enough sunlight would be reflected to keep temperatures and sea ice cover constant. Thus, the outcome (i.e., no melted ice) keeps the original factor (i.e., temperature) stable. Understanding how system components interact through feedbacks is important because it can help explain the outcome of a disturbance on a system that sets feedbacks in motion (e.g., the increase in atmospheric CO_2 concentrations that set the temperature-albedo feedback in motion and causes Arctic ice to melt much faster than anticipated).

The effect of these two types of feedbacks on a system may seem difficult to grasp right now, but many examples are provided to solidify your understanding. These include examples of feedbacks within the climate system, the human body, a forest ecosystem, and in human systems (**Boxes 5.6, 5.7, and 5.8**). As you move through these examples, keep in mind the big picture: *reinforcing feedbacks move the system away from the current state by augmenting the direction of change enacted on them, and stabilizing feedbacks keep a system in its current*

---

shifting dominance
a constant interplay between the two general types of feedbacks in a system that compete with each other for dominance that determines the system's overall behavior at any given time.

reinforcing feedback
a general feedback type that causes a system to change and ultimately shift to a new regime.

stabilizing feedback
a general feedback type that keeps a system in its current regime.

state by resisting the direction of change enacted on them. If reinforcing feedbacks are stronger than stabilizing feedbacks, then the state of a system will change and possibly shift into another regime (Figure 5.22b). If stabilizing feedbacks are stronger than reinforcing feedbacks, then the system will remain in the current regime (Figure 5.22a). In short, reinforcing feedbacks amplify change in a system and stabilizing feedbacks dampen it. It is the interplay and competition between these different feedbacks that determine the behavior, and current and future states, of a system. Thus, it is not only understanding how system components interact through feedbacks that is important. It is also important to understand how different feedbacks interact with each other because this also determines the state of a system and whether it is approaching a threshold. Considering the spatial and temporal scales over which feedbacks interact with each other is also important when determining how feedbacks might regulate system behavior in the future. Finally, it is the constant interaction between the two types of feedbacks that can cause variability and fluctuation in a system, even within a given regime, as contrasted with an actual regime shift as described in Section 5.1.1. The contribution of feedbacks to system fluctuation within a regime will be discussed in more detail in Section 5.2.2.

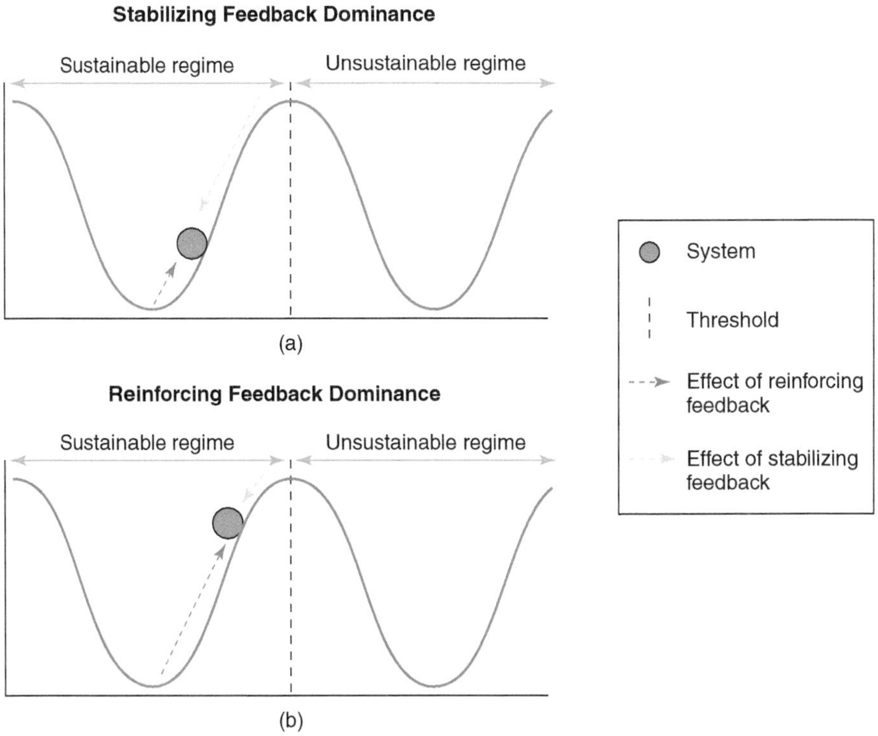

Figure 5.22 When stabilizing feedbacks dominate a system, the system will remain it the current regime (a) but when reinforcing feedbacks dominate then the system will move toward a new regime (b).

The competition among many feedbacks in any complex system renders decision making about, and management of, these systems far from straightforward, at best, and impossible, at worst. Some sustainability scientists studying complex systems argue that management efforts ultimately cause regime shifts when a system becomes overwhelmed by the unintended consequences that result from management actions. (This is related to Characteristic 3 of wicked problems, No End Point, as described in Chapter 2.) Nonetheless, researchers attempt to better understand complex systems using mathematical models, but models are simplifications of reality. When used to make predictions about the behavior of real-world systems to inform policy and other decisions about resource use, care must be taken. How do we know all key processes have been accounted for by a model? How do we know all of the feedbacks are considered? How accurate are the field measurements supporting these models? For example, the chemical weathering that influences global atmospheric CO_2 concentrations involves all rocks on earth (Box 5.6). How can we possibly measure this with sufficient accuracy? Scientists use models and field measurements to make conclusions about climate change, and knowledge is continually increasing. Sustainability scientists integrate human and natural systems in these models to gain a better understanding of the behavior of socioecological systems. However, feedback dynamics, and the immense number of components and interactions among components that must be accounted for, make SESs challenging to understand, and their behavior uncertain. One way to deal with decision making about such systems is to use future scenarios rather than predictive models. The use of future scenarios for this purpose, and many other purposes, is discussed in more detail in Chapter 7.

BOX 5.6 FEEDBACKS IN THE GLOBAL CLIMATE SYSTEM

There are many competing feedbacks present in the global climate system that determine the overall behavior of the system. A subset of global climate feedbacks regulate the quantity of greenhouse gases in the atmosphere over time. Only three of those greenhouse gas regulating feedbacks will be described here.

Stabilizing Feedback: Rock Weathering and CO_2 Concentrations. The first feedback is a stabilizing feedback between temperature and chemical weathering (Figure 5.23). Chemical weathering of rocks by CO_2 dissolved in water depends on temperature, such that more weathering occurs at higher temperatures. When weathering occurs, CO_2 is consumed. Thus, when temperature increases, weathering increases and removes CO_2 from the atmosphere. Removal of CO_2 causes temperatures to decrease and, therefore, chemical weathering

(Continued)

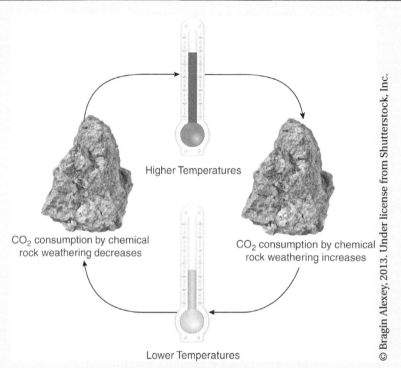

Figure 5.23 A stabilizing feedback between global temperature and chemical weathering of rocks works to keep the global climate system in its current state.

decreases. When chemical weathering decreases, less CO_2 is removed from the atmosphere and temperatures increase. Increased temperatures result in more chemical weathering of rocks. This stabilizing feedback continues as described, keeping CO_2 concentrations in the atmosphere relatively stable over time (unless there is some other disturbance). As you can imagine, chemical weathering of rocks occurs very slowly (millions of years). Thus, this feedback regulates atmospheric CO_2 levels over geologic time scales.

Reinforcing Feedbacks: CH_4 and Permafrost, CO_2 and Temperate Forests. Reinforcing feedbacks in the climate system cause changes in greenhouse gas concentrations. Two reinforcing feedbacks that cause greenhouse gas concentrations to increase have to do with Arctic permafrost and temperate forests (Figures 5.24 and 5.25). Both of these feedbacks ultimately start with CO_2 emissions being released into the atmosphere by humans. CO_2 emissions are causing temperatures in the Arctic to rise. Warmer temperatures in the Arctic result in the melting of permafrost, which results in the release of methane gas (CH_4), stored in the frozen permafrost, into the atmosphere (Figure 5.24). CH_4 is also a greenhouse gas. In fact, CH_4 is 25 times more potent as a greenhouse gas than CO_2. The release of CH_4 results in further increases in the temperature. This results in more permafrost melting, accompanied by more CH_4 release.

The other reinforcing feedback has to do with the shrinking of temperate forest areas (Figure 5.25). Temperate forests remove CO_2 from the atmosphere during photosynthesis and store it in their biomass as organic carbon. Rising global temperatures will result in a shift in global ecosystem distributions, which will cause the area of temperate forests to shrink. With less temperate forests, less CO_2 will be stored in biomass and will instead remain in the atmosphere, causing further warming and further shrinking of temperate forests.

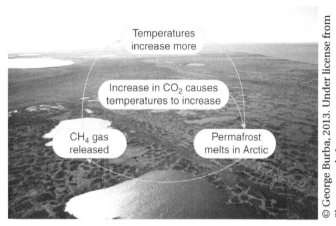

Figure 5.24 **A reinforcing feedback between global temperature and melting permafrost pushes the global climate system toward a new regime.**

Competing Feedbacks in the Climate System. In this example, both of the reinforcing feedbacks that add greenhouse gases to the atmosphere operate on much shorter time scales than CO_2 removal from the atmosphere by the chemical weathering of rocks. Permafrost melts and temperate forests change on time scales ranging from tens to hundreds

Figure 5.25 **A reinforcing feedback between global temperature and temperature forest area pushes the global climate system toward a new regime.**

of years. The chemical weathering process that removes CO_2 from the atmosphere over millions of years simply cannot keep up. Thus, reinforcing feedbacks out-compete the stabilizing feedback because they happen faster. The overall result is that CO_2 concentrations in the atmosphere continue to increase. However, remember from the stabilizing feedback example between temperature and chemical weathering, that there is more CO_2 consumption by weathering at higher temperatures. If greenhouse gas concentrations are increasing as a result of the reinforcing feedbacks of permafrost melting and temperate forest area shrinking, then how much will the resulting temperature rise speed up chemical weathering that removes CO_2? These are complex questions to which we do not have complete answers.

BOX 5.7 FEEDBACKS IN SOCIAL SYSTEMS, POLITICAL SYSTEMS, AND MARKETS

Feedbacks in Social Systems. A reinforcing feedback can influence the behavior of cattle ranchers making decisions about whether to add more cattle to a pasture. If a cattle rancher earns profits from the number of cattle raised each year, then he or she will raise as many cattle as are available to earn a profit. With the profit earned from the previous year, the rancher buys more cattle and adds them to the pasture. This results in more profit that year, and even more cattle for the next year. If this trend continues, and the grass on the pasture is eaten by more and more cattle each year, the system will start to become degraded (Figure 5.26). At some point, it will reach a threshold, beyond which the *grassland supporting many cattle* (Regime 1) shifts to a *desert shrubland that can no longer support the same number of cattle* (Regime 2).

The dominance of a *reinforcing feedback* over a *stabilizing feedback* could ultimately determine whether a city's transportation infrastructure encourages bike commuting. For example, students learn about the climate impacts of CO_2 emissions by cars. As a result, more and more students start biking to work, even though the present bike lane infrastructure on streets that leads from their apartments to campus is not exactly favorable to bicyclists. The city in which the university resides notices this and decides that if student bike commuting continues to increase, then it will invest in additional bike lanes for the city. This results in a reinforcing feedback that could push the system toward a more sustainable transportation

Figure 5.26 A reinforcing feedback between profit and cattle grazing can lead to grasslands degraded by overgrazing.

Figure 5.27 (a) A reinforcing feedback between climate change education and biking could push the transportation infrastructure of a city toward sustainability.

infrastructure that encourages biking (Figure 5.27a). However, at one point during the semester, there is a bad accident between a car and a student biker on a street without a bike lane. The student biker is badly injured, and many other student bikers passing by that day see the injured student. The next day, fewer students bike because they are scared of getting in a similar accident. The city notices that student bike commuting is not increasing. This results in a stabilizing feedback that discourages bike commuting (Figure 5.27b). As the semester goes on, students continue to learn about the climate impacts of CO_2 emissions by cars, and the accident earlier that semester fades in their memory. The next semester, more and more students are biking again. The city notices the increase in student bike commuting and once again commits to investing in bike lane infrastructure if this trend continues. However, there is another biker accident due to a street lacking adequate bike lanes, and student bike commuting declines once again. In this case, two different types of feedbacks are opposing each other (5.27c).

Feedbacks in Political Systems. Most federal regulatory agencies in the United States concerned with the use of public lands, such as the Bureau of Land Management (BLM), have directors appointed by the president of the United States with the approval or consent of Congress. Agency directors guide policies and actions within these agencies, such as leasing BLM lands to private firms for mining or cattle grazing activities. The firms that lease the land profit from its use. Some of that profit is often donated by firms to finance political

(Continued)

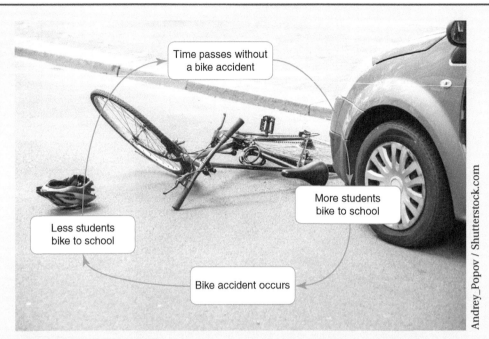

Figure 5.27 (b) A stabilizing feedback between biking accidents and biking could prevent the transportation infrastructure of a city from becoming sustainable.

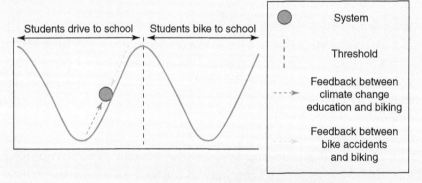

Figure 5.27 (c) The shifting dominance of two opposing feedbacks could ultimately determine whether a city's transportation infrastructure is friendly to bikes.

campaigns for president or for other positions in Congress to those candidates who will appoint regulatory agency directors that favor their private interests. This can lead to more policies and actions within regulatory agencies that favor private interests, and even more profit to fund even more political campaigns (**Figure 5.28**). This *reinforcing feedback* can move the public land system away from *one that serves all citizens* (Regime 1) toward *one that favors private interests* (Regime 2).

A *stabilizing feedback* related to campaign finance is also at play here, which can explain why candidates that do not receive large donations from private firms have a hard time

Figure 5.28 A reinforcing feedback between profit earned by private firms from wood harvested on public land and the donation of some of those profits to political campaigns could lead to a public land system the serves private interests instead of citizens.

being elected to office (Figure 5.29). Candidate A may claim to represent the average citizen, spend countless hours campaigning in many cities across the country, and even raise enough money to run an inspiring advertisement on television that captures the heart and minds of voters across the country. Candidate B, who is funded by wealthy private interests, feels threatened by Candidate A and has many resources at his or her disposal for self-promotion and also to discredit Candidate A. For example, Candidate B could quickly pay for several different television advertisements that reach a diversity of audiences. He could also hire a team of investigators to dig up information about Candidate A that discredits him or her in the eyes of the public. This would crush Candidate A's campaign, and it would take a long time, and a lot of time and resources, to recapture the public's hearts and mind. Once Candidate A's reputation was restored, Candidate B could easily step in and cause political problems for Candidate A again.

Feedbacks in Markets. Many feedbacks found in the market today are reinforcing feedbacks driving rapid economic growth. One *reinforcing feedback* in capitalist economies, in which profit maximization is the underlying goal, is between product prices and wages (Figure 5.30). When prices increase, wages must also increase in order for people to maintain the same standard of living. However, in order for wages to increase, prices also have to increase if the firms establishing the wage increases are to also maintain increases in profits. When firms increase prices, wages will need to rise again to maintain living standards.

(Continued)

Figure 5.29 A stabilizing feedback between campaign effort and monetary resources can lead to a system where only political candidate who receive large sums of money from special interests can win elections.

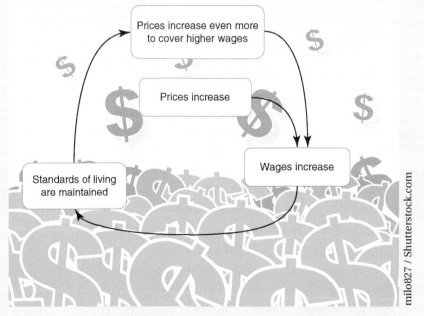

Figure 5.30 A reinforcing feedback between wages and prices can cause prices to continue to increase as wages are adjusted to keep up with the cost of living.

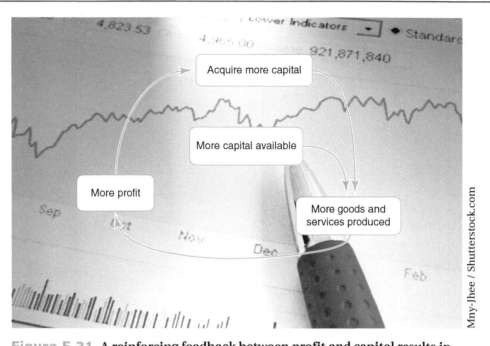

Figure 5.31 A reinforcing feedback between profit and capital results in rapid economic growth.

Another *reinforcing feedback* is between capital and the production of goods and services using that capital to generate profits (Figure 5.31). In economic systems, capital includes factories, equipment, and other infrastructure (manufactured capital); physical labor and mental talents (human capital); and natural resources and services (natural capital), which are all needed for the production and distribution of goods and services. The more capital that is available, the more goods and services can be produced, and the higher the profit. A percentage of this profit can be reinvested to acquire more capital, which will produce more goods and services, and even more profit for reinvestment.

This reinforcing feedback could come to a screeching halt, however, if one of the types of capital just described became limiting. For example, if natural capital were to become badly depleted, then the production of goods and services, and the associated profit from that natural capital, would diminish. Eventually, over long periods of time, natural capital could recover, and economic growth would be able to resume. However, growth would only continue until the natural capital became depleted again and would decline again after that depletion. This describes a *stabilizing feedback* that limits economic growth (Figure 5.32).

Feedbacks in the Human Body. The human body is a complex system regulated by feedbacks. An example of a *stabilizing feedback* in the human body is the regulation of blood pressure by the blood vessels and the heart (Figure 5.33). When blood pressure in your body increases, such as when you become angry or upset, the increase is detected by

(Continued)

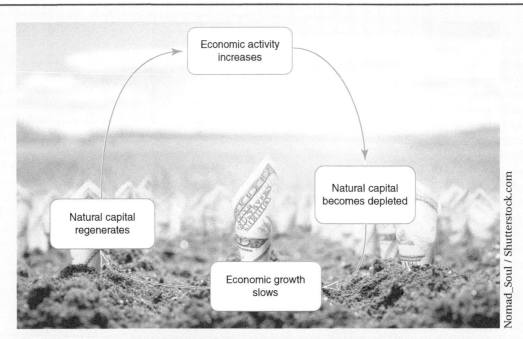

Figure 5.32 A stabilizing feedback between economic activity and natural capital dictates the state of the economy.

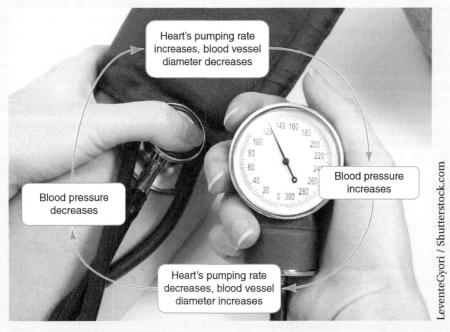

Figure 5.33 A stabilizing feedback between blood pressure and certain physiological processes can keep a person's blood pressure at healthy levels.

receptors in blood vessel walls. These receptors alert the brain, which sends a chemical signal to both the heart to tell it to decrease its pumping rate and to the blood vessels to tell them to increase in diameter. As a result, blood pressure decreases. When blood pressure gets too low, such as when you calm down, the blood vessel wall receptors detect this and tell the brain. The brain sends signals to the heart to tell it to increase its pumping rate and to the blood vessels to tell them to decrease in diameter. This stabilizing feedback keeps blood pressure at about the same level.

An example of a *reinforcing feedback* in the human body occurs during child birth. When a contraction happens, a hormone known as oxytocin is released (Figure 5.34). Oxytocin stimulates more contractions, which stimulates the production of more oxytocin. This feedback continues until the system is pushed into a new regime: the baby is born and the system has moved from *woman with baby in womb* (Regime 1) to *woman without baby in womb* (Regime 2).

Another *reinforcing feedback* in the body occurs to help the body form a clot when you cut yourself with a sharp object (Figure 5.35). Pretend you cut your finger with a knife while cutting onions. The injured tissue releases chemicals into your blood stream. These chemicals activate platelets, which are the clotting agents in your blood. The activated platelets release more chemicals, which activate more platelets. This process continues until a clot forms and no additional clotting agents are needed. At this point, the system has shifted from *body letting out blood* (Regime 1) to *body retaining blood with a clot* (Regime 2).

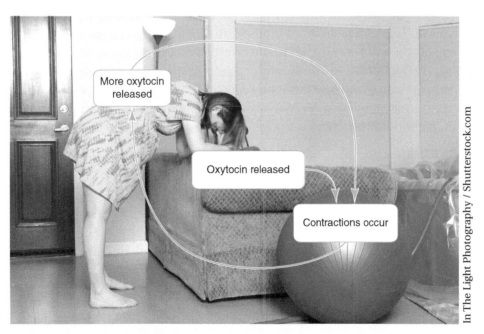

Figure 5.34 A reinforcing feedback between oxytocin production and contractions eventually results in birth.

(Continued)

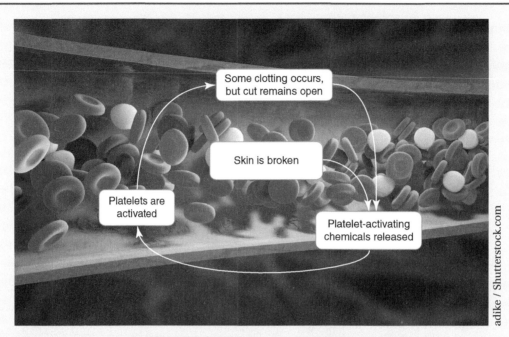

Figure 5.35 A reinforcing feedback between platelets and chemicals in the blood eventually results in a wound fully sealed with a clot.

Temperature Control in a House. When heating a house in the winter, there is a *stabilizing feedback* between the heater and the temperature of the house, as indicated by the thermostat that keeps the house at a constant temperature. For example, if the thermostat is set at 70°F, then the heater will add warmth to the house until the temperature has reached 70°F. At that point, the heater will turn off, and the addition of heat into the house decreases. When the temperature of the house becomes lower than 70°F, the heater will turn back on or be increased. This stabilizing feedback continues to keep the house in its current state with respect to temperature (70°F).

If there were a disturbance that caused a change in the house, such as squirrels chewing a big hole in the roof, then a reinforcing feedback would regulate the system until it shifted to a new regime. In cold climates, snow lying on the roof of a house can act as an insulator that keeps heat in the house. With a new hole in the roof, heat would escape, and some of it would melt the snow surrounding the hole on the way out. This would cause the temperature in the house to drop below 70°F, and the addition of warmth to the house by the heater would intensify until the house temperature rose again to 70°F. This additional heat would melt additional snow surrounding the hole, which would result in more heat added. Eventually, all of the snow surrounding the hole would be melted as the system shifted from *house with no hole* (Regime 1) to *house with hole* (Regime 2). The conditions of Regime 2 result in a higher heating expense for the owners of the house. Can you draw diagrams that represent the reinforcing and stabilizing feedbacks that maintain the temperature of a house?

BOX 5.8 THE TRANSFORMATION OF SWEDEN'S CAR CULTURE: FROM GASOLINE TO BIOFUELS

Sweden is a world leader in weaning the world from fossil fuel energy. In 2008, the country acquired only 30% of its energy from fossil fuels, compared to 77% in 1970. (In 2008, in comparison, the United States obtained 85% of its energy from fossil fuels.) How did this large change happen for Sweden in less than 40 years? One contributing factor was the transformation of Sweden's car culture from vehicles run on fossil fuels to vehicles run on ethanol. Greenhouse gas emissions from corn-based ethanol fuel are the same, or worse, than emissions from gasoline, and using a food source (corn) for fuel drives up food prices. However, ethanol fuel made from sugarcane, cellulose, forestry wastes, or other waste streams emits 85%–90% less greenhouse gases than gasoline and does not drive up food prices. One courageous man, along with a small group of other committed and like-minded individuals, made the Swedish regime shift from gasoline-powered to ethanol-powered vehicles happen.

In the early 1990s, Per Carstedt became passionate about sustainability and about the possibility of bringing flexi-fuel vehicles, which he observed using ethanol fuel while living in Brazil (Figure 5.36a and 5.36b), to his Ford dealership in Sweden. When he proposed this idea, people said the

(a)

Figure 5.36 (a) For more than 40 years, Brazil has used sugar cane and sugar cane waste to produce ethanol fuel at joint sugar cane mill-ethanol manufacturing plants, as shown here, and this has reduced Brazil's greenhouse emissions by more than 60% compared to gasoline-powered cars.

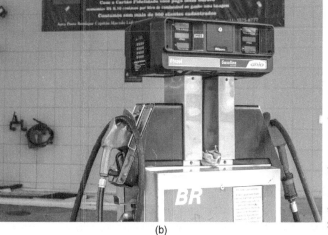

(b)

Figure 5.36 (b) In the mid-1970s, the Brazilian government mandated that gasoline contain 10%–22% ethanol, and by 2015, the percent of ethanol fuel required in gasoline had increased to more than 25%.

(Continued)

cars would not work in Sweden because it was too cold and there was no market for them. These claims eventually proved to be false. A very real problem did exist, however: if flexi-fuel vehicles were brought to Sweden, there was the "chicken-and-egg" problem of no filling stations. In Carstedt's words, as written in Peter Senge's 2008 book *The Necessary Revolution*: "We had no cars because we had no filling stations, and we had no filling stations because we had no cars." This is a stabilizing feedback, which keeps a system in its current state (Figure 5.37). What did Carstedt do to solve this problem? In his words, from Senge's book: "We just got started." In 1995, with the help of the head of Ford's small flexi-fuel program in Detroit, three cars were imported to Sweden. Carstedt thought that this might be enough to "wake up the market," but it was not. Over the next several years, he worked with Ford and the Swedish BioAlcohol Fuel Foundation to form a consortium of Swedish businesses, municipalities, and individual people willing to buy 3,000 cars. Concurrently, Carstedt worked on the filling station problem. By the time 50 flexi-fuel vehicles had been imported, he had persuaded two filling stations to install ethanol pumps, with one being in the capital city of Stockholm. With the help of the Foundation, and using the simple strategy of persuading one fuel retailer at a time, 1,000 filling stations, or 25% of all filling stations in Sweden, had installed ethanol pumps.

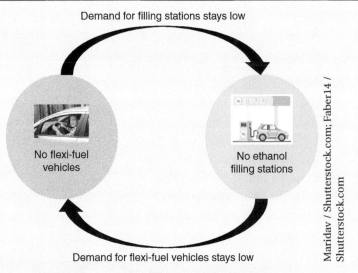

Figure 5.37 A stabilizing feedback between availability of flexi-fuel vehicles and existence ethanol filling stations kept Sweden's vehicle fueling system in an unsustainable state for a long time.

In addition to filling stations, Carstedt worked with ethanol suppliers and car companies. First, he sought out like-minded individuals to buy the company that imported ethanol to Sweden. He did this to ensure that the owner's agenda was to develop ethanol biofuels as a long-term alternative energy system, not just capitalize on a short-term investment. He also sought out like-minded individuals at other car companies, such as Saab, who were interested in developing flexi-fuel vehicles through pilot programs. Finally, he worked with like-minded individuals at universities and other R&D institutions to advance ethanol production technologies, such as using cellulose-based wood chip waste from Sweden's large forestry industry. With a solid network of support for ethanol biofuels in Sweden and public concern for climate change on the rise, reinforcing feedbacks eventually overcame stabilizing feedbacks, as Sweden's vehicle fuel system shifted from a regime dominated by gasoline to one dominated by ethanol (Figure 5.38).

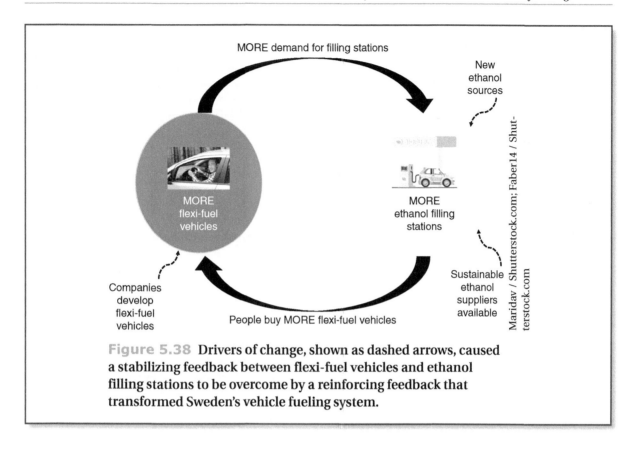

Figure 5.38 Drivers of change, shown as dashed arrows, caused a stabilizing feedback between flexi-fuel vehicles and ethanol filling stations to be overcome by a reinforcing feedback that transformed Sweden's vehicle fueling system.

Section 5.2.2—System Fluctuations and Feedbacks

Key Concept 5.2.2—One major reason for system fluctuations is the continual interaction among system components through feedbacks over time.

An explanation of how feedbacks drive population growth in human societies over time can be used to illustrate how system fluctuations are caused by continual interactions among system components through feedbacks. At a given carrying capacity, a stabilizing feedback between population and arable land can keep human populations from changing too much. (Figure 5.39). As a result of this stabilizing feedback, the population fluctuates around the carrying capacity, rather than remaining exactly at the carrying capacity, as a result of the following interactions among system components: Arable land is the land available for agriculture. As human populations increase (1 in Figure 5.39), they use more and more arable land to grow food. Once populations reach carrying capacity (2), arable land becomes degraded by soil erosion, reducing soil fertility. This results in less arable land and human populations respond by decreasing (3). When the degraded land is not used to grow

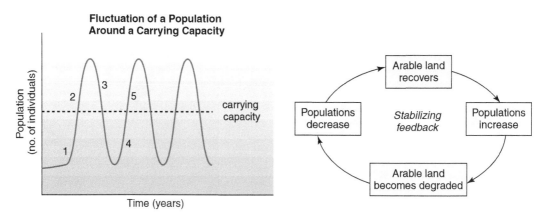

Figure 5.39 A stabilizing feedback keeps human populations relatively stable over time.

food over long periods of time, it has time to recover to become farmable once more, as soil reforms and nutrients are regenerated (4). This allows human populations to increase again (5). As a result of this stabilizing feedback, human populations fluctuate around the carrying capacity over long periods of time (Figure 5.39). These fluctuations are not regime shifts, but only changes in the population size within a given regime. When put into the context of the ball-and-basin model, these fluctuations can be viewed as the ball rolling around within the same basin, but not rolling into a new basin.

Humans have drastically increased their planetary carrying capacity through technology. This is, in part, the result of a reinforcing feedback between human populations and technology (Figure 5.40). When humans first evolved as a species, they had a lower carrying capacity than today (1 in Figure 5.40). Then, humans developed technologies, such as tools for hunting and gathering, that made food acquisition more efficient. More food availability raised the carrying capacity for human populations to K_1 (Figure 5.40). The number of people eventually increased to meet this new carrying capacity (2). More people, with more ideas, and more time for invention with less time spent on food acquisition, resulted in even more technologies, such as the plow. This further raised the carrying capacity to K_2 and human populations again grew to meet it (3). Agricultural systems resulted in people settling down in one place and societies became organized in a new way. People had even more time to innovate. Eventually, technologies developed during the Industrial Revolution, such as sanitation systems, medical advances, and the rise of fossil fuel energy, increased the carrying capacity even more (to K_3). Human populations grew again (4). Each new carrying capacity can be considered a different technological regime for human societies: hunting and gathering (Regime 1), agricultural (Regime 2), and industrial (Regime 3).

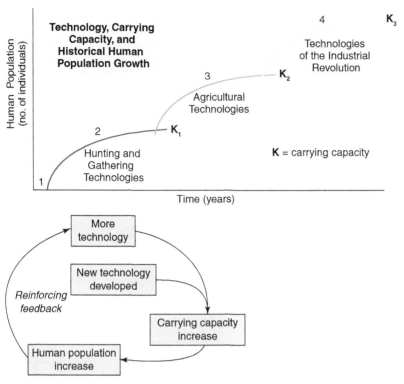

Figure 5.40 A reinforcing feedback has allowed human populations to increase.

Figure 5.41 brings together both of these feedbacks and illustrates their shifting dominance. Both are always present to some degree, but the direction of change, and the behavior of the system at any given time, depends on which feedback is stronger. When the system is stable at a certain carrying capacity, the stabilizing feedback is dominant (blue arrows in Figure 5.41). The stabilizing feedback example between population and arable land just described was focused on human populations reliant on agriculture as their primary food source (at K_2 in Figure 5.40). However, similar stabilizing feedbacks keep populations of hunter-gathers at K_1 and industrialists at K_3. For example, a band of hunter-gathers might live on a plot of land that becomes continually degraded with human waste and depleted food resources until they are forced to move. During their migration, they might lose members of their band to predators or during violent conflict with other roving bands. Once they find a place to settle with appropriate resources, populations can thrive again. There is generally less population fluctuation in industrial societies, due to technologies that promote relatively stable birth and death rates, but there is still fluctuation due to factors such as disease, war, and natural disasters. Whatever the specific situation, a system remains in its current regime when stabilizing feedbacks are stronger than reinforcing feedbacks. Even at stable

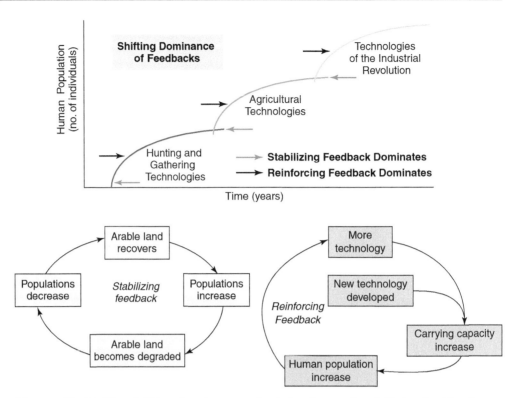

Figure 5.41 The shifting dominance of reinforcing and stabilizing feedbacks has led to periods of population growth and relative stasis.

population levels, humans continually develop new technologies; but it is not until a new development sparks a reinforcing feedback that population change occurs again.

When the system is changing, reinforcing feedbacks dominate (black arrows in Figure 5.41). When these feedbacks dominate, technological innovations overcome the limitations imposed on human population growth by stabilizing feedbacks. This happened with hunter-gathers when they developed, and successfully implemented, technologies for agriculture. They were able to acquire much more food than ever before and populations grew exponentially for a time. However, even during growth phases, there are still population fluctuations. For example, as people settled down into agricultural communities, population densities grew and disease was more easily spread. A major disease epidemic could temporarily reduce population size, even during a period of overall population growth. A natural disturbance, such as drought or an abrupt ice age, could have a similar effect. Despite these possible setbacks, populations eventually reach a new carrying capacity where stabilizing feedbacks take over.

In SESs, this pattern of periods of fundamental change, dominated by reinforcing feedbacks, and periods of stability, dominated by stabilizing feedbacks, is never ending. The fluctuations that occur in SESs throughout both of these stages, as a result of the constant interplay among the two types of feedbacks are also never ending. Thus, using indicators to determine when systems are fundamentally changing to new regimes, and when they are stable, can be difficult due to inherent fluctuations. Figure 5.40 shows how human populations fluctuate around a carrying capacity due to a stabilizing feedback. Figure 5.41 shows how human populations grew to meet each new carrying capacity when reinforcing feedbacks dominated, but population fluctuations disappeared in this graph. Figure 5.42 shows an example of what the solid lines shown in Figure 5.41

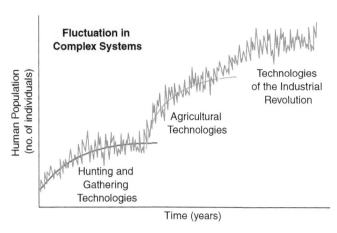

Figure 5.42 Fluctuations, which are always occurring in systems, are distinct from regime shifts, which are not constantly happening.

might look like in reality, with constant fluctuations in the system shown. The fluctuations in the graph are not based on actual data because it is not available in such detail, especially for very early human populations. Rather, the fluctuations are shown to illustrate how SES complexity means that these systems are never still or unmoving. Change is a constant and complex systems are primed for it.

This section on feedbacks will end with a comment about feedbacks between natural and human systems. It is important to understand feedbacks in systems, in order to work with variability to promote resilience and sustainability. It is also important to work to strengthen feedbacks between natural and human systems when trying to move SESs toward sustainability. As stated by Brian Walker and David Salt in their book *Resilience Thinking* (2006):

> *Tightness of feedbacks refers to how quickly and strongly the consequences of a change in one part of the system are felt and responded to in other parts. Institutions and social networks play key roles in determining tightness of feedbacks. Centralized governance and globalization can weaken feedbacks. As feedbacks lengthen, there is an increased chance of crossing a threshold without detecting it in a timely fashion. (p. 121)*

> *A resilient socio-ecological system would strive to maintain, or tighten, the strength of feedbacks. They allow us to detect thresholds before we cross them. Globalization is leading to delayed feedbacks that were once tighter; the people of the developed world receive weak feedback signals about the consequences of their consumption of developing world products. Feedbacks are loosening at all scales. (p. 146)*

Thus, strengthening feedbacks between natural and human systems can make socioecological systems more resilient. One of the case studies in *Resilience Thinking* describes a situation in an Australian river basin called the Goulburn-Broken catchment where this type of strengthening occurred. (This case study was described in detail as part of General Question 3 in the End of Chapter Questions for Chapter 2.) In the 1970s, dairy pastures and many high-value crops in the catchment were destroyed by a flood. Not only did the flood result in soggy soils, but it caused groundwater levels to rise. This was a problem because the naturally salty groundwater harmed vegetation. The flood had a huge impact on the livelihoods of local communities and there was a large response to the crisis. Prior to the crisis, water in the catchment was managed by the state and federal government without local community input. In this governance regime, feedbacks between natural and human systems were weak. After the 1970s flood, water governance institutions were reorganized into regional institutions named Catchment Management Authorities (CMAs) composed of local community groups, governments, and the private sector. This increased the resilience of the river basin by strengthening the feedbacks between natural and human systems. In this case, the crisis caused by the flood was enough to push the system over a threshold from *centralized federal management* (Regime 1) to *local and regional governance* (Regime 2).

Section 5.3: Adaptive Cycles and Resilience

Core Question: How is the resilience of SESs affected over long time scales?

This section concludes this chapter with a brief introduction to *adaptive cycles*, which is another tool for thinking about SES complexity. Adaptive cycles characterize patterns of change that influence resilience. They are a feature of complex adaptive systems (CASs), which are the focus of the next chapter. Thus, many of the underlying mechanisms that govern these cycles are left for Chapter 6. However, one mechanism—the shifting dominance of feedbacks—that was introduced in this chapter, will be used to explain why systems pass through adaptive cycles. The four phases of an adaptive cycle are generally described in terms of how resilience—the ability of a system to absorb impacts, but still maintain the same conditions, function, or identity—is affected as systems pass through each phase.

Section 5.3.1—Introduction to Adaptive Cycles

Key Concept 5.3.1—Complex adaptive systems pass through four general phases that affect their resilience, which influences their response to disturbances affecting the system.

Adaptive cycles describe a sequence of four phases that some, but certainly not all, natural and human systems pass through as they evolve over relatively long periods of time (Figure 5.43). A system exhibits varying degrees of resilience, due to changes in the strength of interactions and feedbacks within the system, as it passes through each phase. Adaptive cycles have been most extensively documented by ecologists, but the idea originally arose from an Austrian economist Joseph Schumpeter's analysis of boom-and-bust cycles in economies. Adaptive cycles have also been applied to other systems, such as to explain the rise and fall of human societies throughout history, and to the SESs of concern to sustainability (See **Box 5.9** at the end of this section for a sustainability case study.). The adaptive cycle concept is illustrated with one ecosystem example and one economic system example in Section 5.3.2, following a general description of each phase in this section.

The phases of an adaptive cycle are characterized by their connectedness and potential (Figure 5.43). Connectedness (x-axis) represents the degree of connectedness, or the "tightness" of connections and feedbacks, among internal system components. Potential (y-axis) represents a system's potential for change, in response to changes in the external environment. For example, in the release phase, there is high connectedness among system components, but low potential for change.

Release Phase (Ω). This phase is often triggered by a disturbance (e.g., forest fire, economic crisis, or war) that is large enough to exceed a system's resilience. This phase is very short in duration as compared to the other three phases. During this phase, the structural complexity of the system is radically reduced, such that the previously strong interactions among system components are broken and the materials, energy, and information contained within these components are released. Feedbacks also weaken during this phase.

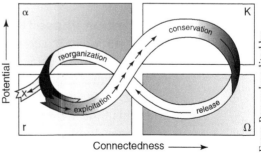

Figure 5.43 Short arrows represent times when the system is changing slowly, whereas longer arrows are periods of relatively more rapid change.

Reorganization Phase (α). After the disturbance, there is a period of chaos and uncertainty that lasts a bit longer than the release phase, but is still quite short relative to the other two phases. During this time, there is very little resistance to new innovations (e.g., new types of vegetation, new designs, or inventive ideas) becoming established for the long term. As a result, small chance events can powerfully shape the future system, such as foreign seeds blowing in from outside the forest and establishing a completely different forest after a fire. The reorganization phase can be viewed as a "window of opportunity" for change following the release phase, which can lead to a regime shift, after which a completely new system is established (compared to the one that existed before). During this phase, transitions to sustainability may occur most readily. However, it is also at this point that a system can completely break down and exit from the adaptive cycle (as represented by an x on the left in Figure 5.43). When this happens, the system no longer has the resources to maintain an organized structure governed by interactions, and feedbacks, among components.

Exploitation Phase (r). This is a rapid growth phase where there is constant change in the features of a system, such as the components, interactions among the components, and feedbacks. During this phase, change is the rule rather than the exception. This is the second longest phase, during which the human actors or biological species, which became established during the chaotic reorganization phase (α), experience rapid growth. The connectedness or interactions among system components remain weak during this phase, as all system components scramble to individually exploit and take hold of all the abundant available resources. During this phase, different system components "seize the day" and jump at opportunities that allow them to become better established.

Conservation Phase (K). There is a relatively slow transition from the exploitation to the conservation phase, as system components gradually become more established and grow. Interactions and feedbacks among system components become stronger and stronger as this phase progresses. It is very difficult to establish new innovations in the system during this time, as system components become more specialized, and certain ones dominate and become "locked in." The system becomes so stable and rigid that change is the exception rather than the rule. However, this also makes the system very rigid and reduces its resilience in the face of external disturbances. As a result, the system is increasingly susceptible to disturbances, such that it is "primed" for change.

Section 5.3.2—Adaptive Cycle Application

Key Concept 5.3.2—Adaptive cycles have been applied to natural systems to understand ecological succession and to economic systems to understand boom-and-bust cycles.

Ecological Succession in a Forest. A disturbance that starts a fire in a forest, such as a lightning strike or a camp fire left unattended, ignites the release phase (Ω). This disturbance must be large enough to exceed the ecosystem's resilience in order to break apart the tight web of interactions among the system's components. When this does happen, the forest ecosystem burns down and comes completely undone. Resources such as nutrients accumulated in the biomass of trees, shrubs, and grasses are released into the ecosystem and made available for new organisms during the reorganization phase (α). During reorganization, small chance events can powerfully shape the future. Seeds from a shrub species, not observed in the forest before the fire, might be brought in by a migrating flock of birds, and could become well enough established so that the new shrub species will dominate a **niche** in the new ecosystem that regrows after the fire. This is also a time when forests are most vulnerable to invasive species, which might have a detrimental impact on the forest. During the exploitation phase (r), species that have established themselves experience a stage of rapid growth during which they exploit the resources left over after the fire disturbance (i.e., nutrients from burned vegetation). As vegetation grows, especially the weeds, grasses, and pioneer species at first, forest biomass begins to accumulate. In the conservation phase (K), biomass begins to accumulate more slowly. In addition to slow accumulation, more and more of it is stored in unavailable forms, such as the heartwood of trees or as dead organic matter and detritus on the forest floor. As this biomass accumulates, the forest becomes less resilient, and once again vulnerable to disturbance by lightning or a negligent camper. At this point, the cycle of release, reorganization, exploitation, and conservation is "primed" to begin again.

> **niche** describes the "way of life" of a given species as defined by the total uses made by the species of the abiotic and biotic resources available in the environment in which it lives.

The Role of Feedbacks in Ecological Succession. The process of ecological succession, and adaptive cycles in general, are driven in part by feedbacks and their shifting dominance through time. During ecological succession, forests repair themselves after a disturbance by passing through several stages defined by distinct biological communities, starting with a pioneer community of grasses and young shrubs, to a mix of shrubs and trees, and finally to a climax community of mature trees with some shade-tolerant shrubs in the understory (**Figure 5.44**). The interplay between stabilizing and reinforcing feedbacks regulates this process, as the forest moves from one distinct biological community

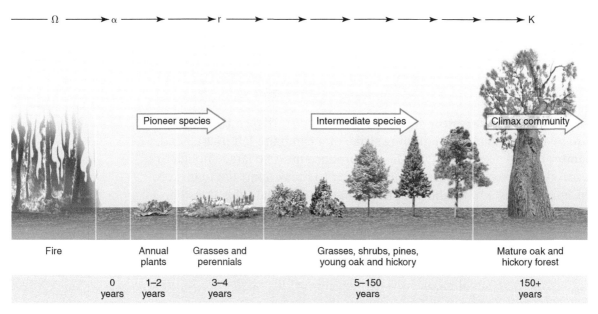

Figure 5.44 Distinct biological communities mark the different stages of ecological succession in a forest.

to the next. Each distinct biological community can be conceptualized as a regime, and the change from one distinct biological community to the next as a regime shift.

When a disturbance, such as a wildfire, leaves a forest floor open and barren, seeds of various plant species begin to drift in on the wind, by water, or with animals (Figure 5.45a). Pioneer species, such as grass, grow very quickly and become the first dominant species in the new ecosystem. Seeds for shrub plants are also able to take root, but they do not grow as quickly as grasses. For years, the ecosystem remains in a regime dominated by grasses with scattered shrubs. These two plant types compete for rainwater and sunlight. Grasses have very shallow roots and can intercept rainwater water as it flows down through the soil, whereas shrubs have deeper roots that miss out on the rainwater that is taken by grasses. The interception of most of the rainwater by grass makes shrub growth extremely slow. Shrubs will eventually grow large enough to shade out grasses, but not until shrub density reaches a critical point, or threshold. Until then, a stabilizing feedback keeps grasses dominant (Figure 5.45b). When a reinforcing feedback comes into play, as the shrub density reaches a threshold (Figure 5.45c) and shrub shading starts to affect grass growth, grass becomes less abundant and intercepts less water. Shrubs have more water to grow, and more shrubs grow to further shade out the grass. This reinforcing feedback continues until shrubs are the dominant species in the next regime of ecological succession.

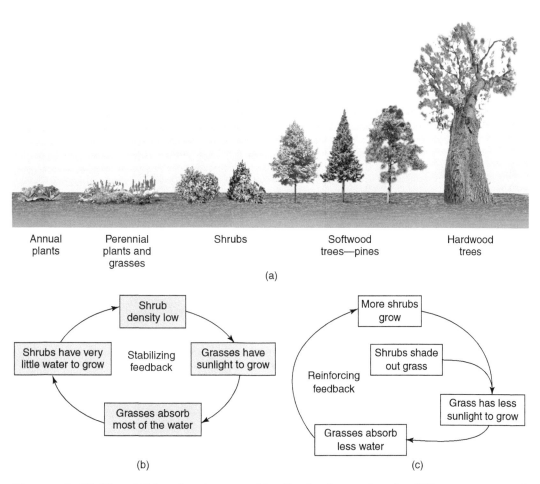

Figure 5.45 The shifting dominance of feedbacks determine the different stages and regime shifts that occur during ecological succession.

Progression of the American Automobile Industry. In the mid-20th century, gas was inexpensive relative to previous periods, and demand for fashionable, gas-guzzling, tail-finned automobiles, such as the Cadillac, was high (Figure 5.46). At the time, the American automobile industry was in the conservation phase (K) and was "locked in" to the production of these types of large cars, allocating most of their resources to the factories and infrastructure required to manufacture and sell the tail-finned automobile. In the 1970s, the system was disturbed by rapidly increasing fuel prices, and demand suddenly shifted toward smaller and more fuel-efficient cars. Along with new and intense foreign competition, the American auto industry fell apart and experienced a release phase (Ω). After this release, the reorganization phase (α) began as chaos and uncertainty now ruled the American auto industry. The system was open to innovations in car technologies

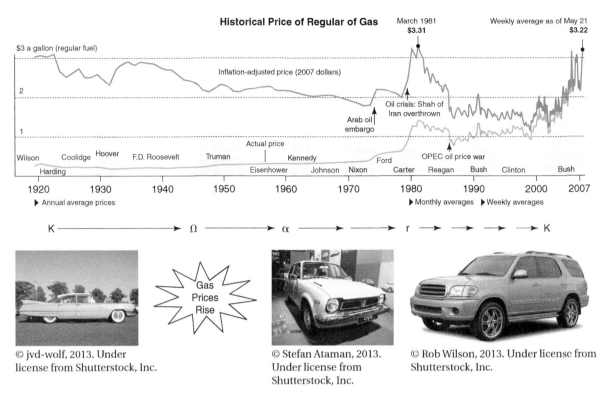

Figure 5.46 The price of gasoline has been a major driver in the progression of the U.S. auto industry.

and ways of conducting business. This "window of opportunity" for change attracted innovators and entrepreneurs to exploit the industry's resources left over from the release phase. Rapid growth and constant changes in car technologies during the exploitation phase (r) became the norm, as innovators and entrepreneurs scrambled to take hold of abundant resources. Eventually, things settled down into a conservation phase (K), during which innovations were only slowly introduced into the rigid and specialized system, if they were introduced at all. The auto industry eventually stabilized into a system dominated by SUVs and other types of "bigger and better" gas-guzzling vehicles. Many believe that the system is once again "primed" for collapse if gas prices rise, consumer demand for smaller more fuel-efficient foreign cars continues to increase, or some other disturbance affects the system in the future.

These two examples illustrate how some types of systems, particularly complex adaptive systems, change and evolve over long periods of time. During the exploitation (r) and conservation (K) phases of the adaptive cycle, resources are gradually accumulated and become more and more specialized. This leaves little room for innovation or major changes to the system, but this type of rigid stability also makes

BOX 5.9 ADAPTIVE CYCLES AND CLIMATE CHANGE ON THE YUCATAN PENINSULA

Researchers Mark Pelling and David Manuel-Navarrete used adaptive cycles, as applied to social systems, as a tool for understanding the lack of action on climate change, and the potential for future action, in two cities located on Mexico's Yucatan Peninsula in Quintana Roo State: Mahahual and Playa del Carmen (Figure 5.47a and 5.47b). The same climate-related threats affect the two cities, including short-term weather-related phenomena, such

Figure 5.47 (a) Mahahual and Playa del Carmen are located on the Yucatan Peninsula south of Cancun, which is a popular Spring Break destination for college students.

Figure 5.47 (b) Both Mahahual and Playa del Carmen are located on the Caribbean Sea and have vacation beach resorts for tourists.

(Continued)

as hurricanes and heat waves, and longer-term threats such as sea level rise. However, the Quintana Roo State Government does not take all of these threats seriously. For hurricanes, a good early warning system is in place, and the state and federal governments together help support evacuation and rebuilding. In contrast, there has not been as much action to address the other threats from climate change or to mitigate climate change itself.

Inaction on climate change appears to be due to the fact that social systems in these cities are stuck in the conservation phase (κ, Figure 5.43). Nothing is changing because the actions of individual people reinforce the actions taken by existing institutional structures, such as government agencies, and vice versa. Individuals could take action, for example, by rising up in protest to demand that the state and federal government, and other institutions, take action on climate change. Formal institutions, such as governments and corporations, could take widespread action on climate change using a top-down approach such as passing laws that influence people's behavior in a way that supports climate change mitigation. Neither the bottom-up actions of individuals, nor the top-down actions of institutions, are adequately addressing climate change due to interactions between individuals and institutions that reinforce their behaviors and keep them from changing.

According to interviews conducted by Pelling and Manuel-Navarrete at the time their study was conducted, most people living in Mahahual and Playa del Carmen valued improvements in their quality of life as individuals, and for their families, over any kind of community action or public goods. They were concerned about the impacts of hurricanes on themselves and their families, but believed it was the government's responsibility to protect them. Along these same lines, their views were that action on climate change, and it's slower moving threats such as sea level rise, was not a local concern and it had minimal influence on their lives. The few individuals that did believe that action was a local concern were not supported by a larger community of people who believed this. Therefore, it was difficult to take action as an individual or community.

There are differences between the two cities, however. Playa del Carmen is a much larger city and large corporations exerted greater control over the local economy (Figure 5.48). The connection between people and place was not as strong, and

Figure 5.48 The many shops found along Playa del Carmen's Fifth Avenue highlight an economy dominated by global corporations.

most did not want to settle in the city for the longer term. They were there mostly to make money. The result was that the city lacked community, and there was a general unwillingness to hold corporations, developers, and the government accountable for action on climate change. In contrast, residents of the smaller city of Mahahual had a stronger connection to place. It was largely a frontier settlement with many immigrants. They also felt more empowered about a local influence on the future of their city, although the focus was on improving infrastructure and promoting the local economy (not mitigating climate change).

Social capital is a required element for building a more sustainable world. It depends on productive social relationships, among different stakeholder groups, that are grounded in mutual trust. Sufficient social capital for action on climate change did not exist among individuals and institutions in Mahaual and Playa del Carmen. In both cities, trust was strongest among family members and groups of friends. These tight social circles were largely closed to community organizers and those with different values and beliefs. This made it difficult to organize for climate change action. In Mahahual, there were hardly any community organizations and those that did exist were very weak. The frontier mentality of reliance on self, and the diverse population that resulted from high rates of immigration, led to general mistrust of community organizations and anyone who stood apart of the masses as a leader. In Playa del Carmen, community organizations did exist, but they were top-down and strongly tied to the government and corporations who maintained the status quo. There was also not a lot of outreach to the public by these organizations, so citizens were not aware of their plans or opportunities for involvement.

Despite the overwhelming dominance of existing social structures, there have been small windows of opportunity for change. When hurricane Dean hit the Yucatan Peninsula in 2009, Mahahual was among the most damaged cities. After the hurricane, established institutions broke down and the system entered a release phase (Ω, Figure 5.43). This opened up an opportunity for transition away from the dominate regime. Residents became willing to organize as a community for the rebuilding effort. People also organized to seek out economic activities that could make up for the temporary loss of their cruise terminal, which was damaged in the hurricane and closed for 18 months. Sometimes natural disasters, such as a hurricanes, can provide a window of opportunity for change. However, the new community organizations in Mahahual were not strong enough to persist and resist the influence of the existing institutions of the dominant regime. They were ultimately "re-absorbed" by existing social structures and things continued as they had in the past.

When promoting change for sustainability, it is important to be aware of adaptive cycles. In some phases of the cycle, such as in the conservation change, systems are less "primed" for change. However, systems that are in the release and reorganization phases often present opportunities for change toward sustainability.

the system increasingly vulnerable to disturbances that will release (Ω) resources, such that they are again "up for grabs" and can be reorganized (α) in new ways by innovators. Thus, an uncertainty about the future of the system, and the introduction of innovation through experimentation, define the release (Ω) and reorganization (α) phases. As a result, these two phases are often referred to as instances of "creative destruction" because of the opportunity to initiate the beginnings of a fundamentally new system.

In the context of SES sustainability, it is during these time frames (Ω and α phases) that human activities can profoundly influence systems in terms of changing a currently unsustainable system to a more sustainable one. This type of thinking about sustainability transitions falls under the "hurry up and wait" adage. This adage applies to situations where there are periods of rushing around and working hard to develop and prepare new innovations, such as technologies and policies that can shape the future, even if at the time it seems unlikely they will be adopted. This rushing around is followed by long periods of waiting for brief windows of opportunity through which change can arise. Thus, using the adaptive cycle framework offers hope for the future in the face of the daunting challenges presented by sustainability problems. On the one hand, actions taken to address these challenges may be ineffective because they are taken at an inopportune time in an adaptive cycle. On the other hand, actions taken at an advantageous time in an adaptive cycle can be extremely effective, making it possible for fundamental change toward sustainability to occur in an instant.

Guidelines for fostering resilience. In 2012, Reinette Biggs of the Stockholm Resilience Centre led a team of sustainability scientists who collectively came up with seven principles for enhancing resilience of ecosystem services, which can serve as a set of guidelines for managing resilience of the complex SESs in which sustainability problems are embedded (**Table 5.2**). These seven principles are briefly described here, but the interested reader should refer to this chapter's bibliography for more detailed information. The first three principles are focused on key aspects of the SES being managed, such as the grassland systems described in this chapter. The ideas behind Principle 1 are touched on under the Diversity and Resilience heading in Section 5.1.3 of this chapter. Principle 2, Connectivity, refers to the strength or "tightness" of feedbacks in a systems, as described in the excerpt from Walker and Salt's book *Resilience Thinking*, which is found in Section 5.2.2. Slow variables, which are part of Principle 3, will be described in detail in Chapter 6 (along with fast variables). In short, slow variables govern the underlying structure, and therefore resilience, of a system over relatively long time scales. Principles four through seven refer to attributes of the governance system under

Table 5.2 Seven general principles for enhancing the resilience of ecosystem services

SES properties to be managed
1. Diversity and redundancy
2. Connectivity
3. Slow variables and feedbacks
Attributes of the governance system
4. Understanding SES as a complex adaptive system
5. Learning and experimentation
6. Participation
7. Polycentricity

which an SES is being managed. Principle 4 refers to the need for scientists and resource managers to recognize key characteristics of complex adaptive systems, CASs, which are the subject of Chapter 6. Briefly, these characteristics include emergent properties, behavior that results from both internal system dynamics and changing conditions outside of the system, and the ability to adapt over time to changing external conditions. Principle 5 advocates for constantly learning about changing conditions through "management experiments," which reveal how an SES might respond to management actions or other disturbances. Participation (Principle 6) denotes engagement of a range of different stakeholders in decision-making processes about SES management, for the reasons discussed in Sections 2.2.3 and 3.3 of this textbook. Finally, Polycentric governance systems (Principle 7) consist of multiple governing bodies at a variety of spatial scales, such as local, regional, and nation. Benefits to polycentric governance include matching the scale of governance to the scale of the problem (e.g., a local issue is addressed by a local governing body), providing redundancy if one or more levels of governance fail (e.g., a local institution might fail to protect a watershed, but a regional institution could succeed), and enhancing opportunities for learning and experimentation (Principle 5) and broad participation in governance (Principle 6).

This chapter described different patterns of change that can occur in SESs in order to help you understand their complexity, and to have better knowledge about how to resolve the sustainability problems embedded in them. The focus of this chapter was on patterns of change that affect a system's resilience, which is its ability to absorb disturbances and still maintain the same fundamental structure, in terms of system components, and functions, in terms of interactions and feedbacks among those components. As described in Section 5.2, a system undergoes a regime shift when reinforcing feedbacks are stronger than the stabilizing feedbacks that keep a system in its current regime. However, as implied by the adaptive cycles concept presented in this section, for some systems, the ability to absorb disturbances also depends on how long-term changes affect its resilience. These long-term changes happen in complex adaptive systems because these types of systems tend to evolve and change over time in ways that help them adapt to changing conditions in the environment external to the systems. Understanding this type of change, and the implications for understanding the complexity of some SESs that exhibit this type of change, is the topic of the next chapter.

Bibliography

Baral, Nabin, et al. 2010. Growth, collapse, and reorganization of the Annapurna Conservation Area, Nepal: an analysis of institutional resilience, Ecology and Society, 15(3), Available online: http://www.ecologyandsociety.org/vol15/iss3/art10/ (Last accessed: May 5, 2016).

Barnosky, A. D., et al. 2012. "Approaching a State Shift in Earth's Biosphere." *Nature* 486: 52–58.

Beier, Colin M. et al. 2009. Growth and collapse of a resource system: an adaptive cycle of change in public lands governance and forest management in Alaska, Ecology and Society, 14(2), Available online: http://www.ecologyandsociety.org/vol14/iss2/art5/ (Last accessed: May 5, 2016).

Biggs, R., M. Schlüter, D. Biggs, E. L. Bohensky, S. BurnSilver, G. Cundill, V. Dakos, T. M. Daw, L. S. Evans, K. Kotschy, A. M. Leitch, C. Meek, A. Quinlan, C. Raudsepp-Hearne, M. D. Robards, M. L. Schoon, L. Schultz, and P. C. West. 2012. "Toward principles for enhancing the resilience of ecosystem services." *Annual Review of Environment and Resources* 37: 421–448.

Chapin, F. S., C. Folke, and G. P. Kofinas. 2009. "A Framework for Understanding Change." In *Principles of Ecosystem Stewardship: Resilience-Based Management in a Changing World*, edited by F. S. Chapin, G. P. Kofinas, and C. Folke, 3–28. New York: Springer.

Chapin, F. S., G. P. Kofinas, and C. Folke. 2009. *Principles of Ecosystem Stewardship: Resilience-Based Natural Resource Management in a Changing World.* New York: Springer.

Chapin, F. S., and G. Whiteman. 1998. "Sustainable Development of the Boreal Forest: Interaction of Ecological, Social, and Business Feedbacks." *Conservation Ecology* **2** (2): 12. Available online at www.consecol.org/vol2/iss2/art12/.

D'Odorico, P., A. Bhattachan, K. F. Davis, S. Ravi, and C. W. Runyan. 2013. "Global Desertification: Drivers and Feedbacks. *Advances in Water Resources* 12: 326–44.

Gunderson, L. H., and C. S. Holling. 2002. *Panarchy: Understanding Transformations in Human and Natural Systems*. Washington, DC: Island Press.

Ludwig, D. 2001. "The Era of Management Is Over." *Ecosystems* 4: 758–64.

Marten, G. G. 2001. *Human Ecology: Basic Concepts for Sustainable Development*. London: Earthscan. Available free online at http://gerrymarten.com/human-ecology/tableofcontents.html.

Meadows, D. H. 2008. *Thinking in Systems: A Primer*. White River Junction, VT: Chelsea Green.

Mitchell, M. 2009. *Complexity: A Guided Tour*. Cary, NC: Oxford University Press.

Pelling, M., and D. Manuel-Navarrete. 2011. "From resilience to transformation: the adaptive cycle in two Mexican urban centers." *Ecology and Society* 16(2): 11.

Raworth, K. 2012. *A safe and just space for humanity: can we live within the doughnut? Oxfam Discussion Papers*, February 2012. Accessed December 16, 2015. https://www.oxfam.org/sites/www.oxfam.org/files/dp-a-safe-and-just-space-for-humanity-130212-en.pdf.

Regime (n.d.). *Merriam-Webster Online*. In Merriam-Webster. Retrieved January 2013, from http://www.merriam-webster.com/dictionary/regime.

Rockstöm, J., et al. 2009. "A Safe Operating Space for Humanity." *Nature* 461: 472–75.

Scheffer, M. 2009. *Critical Transitions in Nature and Society*. Princeton, NJ: Princeton University Press.

Senge, P. 2008. *The Necessary Revolution: How Individuals and Organizations are Working Together to Create a Sustainable World*. New York, USA: The Doubleday Publishing Group, 406 p.

Steffen, W., K. Richardson, J. Rockström, S.E. Cornell, I. Fetzer, E.M. Bennett, R. Biggs, S.R. Carpenter, W. de Vries, C.A. de Wit, C. Folke, D. Gerten, J. Heinke, G.M. Mace, L.M. Persson, V. Ramanathan,

B. Reyers, and S. Sörlin. 2015. "Planetary boundaries: guiding human development on a changing planet." *Science* 347(6223), 1–10.

Walker, B., and D. Salt. 2006. *Resilience Thinking: Sustaining Ecosystems and People in a Changing World.* Washington, DC: Island Press.

End-of-Chapter Questions

General Questions

1. In the two graphs below, circle the areas that represent regime shifts as compared to fluctuations in the system's state within a regime. Explain why you circled the regions that you did.

    a. GDP is commonly used as an indicator of the current state of an economy. An economic recession is often deemed to have occurred when there is a drop in GDP for two consecutive quarters. Assume that a switch from a situation of economic growth to one of economic recession can be considered a regime shift. The graph below shows economic growth in the United Kingdom (UK) from 1986 to 2012.

    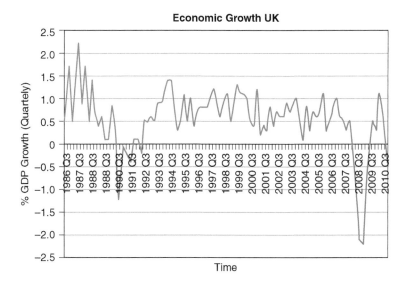

    b. The Pacific Decadal Oscillation (PDO) is a shift in the climate of the Pacific Ocean from a warm regime to a cool regime. One indicator used to detect it is sea surface temperature (SST). When SST is positive and above the average, which is shown as 0 in the graph below, then the PDO is in a warm phase. When SST is negative and below the average, then the PDO is in a cool phase. Each phase lasts about 20–30 years before it shifts to the next one.

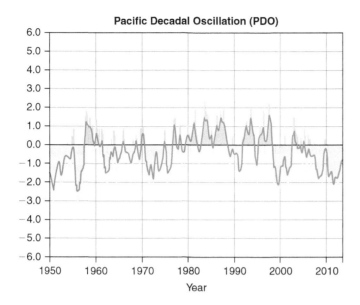

2. Use the ball-and-basin model (as shown in Figure 5.10) to depict how the resilience of the following systems is changing in the hypothetical scenarios below.

   a. Many drivers regulate the state of a coral reef and determine whether it is coral-dominated (Regime 1) or algae-dominated (Regime 2). Two important drivers are nutrient inputs and fishing pressure, such that an increase in one or both of these drivers results in a coral-dominated reef moving closer to a threshold beyond which the reef will shift to an algae-dominated regime. In 1900, a coral-dominated reef experienced an annual nutrient input of 10 mg of nitrogen, and a fishing pressure of 100 fish per month. By 2000, the reef still exists in a coral-dominated regime. However, the annual nutrient input has quadrupled to 40 mg of nitrogen, and the fishing pressure has increased to 450 fish per month.

   b. A nation's governance system currently exists in an unsustainable regime in which the desires of elite special interests, making up only 5% of the nation's total citizenry, are given preference over the majority of the citizens. A shift to a more sustainable democratic regime, in which policymaking is based on the desires of the majority of the citizens, is desired. One driver thought to increase the voice of the majority of the citizens in policymaking is access to education. In 1950, about 70% of citizens had access to education. By 2010, only 50% of citizens had access.

3. Determine whether each of the following is a stabilizing or reinforcing feedback. Draw a representation of each feedback (as shown in many Figures throughout this chapter).

a. When you are hungry, your metabolism slows down so that energy can be conserved. This permits you to continue to survive with less food. When more food is provided, your metabolism speeds up and you require more food to survive. If less food were to become available in the future, your metabolism would slow down once more.

b. When you are late for class, your instructor reprimands you for your tardiness. As a result, you arrive to class 10 minutes early for the next few weeks. Your instructor notices this and rewards you with perfect participation points for those weeks. Once you get these points, you assume that you are now on your instructor's good side and start coming to class later and later each day, until you are 5 minutes late for an entire week. Your instructor notices and threatens to deduct participation points if this happens again. As a result, you once again arrive to class 10 minutes early for the next few weeks.

c. A small snowball rolls down a hill. As it rolls, it picks up small bits of snow and speed. This makes the snowball larger and increases its speed. The larger snowball has more surface area, which allows it to pick up even more snow and move at even higher speeds. The snowball continues to grow larger and larger, and move faster and faster, as it rolls down the hill.

d. A toaster company surveys its customers about the quality of its product. A large number of customers complain that their toaster broke within the first few months of purchase, and swear that they will never purchase a toaster from this company again. As a result, the company changes its toasters so that they will last for more than a year, and the next round of customer surveys is overwhelmingly positive. However, the company's profits will increase if more toasters are purchased. Therefore, once again, it begins manufacturing toasters that break within the first few months of purchase. This profit-generating strategy works for a while until customers begin complaining that they will never again purchase a toaster from this company. As a result, the company once again increases the longevity of its toasters.

e. Jimmy was a C student, until ninth grade. He received a C+ on his first history exam, which he did not study for. After his first history exam, his teacher told him that he was naturally smart to have received a C+ on his first exam without studying, and that he had a lot of potential. As a result, Jimmy decided to study for his second history exam, and earned a B+. His teacher told him that, although he did better on this exam, he could do better still if he just studied harder. On his final exam, Jimmy received an A+.

f. Drought occurs when there is a deficiency in the amount of precipitation that falls in any given year. A precipitation deficiency occurs when there is not enough water vapor in the atmosphere to condense and fall to the ground as precipitation. Plants help to return water in the soil to the atmosphere through a process known as transpiration. Less precipitation results in less water for plants, which causes them to die. When plants die, less water is returned to the atmosphere from the soil through transpiration, and even more plants die.

4. Apply the adaptive cycles framework from Section 5.3 to categorize changes in the systems described in the case studies below. Use the release phase (Ω), reorganization phase (α), exploitation phase (r), and conservation phase (K) to describe the state of the system at any given time.

a. This case study is based on research conducted by Baral, Stern, and Heinen, which was published in a 2010 paper in the journal *Ecology and Society* titled "Growth, collapse, and reorganization of the Annapurna Conservation Area, Nepal: an analysis of institutional resilience." The paper is freely available online at: www.ecologyandsociety.org/vol15/iss3/art10.

From 1950 to 2000, the Annapurna Conservation Area (ACA) in Nepal went through an adaptive cycle in terms of who managed the land in this protected area. Prior to the 1950s, forests were managed by communities using time- and field-tested traditional governance practices. However, when forest degradation occurring during this time, it was blamed on mismanagement of forests by these local communities. As a result, the Nepali government passed the Forest Nationalization Act of 1957. This policy took control of the forests away from local communities and put it in the hands of a national forestry department. A major consequence of this policy was that forest degradation increased rapidly. This was attributed to the fact that, without local communities managing forests, the national government was not able to enforce the rules locally. In addition to the major disruption in forest governance practices as a result of the 1957 policy, a large influx of international tourists in the 1960s contributed to further forest degradation due to development.

After almost three decades of inaction and continued forest degradation, the Nepali monarchy released a 1985 mandate for protection of the forests. The mandate called for a balance between forest conservation and tourism development that benefited local communities. The King Mahendra Trust for Nature Conservation (KMTNC), which was a nongovernmental organization (NGO) established in 1982, seized the opportunity provided by the monarchy's mandate. They conducted studies, in collaboration with local forest communities, which resulted in a pilot project administered by the Annapurna Conservation Area Project (ACAP).

In the early 1990s, the ACAP received the Tourism for Tomorrow award, and the ACAP officially became the ACA. In 1992, the Nepali government granted ACAP/KMTNC legal authority over the ACA until 2002. A central goal of the ACA was that it be governed by local communities. As a result, conservation area management committees (CAMCs) composed of local people were officially recognized in 1996 as decision-making entities, regarding conservation and development in the ACA. The CAMC is a form of co-management composed of a mix of local community members and ACAP staff. Over the 10 year time period from 1992 to 2002, many beneficial programs, focused on natural resource conservation, cultural preservation, tourism and rural development, and the education and empowerment of women, were implemented and continually expanded by this new governance model. As a result of these programs, many problems in the region were resolved, and the system had stabilized by the turn of the century. However, all was not well. The system had become so inflexible and entrenched that it was not able to effectively respond to the Maoist insurgency that began in 2001.

b. This case study is based on research conducted by Beier, Lovecraft, and Chapin, which was published in a 2009 paper in the journal *Ecology and Society* titled "Growth and collapse of a resource system: an adaptive cycle of change in public lands governance and forest management in Alaska." The paper is freely available online at: www.ecologyandsociety.org/vol14/iss2/art5/.

The Tongass National Forest is located in southeastern Alaska and is the largest national forest in the United States. At the turn of the 20th century, there was increasing recognition of the need to balance nature preservation with economic development. The U.S. Forest Service (USFS) was established in 1905 to ensure this goal was met in national forest lands. However, establishing such a system in distant and rugged southeastern Alaska proved challenging. Efforts made during the first half of the 20th century mostly failed. WWII created a demand for timber and, with it, an impetus for launching a forestry system in the Tongass region. From 1950 to 1970, with the help of legal authorities and subsidies from the federal government, the Tongass region saw rapid increases in timber harvesting.

Timber harvesting continued, but in the late 20th century environmental policies and globalization began affecting what had become a well-established system. In addition to increased competition with global timber markets, the U.S. Congress discontinued timber subsides in 1990. These drivers of change resulted in regional job loss and heightened conflict and mistrust among forest managers, federal policymakers, and other stakeholder groups, all occurring in an already degraded forest ecosystem. As of 2010, the situation had not improved.

Project Questions

1. **Defining Regimes.** Complete the following table to define the regimes that (may) exist in your system, name the regimes, and describe why they are sustainable or unsustainable.

	Regime 1	Regime 2
Describe the regime in general terms.		
Identify three defining characteristics that differentiate the regimes.	What is the value of each characteristic in each regime? (e.g., high/low, wet/dry)	
1.		
2.		
3.		
Give the regime a descriptive name using two or three short words		
Is the regime sustainable or unsustainable? (Does it have healthy environmental, social, and economic functioning?) Why or why not?		

2. **Identifying Feedbacks.** Identify the important reinforcing and stabilizing feedbacks affecting your system, which are pushing the system toward a new regime or keeping it in the current regime, respectively. Describe which set of feedbacks—reinforcing or stabilizing—is currently dominating the system, and how this is affecting your system.

3. **Characterize your system's resilience.** Resilience can be hard to characterize exactly, but use some of the ideas from this chapter to help you: the concept of a system's proximity to a threshold and what this implies for resilience (think about your system's thresholds, the drivers as identified in Chapter 3 and Project Questions 1 and 2, as well as the feedbacks regulating your system); the importance of diversity and adaptive capacity to a system's resilience; and how resilience changes as a system passes through different phases of an adaptive cycle. In the next chapter, you will learn about other ways to think about a system's resilience.

Chapter 6

COMPLEX ADAPTIVE SYSTEMS

Students Making a Difference

Sustainability and the Navajo of Northern Arizona, USA

Paul Prosser is deeply inspired by the Navajo people: "To them, sustainability means a lot of different things, but the most important thing was their worldview: everything is connected to everything else, and nothing in nature, including humans, is superior to anything else. Whenever we talk about people, we are afraid of the part that is less science-y, culture and spirituality. I wanted to bring the cultural and spiritual aspects of their lives into the building design. I wanted to learn what it was, that was particular to this group of people, that defines sustainability on their own terms and in accordance with their own worldview." The Navajo governance hierarchy, from largest to smallest, is nation, region, chapter, village/town/

city. Paul worked with the Tonalea chapter, located four hours north of Phoenix, to help them develop plans for a new chapter house using sustainable building designs that were in accordance with their worldview. Their chapter house was condemned and torn down in 2016. A chapter house is an important part of their society. It is where administrative work occurs, and where photos, artwork, and language are preserved to protect their culture and spiritual life. As he worked with the Tonalea, he learned how important culture and spirituality were to the inner workings of life as a Navajo. He also saw how external factors, such as actions and policies of the federal government, affected their lives.

Although he felt welcome by the Navajo, as a white middle-aged male, he was immediately confronted with these affects as he began work with them: "Every white male that steps foot on the reservation is met with some long-standing cultural background from all of the colonization, oppressive government structures that tore their tribe apart. Understandably, they are going to be in doubt and suspect [of me] until I show them otherwise. I don't think I was fully trusted by the time I left." This condition of path dependence, a feature of complex adaptive systems that influences pathways into the future and present-day decisions, arose from past and present interactions with the federal government, and it affected Paul's ability to work with them.

It also affected their chapter house project. According to Paul, it sabotaged the project before it even began. They broke off engagement with him when they felt he was taking too long to collect information: "They have a lot of layers of government that affect them every day. It goes from the chapter to the regional council, to the tribal council, then the BLM, the Department of the Interior, the Bureau of Indian Affairs. All of those people have some finger on what they do. It's very complicated. It causes them to be very cautious and capricious about their decisions because they don't know what comes next. They had some money promised to them for this building, and they were very nervous about not acting right away, so that the funds would not get taken away and applied to something else, before they had the opportunity to act." There was also the approval process, which, in their experience, was complex and often unsuccessful. For example, when people living in Hopi-Navajo disputed territories had a damaged roof, they had to hire lawyers to help them navigate a maze of paperwork and regulation to get it fixed, often to no avail: "It would go into the federal system and nothing would happen. People's homes that had leaks in the roofs would turn into these mold-filled places where they used tarps to keep water from coming in. The Navajo bring these memories of the struggles they went through to fix a roof to trying to fathom how they would be able to build an entire building." Ultimately, due to this path dependency and other factors, the building was not built using indigenous knowledge as part of the design guidelines. Although, Paul did gain insight as to how this could be done in the future.

The behavior of a complex adaptive system depends on interactions between factors internal to a system, such as the cultural and spirituality of the Navajo, and external conditions, such as the federal government. Certain features that can affect the sustainability of a CAS, such as path dependence, often arise from these interactions. Many, but not all, sustainability problems are embedded in a CAS. In this chapter, you will learn about how a CAS behaves and what factors influence it's behavior.

Core Questions and Key Concepts

Section 6.1: Emergent Features and Behaviors

Core Question: How do emergent features arise in complex adaptive systems and how do they differ from other system features?

Key Concept 6.1.1—Certain features and behaviors arise from the individual components of a system, and the interactions among them, that cannot be understood unless the system is viewed holistically.

Key Concept 6.1.2—There is a difference between emergent properties, which are greater than the sum of their parts, and collective properties, which are a simple sum.

Section 6.2: Interactions Between the System and External Conditions

Core Question: How do complex adaptive systems interact with their external environments and how does this affect their behavior?

Key Concept 6.2.1—Dynamic interactions among internal system components and changing external conditions result in a lot of fluctuation in complex adaptive systems.

Key Concept 6.2.2—Dynamic interactions among internal system components and changing external conditions result in changes to the system's stability landscape, and subsequently, its resilience over time.

Section 6.3: Adaptation

Core Question: How do complex adaptive systems adjust to changing conditions over time?

Key Concept 6.3.1—Biological populations adapt their behavior over time by passing on and receiving information in the form of inherited traits.

Key Concept 6.3.2—Human systems adapt over time by passing on and receiving information in the form of inherited traits, but also in many other forms acquired through a process of learning.

Key Terms

- complex adaptive system (CAS)
- biological population
- inherited traits
- biological community
- emergent property
- hierarchy
- carrying capacity
- mortality rate
- autogenic drivers
- allogenic drivers
- slow driver
- fast driver
- stability landscape
- precariousness (P)
- latitude (L)
- resistance (R)
- chaotic
- human agency
- path dependence

> "Nigel Franks, a biologist specializing in ant behavior, has written, 'The solitary army ant is behaviorally one of the least sophisticated animals imaginable,' and, 'If 100 army ants are placed on a flat surface, they will walk around and around in never decreasing circles until they die of exhaustion.' Yet put half a million of them together, and the group as a whole becomes what some have called a 'superorganism' with 'collective intelligence.' How does this come about? ... The mysteries of army ants are a microcosm for the mysteries of many natural and social systems that we think of as 'complex.' ... Similarly mysterious is how the intricate machinery of the immune system fights disease; how a group of cells organizes itself to be an eye or a brain; how independent members of an economy, each working chiefly for its own gain, produce complex but structured global markets; or, most mysteriously, how the phenomena we call 'intelligence' and 'consciousness' emerge from nonintelligent, nonconscious material substances."
>
> —Melanie Mitchell, Complexity: A Guided Tour, 2009

Chapter 5 described different patterns of change in systems—linear and nonlinear, reversible and harder-to-reverse, and adaptive cycles—and how these patterns of change are affected by, and also affect, a system's resilience. However, some questions remained unanswered. Why do some systems have alternative regimes separated by nonlinear thresholds, and others do not? Why do some systems cross relatively reversible thresholds to new regimes, while others undergo critical transitions that are difficult to reverse? This chapter attempts to answer these questions by providing additional tools for thinking about SES complexity using ideas from the complex adaptive systems framework.

This framework is focused on understanding the behavior of a **complex adaptive system (CAS)**. CASs are distinguished from other complex systems because of their tendency to evolve over time in ways that help them adjust to changing conditions. All systems of this type, including some SESs, have three common characteristics, as described by Melanie Mitchell in her 2009 book *Complexity: A Guided Tour*. First, features and behaviors emerge from these systems that cannot be explained by the individual behavior of each system component alone. In other words, the whole is greater than the sum of its parts. Like the army of ants described in the opening quote to this chapter, the behavior of individual components of CASs are not sophisticated on their own, but a "collective intelligence," "consciousness," or "superorganism" arises, seemingly spontaneously, out of the simple and small-scale interactions among system components (Figure 6.1). This happens without a central controller organizing and directing the behavior of each individual component. Second, CAS behavior is determined by both internal system components and processes, and changing conditions

> **complex adaptive system (CAS)**
> a system capable of evolving over time in a manner that helps it adjust to changing conditions and typically in ways that promote its survival

Figure 6.1 Simple interactions among individual army ants, who cannot accomplish much alone, lead to a "superorganism" made up of millions of individual ants that can work together to hunt for and transport prey (*left*) or build a nest out of leaves (*right*).

outside the system, as a result of a two-way exchange of matter, energy, and information. Third, CASs adapt their behavior by passing on and receiving information through processes of learning (e.g., human systems) and evolution (e.g., natural systems). These three major features of CASs are the topic of each of the three sections that compose this chapter.

Section 6.1: Emergent Feature and Behaviors

Core Question: How do emergent features arise in complex adaptive systems and how do they differ from other system features?

Section 6.1.1—Sophisticated Properties from Simple Individual Interactions

Key Concept 6.1.1—Certain features and behaviors arise from the individual components of a system, and the interactions among them, that cannot be understood unless the system is viewed holistically.

The first defining feature of complex adaptive systems is that sophisticated features and behaviors emerge from these systems that cannot be explained by the simplistic individual behavior of each system component alone. CASs seem to take on a life of their own as the simple interactions, among a web of components, culminate in a "collective intelligence" or "consciousness." This arises without a central controller that organizes, directs, and coordinates the behavior of each individual component, and the interactions among those components. In this section, examples of this type of behavior, as exhibited by two distinct CASs—market behavior and biological evolution—will be described to illustrate these ideas.

Individual Behavior and the *Invisible Hand* of Markets. The *invisible hand* is a metaphor coined by economist Adam Smith in the late 18th century to explain how markets work. It is based on the idea that when individuals, such as buyers and sellers, act in their own self-interest in markets, then the hard-to-predict features that will emerge are the market-clearing price and quantity (Figure 6.2). This price and quantity, in turn, guide the behavior of individuals in markets.

Chapter 6 Complex Adaptive Systems **313**

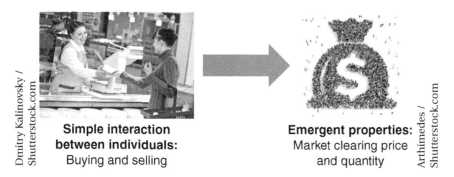

Simple interaction between individuals: Buying and selling

Emergent properties: Market clearing price and quantity

Figure 6.2 The customer and vendor in this picture each acts in her self-interest through the simple interaction of buying and selling a slice of cheese (*left*). A market price for cheese emerges, from millions of similar interactions between millions of other buyers and sellers, which is based on the quantity of cheese that buyers are willing to buy at a certain price and the quantity of cheese that sellers can offer for that price (*right*).

Millions of buyers and sellers are two different types of individual components that interact in markets through buying and selling. Buyers dictate demand, and sellers determine supply. Demand and supply interact to produce a market clearing price, and the quantities of goods and services that are produced, distributed, and consumed (Figure 6.3). In markets, there is no central controller who organizes, directs, and coordinates the decisions of each individual buyer and seller, and the interactions between them. Instead, the market automatically adjusts to changing conditions. Markets do this by combining millions of individual decisions about buying and selling into a market clearing price and quantity. This captures a vast amount of information, which cannot itself be explicitly known or comprehended. Imagine trying to keep track of how much, and what type, of bathroom soap each individual household in the United States prefers in order to establish the price and quantity of soap produced. Markets automatically collect all of this information through their "collective intelligence" or "consciousness."

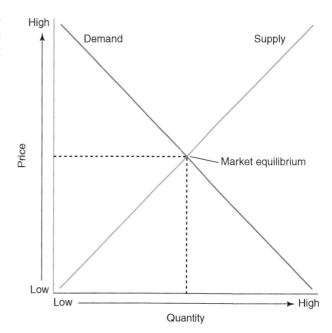

Figure 6.3 The individual decisions and actions of millions of buyers and sellers in markets results in a condition known as market equilibrium, which is where supply and demand meet to determine the emergent properties of market clearing price and quantity.

Individual Selection and Biological Evolution. Biological evolution is the process by which the composition of biological species present on earth changes over time. According to Charles Darwin's theory of evolution by natural selection, it is the differential selection of individual organisms over time that ultimately results in the production of new species. The rules governing the selection of individual organisms are relatively simple. First, the size of a **biological population** is limited by scarce resources in the environment, such that all individuals born in a population will not survive. Second, survival depends on **inherited traits**. Individuals whose inherited traits give them a high probability of surviving and reproducing will leave more offspring. The unequal ability of individuals to survive and reproduce leads to gradual changes in populations over time, with favorable traits accumulating over generations. The sophisticated process of biological evolution, from which new species arise, emerges from the simple interactions among individuals within a population, their **biological community**, and the ecosystem that they inhabit. There is no central controller organizing, directing, and coordinating this process, yet it still happens as biological systems literally take on a life of their own. This process cannot be explained by understanding the millions of interactions among the individual components (i.e., individual organisms) alone. The process of evolution is greater than the sum of its parts. **Box 6.1** provides a case study that illustrates how new species emerge from the process of biological evolution.

> **biological population**
> all individuals of the same species that live in the same geographical location and are capable of producing offspring

> **inherited trait**
> a characteristic that is controlled by genes and is passed from parents to offspring

> **biological community**
> multiple populations of different species living in the same area and interacting with each other in a variety of ways

---

BOX 6.1 EVOLUTION OF THE LONG-NECKED GIRAFFE

Although recognized as incomplete by modern-day evolutionary biologists, Charles Darwin's original explanation for how giraffes evolved to have such awkwardly long necks demonstrates how a new species could emerge from simple interactions among individual organisms and other factors influencing an ecosystem. Giraffes have longer necks than other grazers that inhabit African savanna ecosystems, and their long necks make the essential activity of drinking water awkward and potentially dangerous (Figure 6.4). Given these factors, why do giraffes have such long necks?

Darwin's explanation was that giraffes evolved long necks, as a result of the interactions among organisms living in the African savanna, during a period of prolonged drought (Figure 6.5). In savanna ecosystems, grazers can choose from both ground-level vegetation, such as grasses and other plants, and vegetation that is located higher off the ground in trees. The acacia tree is very common in savanna ecosystems. During periods of drought, there is less water available for vegetation and, therefore, less vegetation available for grazers. Darwin proposed that giraffes developed longer neck over periods of prolonged drought

Figure 6.4 Compared to other grazers inhabiting the African savanna, such as the wildebeest (*left*), giraffes have much longer necks, such that bending down to drink requires splayed legs, and sometimes bent knees, which puts them in a vulnerable position with regard to attack by their lion predators (*right*).

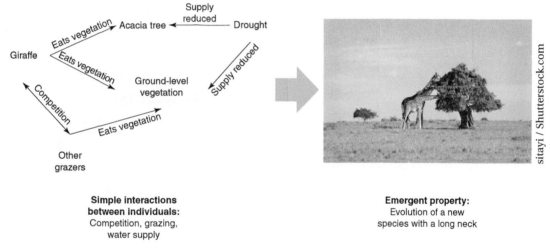

Figure 6.5 Simple interactions between individuals in a savanna ecosystem (*left*). A market price for cheese emerges, from millions of similar interactions between millions of other buyers and sellers, which is based on the quantity of cheese that buyers are willing to buy at a certain price and the quantity of cheese that sellers can offer for that price (*right*).

because giraffes with longer necks were more likely to survive and reproduce than giraffes with shorter necks, such that the longer neck trait was selected for from generation to generation. Eventually, due to simple interactions among individual system components, a new species of longer-necked giraffe evolved.

Section 6.1.2—Emergent Properties versus Collective Properties

Key Concept 6.1.2—There is a difference between emergent properties, which are greater than the sum of their parts, and collective properties, which are a simple sum.

An **emergent property** is the sophisticated feature or behavior that emerges from a system that cannot be explained by only the individual system components, or the simple interactions among them. These properties are called *emergent* because they emerge, or arise, from different levels of organization in a **hierarchy**. Hierarchies are often present in CASs and have been proposed as one of the many possible ways to measure complexity in these systems. The idea behind hierarchies is that a system is composed of subsystems that are, in turn, composed of other subsystems. Each subsystem has its own set of emergent properties that characterize it, but that do not apply to the individual subsystems that compose it. As subsystems combine to become components of the system in the next level higher in a hierarchy, which forms a new functional whole, new properties emerge that did not exist at the level below it (**Figure 6.6**).

> **emergent property**
> a sophisticated feature that emerges from a system and cannot be explained from the behavior of the system's individual components nor the simple interactions among them

> **hierarchy**
> different levels of organization by which items are arranged as a variety of levels relative to each other

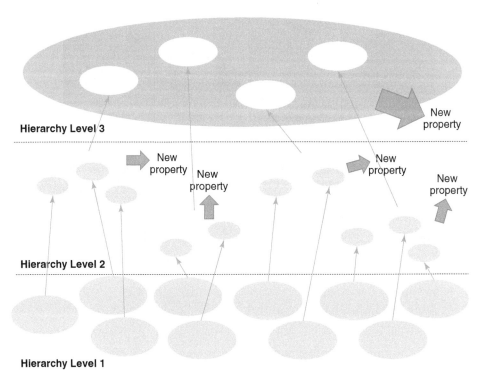

Figure 6.6 Systems at lower levels in a hierarchy become subsystems of a new system, which is formed at the next level up in a hierarchy

The human body is one of the most common examples used to illustrate a hierarchy of organization, and how new properties emerge from each new level of the hierarchy (Figure 6.7). The first level of organization in the human body is the atom. Two atoms present in the human body are hydrogen (H) and oxygen (O). These atoms are each made up of their own subsystems of protons, neutrons, and electrons, which are not shown in Figure 6.7. The two atoms are combined into a molecule to form water (H_2O), which makes up about 60% of the human body. When not combined into H_2O, hydrogen and oxygen exist separately as gases (H_2 and O_2) with very different properties than H_2O. Thus, new properties emerge, such as polarity, when H and O combine into a water molecule. Water molecules then combine together with other types of molecules to become subsystems of organelles (Figure 6.7), such as the mitochondrion. A mitochondrion is capable of extracting energy from food through oxidative metabolism, which is a behavior that emerges when molecules combine to form it, and one that cannot be carried out by the individual molecules alone, nor can it be understood by only studying the individual molecules.

Similar emergent phenomena continue to arise as one moves to other higher levels of organization within the human body, shown

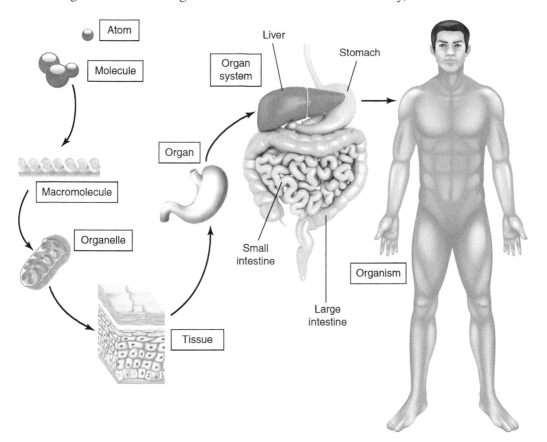

Figure 6.7 Hierarchies of organization, beginning with the atom, combine to ultimately form the human body.

in Figure 6.7, but we will stop there, as the idea has been sufficiently illustrated. There is another important point to recognize about hierarchy, which is related to problem solving. Many different levels of a hierarchy, as well as the features and behaviors that emerge at each of its levels, must be considered in order to fully understand a problem. **Box 6.2** discusses how the tiny spruce budworm larvae contributes to tree death in boreal forest ecosystems, illustrating the importance of understanding the different levels of hierarchy, as each level displays its own emergent property. **Box 6.3** describes a hierarchy of organization for water governance in Bali's traditional sustainable agricultural systems and the properties that emerge from the hierarchy.

Emergent properties should be distinguished from another feature of CASs called *collective properties*. These are still whole-system properties, but are not concerned with new properties that emerge at higher and higher levels of a hierarchy. Rather, they are a simple summation of individual components, and can be understood by studying the individual components. In other words, unlike emergent properties, the whole of collective properties is *not* greater than the sum of its parts. The difference between emergent properties and collective properties can be illustrated through an example of biological populations.

A biological population is one level of the ecological hierarchy of organization. It is made up of many individual organisms of the same species from the level below it and, when combined, make up biological communities, the level above it, composed of many populations of different species interacting. When individual organisms come together to form a population, a property that emerges is the **carrying capacity**, which is the maximum number of individuals that a certain environment can support based on resources available. This affects the number of individuals in the population as a whole because the success of each individual, through reproduction and survival, depends on the number of other individuals in the population, and how close this number is to the carrying capacity. Thus, carrying capacity is a property of populations that emerges because of interactions among individual components of the population. The **mortality rate**, however, is a collective property at the population level. It is simply the sum of the number of individuals that have died over a certain time period (e.g., deaths/month).

There is one more important point to make about emergent properties and why they are useful for understanding and resolving sustainability problems that are embedded in CASs: they can make problem solving easier! The ability to recognize emergent properties, as a basis for understanding, means that each and every individual system component, and the interactions among them, does not need to be fully tracked and comprehended. This would be hard to do and very time consuming! Thus, not only are emergent properties necessary for understanding CASs, they can make it easier.

---

carrying capacity
the maximum number of individuals of a given species that a certain environment can continue to support with the available resource base

---

mortality rate
the sum of the number of individuals in a population that have died over a certain time period

BOX 6.2 MULTIPLE HIERARCHIES IN BOREAL FORESTS

In boreal forests, levels of hierarchy range from a tiny spruce needle, to a forested stand of trees, to an expansive forest-peatland-lake landscape covering a very large area (recall Figure 3.13 from Chapter 3). Spruce needles make up the branches, which, in turn, compose the entire tree. An even higher level of this hierarchy is the forest stand, which is made up of populations of different tree species. Tree populations, along with peatland and lakes, mingle to make up the boreal forest landscape.

In order to understand change in these systems, due to the activity of the spruce budworm, it is important to know which levels of organization, and their emergent properties, should be considered. The spruce budworm plays an important role in the natural life cycle of forest ecosystems, but lately the activity of the budworm, coupled with rising regional temperatures due to climate change, has led to the destruction of large swaths of forest within boreal forest landscapes. Different levels of hierarchy in the forest are considered to understand how budworm larvae contribute to tree death in these forests, as part of their natural cycles, which are really adaptive cycles (Section 5.3).

One level is that of the spruce needle, and the community interactions between the needle and the budworm. After mating, female budworms deposit their eggs on the underside of the spruce needles. Larvae hatch from the eggs in a little over a week and immediately prepare for winter hibernation by spinning silk tents nestled under the bark scales of the tree's trunk. They emerge from their protective silk shelters in springtime, and they are hungry! As a result, they feed on young needles and buds, which they prefer to older needles (Figure 6.8). The system feature that emerges from this interaction is herbivory: spruce budworms devour spruce needles. However, this is not the only issue because a tree is able to regrow its spruce needles after they are eaten by budworm larvae. To fully understand how spruce budworm larvae kill trees, a different level in the hierarchy must also be considered.

The major predators of spruce budworm larvae are insectivorous birds (Figure 6.9). In order for the birds to prey on the larvae, they must be able to find them. When boreal forest stands are young, such that the tree branches are not densely packed together, there are spaces between the branches available

Figure 6.8 Newly hatched budworms prefer to eat the young buds produced by spruce trees in the springtime.

(*Continued*)

for birds to find and eat the budworms. However, as the forest grows and matures, biomass accumulates to such an extent that tree branches are packed very close together. When a forest is in this condition, birds cannot find budworms as easily and budworm populations increase. It is at this point that budworm larvae are able to kill trees by overwhelming their ability to regenerate enough new spruce needles, which are needed for photosynthesis and, ultimately, its survival. Thus, the interaction between budworm and bird populations, at the level of the forest stand, must be considered to understand the problem. The system feature that emerges from this level of interaction is predation, by which bird populations do, or do not, keep budworm populations in check by eating them.

Figure 6.9 Birds living in boreal forests that eat insects, such as the Cape May Warbler shown here, keep budworm populations in check when boreal forest stands are young and not as densely packed as older stands.

BOX 6.3 EMERGENCE OF ORDER IN BALI'S AGRICULTURAL SYSTEM

In his 2006 book *Perfect Order*, anthropologist Stephen Lansing used game theory and field observation to describe how cooperation spontaneously arises from simple and small-scale interactions among rice paddy farmers, and dispersed local institutions, on the island of Bali in Indonesia. Farmers cooperate to coordinate planting and irrigation schedules, and collaborate on pest control. The end result is good rice crop yields for farmers and a system of governance that is self-sustaining, not requiring outside intervention or "top down" management, and adaptive in the face of changing external conditions.

At first glance, based on the physical layout of "water mountains," it seems that the situation in Bali would not lead to cooperation. Bali is an island covered with active volcanoes. Rice paddies lie on terraces constructed on the steep slopes of these volcanoes (Figure 6.10). Water for irrigation originates in volcanic crater lakes and flows through deep ravines into a complex, and often fragile, system of dams, weirs, canals, and tunnels. Rice farming begins as high up on the hillsides as rice will grow. The rice paddies are constructed ponds that need to be flooded during plant growth and dried out when plants are harvested. As such, irrigation schedules are essential to a healthy crop yield. Because of their physical location, upstream farmers could easily cut off the water supply to downstream farmers, keeping more for themselves. However, it is not in their best interest to do this. Water supply is important, but so is pest control. By synchronizing planting schedules with downstream farmers, upstream farmers can prevent pests from destroying their crops: if all farmers, both upstream and downstream, harvest at the same time, pests will lose their habitat all at once and be eliminated.

Figure 6.10 In Bali, rice is grown in artificial pools located on terraces cut into the steep hillsides of active volcanoes.

The actions of thousands of farmers occur within a vast hierarchy of cooperating groups, from which an efficient system of water governance emerges (Figure 6.11). At the lowest level of the hierarchy is each farmer and his rice paddy. All farmers in an area who benefit from the same source of water belong to a Subak, a local water temple that transcends the boundaries of the villages where farmers live. In a Subak, information is shared and decisions about planting and irrigation schedules, and maintenance to irrigation infrastructure, are made by consensus. Any individual who farms a rice paddy using the water from that source is required to attend Subak meetings and participate in work assignments. If they do not, they receive fines and penalties and, at best, have their water supply cut off or get kicked out of the Subak, at worst. Each local Subak elects a leader to attend monthly meetings at a regional water temple, where activities such as information sharing, planting, irrigation, and maintenance are coordinated among Subaks. At the very top of the hierarchy is Bali's supreme water temple, where priests with high spiritual authority have the power to change decisions made by regional water temples or local Subaks.

Given the vast number of irrigation systems within a region, and the constant demands of managing and maintaining them, it would be challenging for a state or national government to manage them as effectively. Actions taken by the Indonesian government in the 1970s, based on advice about how to increase rice yields from the Asian Development Bank, are a testament to this. Instead of following their traditional system of water governance,

(Continued)

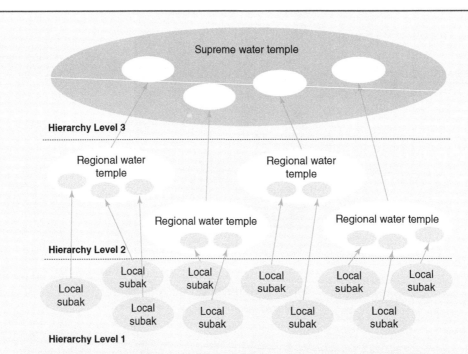

Figure 6.11 Cooperation among rice paddy farmers emerges from a governance system made up of a hierarchy of different interacting components.

where cooperation, and good rice yields, arose spontaneously from a network of Subaks and temples spread across a large region, farmers were encouraged by the government to act in a disconnected way. They were told to plant rice as often as they could, rather than coordinating planting and harvesting schedules with other farmers. The result was disastrous. Pests destroyed crops, sometimes completely, leading to hunger and aerial pesticide spraying by the government. It was not until the traditional governance system was restored that things improved.

Sometimes SESs that behave as complex adaptive systems, if left on their own without outside intervention, can be more sustainable, in many ways, than an SES that is intensively managed from the outside. More resources, such as time, money, and personnel, would be required for an outside agency, such as a state or national government, to efficiently manage an agriculture system like the one found on the slopes of Bali's volcanoes. When the national government did intervene in the 1970s, environmental harms, such as pesticides, and social ills, such as hunger, resulted. Awareness of CASs, and their characteristics and behaviors, can be important for understanding and addressing sustainability problems.

Section 6.2: Interactions between the System and External Conditions

Core Question: How do complex adaptive systems interact with their external environments and how does this affect their behavior?

The second general characteristic of complex adaptive systems (CASs), mentioned at the beginning of this chapter, is the fact that overall system behavior is regulated by both internal system components and processes, and changing conditions outside of a system's boundaries. This regulation is defined by a two-way exchange between internal and external environments resulting from the production and use of matter, energy, and information by both. Two features important for understanding SESs that behave like CASs, and resolving the sustainability problems embedded in them, result from this type of internal-external exchange. First, in addition to the continual interactions among internal system components through feedbacks, as described in Section 5.2.2, changing external conditions also cause systems to fluctuate. Recall from Section 5.1.1 that it is important to distinguish the normal fluctuations of a system *within* a regime from those that indicate shifts to fundamentally different regimes. This feature will be covered first. The second feature will be discussed in Section 6.2.2. Section 6.2.2 is concerned with slow changes in the stability landscape over time that affect a system's resilience. This has to do with both changing internal and external conditions, as well as time scales of change.

The internal and external environments of a system can be conceptualized as follows. Inside the system's boundaries, there are many interconnected components that interact through a variety of processes, including feedbacks (Figure 6.12). Outside of its boundaries, there are external conditions that endlessly disturb internal system components and processes. The dashed arrow pointing from the internal system to the external conditions indicates that systems can also influence and change their external environment to

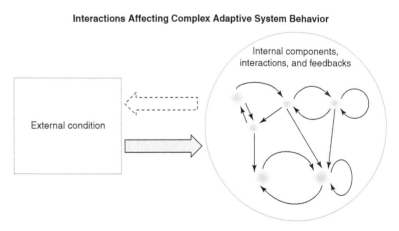

Figure 6.12 The behavior of a complex adaptive system is influenced by internal interactions and feedbacks (right) and changing conditions in the environment external to the system (left).

some degree. This two-way exchange between the internal components and processes, and the conditions in the external environment, ultimately determines the state of a system at any given time, as well as its dynamics over time. Section 6.2.1 focuses on system fluctuations, and Section 6.2.2 examines changes over time that affect system resilience.

Section 6.2.1—Disturbance and Patterns of Fluctuation

Key Concept 6.2.1—Dynamic interactions among internal system components and changing external conditions result in a lot of fluctuation in complex adaptive systems.

Some fluctuations in a system are a result of changing external conditions that affect the system and some, as discussed in Section 5.2.2, are a result of continual interactions among internal system components. If an external disturbance affects a system profoundly, then it can cause a system to shift to a new regime. In other cases, external disturbances do not affect systems enough to cause a regime shift, but they do result in system fluctuations beyond those caused by internal dynamics alone. This type of fluctuating behavior can be contrasted with that of simpler systems existing under more controlled conditions, and how they respond to disturbance. This is an important distinction to make because the concepts used in some disciplines about how systems respond to disturbance, such as in chemistry or engineering, cannot be necessarily transferred to systems exhibiting greater complexity, such as with CASs. In fact, in many instances of resource management, treating complex systems as simple systems has led to unfortunate outcomes. Simple systems do not have as much variability and fluctuation as complex systems, whereas in complex systems variability, fluctuation, and change are the rule rather than the exception. When complex systems are treated as simple systems, and their variability is limited, their resilience can actually be diminished. **Box 6.4** illustrates these ideas with a case study of forest management practices in fire adapted ecosystems.

In order to work with, rather than resist, the natural variability of CASs, system responses to disturbance and patterns of fluctuation must be understood. One way to understand this is to contrast how a very simple system existing under controlled conditions responds to disturbance, as compared to a system exhibiting greater complexity. To accomplish this, a simple chemical reaction occurring in a laboratory beaker is compared with the global climate system in the next few paragraphs.

BOX 6.4 FOREST FIRE SUPPRESSION IN THE UNITED STATES: A TALE OF TWO REGIONS

For centuries, fire was used by Native Americans as a tool for modifying fire-adapted landscapes to improve conditions for foraging and visibility for hunting. In the southeastern United States, early European settlers noticed that burning produced straighter trees for loggers and lusher grasses for cattle, so they continued the native practice. In the western United States, especially in California, higher population densities and, later, stringent air quality standards made fire suppression popular. Additionally, a wave of enormously destructive wildfires in the early twentieth century oriented U.S. Forest Service policy toward suppression. By the mid-twentieth century, political pressures continued to support suppression and educational campaigns, such as Smokey the Bear, became popular. States in the southeast resisted these pressures, whereas western states capitulated. Thus, in western states such as California, in the name of protecting forest ecosystems and people, immense resources have been devoted to actively suppressing fires that naturally occur in healthy fire-adapted ecosystems (Figure 6.13).

Fire suppression was initially justified, in part, by forest managers' misconception that forest ecosystems were healthiest in a mature climax state. This mature climax condition, or "equilibrium" state, of forests was viewed as the end goal or the desired final condition for a forest (such as the hardwood trees in Figure 5.44). However, as later revealed by ecologists and wildfire patterns, suppressing natural fires negatively affects biodiversity and makes forests less resilient to fire disturbance. (Recall that resilience is the ability of a system to absorb disturbances and still remain in the current regime.) When fires are suppressed, lots of highly flammable materials, such as branches and leaves, accumulate on a forest floor over time. Like kindling used to light a campfire, this transforms forests into tinderboxes waiting to go up in flames. Under these conditions, when a natural lightning strike or human activity starts a fire, it is likely to spread rapidly and burn more forest. This makes it harder for forests to recover from disturbance, and increases the chance that forests will shift to another

Figure 6.13 Traditional management practices seek to suppress fires at a very small scale (*left*) before they grow into fires that affect entire large areas (*right*), but this practice does not recognize the complex nature of fire-adapted ecosystems.

(*Continued*)

Figure 6.14 Fire management agencies used prescribed fires, such as this controlled fire burning in a South Carolina pine forest, to accomplish the important function performed by natural wildfires, which is to remove, by burning, highly flammable materials that have accumulated on the forest floor.

non-forest regime. In essence, suppressing fire stabilizes forests, halting the variability, fluctuations, and changes that, if left on their own, are the rule rather than the exception in these fire-adapted ecosystems.

The U.S. Federal Wildland Management Policy, first introduced in 1995, recognized the dangerous tinderbox conditions created by a history of fire suppression in the western United States and supported the reintroduction of fire. Massively destructive wildfires occurring in recent years have moved wildfire management agencies to action. As of 2015, in an effort to alleviate the tinderbox conditions in the unnaturally dense forests of the west, the U.S. Forest Service has greatly increased the number of prescribed fires, also referred to as controlled burns (Figure 6.14). A major challenge to this effort is the growing trend referred to as "rurbanism," where city slickers and suburbanites are moving into rural areas close to wilderness. This trend is also affecting the southeastern United States, where prescribed fire management strategies have long garnered political support, which is waning in the face of "rurbanism." As a result, in both regions, when it comes to prescribed burns for adaptive forest management, there is a strong focus on human safety and community education.

The chemical reaction example is very simple (Figure 6.15). It involves five components in a glass beaker: Chemical A, Chemical B, Chemical C, Chemical D, and water. The interactions among the components are different types of chemical bonds. The external conditions affecting these interactions are temperature, pressure, water added or removed (volume), or other chemicals added to the beaker.

Figure 6.16 represents the four chemicals A, B, C, and D that are part of the reaction with different colored lines. At time = 0, the four chemicals are mixed together in the beaker with water at different individual starting concentrations. After some time, a reaction occurs and the system settles into an equilibrium, such that each of the four chemicals is present at a constant (unchanging) equilibrium concentration. This is indicated by a flattening or leveling off of the four lines. Then, about midway along the x-axis, more of Chemical B is added to the system. Following this disturbance, the system readjusts and eventually settles into a new steady equilibrium. Assuming that all external conditions affecting concentrations in the beaker are held constant following

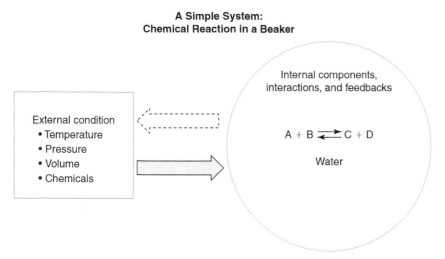

Figure 6.15 Under the controlled conditions in a chemistry laboratory, the small number of components involved in a chemical reaction, and the limited number of external conditions that influence that reaction, lead to a relatively stable system with minimal fluctuation and variability.

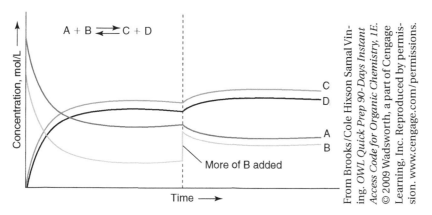

Figure 6.16 The relatively stable concentrations of chemicals A, B, C, and D are a result of the interactions among these chemicals in the beaker, and the external conditions of temperature, pressure, volume, and additional chemicals that influence the system in a predictable way.

this disturbance (temperature, pressure, volume, and additional chemicals), then the equilibrium concentrations will remain at unchanging and constant values. This is how relatively simple systems behave and respond to disturbance. In this case, stability is the rule rather than the exception. This applies to the controlled conditions in a chemistry lab, as well as many engineered systems.

Contrast this with the climate system, which is neither easily controlled nor engineered and where change is the rule rather than the exception! The climate system contains many components. These are generally broken down into five main components according to the Intergovernmental Panel on Climate Change (IPCC): atmosphere, hydrosphere, cryosphere, land surface, and biosphere (Figure 6.17). Each of these internal components can be broken down into subcomponents. For example, the hydrosphere is made up of rivers, lakes, groundwater, oceans, and seas. Each of these subcomponents could be broken down even further. For example, lakes can be broken down into oligotrophic, mesotrophic, or eutrophic based on productivity and nutrient concentrations. Vegetation, which is a subcomponent of land surface, can be broken down into categories ranging from desert to grassland to forest. Microorganisms are a subcomponent of the biosphere that can be further classified based on mode of nutrition: photoautotroph, chemoautotroph, photoheterotroph, or chemoheterotroph. The list could go on and on as subdivision into further components continues. The system is very complex just in terms of the sheer number of components. However, interactions among those components are what really determine system fluctuations, and these must also be considered.

Defining interactions among components can be even more daunting than defining the components themselves. For example, pick just one component from the list shown in Figure 6.17—lakes—and list as many interactions with other internal system components as you can (Figure 6.18). Lakes interact with rivers, which flow into and out of them. They indirectly interact (through rivers) with the ocean, and

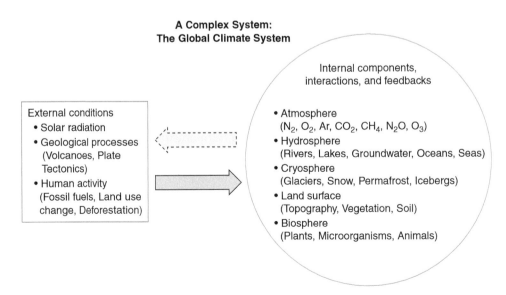

Figure 6.17 The large number of components that make up Earth's climate system, and the different external conditions that can influence this system, lead to hundreds of interactions that cause constant fluctuation and variability.

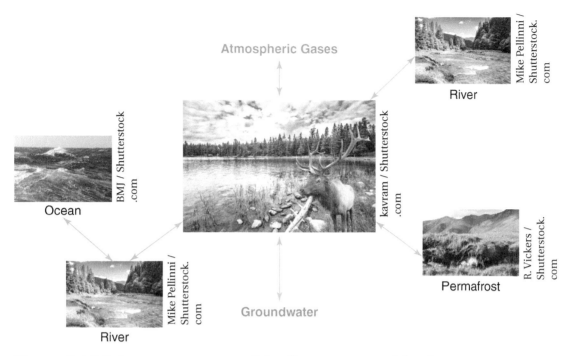

Figure 6.18 Lakes are a component of the climate system and they interact directly with other components, as shown here, leading to system fluctuations and variability caused by interactions among internal components.

influence what flows into the ocean. They interact with groundwater, which may flow into, or out of, a lake, depending on the local hydrology, geology, and soil type. In Arctic environments, lakes interact with permafrost, especially as it melts and creates more lakes. Vegetation grows in lakes, and animals live in lakes. Gases from the atmosphere dissolve in lake water. This is already seven interactions, which doesn't even scratch the surface *for this one component*! The many biological, chemical, and physical processes, through which the components of the climate system interact across many spatial and temporal scales, result in an extremely complex system. The feedbacks among all of these internal system components are incredibly numerous as well. This all results in a lot of dynamic fluctuation occurring from internal system dynamics alone.

Added to that complexity are all of the possible *external* conditions influencing the climate system. Figure 6.17 shows these grouped into three major categories: solar radiation, geological processes, and human activity. Variations in solar radiation occur over very long time periods (10,000s to 100,000s of years). Geological processes, such as volcanic eruptions that spew dust and gases into the air, also affect the amount of solar radiation reaching the earth's surface by blocking it out. CO_2 is one of many gases emitted from volcanoes, and also drives climate change over long time scales. The position of the continents relative to the poles, as determined by plate tectonics, can influence glaciation. Human activities occur over much shorter time scales. Land use changes, such as converting natural landscapes to black asphalt parking lots, result in more solar radiation being absorbed by the earth's surface. Both deforestation and fossil fuel burning add CO_2 to the atmosphere. All of these external changes directly or indirectly affect the internal climate system, which further contributes to fluctuations and variability in the system over time (**Figure 6.19**).

The result of interactions among internal components, and their response to external changes, looks very different in the climate system (**Figure 6.20**) as compared to the chemical reaction in the beaker (Figure 6.16). The climate system is highly variable. Atmospheric CO_2 concentrations are a major indicator of global climate variability; and data records with enough detail to see this variability cover the past 740,000 years (for example, Figure 6.20). This diagram is from the 2007 IPCC Synthesis report, and the top three graphs show changes in the concentration of three greenhouse gases over the past several hundred thousand years: N_2O (green line), CO_2 (red line), and CH_4 (blue line). The data for these records come from direct measurements of these

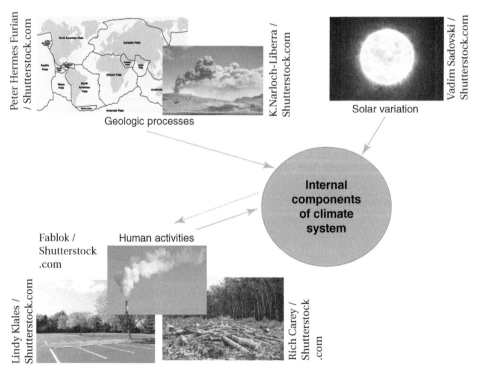

Figure 6.19 Lakes are a component of the climate system and they interact directly with other components, as shown here, leading to system fluctuations and variability caused by interactions among internal components.

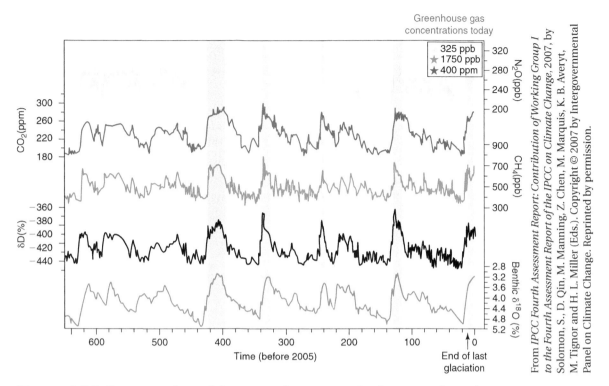

Figure 6.20 Concentrations of three greenhouse gases in the atmosphere shown.

gases in the atmosphere and, before these were available, the gas concentrations in air bubbles trapped in glacial ice from Antarctica.

The large-scale fluctuations over the past 740,000 years are thought to be driven largely by variations in solar radiation caused by the Milankovitch cycles, and they follow a general pattern. A change in this external condition has consequences for many internal components of the earth's climate system. One major change occurs in the cryosphere, which causes glaciations to occur. When greenhouse gas concentrations were much lower in the past, glaciers covered the earth to a much greater extent than today. When concentrations were high, there were interglacial periods with much smaller glaciers. These glacial and interglacial periods are different regimes, and changes from one to the other consitute a regime shift. Within each regime, there is also a lot of fluctuation. However, the drivers of the regime shifts, and the reasons behind the fluctuations, remain as a major unsolved mystery about the earth's climate.

What is clear from the measurements is that atmospheric concentrations of greenhouse gases are higher now than at any time over the past 740,000 years (as noted by the stars in the upper right-hand corner of the graph in Figure 6.20). The earth's climate system has been in the same pattern of shifting between glacial periods and interglacial periods for at least the past 740,000 million years. In fact, geological evidence shows that this shifting between glacial and interglacial periods began about 35,000,000 years ago. Before that time, from about 600,000,000 years ago up until 35,000,000 years ago, the earth was in a hotter condition that did not involve glacial-interglacial periods. Human societies are adapted to the conditions of the current warm interglacial period that began about 11,000 years ago at the end of the last ice age. What exactly will happen to the earth's climate as a result of the human drivers now changing the state of the climate system is uncertain, and is the subject of much research. IPCC scientists have developed several future climate scenarios, which will be explored more in the next chapter. Human activities may be pushing the climate system toward a threshold, or we may have already crossed that threshold, beyond which the Earth's climate system will enter a new regime.

Section 6.2.2—Disturbance, Internal Dynamics, and Stability Landscapes

Key Concept 6.2.2—Dynamic interactions among internal system components and changing external conditions result in changes to the system's stability landscape, and subsequently, its resilience over time.

This section is concerned with explaining how the two-way exchange between internal components and processes, and changing external conditions, shape system behavior over time. As described in Section 5.3, systems advance through pathways of development over long periods of time known as adaptive cycles. At any point along these pathways, external disturbances can affect the progression of a system. To understand how this works, it is helpful to define four new types of drivers: autogenic drivers, allogenic drivers, slow drivers, and fast drivers.

Recall from Section 3.1 that drivers are the governing forces that act on a system, either causing it to change or to remain in its current state. When thinking about internal and external environments of a system, and the dynamics of that system over time that shape resilience, it is useful to distinguish autogenic drivers from allogenic drivers. **Autogenic drivers** are the internal forces that cause a system to move forward in a certain direction along the pathway of development. They are a result of interactions among internal system components. **Allogenic drivers** are the external forces that disturb this pathway of development by either setting the system back to an earlier stage (e.g., a fire burning newly established shrubs at the beginning of the exploitation phase, and setting ecological succession back to the release phase) or causing the system to set off in a new direction (e.g., a forest fire followed by the introduction of an invasive shrub that outcompetes native species for resources to such an extent that the original forest does not regrow).

---

autogenic drivers
the internal forces that regulate a system's behavior and cause it to move forward along a certain directional pathway of development

---

allogenic drivers
the external forces that disturb the directional development pathway of a system by either setting it back to an earlier stage or causing it to move in a new direction

Two other types of drivers, which are important for thinking about system dynamics and resilience over time, are related to temporal scales of change. Allogenic drivers are not always obvious and sudden, such as a forest fire or a gas price spike, but can also be unnoticeable and slow-acting. Such subtle allogenic drivers affect systems slowly over relatively long periods of time. As a result, they are called **slow drivers**. The more obvious and faster-acting allogenic drivers, such as a fire or a price spike, are called **fast drivers**. Both of these types of allogenic drivers shape the **stability landscape** of a system, which is the *character* of a system's internal feedback structure at a given point in time. *Character* refers to the state of the competition among stabilizing and reinforcing feedbacks. Remember from Section 5.2.1 that it is the shifting dominance of a system's internal feedbacks that either keeps a system in its current regime (stabilizing feedbacks dominate) or moves it toward a new regime (reinforcing feedbacks dominate). Changes in the feedback structure result in changes in a system's stability landscape over time, and this change affects resilience. As reinforcing feedbacks become more and more dominant, resilience decreases. As stabilizing feedbacks become more and more dominant, resilience increases.

In summary, allogenic drivers interact with internal system dynamics, which are regulated by autogenic drivers, over different temporal scales to shape a system's stability landscape and resilience. All of these ideas are explained in this section, followed by two case studies of a lake and a grassland, where these ideas were applied to successfully recover ecosystems from hard-to-reverse critical transitions. The ideas presented in this section will help answer some of the lingering questions from the last chapter, such as: why do some systems have alternative regimes separated by nonlinear thresholds and others do not? Why do some systems cross relatively reversible thresholds to new regimes while others undergo critical transitions that are difficult to reverse? The answer to these questions has to do with changes in the stability landscape, and subsequently changes in resilience, over time. Some stability landscapes have thresholds, some do not. Some have more reversible thresholds, others foster critical transitions. Slow drivers gradually shape the stability landscape over time, whereas fast drivers shock it. Both slow and fast drivers can cause systems to suddenly cross thresholds.

Stability Landscapes and Resilience. The ideas of alternate regimes, regime shifts, drivers of change, and thresholds were all illustrated using the ball-and-basin model of resilience in Section 5.1.3. The model is shown again in **Figure 6.21a**. One aspect of this model that was left undefined in the previous chapter is the blue line that defines the boundary of the basins for each regime. This blue line, spanning both regime basins, and including the threshold between them, is the stability landscape. The stability landscape represents the conditions

---

slow driver
a type of driver that influences systems slowly and over relatively long periods of time

fast driver
a type of driver that rapidly influences a system over very short time periods

stability landscape
the character of a system's internal feedback structure at a given point in time as determined in part by the shifting dominance among stabilizing and reinforcing feedbacks

precariousness (P) a measure of the nearness of a system to a threshold

latitude (L) a measure of the maximum amount of change a system can undergo before crossing a threshold

resistance (R) a measure of how easy or difficult it is to change a system

under which a system operates, and its internal feedback structure, and it can change over time. This has consequences for a system's resilience. Moving from Figure 6.21a to Figure 6.21b, the position of the system (the ball) within the sustainable regime is moved closer to the threshold when the shape of the stability landscape changes. A driver (orange arrow) is disturbing the system to the same extent in both Figure 6.21a and Figure 6.21b. However, because the stability landscape in Figure 6.21b has changed to a shallower basin, the ability of the system to remain in the sustainable regime in the face of the same driver disturbance has decreased. In other words, the resilience of the system with the stability landscape shown in Figure 6.21b is less. The stability landscape determines system resilience. In order to anticipate future regimes shifts and changes in a system, information about both the magnitude of the drivers (orange arrows) influencing the system, and the overall resilience of the system as dictated by the stability landscape (blue boundary), must be known.

Now that we have started changing the shape of the regime basins, resilience must be defined in more detail. So far, resilience has been depicted as the horizontal distance between the ball and the threshold (Figure 6.21). This is not the entire picture and was a simplification. To be more precise, resilience is defined in terms of three dimensions: **precariousness (P)**, **latitude (L)**, and **resistance (R)** (Figure 6.22a). Precariousness (P) indicates the nearness of a system to a threshold.

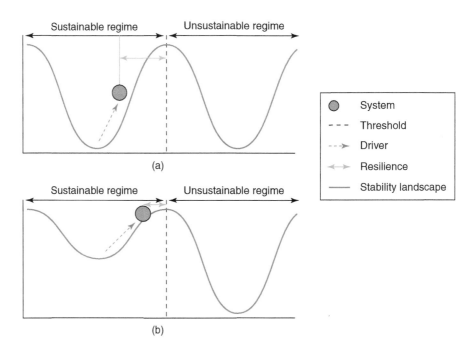

Figure 6.21 When the stability landscape changes, such as from (a) to (b), a driver of the same magnitude can bring a system much closer to a threshold.

The latitude (L) aspect of resilience defines the maximum amount of change a system can undergo before crossing a threshold into a new regime and is essentially the width of the basin. The depth of the basin characterizes the resistance (R), which is how easy or difficult it is to change a system.

These different dimensions of resilience can change independently of each other. For example, a stability landscape could change such that the resistance (R) to change increases, but the maximum amount of change possible for a system (L) decreases (Figure 6.22b). In a grassland system, for example, this type of change might mean that the landscape is more resistant to change caused by cattle grazing. Where there used to be bare spots left behind on the landscape left by 50 head of grazing cattle, there now might be no noticeable impact by the same number of cattle. A driver that might increase resistance (R) to cattle grazing in such as way could be increased soil moisture, which would allow grass to growth back more effectively. At the same time that resistance (R) increases, the latitude (L) dimension of resilience could decrease. In this case, the maximum amount of change a system can handle may decrease. For example, in a system that can graze 50 head of cattle on average, this could mean that the maximum amount of change that could occur in the number of cattle, that would still maintain a sustainable grazing pressure, would decline: if a stability landscape can handle an average of 50 cattle, but 45 to 55 cattle (50 ± 5)

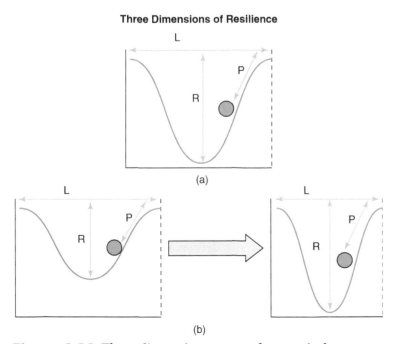

Figure 6.22 Three dimensions are used to precisely characterize resilience, and they can change independently of each other.

is the maximum latitude (L), then a stability landscape with higher latitude (L) may allow the system to handle 30 to 70 (50 ± 20) cattle.

Slow Allogenic Drivers and Stability Landscape Changes. Resilience is determined by the stability landscape, but how is the stability landscape determined, exactly? One way is through the influence of allogenic drivers over time, which can set the overall conditions under which a system operates. In the grassland example, ecological succession is an internal system process that is set into motion when an ecosystem is disturbed. This type of process, and the relevant feedbacks that drive the process, were described for a forest in the last chapter (Section 5.3.2, Figure 5.44). A similar process occurs in grasslands, and in all ecosystems for that matter. As succession occurs, the grassland is slowly changed from young pioneer species to a more mature climax community, and different species come to dominate the physical landscape. This changes the stability landscape as certain species dominate the landscape over time. Different species have different reactions to disturbances, such as fire or drought. Thus, each different biological community presents a different stability landscape for the grassland ecosystem. In a grassland, allogenic drivers that influence the progression of ecological succession, as regulated by autogenic drivers, include grazing pressure, climate change, or globalization.

The allogenic drivers can be slow or fast acting. In other words, they can be slow or fast drivers. Grazing pressure slowly influences the amount of grass, relative to shrubs, present over time. More frequent drought caused by climate change might slowly decrease soil moisture, making it harder for new grass to grow. Globalization increases the rate at which new technologies are spread from industrialized to developing nations. In the grasslands of east Africa, for example, travel in the past was by foot or by animal. Today, safari companies use motorized vehicles that cause erosion and soil compaction on the grassland landscape. Erosion and soil compaction make it difficult for grass to regrow. All of these allogenic drivers result in slow changes in the stability landscape of the grassland, and influence the general operating conditions and internal feedback structures for the system over time.

The ball-and-basin model can be used to illustrate how grazing pressure, which is a slow allogenic driver, might affect the stability landscape of a grassland over time. An indicator used to track the character of the stability landscape over time, as a result of grazing pressure, might be the amount of grass cover (green arrow in Figure 6.23). Figure 6.23 shows how changes in the stability landscape, as indicated by decreasing grass cover, eventually cause the system to shift toward desert shrubland over time. Overall, as the grassland regime basin becomes shallower, and the shrubland basin gets deeper, reinforcing feedbacks increasingly overwhelm stabilizing feedbacks, and the prospect that the system will undergo a regime shift becomes more and more likely. At first, the stability landscape supports only the grassland regime (Figure 6.23a). Under these conditions, it is not possible for the system to shift to desert shrubland because that regime does not exist. This means that the grassland is resilient to changes imposed by other faster-acting allogenic drivers that could shift it to a shrubland, such as a wildfire or a drought event. As grazing pressure increases over time, however, the internal system conditions and feedback structures change, with permanent reductions in grass cover. In other words, the stability landscape changes. As this happens, the desert shrubland regime basin begins to form (Figure 6.23b) and gets larger and larger (Figures 6.23c and 6.23d) until it is the only possible regime (Figure 6.23e). At the same time, the grassland regime basin gets increasingly smaller until it finally disappears. This overall trend in the stability landscape decreases the grassland's resilience to disturbances by other, faster-acting, allogenic drivers (e.g., fire, drought) such that smaller and smaller disturbances will push the system toward a desert shrubland.

Fast Allogenic Drivers and Stability Landscape Changes. Slow acting allogenic drivers, such as grazing pressure in a grassland, are not the only drivers that can alter the stability landscape of a system.

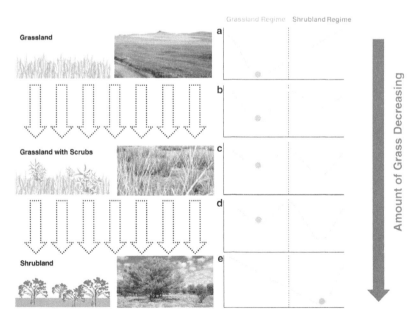

Figure 6.23 As grazing pressure, a slow driver, increases, the amount of grass cover decreases.

Faster-acting allogenic drivers, such as a fire, a meteor impact, or an economic crisis, can also change the stability landscape, if the disturbance is large enough to alter internal system conditions and feedback structures. If the disturbance is not large enough, then fast-acting allogenic drivers will merely change the state of a system within a regime, or cause the system to undergo a regime shift by pushing it over a threshold, but they will not completely alter the stability landscape.

Figure 6.24 depicts how fast allogenic drivers change the state of a system within a certain stability landscape. The fast allogenic driver (orange arrow) influences the state of the system by causing it to "roll around" in the basin, but the disturbances by this driver on the system do not result in a shift of the system to a completely different regime (Figure 6.24a). Such a regime shift is depicted in Figure 6.24b. In both of the cases shown in Figure 6.24 there are two different regimes. Each regime is governed by its own set of interactions among internal system components, and has its own internal feedback structure. When the system is moved from one regime to the next, due to a fast-acting allogenic driver, the internal feedback structure shifts to a new one. Thus, in both cases shown in Figure 6.24, the system's behavior depends

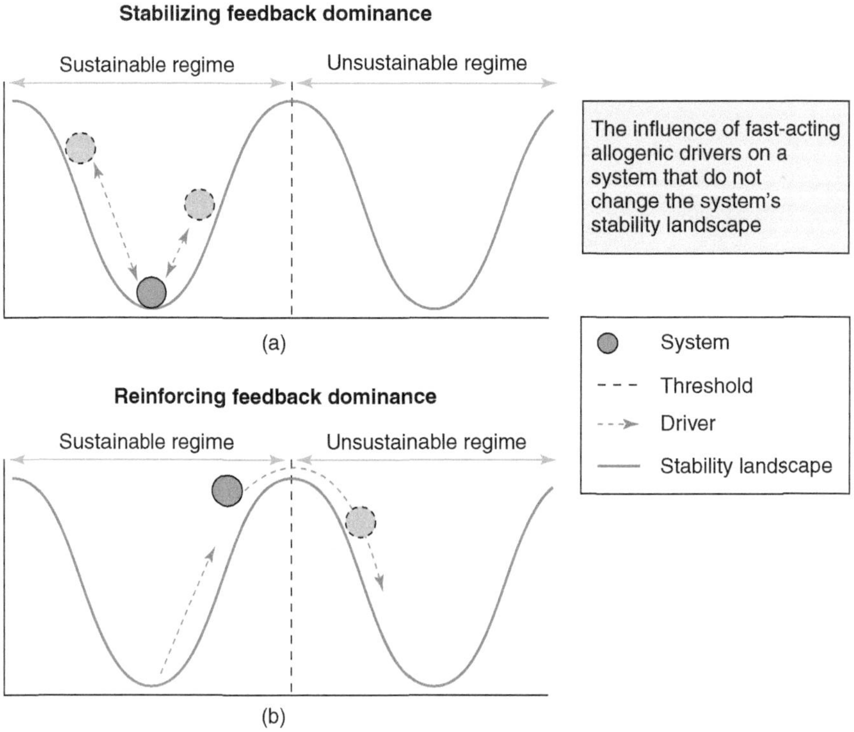

Figure 6.24 When the stability landscape stays the same, fast allogenic drivers influence systems by either causing system fluctuations and variability (a) or by pushing a system into a new regime (b).

on two-way exchanges among internal dynamics and feedbacks (autogenic drivers), and the external conditions (allogenic drivers).

Figure 6.25 shows how a major, high-impact disturbance by a fast-acting allogenic driver can suddenly change the stability landscape of *any* system (a through e), regardless of its stability landscape before the disturbance. When this happens, and the stability landscape goes completely "flat," the interactions among internal system components and the internal feedback structures completely collapse (as represented by the flat ball-and-basin diagram on the right). As a result, the system's behavior is no longer a result of a two-way exchange among internal dynamics and feedbacks on the one hand, and external conditions on the other. Because the system has nonexistent, to very weak, internal dynamics and feedbacks following a major, disruptive disturbance, the system's behavior is governed only by allogenic driver disturbances and can appear somewhat chaotic (as depicted by the random movement of the ball in the right-hand flat ball-and-basin diagram in Figure 6.25). This depiction of what happens to a system following a cataclysmic disturbance by a fast allogenic driver allows

chaotic
the type of unpredictable behavior exhibited by a system that lacks strong internal feedbacks such that the state of a system at any given time is determined largely by random external disturbances

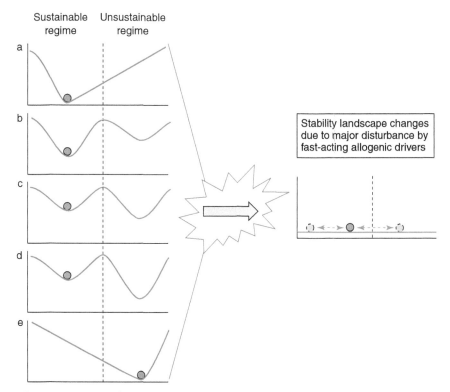

Figure 6.25 A fast-acting allogenic driver that imparts a high-impact disturbance on a system can suddenly change the stability landscape of that system from one with defined regimes, internal system conditions, and feedback structures to one that is chaotic and lacks internal structure.

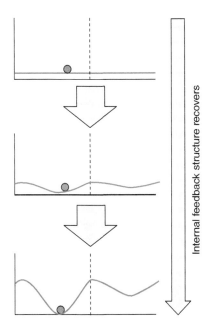

Figure 6.26 After a major disturbance by a fast-acting allogenic driver that renders a stability landscape chaotic, the stability landscape reforms, internal feedbacks restructure, and new regimes are created.

a conclusion to be drawn about internal-external exchanges in CASs: *The stronger the interactions between internal system components and feedback structures, the less influence an allogenic driver will have on the system's overall dynamics.* This is true for both fast allogenic drivers (e.g., a catastrophic grassland fire) and slow allogenic drivers (e.g., long-term grazing pressure). Figure 6.26 shows how a system's internal interactions and feedbacks might recover following a disturbance as the stability landscape is reshaped.

Before moving on to the next topic, one last point should be made in reference to major, high-impact disturbances by fast-acting allogenic drivers that suddenly change stability landscapes. Such disturbances do not always completely obliterate stability landscapes, but can also change the landscapes less drastically. For example, a less severe disturbance might cause the stability landscape to change suddenly from b to e or, alternatively, from d to a (Figure 6.25). Such alterations are severe, in the sense that one landscape is changed to another landscape that is very different, but they are not so drastic that the internal interactions among system components and feedback structures completely collapse. Internal interactions and feedbacks still exist, and they interact with the external environment to determine overall system behavior. This is in contrast to the "flat" ball-and-basin landscape where the system's behavior is governed by allogenic drivers only.

Slow Drivers versus Fast Drivers. How does one know which drivers are slow and which are fast? The short answer is that it depends on the system and the specific time scales over which each of the drivers regulating a system is operating *relative to the others*. Recall from Chapter 3 (Section 3.1.2) that drivers operate over time scales ranging from very short (seconds to minutes) to very long (years to centuries). For example, in a grassland three allogenic drivers might include wildfires, grazing pressure, and soil moisture content. Wildfires in this case would be the fastest driver, because they occur over a time scale of hours to days. Grazing pressure might be the next fastest driver, assuming that animals are grazing on grasslands for decades to centuries. Changes in soil moisture content, which can contribute to seed establishment and other internal processes in grasslands, occur naturally over very long periods of time, such as hundreds to thousands of years, due to changes in a region's microclimate.

Drivers that are slow, such as in the grassland system just described, however, are not necessarily *always* the slowest drivers in *every* system. A driver that is considered slow in one system might be considered a fast driver in another. Whether a driver is labeled slow or fast depends on the *relative* time scales over which different drivers act in a specific system, and also the phenomena being studied. For example, when trying to determine biomass accumulation in a lake ecosystem, both photosynthesis rates and biomass growth are important drivers. Photosynthesis by plankton in a lake is a fast variable relative to the biomass growth of fish that eat the plankton (Figure 6.27a). Photosynthesis rates are on the order of hours, whereas fish growth occurs over days to weeks. Relative to photosynthesis rates and biomass growth, major changes in vegetation surrounding the lake due to regional climatic changes might seem like a slow process and, in this case, climate change would be a slow driver. However, if the goal is to understand how landscape vegetation changes, then regional climate change is a fast driver relative to other important drivers. Regional climate is influenced by

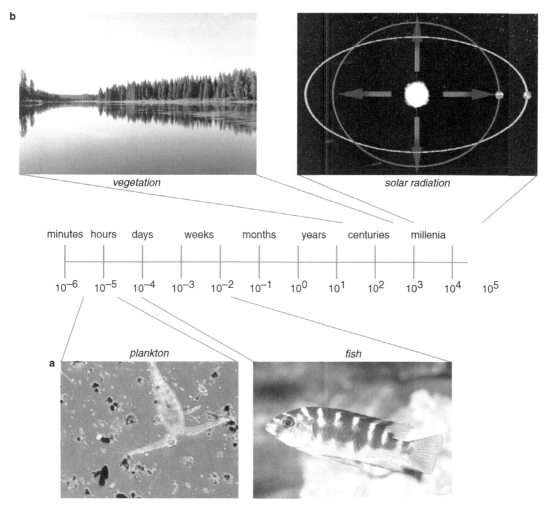

Figure 6.27 Different drivers operate over a variety of time scales.

global climate patterns. Global climate patterns change when there are variations in the amount of incoming solar radiation caused by changes in the position and orientation of the Earth relative to the Sun over 10,000- to 100,000-year time periods (Figure 6.27b). In this case, regional climate change is the fast driver and the amount of incoming solar radiation is the slow driver.

Stability Landscape Changes in Three Dimensions. As always, the ball-and-basin model is a simplification useful for initially visualizing concepts. A more realistic model of a stability landscape is a three-dimensional model (Figure 6.28). In this 3D ball-and-basin model, there are still two regimes: the system is represented by a ball and the threshold is a dotted line. As in the simple ball-and-basin model (Figure 6.21), the system (the ball) is not the only part that moves. The stability landscape, over which the system moves, also changes. In Figure 6.28a, the system is in Regime 1 and its precariousness (P) shows that it is close to a threshold between Regime 1 and Regime 2. Figure 6.28b shows a change in the stability landscapes of both regimes. Regime 1 now has a smaller latitude (L, basin width) and about the same resistance (R, basin depth) as in Figure 6.28a, whereas Regime 2 has a larger latitude (L) and the same resistance (R). Notice that the system (the ball) has shifted from Regime 1 in Figure 6.28a to Regime 2 in Figure 6.28b. This brings up an important point: it is possible for a system to shift into a new regime *only by changes in the stability landscape caused by slow allogenic drivers.* Thus, a regime shift is not always caused by an obvious external shock, such as a wildfire or an extreme drought event. The stability landscape can change imperceptibly over long periods of time, which can eventually cause a sudden and unexpected threshold shift. This is depicted in Figure 6.28 as a change in the resilience dimensions and, as a result, the position of the threshold. In the grassland ecosystem

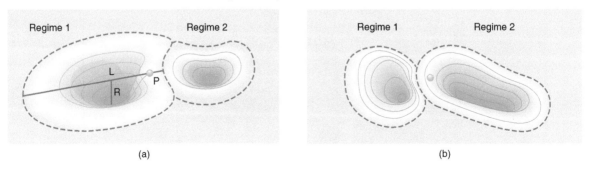

Figure 6.28 A more precise depiction of a stability landscape includes the three dimensions of resilience.

example, a gradual disturbance caused by grazing pressure, a slow allogenic driver, can eventually cause a sudden and unexpected shift to shrubland. The changes leading up to the shift are gradual, and are often unperceived by an observer, as the threshold creeps closer and closer to the system with changes in the stability landscape. The regime shift, in this case, happens without an obvious external shock by a fast-acting allogenic driver, such as a wildfire or a drought, that pushes the system across a threshold.

Regime shifts resulting from slow changes in the stability landscape are sneaky and hard-to-detect because intuition tells us that natural and human systems will undergo major changes *only* as a result of distinct events such as a flood, a tsunami, an earthquake, the election of a malevolent politician, or a stock market crash. However, major change can also occur in an invisible way, through the slow and imperceptible erosion of resilience over time by stability landscape change due to slow allogenic drivers, such as grazing pressure, phosphorus addition to a lake, increasing poverty, or steady deforestation. Intuition is not always accurate, and it is challenging to anticipate this type of change. The implication of all of this for sustainability is that it can be difficult, if not impossible, to detect the development of serious problems until they are a reality. Often, because stability landscape changes happen so imperceptibly over time, managers of natural and human systems tend to assume that they are constant and unchanging system components. However, as illustrated in the previous examples, unnoticed slow variable changes to stability landscapes can move a system toward a critical threshold over long periods of time.

Although the anticipation of sudden and unpredictable shifts of systems from sustainable to unsustainable regimes is frightening, this type of slow change can also give us hope that our sometimes seemingly ineffective actions actually are pushing systems back toward sustainability. Environmental economist Paul Hawken describes a worldwide movement toward sustainability in his book *Blessed Unrest: How the Largest Movement in the World Came into Being, and Why No One Saw It Coming*. (His ideas were described in detail in Box 1.3 of Chapter 1.) He tallies at least 130,000 organizations that exist around the world today that are invisibly working toward environmental and social justice. These range from small businesses to nonprofit organizations, to causes promoted by individual people. One way to view this situation is that these actions could be causing undetected gradual changes in the stability landscape that most do no perceive. In his book, Hawken claims that we don't hear about the slow changes resulting from this movement in the news because politicians and the mainstream media don't pay attention to them. Could this imperceptible worldwide movement one day shift socioecological systems to a more sustainable regime? Let's hope so, and keep making those small changes.

Scheffer Ball-and-Basin Stability Landscape Model. Put on your 3D glasses! This section presents a comprehensive three-dimensional model for visualizing stability landscape change over time, and how this affects whether a system has more than one regime, and also what type of regime shifts are possible. This new model is first generally explained. It is then applied to several natural and social systems to illustrate how all of the ideas come together to understand SESs that behave as complex adaptive systems

In Section 5.1 of the last chapter, you learned *how systems can undergo different patterns of change: linear, easily reversible thresholds, and critical transitions* (Figure 5.3). Now you will learn how the shape of the stability landscape determines which of these types of change can possibly occur in a system. This can be explained by exploring Figure 6.29, which was redrawn from a book by called *Critical Transitions* (2009) written by ecologist Marten Scheffer from Wageningen University in the Netherlands. This figure brings together the concepts related to patterns of change from Section 5.1 of the last chapter, and stability landscapes changes from this chapter. This model is applied to a real world example in **Box 6.5**.

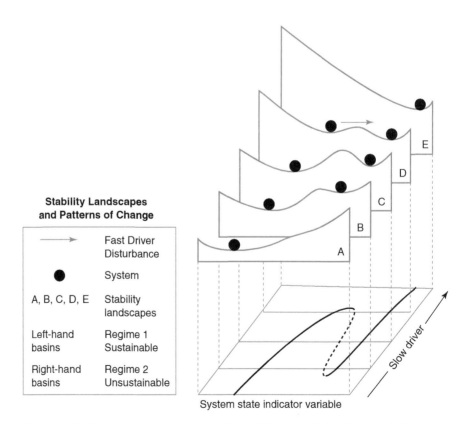

Figure 6.29 A visual representation of Marten Scheffer's stability landscape concept.

In Figure 6.29 there are five possible stability landscapes shown (A through E). The ball represents the state of the system, which is detected using a *system state indicator variable*. This variable literally indicates the state of the system and can be either a system driver or a "passenger" variable. Drivers are variables that regulate the system, like the driver of a car, but passengers move along passively with the changes. (In this chapter, Scheffer's terminology is used. However, his system driver variable is analogous to a "pressure indicator" and his passenger variable is similar to a "state indicator," both of which were introduced in Chapter 4.) In many systems, there are usually a few main drivers governing change and the rest are passengers. However, passengers are not always just riding along with the changes in the system, and they sometimes take the wheel. In other words, passengers can become drivers, and vice versa, as systems change. For instance, recall the example of the feedbacks involved with human population growth in Section 5.2.2 of the last chapter: the amount of arable land available for food production was the driver that stabilized populations over time through a stabilizing feedback. However, as the system changed due to technological innovation, which changed the definition of what land was arable, and how much food could be grown, land available for food production became a passenger, driven by new technology. In the Scheffer ball-and-basin stability landscape model (Figure 6.29), both drivers and passenger variables can serve as *system state indicator variables*. The best indicators, as discussed in Chapter 4, will depend on the specific system and what you need to know about that system.

On the other axis in the Scheffer ball-and-basin stability landscape model is the slow driver, which interacts with the internal system conditions and feedback structure in a two-way exchange. This two-way interaction ultimately causes the stability landscape to change over time. The slow driver in this model can be either allogenic or autogenic. Recall that autogenic drivers are the internal governing forces that cause the system to move forward in a certain direction along the pathway of development. These drivers have not been the focus of this section, but will be discussed in more detail in Section 6.3.

At stability landscape A in Figure 6.29, conditions, as determined by slow variables, are such that there is only one possible regime for the system (as denoted by the system state indicator variable). In stability landscape B, conditions have changed due to the influence of a slow driver, and a second regime possibility develops. At this point, the shift is a reversible threshold shift, not a critical transition. This means that change will occur quickly with only a small perturbation when the system is close to the threshold, but it is not hard to reverse the change to bring the system back to its current regime. To reverse the change, the driver causing the change is reversed to push it back over the threshold, and the system is left to recover. Stability landscape C shows two alternative regimes under conditions that result in a critical transition. Once the threshold is crossed, it is very hard to go back (but there have been success stories as will be described in **Box 6.6**). At stability landscape D, conditions have shifted far enough to the other end of the spectrum that the system tends toward an unsustainable regime in the face of disturbances, rather than a sustainable one. A threshold-shift back to the sustainable regime can occur with enough pressure in that direction. However, because the sustainable regime basin is shallower than in the stability landscapes C and B, smaller fast driver disturbances can shift the system back to the unsustainable condition more easily in stability landscape D. Finally, at stability landscape E, allogenic drivers have changed internal system conditions and feedbacks to such an extent that there is only one (unsustainable) regime. Under these conditions, it is impossible to drive a system back to sustainability. Note that fast drivers can also alter the stability landscape, but more drastically. Thus, a fast driver might cause a "jump" from one stability landscape to another without the slow transition through the five stability landscapes portrayed in this model.

A major conclusion of the Scheffer ball-and-basin stability landscape model is that alternative regimes are not permanent features of systems. They exist only for a subset of possible stability landscapes. Different possible stability landscapes explain why some systems undergo critical transitions (stability landscape C, Figure 6.29), why some cross reversible thresholds more easily (stability landscapes B and D), and why others might get stuck in certain regimes (stability landscapes A and E). It is important to pay attention to overall system conditions, as determined by the stability landscape, when trying to prevent a regime shift from a sustainable to an unsustainable state. For example, if a system is functioning under the conditions of stability landscape A (Figure 6.29), then it is not possible to move a system to an unsustainable regime. However, if the system resembles stability landscape C, then great care must be taken not to push the system over a threshold toward a critical transition from which it would be very difficult to recover.

It is also important to pay attention to stability landscapes when trying to recover systems from unsustainable regimes back to sustainable regimes. For example, if stability landscape D is dictating the conditions under which your system operates, you will need a relatively large disturbance to push your system from an unsustainable regime back to a sustainable one. Additionally, because the sustainable regime basin is shallow relative to other stability landscapes, the system will more easily slip back to an unsustainable regime in the face of ongoing disturbances, following a recovery effort. A better strategy than constantly applying large disturbances to push systems back toward a sustainable regime, or alternatively "walking on egg shells" while in the sustainable regime to prevent relatively small disturbances from shifting the system back to an unsustainable regime, might be to work on changing the stability landscape so that the system has landscapes that resemble B or even A. With stability landscape B, small disturbances tip the system toward sustainability. For stability landscape A, no unsustainable regime exists! In Box 6.5, the Scheffer ball-and-basin stability landscape model is applied to a spruce-fir forest ecosystem to illustrate how this model might be used to understand SESs in general. Box 6.6 provides examples illustrating how the Scheffer ball-and-basin stability landscape model can be used to better inform resilience-based approaches to SES governance. **Box 6.7** applies the ideas in this section to a sustainability case study focused on the indigenous Shimshal society, who live in the furtherest northern reaches of Pakistan.

BOX 6.5 STABILITY LANDSCAPE CHANGES AND RESILIENCE IN SPRUCE-FIR FOREST ECOSYSTEMS

The response of a spruce-fir forest ecosystem to an insect outbreak is a detailed example of how stability landscape changes determine the effect of a fast driver disturbance on a system. As described in detail in Box 6.2, spruce-fir forests, found in boreal ecosystems, experience attacks by the spruce budworm on a regular basis when the larvae eat newly sprouting spruce needles in the springtime. The budworm attack is the fast driver disturbance, and is a natural process in these forests. However, the conditions under which the budworm is operating will determine the overall effect of these outbreaks on the forest ecosystem. These conditions are defined by the stability landscape, which, in this case, is shaped by the foliage density. As spruce-fir forests undergo ecological succession, which is regulated by autogenic drivers, foliage becomes denser. Ecological succession is an internal system process characterized by the continual shifting dominance of feedbacks as forests move from one phase of succession to the next (see Section 5.3.2 to review this idea). This process transitions forests from young spruce trees to mature trees over many decades. Thus, in this case, the stability landscape is shaped by slow autogenic drivers, rather than slow allogenic drivers, such as grazing pressure in grasslands (Figure 6.23).

The budworm is always present, but when forests are young, the low density of foliage allows bird predators to access budworm larvae. This keeps budworm populations low and prevents them from doing too much damage to the trees. However, as forests approach the mature tree stage of ecological succession, foliage density increases. Under this stability landscape, spruce forests are less resilient to budworm attacks because insectivorous bird predators cannot hunt budworm larvae as effectively through the dense foliage. Under high foliage density conditions, budworm attacks (fast driver disturbances) push forests more easily from a *living forest* (Regime 1) to a *dead forest* (Regime 2) (Figure 6.30).

(Continued)

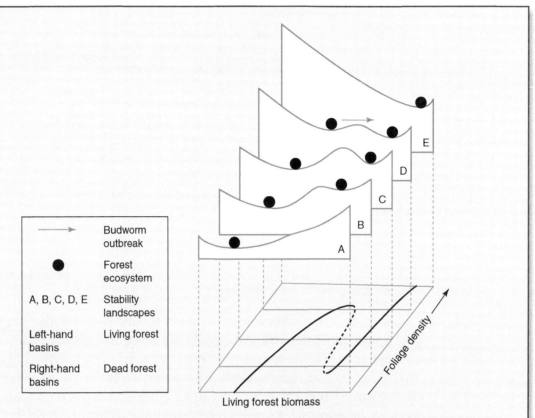

Figure 6.30 The stability landscapes shown here determine the resilience of spruce-fir forests in the face of attacks by the budworm.

Regime shifts are a natural process in these forests, but ecologists are finding that human-caused climate change may be decreasing the resilience of these forests to budworm attacks. The mature budworm is present in these ecosystems in the summer. During this time, they mate and females lay eggs on the underside of spruce needles. Although the larvae hatch in 1–2 weeks, they do not start feeding on the spruce needles until the spring. Instead, they construct shelters on tree branches in which they hibernate and wait out the harsh winters characteristic of the climates in which spruce-fir forests thrive. During the winter, a certain percentage of the larvae die from these harsh conditions. Those that survive the winter emerge from their shelters in the spring to feed on fresh needle growth. However, higher-than-average winter temperatures caused by climate change mean that more larvae survive the winter. Larger budworm populations in the spring, following mild winters, cause more forest damage than smaller populations, which exist following harsh winters with colder temperatures. Thus, human-caused climate change is another slow driver (in addition to foliage density) that causes gradual changes in the stability landscape of these forests, and decreases their resilience to budworm attacks each spring. As a consequence of climate change, the area of dying spruce forests in North America due to yearly budworm attacks is increasing. In summary, both internal autogenic drivers of ecological succession, and external allogenic drivers related to climate conditions, are slow drivers that shape the stability landscape of spruce-fir forests and affect their resilience.

BOX 6.6 RECOVERY FROM CRITICAL TRANSITIONS USING SCHEFFER'S BALL-AND-BASIN STABILITY LANDSCAPE MODEL

This box illustrates concepts introduced in this section and how they have been used, or might be applied, to real-world recovery efforts of systems that have undergone, or may have undergone, hard-to-reverse types of change: the critical transition. The first two examples are of actual recovery efforts that have been successfully carried out in two different types of ecosystems. The first is a lake that has undergone eutrophication. The second example is a type of dryland ecosystem: a woodland. By manipulating the slow drivers that shape the stability landscapes of these systems, and then implementing fast driver disturbance "shocks" at the right time, ecosystem managers have recovered these systems from an unsustainable state. What essentially happens in these cases is that managers find ways to decrease the resilience of the unsustainable regime so that it is easier to implement a fast driver disturbance that pushes the system back to a sustainable regime. In addition to these documented case studies on ecosystems, an example of human system change is also provided.

Recovery of a Shallow Lake Ecosystem from Eutrophication. Shallow lakes generally have two different possible regimes: *clear, vegetated,* and *turbid, unvegetated.* A clear lake is healthy because it contains aquatic vegetation that provides a habitat for many organisms such as zooplankton, fish, and birds. When eutrophication occurs, a lake moves into another regime dominated by microscopic plankton and turbid water. The turbid water does not allow much light to pass through, resulting in a lake with minimal aquatic vegetation and very little habitat for other organisms. The *clear, vegetated* regime supports high biodiversity and can be used by humans. The *turbid, unvegetated* regime supports minimal biodiversity and can be dangerous for humans, such as for fishing, drinking, or swimming, due to growth of toxic cyanobacteria.

Shallow lakes typically undergo critical transitions from *clear, vegetated* to *turbid, unvegetated* when too many nutrients are added to the lake water (Figure 6.31). These nutrients have many sources, including fertilizer, human sewage or animal waste, household gray water, and industrial or mining effluents. The process of anthropogenic nutrient addition to natural water bodies is known as *nutrient loading*. It is a slow allogenic driver. As nutrient loading increases over time, the lake will shift from clear to turbid. These critical transitions are difficult to reverse because nutrient loading must be reduced to levels well below the level just before the critical transition (see Figure 5.3c to review). Reducing nutrient loading to such an extent is often not within the control of human societies because it may be impossible to eliminate nutrient loading due to natural processes, such as nutrients entering the lake from soils, sediments, or the atmosphere. Efforts to recover lakes to healthy *clear, vegetated* regimes have been largely unsuccessful in the past.

However, lake ecosystem managers are starting to pay attention to slow and fast drivers, stability landscapes, and resilience (Figure 6.32). In shallow lakes, recovery efforts that have succeeded involve reducing nutrient loading to extremely low levels (slow driver) and then

(Continued)

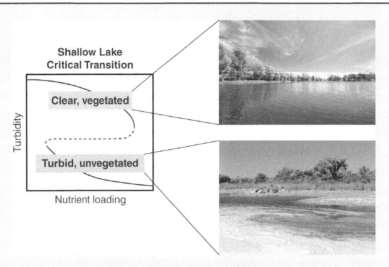

Figure 6.31 When shallow lakes shift from a clear vegetated regime to a turbid unvegetated one, they often undergo a critical transition.

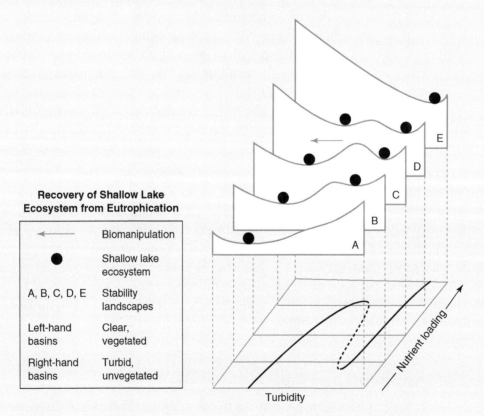

Figure 6.32 The stability landscapes shown here have been used to successfully shift eutrophied shallow lakes back to a healthy clear vegetated regime.

shocking the system by removing all of the fish, adding zooplankton, and planting aquatic vegetation (all are fast driver disturbances). The fast driver disturbance phase of recovery is known as biomanipulation. This may seem strange: if fish are part of lake biodiversity, then why would removing them help biodiversity? It is counterintuitive. However, if you think about biological communities in lakes, this makes a lot of sense (Figure 6.33). Phytoplankton are primary producers that use nutrients for photosynthesis, zooplankton are primary consumers that eat phytoplankton, and fish are secondary consumers that eat the zooplankton. Zooplankton require abundant aquatic vegetation to be present in large numbers because this is where they hide from their predators (fish). In a *turbid, unvegetated* regime, zooplankton populations are low because vegetation is sparse, and this means that phytoplankton populations can explode to process a lot of nutrients and speed up eutrophication. Zooplankton populations can recover if the fish are removed, while zooplankton species are introduced and zooplankton habitats are planted. If zooplankton populations recover, they will limit the growth of their prey: phytoplankton.

Essentially, what happens in shallow lake recovery is that the resilience of the undesired state *(turbid, unvegetated)* is reduced by reducing nutrient loading, and then the system is shocked with a fast driver disturbance (biomanipulation) to push it to a desired state *(clear, vegetated)*. Prior to biomanipulation, the appropriate nutrient loading conditions are set by draining and refilling the lake, or waiting for a low water year when nutrient loading is lower. This type of lake recovery first occurred in the Netherlands and is becoming more common. It is based on knowledge of ecosystems as complex adaptive systems.

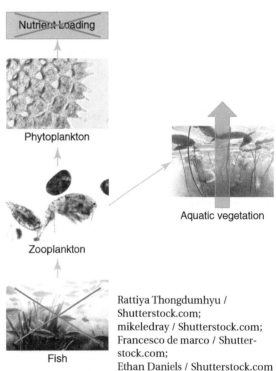

Rattiya Thongdumhyu / Shutterstock.com;
mikeledray / Shutterstock.com;
Francesco de marco / Shutterstock.com;
Ethan Daniels / Shutterstock.com

Figure 6.33 Using biomanipulation to reduce fish populations and increase vegetation cover, while simultaneously reducing nutrient loading, allows zooplankton populations to thrive and, subsequently, graze down phytoplankton populations so that shallow lakes can recover from eutrophication.

Recovery of a Woodland Ecosystem from Desertification. Healthy, *dense woodlands* (Regime 1) can be degraded to a landscape of *scattered shrubs* (Regime 2) as a result of extensive agriculture, wood gathering, and other human activities (Figure 6.34). Healthy *dense woodlands* provide many services to human societies, but one of the only productive human activities that can occur in *scattered shrub* systems is goat husbandry. Recovery of dense woodlands is similar to

(Continued)

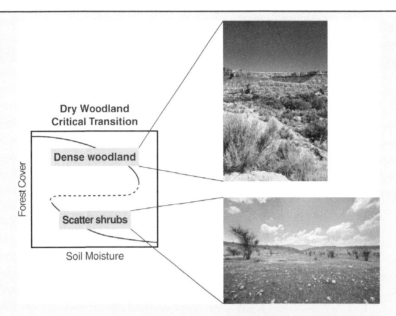

Figure 6.34 When dense woodlands shift to scattered shrub systems, they often undergo a critical transition.

that of shallow lakes in that the correct physical conditions, as dictated by slow drivers (nutrient loading in lakes), are created, and then biological conditions are quickly changed (fast driver disturbance).

In woodland ecosystems, the amount of precipitation is a slow driver that shapes the stability landscape (Figure 6.35). As a result, ecosystem managers wait for a wet year with lots of precipitation, and then shock the system with removal of herbivores and dispersal of seeds (fast driver disturbances). Wet years, which occur during El Niño events in many parts of the world, provide conditions under which new woodland tree seeds can take root in the absence of dense woodlands of mature trees. When mature trees are present, they create localized high soil moisture conditions in an otherwise dry climate through shading. Without mature trees, new tree seeds have trouble establishing in dry conditions, except during wet years. The other challenge to new tree growth is herbivore grazing. When wet years do occur, and trees can grow in scattered shrublands, their saplings are quickly grazed down by herbivores, such as goats and rabbits. To recover these systems, managers wait until a wet El Niño year is predicted. Before rains occur, they remove herbivores and spread plenty of tree seeds across the landscape. These present the right conditions for trees to take root and grow beyond the stage where they are vulnerable to destruction by herbivores. Once mature trees are established, it is easy for new trees to take seed in their protective shade.

Microfinancing and Poverty Reduction. In most cases, the details of change, as portrayed by the Scheffer ball-and-basin stability landscape model, have not been as precisely

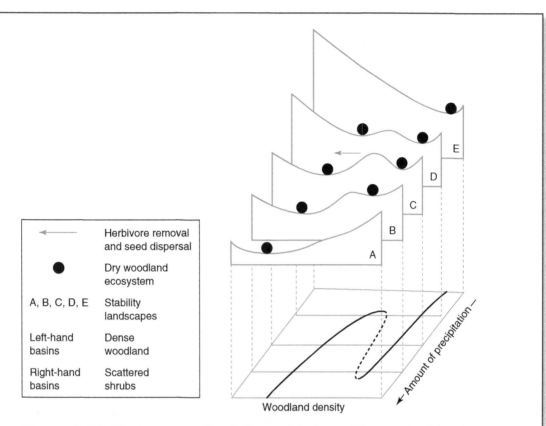

Figure 6.35 To recover woodlands from critical transitions to shrublands, ecosystem managers wait for wet years to remove herbivores and spread seeds over the landscape.

documented and quantified in human systems as they have for ecosystems. However, this does not mean that these types of changes do not happen in social systems. One example is microfinancing as a means of community development for people living in poverty. This idea began with the Grameen Bank in India, for which Professor Muhammad Yunus won the Nobel Peace Prize in 2006. The idea is that small loans of only hundreds of dollars are given to people living in poverty to start small businesses, or other profitable economic activities that will lift them out of poverty. These small loans (fast driver disturbance) can lift people out of *poverty* (Regime 1) to *productive livelihoods* (Regime 2) with improvements in human well-being. The microfinancing model, with a fast driver disturbance in the form of small loans, has been extremely successful in lifting people out of poverty in some cases, but has been unsuccessful under other conditions. The slow drivers that determine the stability landscape conditions under which microfinancing programs succeed are extremely variable from society to society, but they are generally determined by the adaptive capacity of individuals and institutions in each specific case (Figure 6.36). Recall from Section 5.1.3 of the last chapter that adaptive capacity is the ability to innovate, reorganize, and learn when faced with change.

(*Continued*)

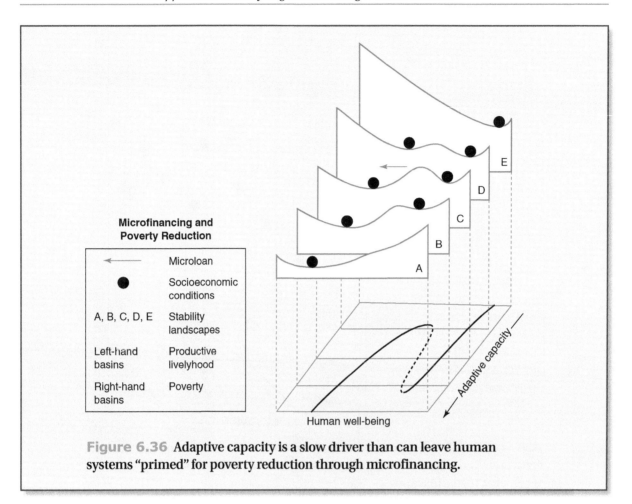

Figure 6.36 Adaptive capacity is a slow driver than can leave human systems "primed" for poverty reduction through microfinancing.

BOX 6.7 STABILITY LANDSCAPES IN TRADITIONAL AND MODERN-DAY SHIMSHAL SOCIETY

"[R]esilience is not only about being persistent or robust to disturbance. It is also about the opportunities that disturbance opens up in terms of recombination of evolved structures, renewal of the system and new emergence."

– *Carl Folke*, Resilience: The emergence of a perspective for socio-ecological systems analyses, *2006*

Indigenous Shimshal society: Internal components, processes, and feedbacks. The village of Shimshal is one of many Wakhi villages located in the Gojal Valley, the rugged mountainous Karakoram region of northern Pakistan (Figure 6.37). The indigenous Wakhi people were farmers, hunters, and nomadic herders (Figure 6.38). They tended herds of yaks, goats, cattle, and sheep in pastures in high elevation summer settlements and near winter residences in the valleys. Their herds are large and their pasture lands vast, with the

Shimshal community exclusively controlling more than 1,000 mi² (2,700 km²) of alpine terrain. They also hunted blue sheep and ibex, which was an important food source during cold winter months. In warmer months, they farmed cereals (wheat and barley), potatoes, peas, beans, apricots, apples, and other vegetables on terraced glacial and alluvial deposits irrigated by melt water streams and, lower down, the larger Shimshal River. More recently, the Shimshal livelihood has expanded to tourism, dominated by trekking, and to employment in cities, where residents migrate for part of the year to undertake seasonal work.

Figure 6.37 Shimshal village, indicated by the red star in the map, is located in far northern Pakistan near the borders with China and Afghanistan.

For centuries, Shimshal was extremely isolated from the outside world. With the exception of support from an Ismaili development agency, the Aga Khan Rural Support Programme, the Shimshal people have been largely self-sufficient due to their isolation. Throughout the 1970s, contact with neighboring Wakhi villages required at least 1 week of walking while carrying a heavy load. It took even longer to reach a town with a sizeable market, where they could earn a supplementary income selling dairy products, yaks, and yak hair carpets. Efforts to improve connection with the outside world began in the 1980s, as villagers initiated various construction projects, such as footpath improvements and bridge building (Figure 6.39). In 1985, the Karakoram Highway (KKH) was completed, following the Hunza River valley and linking China to Pakistani ports on the Arabian Sea. Even with these improvements, Shimshal village was still more than 35 mi (60 km) from the KKH, and even the fastest Shimshalis took almost a day to reach the highway. In 2003, a dirt road connecting Shimshal to the KKH was completed, first initiated by the Shimshal and later completed with the help of the Pakistani government, reducing the journey to 4 hours.

The approximately 1,700 inhabitants of the village are part of the Shimshal culture, which is over 400 years old. The culture is rooted in a deep connection to their place and community in the mountainous Gojal Valley. Until recently, their identity was strongly tied to their nomadic pastoral way of life and the vast alpine ecosystems surrounding them. Infrastructure improvements and other development initiatives, such as the building of bridges, trails, and huts, were traditionally accomplished through collaboration between individual

(*Continued*)

Figure 6.38 For much of their existence, the Shimshal were self-sufficient, as they tended their herd in alpine pastures, such as sheep and goats (*left*), hunted local fauna for venison in winter months, such as the Siberian ibex (*right*), and farmed on a terraced landscape in spring and summer (*bottom center*).

households, who would chose the project and supply necessary resources (materials, food), and volunteers from the community who would actually complete the project. Traditional governance systems were based on collective decision-making, such that a council of household heads would meet to make decisions. These aspects of Shimshali culture demonstrate their indigenous values of self-sufficiency, communal living, and unity.

Khunjerab National Park: An external disturbance. In 1974, George Schaller, an American wildlife biologist, proposed that a national park had to be established. Immediately following this recommendation, the Pakistani government established Khunjerab National Park (KNP) in 1975. The impetus behind the establishment of the park was threefold. First, the government of Pakistan wanted to establish a national park that would become a prestigious World Heritage Site, which is an honor given to places around the world deemed to be "outstanding natural and cultural sites" by the United Nations Educational, Scientific

Figure 6.39 Residents of Shimshal, and other Wakhi villages in the region, traveled using precarious footpaths and dangerous suspension bridges (*left*), small often treacherous dirt roads (*right*), and, once reached via these means, the paved Karakorum Highway (*bottom center*).

and Culture Organization (UNESCO) World Heritage Convention. The neighboring countries of Nepal and India with had national parks with World Heritage Designations, and Pakistan wanted one too. The region encompassed by the proposed KNP would be an ideal candidate for this. Second, international conservation agencies, such as the International Union for the Conservation of Nature (IUCN), were concerned about the many endangered species in the region. These species included the blue sheep (*Pseudois nayaur*) and Siberian ibex (*Capra ibex siberica*), both of which were hunted by the Shimshal for winter sustenance, and also several others, including the Himalayan brown bear (*Ursos arctos*), the snow-leopard (*Panthera uncia*), the Marco Polo sheep (*Ovis ammon polii*), and the Tibetan wild ass (*Equus hemonius kiang*). Finally, some contend that Pakistan wished to create a national park as a way to "stake their claim" to the disputed territory to their north that lay on the Pakistan–China border, which China also wished to claim as their own.

(*Continued*)

Whatever the reason, the establishment of the park was not in the best interest of the Wakhi people living in the area. Government officials, scientists, and international agencies came up with a plan for the park, and designated park boundaries to exclude Wakhi village settlements but include almost 900 mi^2 (about 2,300 km^2) of Wakhi rangeland. Park planners initially sought an IUCN Category II designation for the park, which would forbid grazing and hunting within park boundaries. In other words, grazing and hunting would be prohibited on Wakhi land surrounding their villages. All park planning was done without consulting the Wakhi people. Once planning was complete, the Wakhi were informed of the new park and told that they must abide by park regulations. They were offered monetary compensation in exchange for their grazing rights. All Wakhi villages accepted the park's regulations and the financial settlement, except the villagers of Shimshal. Shimshal was the most remote and isolated Wakhi village and, therefore, most dependent on their way of life, both economically and symbolically. (At this point, neither the KKH nor the dirt road connecting Shimshal to the KKH had been built.)

The Shimshal protested for two main reasons. First, the park would ban hunting, which provided their people with an important source of venison during the cold winter months. Second, a ban on hunting would also mean that they could not cull the wild predators of their livestock, such as wolves (*Canis lupus*), in order to protect their herds. Finally, the ban on grazing would extinguish their pastoral animal husbandry, which was vital to their livelihood, culture, and identity. Two different villagers expressed the seriousness of these restrictions for the Shimshal and their willingness to resist the enforcement of park rules by government officials if necessary: "If they make it a national park, Shimshal will be a tomb" and "First they can kill us, then they can come and make it a national park." Tense and hostile negotiations between the Shimshal and government officials continued for most of the remainder of the twentieth century, as the Shimshal struggled to continue their way of life by both protecting their land from the government and maintaining their traditional practice of nature stewardship.

The Shimshal Nature Trust: A new stability landscape. From the establishment of the park in 1975 until the late 1990s, the Shimshal dealt with the threat of the KNP in mostly reactive ways, through resistance, protest, and outright defiance. During this time, the government, together with international conservation agencies, but without consulting the Shimshal, repeatedly attempted to devise alternate models for the KNP that accounted for local indigenous rights. However, the Shimshal remained suspicious and continued to fear the loss of their livelihood and culture (Figure 6.40). As such, they refused collaboration with the government and continued to oppose the park. At one point, the para-military Khunjerab Security Force established checkpoints within the park, which increased the Shimshal people's feeling of being under siege. As the years wore on, conflict between the Shimshal and the Pakistani government continued. During this time, the Shimshal lived in world of fear, uncertainty, and struggle. It seemed an impasse had been reached, until the Shimshal National Trust (SNT) was established in 1997.

Figure 6.40 Culture is just as important to the Wahki way of life as the natural environment that surrounds them, as illustrated by this elaborate festival to welcome spring in a nearby village.

The Shimshal community created the SNT as a way to better interface with threats from the outside world, such as the KNP, and transform the existing stressful and unproductive situation. The goal of the SNT is to improve quality of life for Shimshal people in a culturally and environmentally sensitive way, while retaining indigenous control of their environment. For centuries, the indigenous Shimshal had interacted with their environment in a sustainable way. There was no evidence of overgrazing. Moreover, scientific studies had shown that the Shimshal's domestic ungulates—their yak, sheep, and cattle—had helped sustain wild carnivore populations and alleviate predation on wild ungulates by these carnivores, thereby contributing to conservation of wild animal populations. However, despite this evidence, international conservation agencies did not trust that the Shimshal could sustainability manage the natural resources in their territory. As such, one purpose of the SNT was to render traditional practices of resource management more transparent by formalizing them.

(*Continued*)

In order to transform their informal traditional institutions into the more formalized institution of the SNT, the Shimshal laid out an official governance system. The SNT is overseen by a 13-member Board of Directors composed of officials elected for 3-year terms by each of Shimshal's eight subclans. The Board of Directors is formally accountable to the council of household heads, which allows the Shimshal to maintain the collective decision-making feature of their traditional society. The SNT Task Force is an additional level of governance composed of about six Shimshali men who have formal education and connections to the outside world, but also a lasting commitment to SNT goals. Most Task Force members do not live in the community, but are vital for maintaining productive relationships between the SNT and actors from the external world advocating for the KNP and other external disturbances. Together, the Board of Directors and the Task Force oversee the operation of six programs, which are maintained and implemented by different village institutions (**Table 6.1**).

Although the new formal governance structure of the SNT has not been formally recognized by the Pakistani government, the Shimshal people have derived many benefits from it. For example, due to the construction of the dirt road connecting Shimshal to the KKH, more children are leaving the village for schooling. This leaves Shimshal mothers without their children, who were a critical part of the traditional summer migration of herds. Using the institutional platform provided by the SNT, mothers of these children are able to hire shepherds to take their place. Another example is the trophy hunt program. With the opening of the road came the inevitable steady flow of foreigners into the community, some of whom wanted to hunt Himalayan animals. Under the SNT, the Shimshal decided, on their own, to ban traditional hunting within their territory. This was replaced with a trophy hunting program, in which outsiders (mostly foreigners) paid fees for hunting licenses. Seventy-five percent of the profits from these fees go to the SNT, with the rest going to the government. Most recently, the Shimshal have installed rooftop solar, giving them a higher quality of life than residents of other surrounding Wakhi villages. Under the auspices of the SNT, the Shimshal were able to negotiate these arrangements and infrastructure improvements.

Table 6.1

SNT Program	Village Institution
Environmental Education	Government Boys' and D.J. Girls' Middle Schools
Self-Help Development Program	Shimshal Volunteer Corps
Shimshal Culture Program	Wakhi Tajzik Cultural Association of Shimsal
Visitors' Program and Mountaineering School	Shimshal Boy Scouts
Nature Stewardship Program	Shimshal Boy Scouts
Women's Development Program	Shimshal Ladies Volunteer Corps & Girl Guides

The SNT is a new stability landscape for the Shimshal people, and it has allowed them to maintain the physical and cultural aspects of their society in the face of disturbances. Prior to the 1970s, Shimshal connections to the outside world were weak. In this case, traditional informal institutions (autogenic drivers) sustained their natural environments and culture. With the opening up of Shimshal society to the outside world, mostly by improvements in transportation infrastructure from the 1970s to the turn of the twenty-first century, the Shimshal people increasingly feared degradation of their alpine environment and loss of their culture. Increased connection brought benefits, but it also brought a constant influx of foreigners and an orientation of Shimshal youth toward urban areas and away from rural ones (allogenic, slow drivers). These slow-acting outside pressures, along with the fast and controversial establishment of KNP (allogenic, fast driver), led them to establish a different governance regime for their society (Figure 6.41). This new formal governance structure has increased resilience for their way of life, in the face of external disturbances. Their newfound resilience has proven to be valuable even in the face of natural disasters, such as the 2010 landslide (autogenic, fast driver) that filled the Hunza valley with sediment. With this landslide, along with other Wakhi settlements, the Shimshal were instantly cut off from the outside world. The Shimshal people, unlike other Wakhi villages located in the same valley,

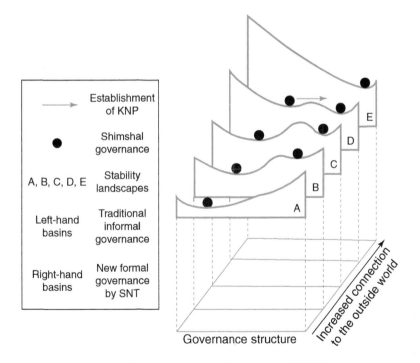

Figure 6.41 The Shimshal's new governance structure is a new stability landscape that allows them to interface with the outside world while, at the same time, protecting their interests.

(*Continued*)

> had fought to maintain their physical and cultural resources and, as a result, were able to revert to a largely self-sufficient and cashless society to weather this disaster.
>
> The ability of Shimshal society to change in the way that it did, in the face of external disturbances, illustrates the third characteristic complex adaptive systems: CASs adapt their behavior to changing conditions over time. This is the topic of **Section 3.3**, the final section of this chapter.

This concludes this section on stability landscapes and system resilience. Stability landscapes are shaped by the two-way exchange between autogenic drivers, arising from internal system interactions and feedback structures, and allogenic drivers, which affect a system due to changing conditions in the environment external to the system. This two-way exchange of matter, energy, and information between internal system components and processes, and the conditions external to the system, is the second characteristic of CASs covered in this chapter. The third CAS characteristic—adaptation—is covered in the next, and final, section of this chapter.

Section 6.3: Adaptation

Core Question: How do complex adaptive systems adjust to changing conditions over time?

The last section focused on how conditions external to a system, by way of allogenic drivers, influence a system's stability landscape and resilience. Another way to view this is that systems are continually adapting their behavior, as determined by internal interactions among components and feedback structures, to changes in the external environment. This adaptation process often plays out in ways that support a system's success and survival into the future. However, this isn't always the case, such as when a societal system collapses when it fails to change its behavior, or when a species goes extinct because all existing populations have failed to adapt to changing environmental conditions.

How does the process of adaptation work? This section will answer this question by presenting the third, and final, characteristic of CASs: *They adapt their behavior over time by passing on and receiving information through the processes of learning and evolution.* This idea of passing on and receiving information for adaption purposes will be illustrated with two examples, one from natural systems (specifically, biological evolution by natural selection) and human systems (cultural evolution by learning). Learning to conceptualize natural and human systems in this way will add another tool to your toolbox for thinking about SES complexity, so that you can use these tools to better understand the types of sustainability problems that are embedded in them.

Section 6.3.1—Biological Evolution by Natural Selection

Key Concept 6.3.1—Biological populations adapt their behavior over time by passing on and receiving information in the form of inherited traits.

Biological evolution is a process by which the composition of species on earth changes over time (Figure 6.42). According to Darwin's theory of evolution by natural selection, it is the differential selection

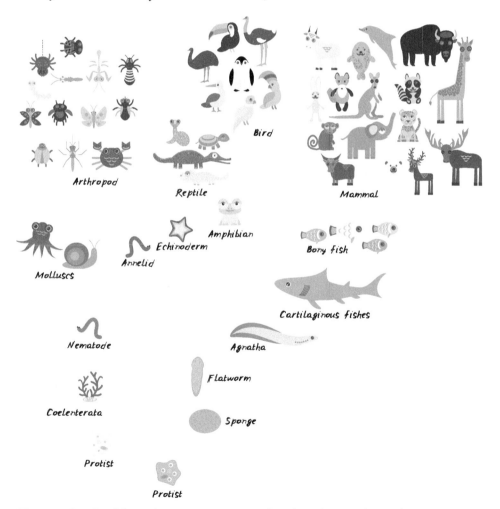

Figure 6.42 Although it misrepresents the abundance of vertebrate species on earth relative to other types of organisms, this flow chart depicts the evolution of new species, beginning with single-celled organisms such as protists.

of individual organisms over time that results in the production of new species. New species arise because existing species are continually adapting to changing conditions in their environment. Those that do adapt and survive have inherited traits from their parents that give them a high likelihood of surviving and reproducing. These inherited traits are the genetic pieces of information, encoded in DNA, passed on from parents, and received by offspring, that support the adaptation of biological populations to their environment. If all populations of a species fail to adapt, based on their inherited traits, then the species will go extinct. If the population does adapt in some way, it will either continue on as a slightly different version of that same species, or it will undergo a regime shift to a new species, with new internal components, interactions among those components, and feedback structures. In any case, the information contained in the DNA of those species that did not survive is not passed on.

Section 6.3.2—Cultural Evolution by Learning

Key Concept 6.3.2—Human systems adapt over time by passing on and receiving information in the form of inherited traits, but also in many other forms acquired through a process of learning.

Biological evolution by Darwinian natural selection involves the passing on of inherited traits from parents to offspring. One of the keys to distinguishing cultural evolution from biological evolution is that, in the latter, the traits are inherited through encoded DNA, which is not altered during the lifetime of an individual (although scientific research in the field of epigenetics has recently led to questioning of this assumption). Thus, traits are passed on to an individual's offspring in the same format in which they were inherited. Cultural evolution is different because humans inherit traits not only from the DNA of their parents, but also by learning from their parents and peers, as well as from information provided by those that came before them, such as in books and many other sources of information. This passing on of information affects the way that human cultures evolve: humans use both the process of learning, and the characteristics acquired through inherited traits encoded in DNA, to adapt to continually changing external environments.

Humans have another characteristic that sets them apart from other species: **human agency**. This is the ability of humans to make decisions that influence SES behavior, and ultimately the capacity of SESs to adapt to changing conditions over time. **Path dependence** has to do with the fact that both past occurrences and present activities lay the groundwork for SES behavior in the future, and what is possible in the

---

human agency
the ability of humans to make decisions that influence the behavior of socioecological systems over time

path dependence
a condition under which past events and present actions lay the groundwork for socioecological system behavior into the future by constraining possible development pathways and influencing present-day decisions

future. *Path* refers to the past and future course or direction that a natural, or human, system has taken or will take. *Dependence* refers to the fact that actions taken in the past or present will influence the actions that can be taken in the future. A common example of path dependence is transportation infrastructure and urban sprawl (Figure 6.43). Once cities become low density and spread out in space, they tend to become more dependent on cars. Having low density sprawl as the foundation for future transportation options can make transitioning to more sustainable options difficult. Installing a light rail over such a large area can be prohibitively expensive. The cost of a typical mile of light rail ranges from ten to hundreds of millions of dollars, depending on the specific context. In contrast, freeway expansion is much less expensive, averaging a bit more than $2 million per mile, per lane. In these times of tight government budgets, the decision is often made to add more lanes of highway. In this case, the future of transportation in a city is dependent on the past development path that led to the present infrastructure. This influences the decisions that can be made, and are made, today.

The idea of path dependence applies to human influences on ecosystem development as well. For example, grasslands from the same ecosystem grazed hundreds of years ago (grassland A) have a different species composition than those that were never grazed (grassland B). If a wildfire burns both grasslands, then the future path of ecosystem succession and recovery after the fire may depend on the species present before the fire. If cattle in grassland A overgrazed grass species whose seeds typically survive wildfires, but not species whose seeds cannot survive a fire, then this grassland is unlikely to regenerate quickly after a fire. Ungrazed grassland B may have a more even balance of both species and regenerate well after the fire. In this case, the grazing

Figure 6.43 The path dependencies created by urban sprawl often lead to a transportation infrastructure centered on highways, such as in Los Angeles, CA (left), due to the high costs of installing a vast light rail network, but this is not always the case and some cities, such as Curitiba in Brazil (right), have found innovative ways to use existing road networks to create sustainable bus-based transportation systems.

history of a grassland determines the path taken by the grassland as it is re-established after a wildfire.

It is important to be aware of the path dependent nature of SESs, not only to build an understanding of what is possible today, based on the legacies of the past, but also to take note that actions taken today *will* influence the future. (**Box 6.8** provides a case study that illustrates this.) Thus, it is possible for a variety of different futures to play out as a result of both past influences and today's decisions and actions. Path dependence is a feature of many SESs, and it results in a high level of uncertainty about future system conditions. Oftentimes, all the necessary information about past events and present conditions is not available to predict, with certainty, their influence on the future path of a system. How to deal with this is the topic of the next chapter, both in terms of exploring plausible future scenarios and collectively visualizing a desirable future aligned with sustainability.

BOX 6.8 UNDER THE SEA: ADAPTATION AND PATH DEPENDENCE IN THE NETHERLANDS

The land surface lies below sea level in some western regions of the Netherlands, near the Rhine River, but it was not always this way (Figure 6.44). When human settlers first moved into the region, the land was a peat marsh protected from inundation by the North Sea by sand dune levees. People built their homes on small elevated clumps of peat, planted crops, and raised livestock. Occasionally, high tides or storms prevented water from flowing out of the Rhine River and into the North Sea, such that the river water piled up on land, flooding their homes, crops, and grazing land. In response, settlers dug drainage ditches around their homes and began building them on man-made mounds called terpen. They also used ditches to drain the land used for crops and grazing, but dry peat was easily eroded and the new exposed land surface dropped in elevation. This made the flooding worse. Eventually, an extensive network of drainage ditches and levees (dikes) were required to keep pace with the sinking land surface. To accomplish this, individual households began to collaborate with each other.

Around 1150 A.D., North Sea currents pushed a huge quantity of sand to the mouth of the Rhine River so that it could no longer flow into the sea at its original outlet. As a result, even more frequent flooding occurred. In response, the Rhine River was dammed upstream and even more canals, this time with locks to prevent flooding if the canals themselves overflowed, were built (Figure 6.45). The lakes created by the dams posed a flooding hazard, so another dam was built. By this time, the western Netherlands had been divided into different political entities with distinct rulers, such as counts and bishops. In order to build the system of dams, locks, and canals as larger and larger spatial scales, collaboration among these entities was needed. Eventually, a water authority with power over the territories of all entities, when it came to water issues, was established.

The new regional drainage system exposed a good deal of land, which became intensively used for agriculture. As the exposed peat land dried out and eroded, the land surface continued to subside until it reached the water table. Once this happened, pumps were needed to move water from smaller drainage ditches up to major canals. At the time, in the early

Figure 6.44 The region of the Netherlands that is the focus of this case study generally falls within the present-day provinces of North Holland, South Holland, and Utrecht.

(*Continued*)

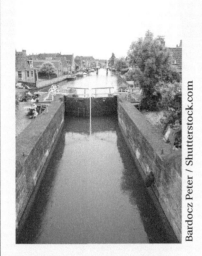

Figure 6.45 Locks, such as those shown here from a small Dutch village, were built to regulate water levels in the canals that drain the land surfaces.

Figure 6.46 Windmills were used to power pumps, which drained the land of water, and are an icon of the western Netherland's battle for dry land.

1400s, watermills driven by wind and horse power were used (Figure 6.46). As cereal crops became easier and cheaper to obtain from other areas, rural people shifted to milk, butter, and meat production or moved to cities to find work. By the sixteenth century, urban areas were growing, trade with faraway places was expanding, and industrialization was ramping up. Dried peat became an abundant local energy source that fueled new industries. As more and more rural people abandoned farming and animal husbandry, and took to digging up their land to sell peat for fuel, large areas of standing water appeared and made the region even more susceptible to flooding events.

The battle against water continues in the Netherlands today, especially in the face of climate change and rising sea levels, but the story will stop here for now. (The interested reader should consult this chapter's bibliography to read about the story of the western Netherlands to the present-day.) The SES described above continually adapted by way of internal interactions among components. For example, a reinforcing feedback developed between flooding, on the one hand, and technologies and institutions designed to reduce flooding, on the other hand, which ultimately made the flooding even worse. The SES also adapted in response to changing external conditions. For example, broader external trends toward industrialization and trade pushed the system toward a new economy of digging up rural landscapes for fuel rather than farming them. Path dependencies were also created when past activities laid the groundwork for SES behavior in the future. The most obvious of these is the constant "battle against water" that defines life in this region of the world. It is hard to say whether the adaptations made in the western Netherlands will support the success and survival of this SES, especially in the face of climate change and rising sea levels. Only time will tell.

Bibliography

Abidi-Habib, M., and A. Lawrence. 2007. "Revolt and remember: how the Shimshal *Nature* Trust develops and sustains socio-ecological resilience *in Northern Pakistan*," *Ecology and Society 12(2):Article 35*. Accessed December 12, 2016. http://www.ecologyandsociety.org/vol12/iss2/art35/.

Ali, I., and D. Butz. 1997. "Report on Shimshal Nature *Trust (SNT) Ghojal, Northern Areas, Pakistan. Brock University, Canada."* Accessed *D*ecember 12, 2016. https://brocku.ca/webfm_send/13297.

Barnosky, A. D., et al. 2012. "Approaching a State Shift in Earth's Biosphere.*" Nature 486: 52–58.*

Bliss, L. 2016. "What California can learn from how the South manages wildfires," *CityLab, The Atlantic,* July 1, 2016. Accessed December 5, 2106. http://www.citylab.com/weather/2016/07/california-wildfires-south-prescribed-burning/489539/.

Cavalli-Sforza, L., and M. Feldman. 1981. *Cultural Transmission and Evolution: A Quantitative Approach.* Princeton, NJ: Princeton University Press.

Chapin, F. S., and G. Whiteman. 1998. "Sustainable Development of the Boreal Forest: Interaction of Ecological, Social, and Business Feedbacks." *Conservation Ecology 2 (2): 12. Availabl*e online at www.consecol.org/vol2/iss2/art12/.

Chapin, F. S., C. Folke, and G. P. Kofinas. 2009. "A Framework for Understanding Change." In *Principles of Ecosystem Stewardship: Resilience-Based Management in a Changing World,* edited by F. S. Chapin, G. P. Kofinas, and C. Folke, 3–28. New York: Springer.

Chapin, F. S., G. P. Kofinas, and C. Folke. 2009. Principles of Ecosystem Stewar*dship:* Resilience-Based Natural Resource Management in a Changing World. *New York*: Springer.

Cook, N., and D. Bu*tz. 2011. "Narratives of accessibility and social change* in Shimshal, Northern Pakistan." Mountain Research and Development 31(1): 27–34.

D'Odorico, P., A. Bhattachan, K. F. Davis, S. Ravi, and C. W. Runyan. 2013. "Global Desertification: Drivers and Feedbacks. *Advances in Water Resources 12: 326–44.*

Folke, C. 2006. "Resilience: the emergence of a perspective for socio-eco*logical systems analyses."* Global Environmental Change 16(3): 253–267.

Gladwell, M. 2002. *Tipping Points: How Little Things Make a* Big Difference. New York: Bay Back Books.

Gunderson, L. H., and C. S. Holling. 2002. Panarchy: Understanding Transformations in Human and Natural Systems. Washington, DC: Island Press.

Hawken, P. 2007. Blessed Unrest: How the Largest Movement in the World Came into Being and Why No One Saw It Coming. New York: Penguin Group.

Henson, P. 2011. "[Opinion] Shimshal Pakistan; a resilient community," Parmir Times, July 1, 2011. Accessed December 12, 2016. http://pamirtimes.net/2011/07/01/opinion-shimshal-pakistan-a-resilient-community/.

Johan Rockström, J., et al. 2009. "A Safe Operating Space for Humanity." Nature 461: 472–75.

Khan, S. R. 2012. "Linking conservation with sustainable mountain livelihoods: a case study of Northern Pakistan." PhD diss., University of Manitoba, 2012, 385 pp.

Knudsen, A. 1999. "Conservation and controversy in the Karakoram: Khunjerab National Park, Pakistan." Journal of Political Ecology 56: 1–30.

Lansing, J. Stephen. 2006. Perfect Order: Recognizing Complexity in Bali. Princeton, NJ: Princeton University Press, p. 225.

Ludwig, D. 2001. "The Era of Management Is Over." Ecosystems 4: 758–64.

Marten, G. G. 2001. Human Ecology: Basic Concepts for Sustainable Development. London: Earthscan. Available free online at http://gerrymarten.com/human-ecology/tableofcontents.html.

Meadows, D. H. 2008. Thinking in Systems: A Primer. White River Junction, VT: Chelsea Green.

Mitchell, M. 2009. Complexity: A Guided Tour. Cary, NC: Oxford University Press.

Mock, J.H. "The discursive construction of reality in the Wakhi community of Northern Pakistan." PhD diss., University of California, Berkeley, 1998, 578 pp.

Odum, E. P., and G. W. Barrett. 2005. Fundamentals of Ecology, 5th ed. Independence, Kentucky: Brooks/Cole.

Pamiri, N. 2016. "How the residents of Shimshal are setting a shiing example for Pakistan." Dawn, July 30, 2016. Accessed December 12, 2016. http://www.dawn.com/news/1232261/

Scheffer, M. 2009. Critical Transitions in Nature and Society. Princeton, NJ: Princeton University Press.

United States Environmental Protection Agency. 2016. "Climate change indicators: atmospheric concentrations of greenhouse gases." Accessed December 6, 2016. https://www.epa.gov/climate-indicators/climate-change-indicat*ors*-atmospheric-concentrations-greenhouse-gases

van der Leeuw, S. 2013. "For every solution there are many problems: the role and study of technical systems in socio-environmental coevolution." Danish Journal of Geography 112 (2): 105–116.

Walker, B., and D. Salt. 2006. Resilience Thinking: Sustaining Ecosystems and People in a Changing World. London: Island Press.

Walker, B., C. S. Holling, S. R. Carpenter, and A. Kinzig. 2004. "Resilience, Adaptability and Transformability in Social–Ecological Systems." Ecology and Society 9 (2): 5.

End-of-Chapter Questions

General Questions

1. For each example below, list each different subsystem that is part of the overall hierarchy described. List these from the lowest to the highest level of the hierarchy. Then, identify any emergent properties mentioned, and from which level of the hierarchy they emerged.

    a. All of the people sitting in a football stadium are made up of organ systems that allow them to move. These organ systems are composed of individual organs. When each person in the stadium follows this simple rule, they carry out a wave, and this propagates through the entire stadium: when the person to your right stands up and throws his or her hands into the air, you immediately do the same.

    b. Shoppers interested in buying a fuel-efficient vehicle sit down at their computers and type the following into an Internet search engine: "fuel efficient vehicle." They are presented with several options, and with several different places to buy. On each of these pages, satisfied (or dissatisfied) customers have written reviews of the vehicles that they decided to purchase. After a good deal of research on the Internet, shoppers narrow down their options and decide to visit three or four different car dealerships, and eventually end up purchasing a vehicle.

    c. You are sitting in a classroom taking part in a discussion with your instructor and classmates. For each person to be able to speak, he or she requires lungs, vocal cords, and a mouth complete with a tongue, lips, and jaw. Your lungs provide the air required to make sounds, your vocal cords vibrate and make a noise as the air from your lungs passes through them, and the different parts of your mouth help to make a variety of sounds. These sounds are interpreted by the person to whom you are speaking. When one or more people speak to each other, a conversation develops.

   d. Water molecules, composed of hydrogen and oxygen atoms, evaporate from the surface of the Atlantic Ocean at about 15°N latitude. The water molecules are attracted to each other and to other particles in the atmosphere and, when temperatures are low enough, form clouds. Due to forces created by the Earth's rotation, the clouds move and rotate together. When atmospheric pressure near the surface is low enough, a hurricane forms.

2. Using the examples provided in (a) through (d) of the last question, determine which of the following is an emergent property, and which is a collective property.
   a. Number of people in the stadium; the wave created by the people in the stadium
   b. Decisions made about which car to purchase; purchasing rate of a specific type of vehicle
   c. Sounds that come out of your mouth; average number of times a heated disagreement develops
   d. Mass of total water molecules in a hurricane; geographic path that a hurricane takes

3. For each of the following examples, the internal components and processes for a system, and the external conditions that might affect that system, are described. Draw a diagram similar to that shown in Figure 6.12 that depicts internal components and processes, and external conditions. Then, for each example, determine whether a change in external conditions would result in internal system behavior and fluctuations more similar to that shown in Figure 6.16 or that shown in Figure 6.20. Explain why you picked the answer that you did.
   a. A bicycle is composed of pedals, a chain, tires, and tubes. When a person uses energy to push down on the bicycle's pedals, all of these parts, and others not mentioned here, interact with each other to move the bicycle forward. In order for the bicycle to function correctly, there must be air in the tubes at a certain air pressure. Several factors can affect air pressure in the tubes, such as the temperature outside on the street on the day the bicycle is ridden, or a nail on the street that could puncture the tube.

 b. Tornados are thought to form when a relatively cold thunderstorm supercell above the earth's surface meets a relatively warm updraft air mass originating from the land surface. The thunderstorm supercell and updraft air mass are composed of several different molecules, such as nitrogen, oxygen, argon, carbon dioxide, and water, in addition to particles, such as dust and pollen. Different possible features on the earth's surface that a tornado might run into as it moves along are a hot asphalt parking lot, a cooler large flat wheat field, or a high elevation mountain.

4. In the following examples, identify the autogenic and allogenic drivers regulating the system described.

    a. Our bodies are composed of cells. Throughout our lives, the cells in our bodies die and are continually replaced with new ones. As we age, the rate of cell death begins to exceed the rate at which cells can be replaced. This results in aging as some systems, such as our immune and nervous system, become compromised and affect our health. In addition to this "natural" aging process, other factors can affect how we age, such as how much we exercise, whether we smoked, or what types of chemical toxins we may have been exposed to during our lifetime.

    b. A self-fulfilling prophecy is a strong belief that becomes reality. Brian may not feel that he is smart enough to earn an aerospace engineering degree. In fact, he has convinced himself that he will fail his first test. As a result of this belief, he actually does fail his first test. This outcome reinforces his belief that he will fail, and he continues to fail his engineering exams. After his first semester, he shows his parents his grades and they are shocked because they know he is capable. They are convinced that his failure is a psychological problem and send him to a counselor before he returns to college the next semester. As a result of a change in his thinking, which arose due to his counseling sessions, he passes all of his engineering exams with flying colors the next semester.

5. In each of the following examples, identify the fast drivers and the slow drivers regulating the system described. List the drivers in order from fastest to slowest.

    a. Many food company executives are wondering what factors will have the greatest effects on their business over the next few years. Changes in human values can affect the types of food they are willing to purchase. In general, values are very deeply entrenched and slow to change. Another recession, coupled with increased unemployment rates, could affect the ability of consumers to buy certain types of foods. A worst-case possibility is a nuclear disaster at the nuclear power plant located close to the fields where their crops are grown.

b. The nature of higher education is constantly changing. The cost of higher education is rising, and this can affect the type of student who is able to attend college. The pace of technological changes results in new technologies in the classroom, ranging from laptops and Facebook, to classroom developments, such a clickers and online discussion boards. The expectations and aspirations of students also change from generation to generation. For example, your grandparents likely had different expectations and aspirations than you do.

c. Many factors can cause the earth's climate to change. The albedo of the earth's surface is one factor. If a forest is replaced by an asphalt parking lot, then more heat will be absorbed. Over time, the buildup of more and more CO_2 in the atmosphere, due mostly to fossil fuel combustion and deforestation, can result in more and more heat being trapped in the atmosphere. Finally, the Milankovitch cycles affect the amount of sunlight that reaches the earth's surface. These cycles change this over tens of thousands to hundreds of thousands of years.

6. In the hypothetical scenarios below, describe how the resilience of each system has changed in terms of precariousness, latitude, and resistance from situation A to situation B.

a. When a person is not employed at a full-time job, then expensive health insurance must be purchased directly by that individual. By working one or more part-time jobs, it is possible to earn enough income to pay monthly insurance premiums. When a person does not have health insurance, any type of expensive healthcare need can quickly lead to loss of savings and, eventually, bankruptcy.

Situation A: Sue works a part-time job as a cashier at a small skiing and mountain biking resort nestled in the mountains. In most months of the year, she earns $1,000 per month. However, right after the Christmas holiday, as well as in May and November when there is too little snow for skiing and too much for mountain biking, there are fewer tourists visiting the resort. During these months, there is not as much work available, and she earns only $800. When her monthly income reaches a lower limit of $800 ± 100 per month it becomes difficult to afford health insurance in addition to housing, food, and other basic necessities.

Situation B: Sue works a part-time job as a cashier at a small skiing and mountain biking resort nestled in the mountains. In most months of the year, she earns $1,000 per month. However, right after the Christmas holiday, as well as in May and November when there is too little snow for skiing and too much for mountain biking, there are less tourists visiting the resort. During these months, there is not as much work available and she earns only $800. When a person earns less than $1,500 per month, the state in which she lives provides a $400 per month health care subsidy to that person to assist in paying for monthly health insurance premiums. Due to this subsidy, her monthly income can now fall to a lower limit of $400 ± 100 before she is at risk of not having healthcare coverage. This is reassuring because, in the event of an economic recession or some other life-altering event, she will be able to handle less available work and still pay for her health insurance.

b. Ozone in the air of urban environments causes smog and human health issues, such as as asthma and other respiratory diseases. Trees and other types of vegetation can help clean the air and decrease ozone (O_3) levels in urban areas.

Situation A: Under current land-use patterns, safe outdoor ozone (O_3) levels in a city are considered to be 50 ± 5 ppb. Beyond this, human health issues can develop. A city's air quality data shows that O_3 levels throughout the city are currently 40 ± 10 ppb.

Situation B: A city decides to improve its air quality by planting trees in parks, along streets and sidewalks, and in parking lots. Before planting trees, safe outdoor O_3 levels in the city were considered to be 50 ± 5 ppb. After planting trees, safe average levels were deemed to be 50 ± 25 ppb on average because trees now clean a lot of the ozone out of the air.

7. Use the Scheffer ball-and-basin stability landscape model, as depicted in Figure 6.29, to describe resilience in the following hypothetical situations. Use the model to depict how allogenic drivers, both fast and slow, would change the shape of the stability landscape, and the position of the system (the ball) within the stability landscape. Also, note which indicator variable you would use to track changes in the system.

a. Type 2 diabetes in children is on the rise. Incidence of this disease in children is strongly correlated with obesity, and is thought to be caused by many factors, including level of physical activity, diet, and genetics. Although genetics cannot be changed, it is possible to alter levels of physical activity and diet. A child's parents are concerned about their child's weight gain. As a result, they decide to send their child to a health-focused summer camp. At the camp, the child gets plenty of physical activity and healthy food. By the end of the summer, the child has dropped 25 pounds. However, once school starts again, the child is back to eating unhealthy cafeteria foods, and sitting at a desk all day because afternoon recess was done away with a few years ago. By Christmas time, the child has gained 30 pounds. The next year, the parents enroll the child in a private school focused on health. The child loses weight again due to physical activity during two short recesses per day, and the healthy organic whole foods served in the school cafeteria. However, outside of school, when playing with friends, at birthday parties, and during sporting events, the child still seeks junk food. In addition to keeping the child enrolled in the healthy private school, they also enroll the child in weekly health education classes. Over time, these classes change the child's perceptions about eating healthy food, and now, even outside of school, the child would prefer to eat healthy food, over junk food. When the child goes away to college, and is no longer under the supervision of his parents, he will tend to prefer healthy foods over junk foods.

b. Many food crops need nutrients to grow and nutrients must be available to them in the soil. If enough nutrients are not available in soils, then a crop will fail. Growing a crop and then removing the biomass from that crop to another location, such as a grocery store where it will be bought by a household who will put the leftovers in a garbage disposal or trash can, depletes the soil of nutrients. Plowing and tilling a field before planting a new crop destroys the integrity of the soil and reduces its capacity to replenish nutrients through microbial activity. As a result, farmers need to add nutrients to the soil each time a crop is planted. This is the way farming has been done for the past several decades. Organic farming often involves *in-situ* composting of crop residue to keep some nutrients in the soil, and no-till methods that allow microbial communities and other soil organisms to survive and replenish nutrients in the soil between planting. Farmers who want to try organic methods, but are not knowledgeable about them, tend to continue with traditional, less risky farming methods. Once knowledge of, and experience with, organic methods increases, however, farmers see the benefits of organic farming to soil health, and switch farming methods entirely. These methods keep soils healthy and reduce the need for nutrient inputs, while still gaining the same crop yields from year to year.

Project Questions

1. **Identifying Drivers.** Go back to the list of drivers that you developed for Project Questions 1 and 2 in Chapter 3. Determine which of these are autogenic and which are allogenic. Also, determine which are fast drivers and which are slow drivers. After learning more about different types of drivers in this chapter, you may need to add some drivers to your already existing list, or modify your list.

2. **Thinking More about Resilience.** In Project Question 3 of Chapter 5, you began characterizing your system's resilience. This chapter introduced two additional aspects of resilience: the 3D model of resilience and stability landscapes. Use these two new tools to further characterize your system's resilience.

    a. First, use the 3D model to describe the resilience of your system in terms of precariousness, latitude, and resistance, and how certain key drivers might be changing any of these three dimensions of resilience in your system.

    b. Next, use the drivers that you identified in Question 1 of this Project Question section to apply the Scheffer ball-and-basin stability landscape model, as depicted in Figure 6.29, to your system. Where is your system now? What drivers are affecting your system? How are they affecting your system in terms of both the shape of the stability landscape, and the position of your system (the ball) on the stability landscape?

    c. Recall the system recovery efforts discussed in Box 6.6. Assuming that your system has undergone, or will undergo, a critical transition to an unsustainable regime, how might you manipulate the stability landscape and the position of your system (the ball) on that landscape to bring your system back to a sustainable regime?

CHAPTER 7

THINKING ABOUT THE FUTURE

Students Making a Difference

Photo courtesy of Jathan Sadowski.

Futuring for the people and by the people

Jathan Sadowski is a scholar-activist with an interest in technology, specifically with regard to who is left out when new technologies are designed and deployed. Much of the time, new technologies for sustainability, such as solar panels, electric cars, and a range of other innovations, are made to meet the desires and needs of an elite group that has the power to say what gets developed. According to Jathan, this is a social equity issue because it leads to sustainability for only a small subset of people and excludes marginalized groups: "We should be building technologies for the most marginalized people, and with them, and giving them the capacity to build it for themselves." His interest in technology, and in a participatory approach to sustainability, led him to be a team member on the Futurescapes City Tours (FCT) pilot project in Phoenix.

The project involved three meetings, over a 3-week period, with 20 volunteers who represented Phoenix's demographics. At the first meeting, they asked participants about their concerns when it came to sustainability and the future of Phoenix. Three major issues surfaced: solar energy, transportation, and water. Then, behind the scenes, the FCT team planned a walking tour to four locations that were relevant to these issues: solar panels on the roof of a local high school, a public parking garage with "stay cool" surfaces, a water cooling facility that keeps buildings in Phoenix's Central Business District cool during the day, and a local water canal. At each location, they brought together a mix of different experts, including academics, city employees, and the private sector. The second meeting was the walking tour, during which participants talked with these experts and were also asked to take photos of places where they saw the past, present, and future come alive. For example, in the parking lot with a "stay cool" surface, which had been donated by a nanotechnology company, they saw the future. In that same place, they also saw the past and present: the surface was eventually paved over with asphalt due to the city's already-existing maintenance schedule for asphalt surfaces. In the third meeting, people talked about what they saw, which brought out the diverse perspectives and values present and uncovered the tensions that exist between the past, the present, and the future.

The FTC visioning process was truly participatory. The participants set the agenda right from the start, such that the FTC team did not know ahead of time where the walking tour would go and who else needed to be involved. According to Jathan, this created a lot of work due to all the last-minute planning and coordination: "There were so many long nights and long hours put into actually making it happen. That was the hard part." Despite this, he feels it was worth the effort: "Getting into the weeds—actually talking to people, actually learning about the different kinds of intersections, the different communities, the different social groups—I think is really important to building a vision of sustainability that is for everybody, and not just for a small group of people."

The FTC project was ultimately focused on capacity building. They wanted the participants to come away with new ways of thinking about the city: the idea of interconnectedness, the fact that technology is human made and has values embedded in it, and the notion that multiple plausible futures exist that, although constrained by the past and present, are influenced by actions taken today. They also wanted participants to feel empowered by realizing that they have the ability to engage with what is happening in their city and also have a right to do so. During the sessions, participants exchanged business cards and networked with each other, the FTC team, and the experts on the walking tour. Many found common ground and interests that they did not know existed. Ultimately, the project left many feeling inspired. Excitement continued and many people stayed in touch. In 2014, the FTC method was used in four other cities around North America. There are many different reasons for future thinking in sustainability, as you will learn in this chapter. Sometimes it is for capacity building, but at other times the future thinking is used explicitly to guide decision making, or to help prepare for the future, or to just generally explore the possibilities for the future.

Core Questions and Key Concepts

Section 7.1: Scenarios and Future Thinking

Core Question: How can we determine what futures could plausibly arise?

Key Concept 7.1.1—Socioecological systems exhibit five general characteristics that make predicting their future challenging.

Key Concept 7.1.2—Scenario analysis is a powerful tool for exploring a range of plausible futures that could arise from socioecological systems.

Key Concept 7.1.3—Scenarios are developed for exploratory purposes or to support decision making and are created using a variety of methods ranging from qualitative and intuitive to more technical and formal.

Section 7.2: Visioning and Future Thinking

Core Question: How can we determine what a desirable and sustainable future should be?

Key Concept 7.2.1—Visions promote movement toward sustainability by motivating and inspiring, guiding and directing change, building capacity and social capital, giving solutions "staying power," and promoting long-term thinking.

Key Concept 7.2.2—Vision development involves building consensus around questions related to defining human progress, needs, and ethics and the roles of technology and place in sustainability.

Key Concept 7.2.3—Ten criteria can be used to guide visioning in order to distinguish efforts that will be effective for addressing sustainability problems from those that will not.

Key Terms

desirable future
probable futures
plausible futures
input variable
system relationships
proxy variable

To a large degree, sustainability is a challenge to think about the long-range future and, in so doing, to re-think the present. Sustainable development brings the question of the future to the strategic forefront of scientific research, policy deliberation, forward-thinking organizations, and the concerns of citizens.

—Global Scenarios in
Historical Perspective,
Millennium Ecosystem
Assessment

As first defined in Chapter 1, sustainability is fundamentally about preserving the conditions necessary for the survival of prosperous human societies. Inherent to this definition is the fact that these "conditions" are being preserved for societies that will exist *decades to hundreds of years from now*. In sustainability, it is necessary to take a long-term perspective and consider the influence of our present actions on the future. The opening quote for this chapter reflects the necessity of doing so, but this process is not straightforward. In fact, it can be profoundly difficult.

Up until now, this book has been about understanding present-day socioecological systems. This book will now move away from the current state toward thinking about the future. This chapter introduces two tools to help answer the questions: *where are we headed?* (future scenarios) and *where do we want to go?* (visioning) (Figure 7.1). Future scenarios are used to consider the ways that a sustainability problem may change and develop in the future. Visioning allows stakeholders to collectively imagine a desirable future. The results of future scenario and visioning activities are used to inform transition strategies (Figure 7.1), which are the focus of the next chapter. This chapter begins with an overview of why it is difficult to predict the future, especially for sustainability problems that are embedded in complex systems. Following this, the usefulness of future scenarios as a tool for thinking about the future will be explained, and different types of future scenarios described,

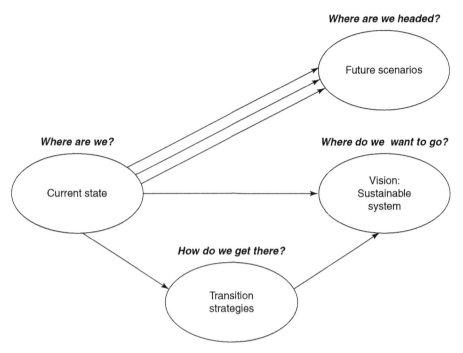

Figure 7.1 TSR Framework.

along with examples. The second section of this chapter is focused on visioning, which is the process of creating pictures of a sustainable future where society would like to go.

Section 7.1: Scenarios and Future Thinking

Core Question: How can we determine what futures could plausibly arise?

Men, forever tempted to lift the veil of the future—with the aid of computers or horoscopes or the intestines of sacrificial animals— have a worse record to show in these "sciences" than in almost any scientific endeavor.

—Hannah Arendt, The Life of the Mind, Vol. 2: Willing, Ch. 14, 1978

Thinking about the future is a unique and challenging problem that has persisted through human history (Figure 7.2). As noted in the opening quote to this section, a variety of methods have been used to predict the future, including abnormalities in animal entrails after a sacrifice, the position of planetary bodies in astrological forecasts, and outputs from computer models based on translation of real-world phenomena into mathematical relationships. For much of human history, people believed the future was already set out according to a divine plan and that our destiny was "written in the stars." With the advent of modern science in the 17th century, thinking shifted. People began to believe that future events are not predetermined, but instead depend on past and present actions and circumstances. Using the terminology of this book, people began to realize that actions taken by people, or the occurrence of chance events, influence the behavior of SESs and their future.

Beginning with the Scientific Revolution of the 17th century, and continuing to the present-day in many forums, quantification and predictive power (often through mathematical modeling) became the hallmark of legitimate knowledge about the future. This tradition began in the natural and physical sciences and has extended into social sciences such as economics. Such predictive approaches are used to produce long-term forecasts meant to inform decision making for sectors such as transportation, energy, and fisheries (recall MSY from Chapter 4). Although predictive modeling has its advantages, including reducing human bias and enhancing human capacity to manage complex information, it is not, by itself, sufficient for addressing the future of sustainability problems. Thinking about the future of SESs is a challenge for several reasons, which are listed in Table 7.1 and are the topic of the next section.

Figure 7.2 It is difficult for anyone to predict the future, no matter what method is used, such as an oracle using a crystal ball to divine the future (*left*) or a meteorologist using a science-based computer model (*right*).

Table 7.1 Characteristics of Socioecological Systems Relevant to Thinking about the Future

SES Characteristic	Literature Terminology	Example
1. Unanticipated events	Discontinuity	Lottery ticket, car accident
2. Unavailable information	Ignorance; Limited understanding of ecological and social processes	Rising atmospheric CO_2 concentrations and ocean acidification
3. Complex system behavior	Surprise; Intrinsic Indeterminism	Feedback, non-linear behavior, cascading effects (Chapter 5)
4. Human actors	Volition; Reflexivity	Choice to buy a Ferrari or a Porsche
5. Long-term challenges	Inertia; Path dependence	Long-term weather forecasts

Section 7.1.1—Challenges to Future Predictions about SESs

Key Concept 7.1.1—Socioecological systems exhibit five general characteristics that make predicting their future challenging.

Characteristic 1: Unanticipated Events. When thinking about the future, three types of futures are distinguished: probable, plausible, and desirable. Desirable futures (visions) are dealt with in Section 7.2. Here, the ideas of probable and plausible futures are used to explain one reason why predictive models are not sufficient for dealing with many sustainability problems. Probable futures are likely to happen. They can be revealed through modeling based on quantified relationships among SES components. In these models, probabilities are assigned numerical values and the chance of an event occurring is calculated. However, qualitative methods are also used, such as intuition based on practical knowledge and experience or from stories passed from generation to generation. Both can point to probable futures, and rely on knowledge of past system behavior, to predict the future. Using probable futures to predict the future is most effective when uncertainty is low, the system is well understood, and the issue at hand is not extended too far into the future. This is often not the case for SESs of concerned to sustainability. Plausible futures are those that could possibly occur, but are not necessarily the most likely and can arise from unanticipated events. Although all probable futures are plausible, not all plausible ones are probable. Both are considered when dealing with sustainability problems.

desirable futures futures that are preferred.

probable futures futures that are deemed likely to happen based on knowledge of past systems.

plausible futures futures that could possibly occur but have not been deemed the most likely because they arise from unanticipated events.

The difference between probable and plausible futures is best illustrated with an example. If you continue to work hard in school, it is likely you will graduate with a university degree. This is one plausible future that is also likely or probable. However, there are other plausible futures for you, however likely or unlikely they may be. You might hit a multimillion-dollar jackpot with that lottery ticket your great aunt gives you for Christmas next year. As a result, you decide to drop out of school and start a nonprofit for after-school literacy programs. Another plausible, but hopefully improbable future, is that you may be in a tragic car accident while driving home from a party on New Year's Eve that leaves you paralyzed. This can completely change the trajectory of your life. All three of these futures—graduate on time, leave school to start a nonprofit, and get in a tragic car accident—are all plausible, but they are not all expected (Figure 7.3). The lottery ticket and the car accident are unanticipated events or "wildcards".

When tackling sustainability problems, it is helpful to consider the future as both a probable extension of the past and a plausible alternative that can be greatly influenced by unanticipated events. Thus, predictive models, based on known relationships and observed historical trends, are not sufficient on their own because they cannot deal well with unanticipated or unimagined events. The introduction of new technologies into SESs often brings unanticipated events. For example, people living in 1910 could not possibly have imagined the far-reaching changes in agricultural practices during the Green Revolution (nor the unintended consequences). Today, it is similarly hard to imagine what the future may bring.

> **input variables** information about a specific system that is measured in some way and used as data for models.

> **system relationships** interactions among different system components that are used in quantitative models when they can be defined mathematically, such as in equations.

Characteristic 2: Unavailable Information. A second reason predictive modeling is not sufficient on its own is related to the lack of complete information surrounding many sustainability problems. There are two types of information required for predictive modeling: input variables and system relationships. For simple systems, it is possible to obtain accurate information for both of these. One illustration of a simple system is a beaker of water, containing only a few chemicals, existing under controlled laboratory conditions (described in Chapter 6). An example that contrasts this simple system with a more complex system, the Earth's oceans, to illustrate why input variables and system relationships are harder to determine for complex systems, is given in Box 7.1.

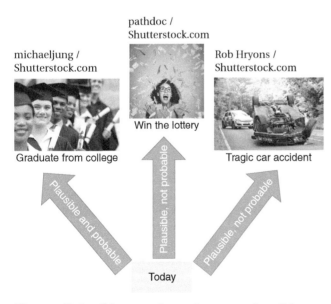

Figure 7.3 All futures shown here are plausible, but not all are probable.

BOX 7.1 CO_2 IN A BEAKER AND IN THE GLOBAL OCEANS

Rising CO_2 Concentrations in a Beaker. In the beaker, there are water molecules and a few chemicals dissolved in the water. External conditions affecting the beaker are temperature, pressure, volume, and other chemicals added to the beaker. The beaker is open to the atmosphere, so gases can enter and exit the water (Figure 7.4a). If gas pressure in the atmosphere outside of the beaker (an external condition) increases, then this will push more gas into the water. For example, if the CO_2 concentration in the atmosphere doubled, this would drive more CO_2 into the beaker (Figure 7.4b). Predictive models can be used to think about the future of this beaker because chemists can measure the input variables and know system relationships. In this example, system relationships are expressed mathematically using an equation known as Henry's law (Figure 7.4c). This law states that the amount (concentration) of a gas (in this case, CO_2) in a liquid (in this case, water) depends on the pressure of that gas in the atmosphere. The pressure of CO_2 gas in the atmosphere and the constant k, two input variables, can be measured and the amount (concentration) of CO_2 in the water can be easily calculated. It is not necessary for you to fully understand this equation and the input variables. The point here is that system relationships and input variables are known and can be used together to predict the future.

Global Oceans and Rising CO_2 Concentrations. When it comes to the SESs in which sustainability problems are embedded, system relationships are not as well understood and can change over time. This is a result of their complex behavior, as described in Chapters 5 and 6. Also, input variables may not be known precisely, or at all. The global ocean is a good example of a complex system. Ocean acidification results from increased atmospheric CO_2, which is due to human activities, and is detrimental to marine organisms, especially those with shells made of calcium carbonate. When CO_2 dissolves in water, it reacts to form and acid, leading to more acidic ocean water. This makes it hard for organisms that build their shells with calcium carbonate, such as corals, clams, plankton, and many others, because calcium carbonate dissolves under acidic conditions. CO_2 dissolves in ocean water in a similar way as it does in the beaker, following Henry's law, but the situation is more complicated because there are more system components and more variation in external conditions. This makes system relationships harder to define, and translate into equations. It also means there are more things to measure, such that not all input variables are known precisely.

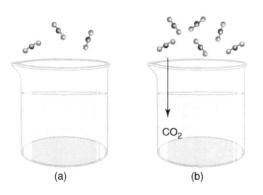

Henry's Law: Amount of CO_2 in water = $k \times$ Pressure of CO_2
(c)

Figure 7.4 A beaker of water is a simple system with known input variables and system relationships.

(Continued)

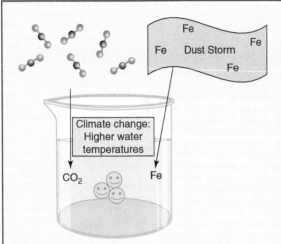

Figure 7.5 The ocean is a complex system and many input variables and system relationships are unknown.

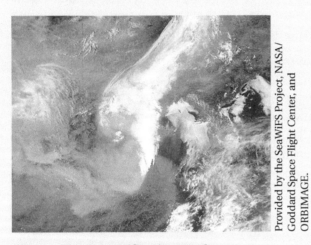

Figure 7.6 Desertification produces massive dust storms that supply extra iron to ocean waters.

Figure 7.5 shows the ocean as a beaker. The same components as in Figure 7.4 are shown, but others have also been added. Plankton are a component of ocean systems not present in the simple beaker of Figure 7.4. Plankton use CO_2 for photosynthesis and, therefore, affect ocean acidity. They also need sunlight and nutrients. One nutrient that they need is iron (Fe). In the open ocean, Fe is a limiting nutrient, such that more available Fe leads to higher photosynthesis rates. (This is similar to nitrogen and phosphorus as limiting nutrients in lakes or coastal waters that, when present in excessive amounts, cause eutrophication of these systems as discussed in Chapter 3.)

Dust storms add Fe minerals to the ocean and are expected to intensify with increasing desertification, a process that was described in Chapter 5. Figure 7.6 shows the type of intense dust storms that arise over Asia and Africa as a result of desertification. Another change in external conditions is increased ocean temperature due to climate change, which will reduce the amount of CO_2 in ocean water. (The solubility of a gas in water decreases as temperature increases.) As you can see, adding just one additional component (plankton), and changing two external conditions (Fe, temperature), rapidly increases system complexity. System relationships in Figure 7.5 are not well-defined and many variables are not precisely known. Although Figure 7.5 barely scratches the surface when it comes to complexity in ocean systems, it does illustrate the difficulty of using predictive modeling to think about the future of complex systems.

Characteristic 3: Complexity Results in Surprise. A third characteristic of SESs that makes predictive approaches to future thinking a challenge is that the inherent behavior of complex systems results in surprise. The features of complexity often exhibited by SESs, covered in detail in Chapters 5 and 6, and the resulting behaviors, are highlighted only briefly here. Reinforcing feedbacks cause systems to undergo rapid "runaway" changes, such as the melting of Arctic sea ice and permafrost in northern latitudes as described in Section 5.2 of Chapter 5. Stability landscape changes over long periods of time can imperceptibly alter the shape of regime basins (Section 6.2.2 in Chapter 6), such that systems suddenly cross thresholds and shift into new regimes. Such nonlinear regime shifts, and critical transitions, are often surprising events. Once a system crosses a threshold, different internal processes and feedback structures govern the system under the new regime and, as a result, the system's response to external change will be different. This shift is hard to foresee and capture mathematically with predictive modeling, although progress is being made in many areas of complexity research. Multiple system components are interconnected, such that intervention in one part of the system (e.g., removing a keystone species from an ecosystem) can lead rapidly, and often unpredictably, to changes in other parts of that system. Such cascading effects are hard to quantify and describe in the exact language of mathematics. Thus, capturing these chaotic, and largely unpredictable, surprises due to the normal behavior of complex systems in predictive models is a challenge. This *intrinsic indeterminism* of complex systems can be contrasted with the second SES characteristic shown in Table 7.1, which is more a result of limits to human understanding of ecological and social processes.

Characteristic 4: Human Actors. The fourth characteristic has to do with the fact that SESs have human actors as internal system components. This is in contrast to the physical and ecological systems just described—water in a beaker, the world's ocean water—that do not explicitly include humans. Just as the future acidity of oceans is determined, in part, by CO_2 concentrations in the atmosphere, the future of an SES is determined, in part, by human decisions and actions not yet made. However, the consequences of decisions and actions cannot be predicted in the same way that the effect of increased CO_2 concentrations on the acidity of water can be predicted. This is because human decisions and actions are driven by values and goals, whereas the behavior of a CO_2 molecule is not. A CO_2 molecule dissolves in water, but it does not have a goal that drives this process. Rather, this "action" is governed by physical and chemical processes and is predictable, assuming all necessary information is available.

In contrast, human behavior is governed by factors, such as emotions and beliefs, that are largely unpredictable. A unique feature of humans is a quality known as reflexivity, which refers to the fact that the very act of *thinking about* the future effects of decisions and actions today can influence the actual decisions made and actions taken. For example,

reflexivity a quality unique to the human components of socioecological systems, as opposed to their other biophysical components, that results in the actual decisions made and actions taken being influenced by the very act of *thinking about* the future impacts of decisions and actions.

when buying a car, your first inclination might be to buy a Ferrari because you like the speed and style. You notice that the fuel efficiency is 10 mpg. After taking a sustainability class, during which you've considered the effect of your decisions and actions on the future, you decide to opt instead for a Porsche 911 with fuel efficiency of 20 mpg and, therefore, lower CO_2 emissions. A CO_2 molecule, on the other hand, does not exhibit reflexivity. It is not able to consider the effect of its "actions" on the ocean's future and then change the actual "action" it takes.

Characteristic 5: Long-Term Challenges. The fifth SES characteristic brings together all other characteristics to illustrate how thinking about the long-term future, which is a requirement for sustainability, is challenging. As a result of phenomena such as temporal inertia and time lags between cause and effect, the predictive power of any model decreases as the time period lengthens. For example, short-range weather forecasts made for time horizons of up to 48 hours are more reliable than long-range forecasts intended to predict weather seven days or more into the future. This is because the Earth's atmosphere is a complex system, with many interacting components and internal feedback structures (characteristics of SES complexity discussed in Chapters 5 and 6). Meteorologists do not have enough information to build a perfect representation of system relationships in computer weather models (SES Characteristic 2 in Table 7.1) and the atmosphere is a complex system that produces surprises as part of its normal behavior (SES Characteristic 3). Unanticipated events (SES Characteristic 1) can influence the weather system, such as the eruption of a volcano (Figure 7.7). When Mt Pinatubo erupted in the Philippines in 1991, there was a marked decrease in temperature and precipitation on land. Finally, it is hard to imagine how people in the future will make decisions based on the values and beliefs of their time period (SES Characteristic 4). It seems strange to us today that people hundreds of years ago based decisions on beliefs about what abnormalities in animal entrails tell us about the future. It might seem strange to people living hundreds of years in the future that we currently make decisions by sitting in front of a white glowing square box (computer), pushing on little squares with our fingers (keyboard), until future predictions are made.

Figure 7.7 When volcanoes erupt, they fill the atmosphere with dust and other particles that can block out sunlight, which decreases temperature, and cause more rain to fall.

Yana Sutina / Shutterstock.com

Much of the challenge in dealing with the future, when it comes to sustainability problems, stems from the SES characteristics discussed here and summarized in Table 7.1. Despite these challenges, decisions must be made, and actions must be taken, if we are to move toward a more sustainable future. But how do we move forward? Future scenarios and visioning are two tools that can help us cope with these challenges, make decisions about a course of action, build capacity to adapt to changing conditions over time, collectively decide what the future should look like, and develop transition strategies (Chapter 8) to move toward a more desirable and sustainable world.

Section 7.1.2—Introduction to Future Scenarios

Key Concept 7.1.2—Scenario analysis is a powerful tool for exploring a range of plausible futures that could arise from socioecological systems.

> *Predictive modeling is appropriate for simulating well-understood systems over sufficiently short times. But as complexity increases and the time horizon lengthens, the power of prediction diminishes. Quantitative forecasting is legitimate only to the degree that system state can be well specified, the dynamics governing change are known and persistent, and mathematical algorithms can be devised to validly represent these relationships. These conditions are violated when it comes to assessing the long-range future of socio-ecological systems—state descriptions are uncertain, causal interactions are poorly understood and may change by unknown ways in the future, and nonquantifiable factors are significant. . . . As an alternative to prediction, scenario analysis has emerged as a key methodology for exploring alternative futures, identifying critical uncertainties, and guiding action.*
>
> —Global Scenarios in Historical Perspective, Millennium Ecosystem Assessment, 2005

Scenario analysis is a powerful means for examining the range of plausible future trajectories that SESs may follow, in the long term, when there is a chance for unanticipated events, a problem of unavailable information, occurrence of surprise due to complex system behavior, and inclusion of human actors as system components. Scenarios are different from predictive models because they acknowledge and work with the uncertainty that inevitably arises from SES characteristics (Table 7.1). Rather than relying on precise predictions, scenario analysis systematically compares and contrasts a set of coherent alternative futures in order to explore how different drivers of change, and assumed system relationships, may result in different futures. Scenarios examine how different natural and human drivers may cause systems to change in the future, so that the effects of present actions on the future can be explored. In addition to using system relationships, such as scientific laws for natural systems (e.g., Henry's Law in Box 7.1), scenarios incorporate multiple legitimate, value-based human perspectives. Scenarios are used as a tool for critically examining assumptions behind the causal chains that define SES system relationships, in light of these diverse perspectives. Thus, the process of constructing future scenarios includes asking questions and learning, as well as analyzing possible futures to determine the best decisions and actions for today.

Scenarios merge analysis with imagination by blending knowledge of the present and past with creative stories of the future. Present conditions influence the future and are derived from a current state analysis (Chapter 3). It is also important to consider the past due to path dependence, which refers to the fact that the future depends on past events (described in Section 6.3 of the last chapter). Past events can either constrain future actions or leave options open. For example, if you told your professor to "stick it where the sun don't shine" after your sustainability class was over last year, then when you need a letter of recommendation for a job application in the future, you will be constrained by your past actions. If you instead gave her a Christmas present before leaving for the semester, and said that you loved taking her class, then you have left options open for letters of recommendation in the future. In addition to using information about the present and past, scenario analysis involves creatively imagining unanticipated events. The future is partially influenced by the present, and dependent on the past, but it is not completely determined by these factors.

Future scenarios provide a lens through which to see how choices made today may influence and play out in the future. The insights gained from this process can be used by decision makers to guide policies and actions. By directing attention to aspects of the future that may have otherwise been ignored, scenario analysis provides a virtual visit to the future that can increase the capacity of decision makers to respond to unanticipated future events. Although scenarios do not precisely predict the future, they provide an opportunity to imagine events that may occur, investigate SES resilience to these events, recognize indicators that may signal when these events are about to occur, and examine sources of uncertainty. This allows decision makers to be better prepared for surprises by stimulating debate about appropriate responses to surprise events, should they occur. People can rehearse their response to plausible future events. By thinking through the future before it happens, scenario construction promotes rational and informed decision making in the face of crisis. Thus, scenarios are not about predicting the "right," or most probable future, but instead can be used to raise awareness through education and build capacity to deal with surprise and crisis.

Scenario analysis as a means of thinking about the future has a long history (Figure 7.8). After WWII, the military used scenarios to explore potential consequences of nuclear proliferation and devise alternatives to this future. The hypothetical nuclear war futures constructed by this process were used to explore how different human choices, and assumptions underlying system relationships, led to distinctive outcomes. Shell Oil was the first company to use scenarios in the private sector for corporate management and strategic planning. In the 1970s, Shell developed its *Year 2000 Study* to prepare managers for future uncertainties in global development pertinent to the company's long-term survival. As a result of this forward thinking, Shell successfully negotiated the 1970s oil crisis while other companies floundered. Through scenario analysis, Shell developed the capacity to respond to crisis situations with rational and informed decision making. Other companies observed the value of such adaptive capacity (first defined in Section 5.1.3), and scenarios have been widely used in the business community since. More recently, scenarios have been used in political arenas to inform decision making on challenging national issues. South Africa was the first country to use scenarios to promote debate about the nature of a post-apartheid society. The Latin American World Model scenarios were used to consider how political and social drivers regulate development in these countries.

Following the 1987 Brundtland Commission report (*Our Common Future*) scenario efforts aimed at sustainability challenges were

Figure 7.8 The U.S. military has used scenarios to think about a future with nuclear war (*top*) and the Shell Oil Corporation used scenarios to anticipate oil crises, which lead to gasoline shortages (*bottom*).

adaptive capacity the ability of a system to innovate, reorganize, and learn when faced with change.

launched. Four of these efforts are described next in Section 7.1.3. The application of scenarios to sustainability problems is relatively new and continuously developing. As a result, there is no one correct or widely accepted procedure for constructing future scenarios for sustainability. However, the undefined nature of scenario analysis is an asset, as the diversity and uniqueness of sustainability problems demands a creative and customized approach. But scenarios must be broken down and classified in some way in order to understand their features and usefulness for addressing sustainability problems. Thus, as an introduction to scenarios, and how they might be used in sustainability, this chapter classifies scenarios into four general types based on a 2003 scenario typology developed by Dr. Philip van Notten and others at the International Centre for Integrative Studies (ICIS) in the Netherlands. In this typology, scenarios are classified into three broad categories based on scenario content, project goal, and process design. The next section contains a general description of the typology. Then, the sections that follow provide case studies of scenario projects focused on sustainability that fall into four different regions of the typology.

Section 7.1.3—Scenario Typology for Future Thinking

Key Concept 7.1.3—Scenarios are developed for exploratory purposes or to support decision making and are created using a variety of methods ranging from qualitative and intuitive to more technical and formal.

The first of three criteria for classifying scenarios is scenario content, which can range from simple to complex. Sustainability problems lend themselves to complex scenarios for several reasons (Figure 7.9). Many SESs of concern to sustainability are complex systems, with a high degree of interconnection among system components and complicated causal chains. Complex scenarios deal with systems exhibiting complex behavior as the rule rather than the exception. The level of uncertainty, and the number of human perspectives involved, also distinguish simple from complex scenarios. SESs again land on the complex end of the spectrum here, as uncertainty is high and diverse perspectives abound. Scenario content also categorizes the difficulty of the decision-making process. The more complex a situation is, the more difficult decision-making will be. Decisions about a problem are harder to make when that problem is set within an entangled web of numerous other

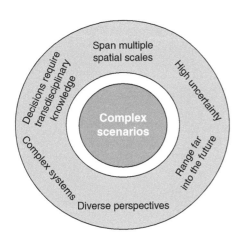

Figure 7.9 Complex scenarios are used for sustainability problems, as a result of several characteristics of these types of problems.

problems. Decisions are also difficult when problems require transdisciplinary knowledge to inform decision making about solutions, and range across multiple scales from local to global, and far into the future. Based on the difficulty of decision-making, SESs are once again placed into the complex scenarios category. Because sustainability problems range closer to the complex (as opposed to simple) end of the spectrum, this chapter will explore complex scenarios only.

Project Goal and Process Design are the two other criteria used to classify sustainability scenarios. Project goal is aimed at the "Why?" aspect of scenarios, which ranges from Exploratory to Decision Support (y-axis in Figure 7.10).

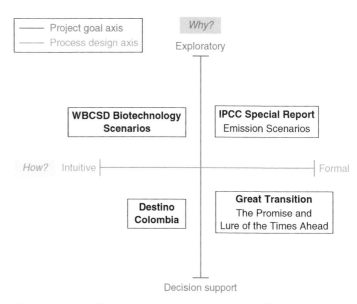

Figure 7.10 Four general types of scenarios are used for thinking about the future of sustainability problems.

Scenario projects with an exploratory goal promote awareness about issues, encourage learning by the general public and expert communities about SES processes, and inspire creativity for imagining unanticipated events and their associated futures. These scenarios are generally more descriptive, open-ended, and focus on the longer term. Possible futures are described and explored, and scenarios are developed, but the developed scenarios are not intended for use in later decision-making process (although they sometimes are used). In exploratory projects, the benefits gained during the process of scenario construction are just as important, if not more important, than the end product. For example, the process itself builds capacity among stakeholders to deal with the future by increasing knowledge and social capital. Decision Support scenarios, on the other hand, scan plausible futures, ranging from most desirable to least desirable. These scenarios are intended to address actual problems in the short term, so they are produced to support later decision making and strategic action.

Exploratory and Decision Support scenarios are not mutually exclusive. Exploratory projects often lead to more refined scenarios that produce a product useful for decision support and strategy development. A benefit that can arise from both of these types of scenario projects is rational and informed decision making in crisis situations, such as when elements of SES futures surprise us. In these situations, there is no time to systematically consider alternative choices and their consequences or to engage all relevant stakeholders. Rather, hasty and ineffective choices are often made, resulting in resource waste and a loss of public confidence in leadership and decision-making organizations.

The second criteria for scenario classification is Process Design, which ranges from Intuitive to Formal (x-axis in Figure 7.10). Process design is concerned with the "How?" aspect of scenarios, such as the methods, processes, and information types used to construct them. Intuitive scenarios are based primarily on qualitative information and understanding. To ensure accurate and complete qualitative knowledge, intuitive scenario construction often occurs in collaborative group sessions involving a diversity of stakeholders. Thus, such scenarios use a participatory approach (as defined in Section 2.2.3). Stories or narratives are the primary techniques for intuitive scenario construction and presentation, and are often enhanced with imaginative artifacts such as pictures, illustrations, or contrived newspaper articles. In many cases, scenario construction by this process is more of an art than a formal analysis.

On the other end of the process design spectrum are formal scenarios. These are predominantly quantitative in nature and require specialized technical knowledge, which make construction of these scenarios more dependent on *experts* (as defined in Section 2.2.3) as opposed to a more participatory process. Scenarios are constructed using computer models, which are based on quantifiable ecological, economic, and social processes. However, not all information needed to construct SES futures can be quantified, such as values and other cultural aspects. Integrated scenarios bring quantitative and qualitative knowledge together, to varying degrees, to complement each other. Qualitative scenarios add texture and richness to quantitative information and enhance communication of technical model outputs to nontechnical audiences. Quantitative analysis adds rigor and consistency to qualitative aspects.

The remainder of this section describes four scenario examples aimed at sustainability challenges ranging across the spectrums of project goal and process design (Figure 7.10). These are:

Example 1. Intuitive–Exploratory: World Business Council on Sustainable Development (WBCSD) Biotechnology Scenarios

Example 2. Intuitive–Decision Support: Destino Colombia

Example 3. Formal–Exploratory: Intergovernmental Panel on Climate Change (IPCC) Special Report Emissions Scenarios (SRES)

Example 4. Formal–Decision Support: Global Scenario Group's (GSG) Great Transitions

Example 1. Intuitive–Exploratory: World Business Council on Sustainable Development (WBCSD) Biotechnology Scenarios. The WBCSD is a CEO-led global alliance founded during the 1992 Rio Earth Summit, which was dedicated to promoting a sustainable future for business, the environment, and society, within the business community. The council collaborates with international organizations and academic institutions to devise solutions to sustainability challenges through eco-efficiency and corporate social responsibility. In 2000, the council published a report, *Biotechnology Scenarios: 2000–2050 Using the Future to Explore the Present,* based on a collaborative project aimed at developing exploratory scenarios using an intuitive process design (Figure 7.10). This scenario project raised awareness within the business community regarding sustainable development and challenges this may pose for companies. To this end, scenarios were constructed to explore possible future worlds and drivers that might shape such worlds. In addition to raising awareness, the project promoted learning so that corporate managers would understand driving forces, and causal relationships, shaping the world today and into the future. Through this process, it was intended that business leaders develop new ways to view the present.

The project was also meant to build capacity for businesses to deal with future surprises. This is especially important for biotechnology, where rapid technological innovation leads to unintended consequences when implemented in society (Figure 7.11). By creating a safe space outside of the boardroom for business stakeholders to "think outside the box," they were able to imagine future surprises and unexpected events. The virtual trips to the future provided by the WBCSD exploratory scenario project allowed corporations to build a capacity for dealing with future uncertainties that biotechnology may bring. In a similar way that Shell Oil used foresight from scenarios to negotiate the 1970s oil crisis, biotechnology companies can prepare for the future using scenarios to ensure long-term survival.

The WBCSD Biotechnology scenarios were designed in a series of workshops composed of participants from the global biotech business community such as Monsanto, Dupont, Hoechst Schering AgrEvo, and Rhône-Poulenc. Additional knowledge to inform scenario construction was gathered from interviewees and resource people from outside of the biotech corporations, such as from the World Bank, the International Union for the Conservation of Nature (IUCN), and U.S. universities. The process was participatory in the sense that it involved over 35 organizations, although the dominant participants were biotech corporations. In the workshops, participants first defined their system by laying out assumptions about the way the world

Figure 7.11 Potatoes have been genetically modified to synthesize insecticides, but an unintended consequence is the potato's reduced ability to produce chemicals in their leaves, called glycoalkaloids, which naturally ward off pests.

works. Specifically, they asked the question: *What are the "givens" that will happen no matter what?* They collectively identified six "givens" or assumptions:

1. New technology will continue to emerge.
2. Population growth will continue.
3. Connectedness of telecommunications and global financial systems will continue to develop.
4. Biotechnology is here to stay.
5. There will be unintended consequences from biotechnology.
6. People are fearful and anxious of the unknown when it comes to biotechnology.

After making these assumptions, the workshop participants identified three major external drivers (i.e., allogenic drivers, as defined in Section 6.2.2) that would shape the future success or failure of the biotechnology industry:

Driver 1. Public fear of biotechnology unknowns.
Driver 2. Consumer choice regarding whether to buy biotechnology products.
Driver 3. Opportunities taken (or not taken) by biotechnology companies to build public trust.

Based on these assumptions and drivers, participants constructed three future scenarios in narrative form, gave each a catchy title, and included accompanying illustrations. The scenarios describe plausible trajectories into the future through stories unfolding in a chain of events, from the present day to 50 years in the future. Each is dominated by one of the three drivers: Domino Effect Scenario (Driver 1), The Hare and the Tortoise Scenario (Driver 2), and Biotrust Scenario (Driver 3). Domino Effect was subject to an unexpected event (i.e., fast driver, Section 6.2.2) followed by a surprise, resulting in cascading effects through the SES relevant to this scenario. The other two scenarios highlight more gradual changes (i.e., slow drivers, Section 6.2.2), such that future events follow more linearly from present-day conditions. The three scenarios, taken directly from the *Biotechnology Scenarios: 2000–2050 Using the Future to Explore the Present* report, are presented in Box 7.2. The illustrations shown in Box 7.2 that accompany each scenario are not those used by the WBSCD, and the interested reader should consult the *Biotechnology Scenarios* report to see the actual artistic illustrations used. The illustrations included in Box 7.2 are shown to give the reader an example of what a scenario narrative, with accompanying illustrations, might look like.

After developing the scenarios, workshop participants concluded that the future of biotechnology markets would depend largely on public trust in their products, which ranged from very low (The Domino Effect) to high (Biotrust). This heightened awareness in the biotech industry may already have led to preemptive measures. For example, following the development of these scenarios the Council for Biotechnology Information (CBI), an industry group, with membership from corporations such as Monsanto and DuPont, published an activity book for children called *Look Closer at Biotechnology*. There is much controversy surrounding this book because it focuses only on positive aspects of biotechnology—increasing crop yields, decreasing resource use, and feeding more people with healthier vitamin-enhanced foods—and omits possible unintended consequences, such as biodiversity loss and organ damage, that scientists are just starting to uncover. Nonetheless, although it is hard to directly link this book with the *WBCSD Biotechnology Scenarios* project, it is an action taken based on industry awareness of the importance of public trust.

BOX 7.2 WORLD BUSINESS COUNCIL ON SUSTAINABLE DEVELOPMENT SCENARIOS: EXPLORATORY AND INTUITIVE

Scenario 1: The Domino Effect

In The Domino Effect, biotechnology continues to make steady progress until 2010, when a curious incident happens that no one much notices at first. A number of people who had received gene therapy are exhibiting debilitating, AIDS-like symptoms, and 25 of them die over a period of two years. While recipients of gene therapy had died in years past, the cause of these deaths had been addressed, it was thought. In any case, this seems to be a different set of symptoms. In at least half the cases, spouses also appear to suffer similar symptoms, and two children of these gene therapy patients also die. Officials at the U.S. CDC (Centers for Disease Control and Prevention) cannot, at first, determine what the cause is or even whether the deaths have anything at all do with the gene therapy. They also can't explain why the spouses and the two children seemed to exhibit many of the same symptoms as the recipients of gene therapy. When the news of this new "plague" reaches the public, reaction is immediate and intense. Many commentators remark that once again, officials have claimed something is safe when it turns out not to be. The public is especially worried about BBIs (biotechnology-based innovations) because if anything goes wrong in this area, the implications could be profound and far-reaching. Industry representatives counter these accusations by insisting that the gene therapy in question is safe and that the symptoms

(Continued)

are most likely associated with some other factor these patients have in common. Whatever is going on has nothing to do with BBIs.

Blame Biotech. A health-related Internet magazine picks up the story, and several NGOs make this strange new disease a key issue in the upcoming U.S. elections—candidates must pledge to "do something" about biotechnology. This pledge is easy enough to make because it's vague and doesn't offend very many voters, so all but one of the presidential candidates and most of the Congressional candidates go on record as being "against"

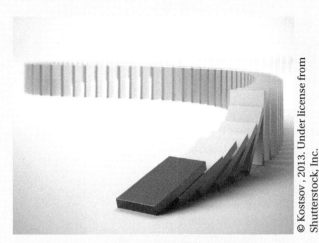

Figure 7.12 The deaths of a significant number of gene therapy patients start a domino effect that ultimately leads to the downfall of biotechnology.

biotechnology. Although industry analysts don't know what "against" means, exactly, one analyst, attempting to make a name for himself, attaches himself to this issue and speaks out on a number of popular talk shows about the dangers of investing in biotechnology—liability issues being the chief of the dangers. The two-tiered agricultural market—where non-BBI-based products fetch higher prices than BBI-based offerings—looks as if it is poised to develop in the pharmaceutical industry as well. A consortium of NGOs organizes a movement for a moratorium on all BBIs as well as certain forms of gene research. Industry spokespeople point out that the surplus food enjoyed in the United States and the relatively low prices of basic foods owe a lot to BBIs. But European farmers choose this moment to offer their non-BBI based crops for export—at a price—to the United States. Pharmaceutical companies join in the debate, arguing that their BBIs are crucial for many lifesaving treatments and that nothing has been proved yet about the link between the BBIs and the mysterious deaths. But the newly elected U.S. president, pressed to make good on his pledge, proclaims the moratorium, meanwhile promising industry representatives in private that the moratorium will be short-term. News of the president's private reassurance to the biotech industry leaks out, and angry activists stage noisy protests against the biotech industry in general. Growth in the industry slows down further, and many companies are thrown into relative confusion, with some choosing to move their operations out of OECD countries altogether.

The Revolution That Doesn't Happen. The 21st century was supposed to be the biotechnology century. But by the end of the first decade of this new century, the large bets made by many companies have not paid off. Several Congressional investigations and a growing number of lawsuits have soured the high expectations of many U.S. investors. And in the United States and Europe, anti-BBI legislation fences in biotechnology companies with many onerous "safety" requirements and restrictions. The best potential employees do not want to work for biotechnology companies, and daily life in the biotechnology industry is more about fighting fires than moving forward. Pharmaceutical companies are doing somewhat better than their colleagues in agriculture, but many come under the same onerous and capricious regulations that afflict the agricultural sector. Lawmakers seem not to make fine distinctions among BBI applications, and the increased use of non-standard Internet news sources means that the public is awash in misinformation about BBIs. Every rumor seems to be magnified. Even when misinformation is corrected, the public seems to be suffering a kind of millennial "hangover," prey to the fear aroused by Armageddon-like prophecies and any health warning that comes along, however tenuous. Numerous self-appointed watchdog organizations offer a seal of approval to products that are BBI-free. Thousands of school boards in the United States follow the example set by Berkeley in the 1990s and insist that school lunches be guaranteed BBI-free. Somewhere in all these fear-arousing rumors, a small news item appears: The mysterious deaths and AIDS-like symptoms suffered by some of the gene therapy recipients are caused by a new variety of flu—serious enough, but probably nothing to do with gene therapy. But this information receives scant attention, and no one cares anyway, because anti-BBI sentiment seems to have a life of its own. By 2020, companies whose profits were solely dependent on BBIs are suffering. It has proven impossible to obtain insurance for many BBIs, and many class-action suits pending in the U.S. courts threaten even the few who are hanging on to life by a thread. Most of these suits, it is clear, will be thrown out of court, but the time and expense of dealing with them is enormous.

The Domino Effect. BBIs in agriculture are also suffering from the domino effect. The effectiveness of the WTO is threatened by a long-term impasse over the EU's refusal to import genetically modified agricultural products and the U.S. response in restricting the import of EU agricultural products. Industry spokespeople point out that the surplus food enjoyed in the United States and the relatively low prices of basic foods owe a lot to GMOs and that third-world countries will suffer if this impasse is not resolved. But anti-BBI feeling continues to develop, and now begins to spill over into the area of biobased polymers, where the fear of toxic waste is being replaced by the fear of genetic pollution. By 2030, it is common for any unusual upset in the ecological balance of a region or any new seasonal strain of flu to be blamed on BBIs—the so-called "Unintended Consequence Effect." One example of the unintended consequence effect has occurred with salmon. A gene modified in salmon to help them breed more productively seems to have resulted in an excess salmon population. When salmon show up where they don't normally occur, environmentalists claim the eco-balance is upset. Nothing is proved, but it doesn't seem to matter. Industry spokespeople claim that the unintended consequence effect is more myth than substance, and that what is really happening is not a material effect at all, but a "domino effect" in which one bad news story creates another, bringing down one product or company after another. Like a giant game of "gossip," a rumor gets more and more exaggerated until after only a few rounds of Internet traffic, the distortion has made the original report unrecognizable.

(Continued)

Every new product and every event seem to be part of some vicious circle. Introducing new BBIs seems almost impossible, and many BBIs already in the market are withdrawn. The FDA and the EMEA recall a number of pharmaceutical BBIs for "further testing," in spite of protests by those consumers who feel they benefit from certain BBIs. Emergency exceptions are made—for insulin, for example—but competitors with conventional medications use this opening to make a case to go back to certain "safer" drugs. Drug policies in the developed nations diverge from each other and from policies of less developed nations, and even within individual countries, many regulations fluctuate wildly from year to year. Pharmaceutical BBIs seem to be approved or not approved almost on a whim. No common standards emerge, either among developed nations or from administration to administration. Large companies begin to spin off individual business units as a way of avoiding liability. At this point, some third-world entrepreneurs begin to buy up these vulnerable little spin-offs, selling new products as cheaply as possible to third-world and even first-world customers, often over the Internet. Although many such drugs are by now illegal in Europe and the United States, suppliers can usually manage to work around these legal restrictions, especially given the increasing number of Internet transactions and the new delivery companies that are formed almost daily. In addition, a number of unscrupulous entrepreneurs set up factories with limited quality, safety, and environmental standards, and observers feel it is only a matter of time until a serious disaster occurs. The Domino Effect, by increasing fear, has also, inadvertently, increased the very dangers it feared.

Scenario 2: The Hare and the Tortoise

In The Hare and the Tortoise, progress in biotechnology and its practical applications comes to a virtual standstill, not because of protest movements or government regulations or an incident that increases fears of biotechnology, but because consumers and investors choose other, non-BBI alternatives. In a number of areas, "classical" R&D delivers solutions with better performance and higher profits—in part because most consumers, given a choice, follow the precautionary principle of "better safe than sorry." Low-tech, holistic health practices emphasizing prevention and sustainable, non-BBI agriculture prove to be much more popular with consumers in developed nations. In less developed nations, BBIs are often too expensive to be adopted. In this scenario, there is still a niche market for BBIs as analytical tools. But the older, more classical approaches are like the slow tortoise in the old fable, who has a kind of patient persistence that, almost unnoticed, wins the race in the end over the swifter but more erratic hare.

The Hare Takes Off. In the last decade of the twentieth century, biotechnology stocks are among the hottest things going. Patent applications for life forms are being submitted at a breathtaking rate, and in the United States, especially, new BBIs are favored with what many perceive to be a fast-track approval process. Some critics point out the relative lack of performance in relation to investment—but most argue that a virtuous circle is developing: more investor money results in more research, which results in news of more potential products, which leads to new investment, more research, and so on.

Emergence of the Tortoises. But other contestants are lumbering along in the same race—alternatives based on high-tech but non-BBI approaches, such as traditional farming techniques and holistic health remedies. BBI advocates point out that traditional farming practices have serious environmental consequences, such as lost topsoil and lost biodiversity due to the habitat encroachment caused by expanding land use for producing food. Where that is true, critics respond, the ill effects can be reversed through better practices—a reversal not so easy to produce when it comes to "genetic contamination." "Better safe than sorry" is the motto of these "tortoise-like" contestants.

Figure 7.13 Biotechnology (the hare) ultimately loses the race to the tortoise, which represents alternatives to biotech chosen by consumers.

The Consumer Chooses. Most consumers just cannot see a clear benefit in BBIs. By 2010, a two-tiered market has developed, with higher premiums being paid for non-BBI-based food and drugs. Public opinion is firmly grounded in the precautionary principle, pointing to the unexpected consequences of the industrial revolution, for example, or of introducing new species into local habitats. It's very obvious that a large part of the next millennium will be spent paying the bill that has become due as a result of the industrial revolution. The cost of global warming and the numerous environmental cleanup chores of polluted air, water, and soil is higher than expected. Many wonder what bills will land on the doorstep of the next generation as a result of the equally dramatic revolution in biotechnology. A consensus begins to emerge in many Western nations that in relation to biotechnology or any other technology that affects society as a whole, it is best to encourage those choices that leave other options open. In The Hare and the Tortoise, a second principle arising from the biotechnology debate is that where a conventional alternative will do just as well, there's no reason to encourage the use of BBIs. Reflecting this development, governments jump on the bandwagon by creating programs to support conventional agriculture and pharmaceuticals: tax breaks and subsidies for non-BBI farming; tax policies favoring unmodified seeds and other "natural" practices; grants and subsidies for development of new pharmaceuticals that don't use gene-based technology; and WTO exceptions to fair trade policies in the case of genetically modified products that importing countries want to resist.

Health Consciousness. The "tortoises," or the non-BBI alternatives, are emerging from many directions, in part because of movements in lifestyle and health. The more affluent and better educated consumers entering the marketplace are highly aware not only of the quality of the products they buy, but also of the circumstances of production. In addition, they are extremely health conscious and fitness oriented, and many of them scan labels

(Continued)

thoroughly, using their personal shopping scanners to compare the information embedded in the bar-codes with their personal ethics and nutritional value profile. When it comes to food and health, consumers are becoming much more personally involved, taking direct responsibility for choosing healthy food and practicing preventive health care. Food retailers are increasingly interested in pleasing consumers, especially because the food crises in Europe at the turn of the century have led to an even greater skepticism about the ability of governments to guarantee food purity and safety. A number of private food certification agencies spring up, and most of the popular retail brands voluntarily follow their guidelines. In addition, new labeling laws and aggressive "natural" food and drug marketing make it difficult for genetically altered products to compete. Consumers seek more local food and gradually become more willing to accept seasonal variations in their diet. By 2010, market researchers notice a significant trend toward the consumption of more organic produce. In Europe, habits of eating have changed with the new generation, who remember the mad cow disease scare and other meat contamination incidents from their childhood. Vegetarianism is becoming more and more popular for ethical as well as health reasons, in part because young people are increasingly supportive of "animal rights" and other related issues. In the global marketplace, consumers have the luxury of a wide choice of products—and they tend to choose on the basis of a company's social responsibility profile as well as the quality of its products. The aging populations in OECD countries are also interested in health, and many people shift their eating preferences lower down on the food chain. Product purity for these consumers means no additives and no BBIs in their food and drugs. Both older and younger consumers now expect to devote a higher percentage of disposable income to high-quality food and wellness. Holistic health, which features exercise, non-BBI food, vegetarianism, and homeopathic rather than allopathic remedies, becomes more popular as the conventional health care safety net in developed nations becomes less able to function. People want to take charge of their own health and are willing to experiment with alternative medicines before turning to BBIs. Pharmaceutical companies still do well with BBIs in certain niche markets, but profits in this sector begin to flatten out. Many new drugs that were expected to be profit sources end up, for various reasons to do with politics and patents, in the public domain. Because expectations were so high, share prices begin to slide, and investor attention turns to those companies that specialize in products derived from new plant discoveries in the Amazon and elsewhere. These new herbal products seem to work—"mostly through the placebo effect," says a noted skeptic. But new discoveries in the powerful effect of the mind on the body lead many consumers to embrace the placebo effect—so-called "self-administered homeopathic drug therapy." New markets develop in products that are innovative, but based on traditional medicine, a paradoxical combination that seems to suit the spirit of the times. As with other aspects of the world of The Hare and the Tortoise, people are not worried about BBIs—they just are not that interested.

Feeding the World. What begins as a health and lifestyle feature quickly becomes a cause. Just as high ideals led to the eradication of polio and smallpox, so now many become convinced that famine, too, can be eradicated. Conceiving of this as a realistic though difficult goal, private foundations work toward food self-sufficiency in every community. This is more than just sorting out the distribution and transportation problems, but has to do with growing small amounts of non-commercial produce in a small space, saving seeds for next year's garden, and practicing low-pesticide farming. Computer technology and labor-saving devices have reached even the most remote parts of the globe, so people have both the time and the information necessary to develop "small plot" permaculture strategies. In the United States, the "sprout scout" movement has taken hold, so that by the time children graduate from high school, they know a great deal about urban gardening. Gardening for food is popular, partly in reaction to the increasing sense of vulnerability people feel because they are so interconnected and interdependent. The more sophisticated and "networked" the lifestyle, the more popular food gardens become. This cultural shift is also felt by farmers, who dislike the way that farming based on BBIs makes them feel less self-reliant and more vulnerable to external factors. What has helped this movement enormously is the World Bank Initiative to make high-yield corn, rice, and wheat seeds and plants available to all small-lot farmers. A consortium of multinationals has formed a group that has taken on the daunting challenge of coordination and distribution, a challenge aided by the increasing numbers of people in remote villages in Africa, India, and China who have received cell phones through international aid agencies. All these efforts are greatly aided by breakthroughs in high-tech, non-BBI farming and breeding techniques. In this developing culture of high-value agriculture and self-sufficiency, the goal of ending world hunger seems a real possibility. In addition to exporting new agricultural methods and the principles of simple water purification and permaculture, UN agencies, private foundations, and NGOs all over the world begin to look more closely at the whole issue of distribution. In this climate, the argument that world hunger can be eradicated only through BBIs looks less convincing, especially because the distribution problems associated with conventional large-scale agriculture are still apparent in BBI-enhanced agribusiness. The problem is not the growing method, it's the will to feed to world, argue the reformers. By 2015, biotechnology has lost the sustainability argument, which, critics say, it never made convincingly in the first place. The remarkable increase in the sales of non-BBI foods lures many smaller farmers back to the farm because the profit margins on such produce are very enticing. In addition, recent research has produced high-tech innovations for conventional farming. New measuring devices allow much more precise inputs of nutrients and water, and discoveries about the interaction of soil, natural fertilizer, light, and plant chemistry lead to significant increases in productivity. These increases are relatively high because information technology allows even poor farmers access to sophisticated analysis of crop needs. Biotechnology becomes an important tool for understanding, but not as important for commodity production. The general attitude is that where possible, it's best to optimize the ecological system as a whole rather than to optimize the seed, a narrow approach which might, in fact, damage the system as a whole.

(Continued)

Diagnostics—A BBI Niche Market. In addition to an increasing sentiment favoring a "go slow" approach to what was supposed to be the "age of biotechnology," another counter-trend is developing: A surprising number of people are not as eager for gene-based diagnostic tests as many had expected them to be. One reason is that the capacity to diagnose disease susceptibility far outstrips the ability to cure the disease. Consumers quickly understand that diagnosis offers them not many upsides and quite a few disadvantages. They don't want the "genomic depression" diagnosed in some people who have been told they will develop an incurable condition at age 35 or 40. Also, they don't trust the ability of diagnosticians to keep confidential information secure—and U.S. citizens, at least, don't want to risk losing employment opportunities or health insurance coverage if a negative diagnosis is discovered. Thus, expensive and sophisticated diagnosis for congenital conditions turns out to be a product without a market.

The Tortoise Wins the Race. While governments encourage conventional agriculture and health practices, they do not attempt to stop the development of biotechnology. But lack of consumer demand and other factors redirect innovation away from BBIs. The cost and uncertainty of patent protection and the perception that biotechnology is a high-risk business lead many companies to invest in alternatives to biotechnology. These investments increase the pace of important innovations in conventional agriculture, such as a joint venture of two startup companies that introduces environmentally friendly insecticide and herbicide based on conventional technology. As both producers and consumers increasingly choose non-biotech alternatives, and the holistic health and "safe" foods sectors grow, investors become aware that BBI-based enterprises are underperforming. In both agriculture and pharmaceuticals, the race to patent life forms is beginning to be questioned. Some critics denounce this view of life as a utilitarian set of raw materials. It's one thing to prospect for gold and oil in specific geographical regions; but it's quite another, they argue, to "bioprospect" in the material that is common to all. Gradually, a consensus begins to emerge on two fronts: first, that contained risks, such as those associated with the use of enzymes in paper manufacturing, are supportable, but uncontained risks, such as vaccine production crops, should not be allowed; and second, that process patents make sense, but that product patents on life forms should not be allowed. By the time products appear, however, even process patents have usually expired. Just as product patents begin to be challenged by courts, the wide sharing of genomic information makes the protection of intellectual property in this area extremely difficult. The human genome sequence is in the public domain, for example, as are SNPs, the places in the genetic code where individual variations take place and that hold the clues to genetic predispositions for various diseases. In The Hare and the Tortoise, the biotechnology industry, like the nuclear industry before it, continues to hold a small niche share of the market—but it doesn't produce the expected revolution.

Scenario 3: Biotrust

A Narrow Window for Success. Early in the new century, biotechnology is growing in its number of applications—pharmaceuticals, drought- and disease-resistant seeds, large-weight fishing stock, fast-growing and disease resistant forestry products, genetically modified bacteria-based mining applications, bioinformatics, bio-plastics and enzyme manufacturing, tissue engineering, bio-remediation, biocomputers, bio-sensors, therapeutic vaccines, and many others. But when it comes to public opinion, the industry is losing ground. Citizen groups, especially in Europe, have organized stiff resistance to BBIs in foods, and this anti-BBI sentiment has spread into the medical arena, even though the issues there are very different. But in this increasingly emotional debate, fine distinctions and scientific arguments are overshadowed by politics and a very effective anti-BBI media campaign. Some U.S. analysts argue that the funding for the campaign has come, in part, from businesses that have a vested interest in keeping U.S. agricultural products out of European markets, and some Congressional representatives from agricultural states propose retaliatory trade measures. The International Council for Genetics, an institution that has grown out of individual, country-based initiatives, proposes much stricter oversight of pharma research and much broader distribution of what they refer to as "the health technology of the rich." Distrust grows: NGOs distrust the companies; companies distrust the media; Europe and the United States distrust each other's intentions in the ongoing BBI debate; citizens distrust what their governments say about the safety of BBIs, especially in relation to food; and third-world countries distrust both OECD governments and the large biotech firms.

Building Biotrust. In this climate, a number of biotechnology companies along with NGOs, patients' rights groups, and other stakeholders join together in what becomes known as the "Biotrust Project." The aim of this project is to create a common meeting ground on which to build trust among all stakeholders in the short time remaining before any major products based on genome sequencing come to the marketplace. If the 21st century is to be the century of biotechnology, they argue, then all stakeholders must have a part in its development. The process of building stakeholder involvement is difficult and at times threatens to break down. Media representatives tend to emphasize the David-versus-Goliath aspects of any situation, while representatives from NGOs and other citizen groups are wary of being used as part of a "window-dressing" campaign. But after serving on the Biotrust panel for a year, they tend to modify their positions somewhat, as do the industry representatives—even though the change in stance sometimes creates friction with the individual sponsoring groups. In spite of the difficulties, participants eventually reach agreement in eight areas of concern: (1) transparency, (2) ongoing stakeholder involvement, (3) ground rules for risk-benefit analysis, (4) a global system of safety standards, (5) inclusion of third-world nations

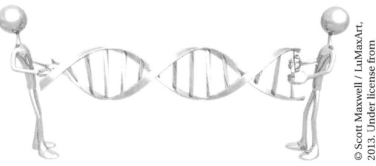

Figure 7.14 By building public trust, biotechnology ultimately succeeds.

(Continued)

in the benefits of biotechnology, (6) data protection, (7) guidelines for patenting and licensing, and (8) responsibility for externality costs and other liability issues. What makes the agreement more effective than many had initially expected is that it is based on social values as well as science and also, that other groups are involved in similar initiatives. The Biotrust Project is usually considered to have more significance than many of the others because it involves major players on both sides, and it rapidly initiates projects that make a difference. For example, one early project is the setting up of a series of regional seedbanks and genetic databases for public use, designed to eventually become the basis for "libraries of life" in every region in the world. Pharma companies follow the lead of agricultural companies and in conjunction with the World Bank, set up a global trust fund for developing countries that allows them to purchase new drugs and therapies at a price that provides a rate of return competitive with successful product launches in the United States and Europe. In addition, not-for-profit virtual drug companies bring together private- and public-sector research to create new vaccines and drugs for diseases found in developing countries. These projects, along with agricultural initiatives, lift more than a billion people from "poverty status" to "customer status."

Transforming an Industry. The Biotrust Project has a profound influence on the development of the industry, helping to guide the biotechnology revolution through the wishes and needs of many people, not just the holders of capital or consumers *of* products. Some "fundamentalist" capitalist theorists argue that this approach is creating an anomaly in the free-market system. But others argue that the real capital here is the knowledge of the genome itself, which, it could be argued, belongs to every living creature on the planet. In addition, if biotechnology is to benefit the world and not just the companies that hold patents, human societies need to produce more creative thinking about access to food and medicine. In this climate of public opinion, companies compete for the trust of consumers by competing for the highest standards and for the reputation of being socially conscious, committed to the environment, and transparent in their financial practices as well as in the basis of their risk-benefit analyses. Business learns to adopt its full, comprehensive role in society to help form the new game rules necessary for sustainable development.

The Biotrust Revolution—A Larger Vision. By 2030, the world is transformed. Most infectious diseases have entirely disappeared. A vaccine for Alzheimer's has greatly increased the chances that older people will live an active life right through to the end. And average life expectancy in the west is 120 years, thanks to new diagnostic techniques and cures—including cures for various cancers—that are genomically specific to individuals. Since 2020, 95% of human body parts have been replaceable with laboratory-grown organs, and the costs have continued to come down every year. In farming, biotechnology means that much less energy and water are used, and food is cheaper, more nutritious, and more plentiful. In Biotrust, crops are genetically engineered to be optimally adapted to local growing conditions. Genetic diversity of crops actually increases as plants are more carefully "genetically tuned" to the local environmental conditions. Rather than losing species, the count of known species in 2030 is higher than it was in 2000. For the first time in modern history, the ecosystem is healthy, resource usage is down, the quality of life is high, and poverty is only relative. The seedbank project has grown into a worldwide foundation that has brought biotechnology to third-world countries. Food is now grown much nearer local markets, and so the old disparities between food production and food need are much reduced. Even the water wars of the early 21st century have disappeared, in large part because of the biotech revolution in drought-resistant plants and the cleanup of polluted water, which has reduced water usage by 25%. By 2040, the world has largely completed the process of shifting from non-renewable energy and chemical feedstocks—coal and natural gas, for example—to biological renewable resources that support a relatively decent quality of life for the global population—now 8.5 billion and soon to be 10 billion. This and all future generations are considered to have the food, fiber, energy, and chemical feedstocks they need, and they receive this bounty from just 30 to 40% of the earth's surface. This development is just in time, as the compelling need to reduce greenhouse gases has been codified, and the resources of fossil fuels have passed from economic recovery. Human creativity has not only muddled through once again, with a solar-powered, renewable-resource-based economy, but future generations of humans have the opportunity to turn their full attention to the age-old questions of tolerance, compassion, beauty, and civility that are within the human potential, knowing that with the management of biological resources, the basic material needs of food, housing, and energy can be met for all. Just as the social order of hunter-gatherers was replaced by a new way of life associated with crop cultivation and animal husbandry, so, too, the information age has been followed by the century of biotechnology, which has brought new rules to society and a better way of life. The World Bank, in conjunction with pharmaceutical corporations, has created a fund that makes health care developments accessible to all nations and cultures. The importance of "world health" is seen by most societies, as is the importance of genetic diversity, which is assured by global legislative developments initiated by the UN and the WTO. Business practices, too, have undergone a transformation. So-called "stakeholder business design" has taken hold in most of the developed world, with its emphasis on stakeholder involvement in decision making, transparency, and commitment to sustainable development. And within this new capitalism, the biotechnology industry has taken the central place as the largest sector of the global economy, not only because most new significant products and services are based in part on BBIs, but also because the companies themselves are quite profitable. Many of them attribute this to turning the corner from being product and technology oriented to becoming market oriented and from emphasizing short-term shareholder value to long-term stakeholder value. Employees in the industry pride themselves on being both "scientifically adept and socially enlightened"—in the forefront, they claim, of the move from an ethos of "ownership" to one of "stewardship." And even though the gap between the rich and the poor is growing, the poor are relatively better off than ever before, and social mobility is increasing in most countries.

A Transformed Society. From the outside, this new society of 2050 looks like a Garden of Eden. But many of the old folks are a little nostalgic for the good old days. They are grateful

(Continued)

for living longer, but feel sorry for young people, who do not have the wonderful illusion of being able to do anything they want to do. Much of what a young person thinks about his or her future is related to the individual genomic profile provided at birth. As each child grows up, part of the ethical and even religious training is focused on what someone with a particular profile is best suited for—and what the ethics are of going against the picture of the future suggested by the profile. Courtship and marriage are often overshadowed by issues of disclosure and genetic compatibility, not for offspring, who can be "fixed" in the womb, but for emotional characteristics suggested by different chemical and neurological profiles. Parenting involves an endless round of gut-wrenching decisions about what genes to fix and when in each child, and children themselves often fault their parents for not providing a better genetic base for intellectual and physical development. Courts both in Europe and the United States appear to be increasingly willing to hold parents accountable in cases of gross negligence involving the genetic welfare of an unborn child, such as neglecting to modify the gene associated with schizophrenia. While procedures of genetic modification are getting less expensive all the time, they still assume a significant proportion of a middle-class budget, which means that while absolute poverty is decreasing, the gap between the rich and the not-rich is growing at an even greater rate than it did in the late 1990s. For many, the developed nations look like the "commercially driven eugenics civilizations" biotechnology critics had prophesied. In countries where the technology is still too expensive to be used extensively, there are fewer "superchildren"—but a higher proportion of children in relation to the numbers of older people. Most couples in developed nations have only one child because of the cost of gene therapy in utero and before the fifth year. Thus, global demographics take an entirely novel turn. On the one hand are the less developed nations with many young boys (gender selection in the womb is easy to obtain through over-the-counter drugs to be taken at a certain periods in gestation). On the other are the "geriatric" nations with well-preserved 120-year-olds, many on their third or fourth organ transplant. These nations spend an enormous amount on health in relation to education. Meanwhile, most of the new innovations in science and technology are beginning to come from those nations with younger populations, and observers note a shift of knowledge production from west to east and north to south. In richer nations, where genetic modification plays a larger role, most employers publish specific genetic profiles of the kind of employee they want, and prospective parents, aware of these ideal gene profiles, tend to demand genetic modifications that will give their offspring the edge in desired fields. Just as in the 20th century, certain names would be popular in cycles ("John" one year and "Michael" the next), in the 21st century, certain genetic profiles are in fashion in particular years. The similarity among children in certain age cohorts is sometimes uncanny. One curmudgeonly social critic complains that you have to go to a remote Pacific island to find an eccentric these days. A deeper complaint is that "modification" of children toward an ideal norm leads to prejudice against those who are even the least bit different—a prejudice that is subtler and more insidious than the old prejudices of race and gender. These trends are disturbing only to the elderly, however. The young take them in stride, accepting the genetic modification of their children in the same way that their grandparents, liberated from the need for large families by labor-saving technology and antibiotics, accepted the downsizing of families from a dozen children to one or two. Every technology produces new challenges. As in ages past, human beings adapt themselves to whatever world their technology has created. And the world of Biotrust, for most observers, is the best world created yet.

Excerpt from *Biotechnology Scenarios: 2000–2050 Using the Future to Explore the Present*, 2000. Reprinted by permission of World Business Council for Sustainable Developement.

Example 2. Intuitive–Decision Support: Destino Colombia.
The country of Colombia has the longest-standing democracy in the world, which is 180 years old. It is also one of the most diverse countries on the basis of ideological, religious, sociopolitical, geographical, ethnic, and linguistic grounds, but this diversity has not blended together harmoniously. Following two brutally violent civil wars in the early 20th century, the country has become increasingly polarized politically and socially. Since the 1960s, violence has escalated among discordant warring groups such as the national military forces, right-wing paramilitary vigilantes, drug traffickers, and left-wing guerrilla factions of the National Liberation Army (ELN) and the Revolutionary Armed Forces of Colombia (FARC). At the time that the *Destino Colombia* scenario project began in 1997, Colombia had the highest rates of kidnapping and murder worldwide relative to its population size (Figure 7.15). Violent conflict was so extreme that the academic discipline of violentology was born there. Economic and government institutions were awash in crisis, including a scandal in

Figure 7.15 In Colombia, the military patrols areas where violence is common (*left*) and citizens march in the street for peace and the end of violence (*right*).

which a government official was accused of spending drug money to fund a campaign. Unemployment, inequity, and dismal educational opportunities were becoming the norm.

In response to this dire situation, in 1997 a business group from the private sector funded a series of scenario workshops held at a farm set in the rolling pastoral landscape outside of the Colombian city of Medellín. With this, Colombia became the first Latin American nation to use a scenario process to try to bring peace to the country and inform strategic action. The *Destino Colombia* process design was intuitive, similar to the WBCSD Biotechnology project (Figure 7.10), in that the scenarios were predominantly qualitative in nature and presented in narrative form. The knowledge base for constructing the scenarios, however, came from all sectors of Colombian society, and was both practical and expert in nature. In this sense, the *Destino Colombia* process was more participatory than the WBCSD scenarios. The project goal of *Destino Colombia* was decision support. The hope was that the scenarios produced in the workshops would be used by the Colombian government to inform policy, and by other sectors of the Colombian society to inform strategic actions aimed at resolving the country's problems.

In July 1997, forty-three participants arrived at the farm for the first of three workshops held throughout that year. They came from diverse groups spread across Colombian society: indigenous groups, businessmen, policy analysts and politicians from both sides of the political spectrum, members of the church, NGOs, peasants, trade unions, popular movements, young people, journalists and the media, and academics and other intellectuals. Even members of the ELN and FARC joined in by phone, with some calling as political prisoners, one from a maximum security prison and another in hiding at an undisclosed location in Costa Rica. The only stakeholder group excluded was the drug traffickers. The goal of the workshops was to develop scenarios laying out plausible future directions for the country over the next 16 years, examine the consequences of those futures, and realize what actions taken today might lead toward or away from those futures.

The initial phase of scenario construction, dubbed the divergence stage, involved gathering information from all stakeholders regarding their diverse perspectives on present-day challenges facing Colombia. This process was carried out via a dialogue among stakeholders that required open communication, which was challenging for several reasons. Some people did not want to open up for fear of saying the wrong thing, which they felt might lead to being viewed as ignorant or ill-mannered, unlikable, or even being arrested, fired from a job, or killed by a guerilla in retribution for something said against them during the workshop. Others were hurt and resentful, as members of their family had been murdered, kidnapped, or assassinated by other stakeholder groups. In the midst of fear, grief, and general uneasiness, it was necessary to create a space for open communication. The group did this by collectively developing and agreeing on "rules of the game" to guide conversation (Figure 7.16). This method worked and trust was built as members talked and listened to one another.

After a one-month break following intense open dialogue, participants convened for the second workshop in August 1997. Having gathered information from participants during the first workshop, experts were brought in for the emergence phase to add to the knowledge that would be used to construct scenarios. The purpose was to widen the knowledge base of participants by educating them about the economic, political, and social dimensions of national and international issues relevant to facing Colombia's challenges. The next and final step, the convergence phase, was spread over the second and third workshops. During this phase, participants constructed a plethora of preliminary scenarios. Then they collectively agreed on, and refined, the scenarios into four alternative futures. Similar to the WBCSD, these scenarios were shaped into narratives with illustrations, which are presented in Box 7.3 in an abbreviated form (the interested reader should consult the *Destino Colombia* project documents to see the illustrations and for the full scenario narratives). Economic structure, political organization, social policy, and the international situation were all considered as drivers leading to the various futures conveyed in the scenario narratives. These drivers, and the actions that led to the various future scenarios, are summarized in Table 7.2.

- Express differences of opinion without irony. No stigmatizing; no personal attacks.
- Be sincere. Assume that others are speaking and acting in good faith.
- Exercise tolerance.
- Observe discipline and punctuality.
- Respect others' right to speak.
- Be concise. No repeating ideas.
- Be inclined and willing to learn.
- Exercise confidentiality and discretion when citing others' opinions.
- Be prepared to "go back to the drawing board."
- Put forward proposals on the basis of an agreement or plan.
- Be disposed to arrive at consensus.
- Call things by their names.

Figure 7.16 **Rules of the game were developed by stakeholders and used to guide conversation during Destino Colombia.**

After the scenarios were finalized, the workshop participants were asked to reflect on the desirability of each scenario, and propose actions that might be taken to change things by focusing on the four following questions:

1. What are the benefits and costs of each scenario for your family, for you personally, for the work you do, and for the nation?
2. Which scenario are you helping to build through actions that you take today?
3. Which scenarios do you prefer? Which seem undesirable to you? How would you alter them?
4. Are there possible scenarios, other than the four finalists, that you can see for Colombia?

The scenarios were then distributed to the Colombian citizenry at large, via a magazine, a national television program, and a video, and they were asked to reflect in the same way.

As with all decision support scenarios, the ultimate goal of *Destino Colombia* was to produce scenarios that could be used to inform policy

Table 7.2 Comparative Analysis of Destino Colombia Scenarios Based on Different Drivers

	Economic Structure	Political Organization	Social Policy	International Situation
Scenario 1 *When the Sun Rises, We'll See*	• Conflicting, ambiguous relations between the state and the productive sector • Improvisation, uncertainty • Economic concentration • Illegal economic activity across national boundaries • Stagnation, instability	• Government mismanagement and corrupt institutions • Continuous policy change from national and international pressure • Collapse of the state • Tendency toward feudal power distribution	• Basic public services deteriorate (education, security, health, justice) • Social fragmentation • Violation of human rights • Increase in poverty • Hopelessness & mistrust dominate interactions	• Censure, rejection, loss of prestige; isolation and international voice lost • External pressure on internal affairs leads to greater intervention • National sovereignty weak and vulnerable
Scenario 2 *A Bird in the Hand is Worth Two in the Bush*	• Mixed economy with state, private, and joint ventures. State provides basic goods & services. Reconstruction of productive infrastructure occurs, with emphasis on strategic sectors.	• Solution comes from compromise among armed elements and other social, economic, political & cultural forces. Power redistributed through negotiation.	• Participatory democracy with redistribution of wealth • A centralized state, with the social structure as its focal point	• Initial uncertainty and expectation, followed by cooperation and support. Eventually, international involvement and representation is strong.
Scenario 3 *Forward March!*	• Private property favored • State promotes enterprise without playing a direct role • Confidence in markets • Reduced government regulation • High taxes finance military, low taxes finance growth	• Authority imposed with emergency measures • War ended using force • Government centralized • Limits on civil freedoms result in discontent from opposition groups	• Not a prime concern at start; priority to military • Government welfare assistance directed to most vulnerable sectors • General improvement; but social rift remains	• Initial international isolation • Gradual acceptance by international community, which maintains pressure in areas such as human rights
Scenario 4 *In Unity Lies Strength*	• Mixed economy • State regulates efficient delivery of goods & services directly or through private sector to guarantee fairness and efficiency. • Economic reforms move Colombia to world economy	• Public peace consensus includes armed groups • Society learns to manage its own conflicts • Government decentralized and local regions empowered • Democratization of power	• Individual need for social services emphasized; demand-based benefits • Deep reforms and social transformations • State meets demand through social compromise	• Full acceptance into the new international order, to the country's benefit • Recovery of prestige and dignity at international level

decisions, and strategies for action aimed at resolving an existing problem in the near term. In this way, this project was different than the WBCSD Biotechnology scenarios, which had more of an open-ended and exploratory goal focused on anticipating longer term challenges, capacity building, and raising awareness. Although the *Destino Colombia* project did raise awareness, promote education, and build capacity, it went beyond this. Its aim was to build consensus for future change and adaptation by using knowledge gleaned from the scenarios to inform strategies for policymakers or for actions taken outside of government arenas. To do this, the scenarios were designed to demonstrate that present-day actions and choices actually do lead to different futures. Each scenario described a path into the future, the consequences of that path, and how that future was reached by daily actions. As far as actual effects on Colombian society, the 43 workshop attendees brought what they learned into their respective organizations, social groups, or sectors. Thus, by changing their own actions, they have affected parts of society. One participant from the private sector started the Ideas for Peace Foundation. The workshop has affected the writing of journalist participants; and one academic brought the scenarios into classrooms for a similar reflective process. The process has spread among indigenous leaders who want to try these scenarios as a means to promote peace. The Colombian government, unfortunately, has not expressed interest in using insights from the process to inform policy decisions at a national level. The spread of insight to regions deeply embedded in conflict has also not been possible.

However, increased social capital resulted from this highly participatory process. As noted in Section 1.3.3 of Chapter 1, good social relations (or social capital) are a required element for building a more sustainable world, and depend on trust, understanding, and mutual respect among different stakeholder groups. A group of stakeholders with high social capital have the capacity to find common ground, talk and listen to one another, and generally work together to address sustainability challenges. Social capital was increased during the *Destino Colombia* scenario process, as barriers of fear and resentment broke down. At one point, a guerilla participant who had phoned in vowed not to kill anyone for anything said during the workshop. As relationships strengthened and trust increased, one landowner with a history of conflict with guerrillas was able to say without fear that he thought the only solution was to intensify the military effort against the guerrillas. By the end, people who would never have interacted in Colombian society joked together. This excerpt from a book called *Solving Tough Problems: An Open Way of Talking, Listening, and Creating New Realities* by Adam Kahane, who was the facilitator hired for the *Destino Colombia* scenario project, describes his reflections on the process:

During the breaks in the meetings, people now started to huddle around the speakerphones, continuing to talk with the guerrillas. People worked hard all day, then talked and laughed and played guitar in the bar until late at night. I was deeply touched by their heartfelt commitment and communication. The team joked about dynamics that were very close to the bone. One morning the representative of the Communist Party overslept after a long evening of singing duets with the retired army general. When he did not show up on time for the meeting, there was a lot of wisecracking about what might have happened to him. "The general made the communist sing," one person said. Then the representative of the rightwing paramilitary said, mock-threateningly, "I was the last one to see him." I was relieved when, a few minutes later, the communist walked into the room.

Despite the success of the project in building social capital, and even creating hope for cooperation so that Colombia might have a better future, good social relations can be fragile. They are slow to build, but deteriorate quickly in the face of conflict. In the end, civil armed conflict and violence continue to fragment Colombian society. In 2002, when Álvaro Uribe became president, the *Forward March* scenario actually began to play out when the administration concluded that force was the only answer, because diplomatic means had failed to work with the guerilla factions. As a result, the military campaign against them was intensified. The actions, social capital, and hope generated by *Destino Colombia* are a testament to the strength of the scenario process. The 1997 workshops were one of the only instances in Colombia at the time where enemies participated in an open dialogue for peace. However, having the resources and political will to implement such a process at the national level through formal government institutions remains a challenge. Imagine what a difference this could make! Perhaps a sustainable future includes government bodies, such as the Social Capital Protection Agency (SCPA), to carry out stakeholder engagement processes much like existing government offices, such as the Environment Protection Agency (EPA), carry out ecosystem stewardship.

BOX 7.3 DESTINO COLOMBIA SCENARIOS: DECISION-SUPPORT AND INTUITIVE

Scenario 1: When the Sun Rises We'll See

The country collapsed into chaos. The lack of will to confront necessary changes had left us without the ability to act—because the worst thing people can do is do nothing! Weariness, laziness, or inability to face problems are all justified by the phrase, "When the sun rises we'll see." The darkness of night turns into a pretext for dreams and apathy, but the clear light of dawn, rather than inspiring important decisions, simply gives rise to a new period of uncertainty. In the face of the country's crisis, this irrational confidence in unexpected miraculous outcomes, this recourse to halfway solutions, this generalized tendency to put off basic actions until later have combined to the point of becoming a collective alternative.

Scenario 2: A Bird in the Hand is Worth Two in the Bush

Following 10 years of bloodshed, and under continuing pressure from armed groups, the state and society decided that it was time to enter into a dialogue and come to serious agreements. Rather than losing it all, everybody gained something—because any settlement is better than continuing a bad lawsuit. Given the distinct possibility of losing everything through violence and armed conflict, we decided to save as much as we could. In the end, no group got everything it wanted. On the other hand, they didn't lose everything, either. Seen from this angle, our solution—based on the realities of the situation we were facing—can be expressed by the popular proverb "A bird in the hand is worth two in the bush."

Scenario 3: Forward March

Seen from a distance, we were experiencing such grave troubles that the measures we adopted presented the only possible response. We had to elect a strong military government to impose force because "harsh problems require harsh solutions." This solution cost us a great deal, but we had to accept it as the price of the many errors and omissions we had allowed to occur. To rebuild a broken nation and mend the lacerations in the country's social fabric before other attempts to achieve peace could be frustrated, people elected a government that proved strong enough to impose order and put an end to institutional chaos.

Scenario 4: In Unity Lies Strength

The nation changed an old way of life and the source of many troubles: the inclination to work against each other. They discovered what could be achieved through respect for differences and the strength of unity. Through united efforts of rural and urban groups, Colombia opened roads and built airports, sewer systems, schools, churches, and health centers. Colombians discovered strength from identifying shared interests and lending many hands to a project—that societies draw lifeblood not from weapons, money, or laws but from dreams, projects, and accomplishments held in common. They depend on developing strength through unity.

Example 3. Formal–Exploratory: Intergovernmental Panel on Climate Change (IPCC) Special Report Emissions Scenarios.
The IPCC was established by the United Nations Environment Programme and the World Meteorological Organization to build an understanding of the risks associated with anthropogenic climate change. The role of the IPCC is to evaluate the scientific, socioeconomic, and technical information relevant to this understanding. In 2000, the IPCC

released its Special Report Emissions Scenarios (SRES). (These scenarios have since been followed by the Representative Concentration Pathways, RCPs, scenarios published in 2014 by the IPCC.) Like the WBCSD Biotechnology Scenarios, the goal of the IPCC scenarios was exploratory, with the purpose of increasing awareness, learning, and building capacity to deal with plausible climate futures (Figure 7.10). The IPCC scenarios were aimed at exploring the uncertainties surrounding both the drivers of human-induced climate change and the models used to project future CO_2 emissions. The idea was to promote awareness and learning about uncertainty within the scientific community and to assess impacts, adaptations, and mitigation. Unlike the *Destino Colombia* scenarios, the IPCC scenarios were not meant to support decision-making. Exploratory projects are more descriptive, open-ended, and focused on the long term for the purpose of capacity building and raising awareness.

Unlike the WBCSD Biotechnology and *Destino Columbia* projects, the IPCC process design is formal (Figure 7.10). Rather than being based on qualitative data and formed into narratives through a participatory process, the IPCC scenarios are quantitative in nature, and require technical knowledge for scenario construction with computer simulation models. Thus, alternative futures built by this means are based on quantifiable ecological, economic, and social processes. However, the IPCC process did not rely solely on quantifiable information. A broad range of qualitative storylines about the future were developed and then linked into quantitative models based on assumptions about the main climate change drivers.

The process of IPCC scenario construction involved several steps. First, climate change scenarios already existing in the literature were surveyed for major assumptions about systems relationships and drivers of change. Five major drivers of anthropogenic climate change were identified: demographic, economic, social, technological, and environmental. Next, storyline narratives were formulated to describe alternative futures for the year 2100 based on the differences between the extent of influence and trends of these drivers (Figure 7.17). The A1 storyline assumes population growth cresting in 2050 and then dropping off into 2100, very fast economic growth, increasing regional equity, and rapid technological innovation. The globalization aspect of this scenario envisions a more interconnected and homogeneous world. This storyline has three substorylines that imagine

	Globalization	Localization
Economic growth emphasized	**A1 Storyline:** Rapid economic growth	**A2 Storyline:** Regional economic growth
Sustainability emphasized	**B1 Storyline:** Global sustainability	**B2 Storyline:** Local sustainability

Figure 7.17 Four qualitative stories about the future were used to constructive quantitative models.

different technological innovations: fossil fuel intensive technologies dominate (A1F1), technologies for non-fossil fuel energy dominate (A1T), and balance between fossil and non-fossil sources (A1B). The B1 storyline maintains the globalization aspect with similar population growth patterns. However, economic growth is less resource intensive, with a shift toward an information and service economy. With this shift, technological innovation is focused on sustainable and efficient technologies. Equity improves globally, as income gaps becomes smaller.

The A2 and B2 storylines describe a more localized world that is heterogeneous and less interconnected. These futures exhibit the highest rates of population growth, with A2 being higher than B2, and equally lower levels of economic growth. In the A2 storyline, equity in per capita income is low due to the fragmented nature of the world. B2 equity is higher than A2, but not as good as the globalized A1 and B1 futures. Due to the inevitable regional inequities that come with fragmentation, B2 has a local sustainability focus that A2 does not have. Technological change for A2 and B2 is slower due to the slow spread of new technologies in a fragmented, localized world. There is more fossil fuel-based technology in the A2 future, which is less focused on renewable energy technologies than the more sustainable B2 world.

Next, storylines were translated into proxy variables used as input variables for quantitative models (Table 7.3). In quantitative modeling, a proxy variable is a quantifiable variable that can be used in a computer model to represent a qualitative variable that cannot be directly represented in a model. For example, one social driver of

proxy variable
a quantifiable variable used in a computer model in place of a qualitative variable that cannot be directly represented in a model.

Table 7.3 Summary of Driver Proxy Variables in 2000 SRES IPCC Scenarios

	2100 Population (billions)	2100 GWP (US$/year)	2100 Per Capita Income Ratio	Energy Intensity (10^6 J/US$)	No carbon energy (% total)	Land use Change
A1F1	7	25× 1990	1.5	18% of 1990	31	Model specific
A1B	7	25× 1990	1.5	20% of 1990	65	
A1T	7	25× 1990	1.5	14% of 1990	85	
A2	15	10× 1990	4.2	35% of 1990	28	Model specific
B1	7	15× 1990	1.8	8% of 1990	52	Model specific
B2	10.4	10× 1990	3.0	24% of 1990	49	Model specific

human-induced climate change from the storylines is equity. However, this is a qualitative variable and cannot be used in computer models. The IPCC represented equity in its quantitative models with the proxy variable *2100 per capita income ratio* (Column 4 of Table 7.3). This is the per capita income in developed countries, and those with economies transitioning toward developed, divided by the per capita income in developing countries. The lower the ratio, the higher the equity between developed and developing nations. Similar proxy input variables are used for the other drivers mentioned in the storylines:

1. **Demographic driver:** Population in the year 2100 is the proxy variable for this driver. The actual numbers are derived from projections by the United Nations (UN) and the International Institute for Applied Systems Analysis (IIASA). Projections by these organizations range from lowest to highest based on the assumptions made about drivers of population growth itself. The numbers used for this driver in each quantitative scenario is shown in Column 2 of Table 7.3.
2. **Economic driver:** The proxy for economic growth is the gross world product (GWP) in 2100. The GWP is the sum of all the gross domestic products (GDPs) for every country in the world. The GWP was assumed to be different for each scenario and is expressed as a multiple of the 1990 GWP. The actual numbers for this driver are shown in Column 3 of Table 7.3.
3. **Social driver:** The quantitative proxy for equity has already been explained, and the ratios used as inputs to the different quantitative scenarios are shown in Columns 4 of Table 7.3.
4. **Technological driver:** There are two proxies used for technological innovation. One is a measure of the intensity of energy use worldwide and is expressed in units of joules used/US dollar spent (Column 5 of Table 7.3). In all scenarios, energy usage in 2100 is assumed to be lower than in 1990. Thus, these values are expressed as a percentage of 1990 usage in Table 7.3. The other proxy is the percentage of energy derived from non-greenhouse gas sources (Column 6, Table 7.3).
5. **Environmental driver:** Land use change is the environmental driver of climate change for these scenarios. This is because land use changes, especially deforestation and intensive agriculture, contribute anthropogenic greenhouse gas emissions to the atmosphere (Figure 7.18). In these models, deforestation is a function of population and income growth. Agricultural land use is determined by food demand and dietary preferences, and also a variety of technological, institutional, social, and economic factors. Each of the six models represents system relationships among all of these factors, and thus land use change, slightly differently. As such, the land use change driver depends on each specific model.

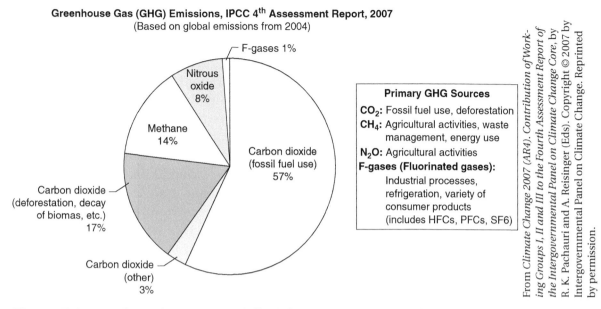

Figure 7.18 Land use changes, especially deforestation and agriculture, contribute almost 40% of anthropogenic greenhouse gases to the atmosphere.

For the IPCC scenarios, a multi-model approach was used (six different models) to learn about the current range of uncertainties surrounding climate change. This was done because each model assumes slightly different system relationships and this will affect the model output, even if the same input variables are used. Box 7.4 provides an illustration of how different system relationships in different models result in uncertainty about the climate futures predicted by these models by returning to the example of the beaker given in Box 7.1.

BOX 7.4 UNCERTAINTY AND THE FUTURE OF GLOBAL CLIMATE

Recall that Henry's Law is a system relationship used to figure out how much CO_2 would be present in the beaker if CO_2 concentrations in the atmosphere were to double. This equation is a quantitative model (albeit, very simple one) that describes the behavior of CO_2 in the beaker based on a defined system relationship. To use Henry's Law, if CO_2 concentrations in the atmosphere doubled, you would have to know k and the pressure of CO_2 in the atmosphere (P_{CO_2}) to predict the new CO_2 concentration in water ($[CO_2]_{water}$) (Figure 7.19). Now, pretend there are three different models representing this system's behavior—Model A, Model B, and Model C (Figure 7.19)—because the scientists who developed these models are unsure of how to exactly characterize this process using equations. (In reality, Henry's law defines simple systems very well.) Let's say the input variables have the following numerical values: $k = 3.7 \times 10^{-2}$ *mol/L atm* and $P_{CO_2} = 1$ *atm*. If these same input variables are used for all three models, slightly different model outputs will result (Figure 7.19). Thus, the range of uncertainty when predicting new CO_2 concentrations in water ($[CO_2]_{water}$) when atmospheric CO_2 concentrations double is 1.9×10^{-2} to 5.6×10^{-2} mol/L. In other words, when predicting the future CO_2 concentration in water using this model, you can say that it is likely that $[CO_2]_{water}$ will fall between 1.9×10^{-2} mol/L and 5.6×10^{-2} mol/L in the future. But we can't be more certain than that range.

Compare this simple situation to the IPCC's situation, when constructing future scenarios to predict atmospheric CO_2 concentrations using its models, and you can see why the IPCC's situation is profoundly more complex. There are many more input variables for these models (Table 7.3), which are themselves uncertain because they are not precisely known (for example, population growth is based on demographic models developed by the UN and IIASA, and these models have their own uncertainties). The six different IPCC models also use slightly different system relationships to construct climate futures (just like the three Henry's Law model varieties in Figure 7.19).

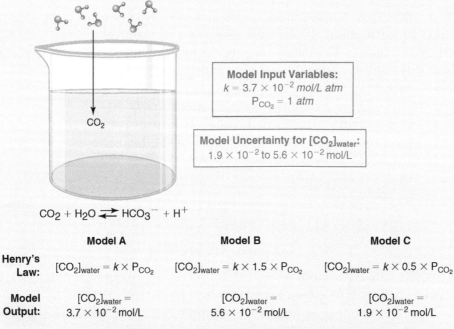

Figure 7.19 Three different representations of Henry's law lead to three different predictions for the CO_2 concentration in water, even when the same input variables are used.

As shown in the summary figure (Figure 7.20), the SRES scenarios were constructed by creating four qualitative storylines, translating each storyline into a quantitative model using quantifiable proxy variables, and then constructing each of the scenarios from computer simulations using all six model varieties *for each storyline*. The quantitative results of this formal scenario process are presented as graphs of CO_2 emission rates over time (Figure 7.21). As you can see, each scenario has an uncertainty range spanning the outputs of the six models used to construct each scenario. For example, using the same driver input values (Table 7.3), but six different model outputs to predict global carbon dioxide emissions in 2100, the A2 scenario shows that in the year 2100 global CO_2 emission rates due to human impacts will be somewhere between 23 and 34 GtC/yr. According to these scenarios, it is likely that emissions will fall in this range, but it is uncertain exactly what they will be.

The results of SRES scenarios are shown together in Figure 7.22 to illustrate the overall uncertainty about future CO_2 emissions, due to variations in the driver inputs. (This is different from the uncertainty due to variations in the system relationship representations among the six different models, as shown in Figure 7.21.) Based on assumptions about the five different drivers of human-induced climate change (Table 7.3), total cumulative CO_2 emissions in the year 2100 range over two orders of magnitude (from about 700 GtC to 2,500 GtC). Scenario A1F1 results in the highest emissions, whereas B1 results in the lowest. Based on these results, different drivers have vastly different effects on future emission trends. Thus, in the end, the IPCC SRES project was used to explore how different drivers influence future emissions based on quantitative model outputs. There were two types of uncertainty regarding future CO_2 emissions: the uncertainty associated with different model representations (Figure 7.21) and the uncertainty associated with the assumptions of how human-induced drivers of climate change might develop into the future (Figure 7.22). In

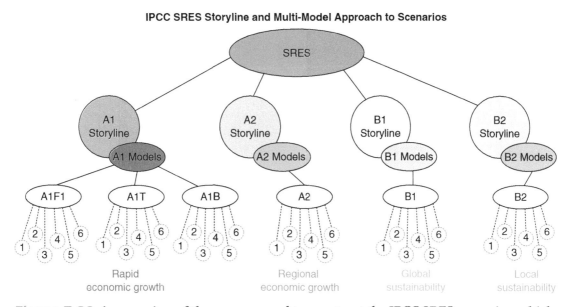

Figure 7.20 An overview of the process used to construct the IPCC SRES scenarios, which ultimately resulted in 36 different model outputs (6 for each scenario).

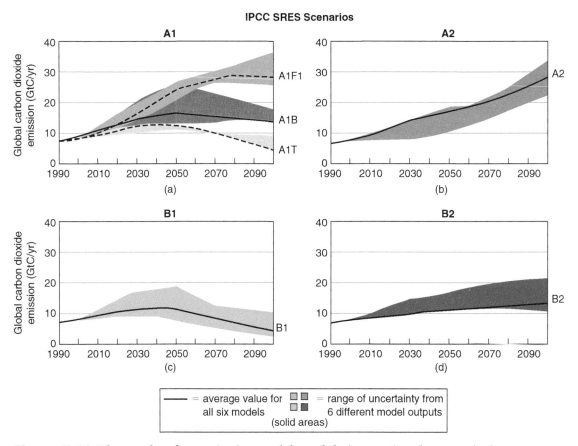

Figure 7.21 The results of quantitative models and their associated uncertainties.

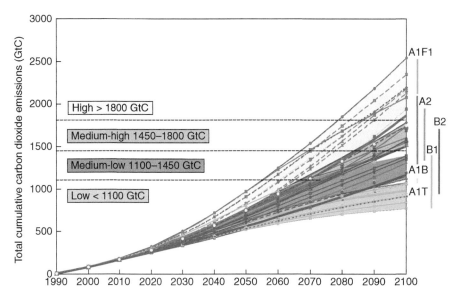

Figure 7.22 The results of all scenarios together show the uncertainty in future CO_2 emissions due to different driver representations.

the report, none of the scenarios are deemed more likely to occur than others, as they are not appointed probabilities. As such, all can be considered equally plausible. One important caveat is that these plausible futures do not consider unexpected events and surprises. Occurrences, such as volcanic eruptions, nuclear winter due to nuclear armageddon, or the slowing of the ocean conveyor belt, can happen and have drastic effects on future climate. (The slowing of the ocean conveyor belt was portrayed, albeit inaccurately, in the blockbuster movie *The Day After Tomorrow*.) Because of this, and also the general difficulty of defining system relationships for models of the complex climate system, scientists often refer to the IPCC SRES results as *projections* rather than *predictions*.

This section will close with some insights about quantitative modeling. In the type of model used by the IPCC SRES project, which extrapolates from present conditions into the future, predictive power diminishes the further into the future you go. Just as weather forecasts don't perfectly predict the weather a week in advance, there is uncertainty associated with predicting the climate many decades, to hundreds of years, into the future. This type of modeling also does not incorporate unpredicted events and surprise. Also, the scenarios resulting from this type of modeling are only as good as the understanding and representation of the system relationships and the input variable data. A common saying when it comes to quantitative modeling is: "Garbage in, garbage out." Because of all of this, such scenarios should be used cautiously to inform policy decisions or actions in the real world, as warned by the climate scientists who wrote the report. The scenarios were not intended for decision support, but to gain insight about uncertainties surrounding the future climate. Assumptions behind modeling should always be clearly stated. The upside of formal scenario activities is that additional scenarios can be developed and explored relatively quickly by changing the input variables. This is in contrast to lengthy participatory stakeholder processes, as described for *Destino Columbia*. Quantitative models are also beneficial in helping the human mind deal with, and manage, complex information, which it is not naturally good at doing.

Example 4. Formal–Decision Support: Global Scenario Group's Great Transitions. This is the final example of scenarios in this chapter, and it will not be described in great detail. After reading the previous three scenario examples, the reader has a good idea of the purpose of exploratory projects, as compared to decision support (project goal), as well as the differences in how scenarios are actually constructed using different methods, processes, and information types in intuitive versus formal projects (process design). The final scenario project example is called *Great Transition* and was created by the Global Scenario Group (GSG). The GSG was assembled by the Stockholm Environment Institute in 1995 to employ an international cohort of distinguished thinkers to investigate possibilities for 21st century global development. The group developed a scenario analysis called *Great Transition: The Promise and Lure of the Times Ahead*. The framework they developed has been used by many groups worldwide to examine the future of specific nations and regions. The *Great Transition* project considers how historical dependencies, present conditions, and future challenges might steer the world toward six different plausible futures. The process design for these scenarios is formal, such that they rely on a combination of qualitative storylines and quantitative modeling, much like the IPCC SRES project (Figure 7.10). Unlike the IPCC SRES project, the *Great Transition* project is aimed at decision support. Scenarios were produced by this project with the intent that they will be used to inform decision making and strategic action for sustainability.

Following an examination of historical trends, three central scenario classes were developed using storylines: Conventional Worlds, Barbarization, and Great Transitions. In Conventional Worlds, unexpected events and surprise are left out, such that the drivers presently shaping the world continue to shape it into the future. Specifically, as environmental, social, and economic challenges crop up, they are addressed through policy adjustment or market reforms just as they are today. In Barbarization, environmental, social, and economic challenges still crop up, but they are not addressed and left to fester. Challenges accumulate as a result, ultimately leading to effects that cascade into crises made worse by reinforcing feedbacks among all the problems. The world descends into tyranny and anarchy. Great Transition portrays deep and meaningful value-based societal transformation. In contrast to present conditions regulated by current value systems, people in the future value quality of life (over material possessions), unity and equity (rather than competition), and environmental sustainability. Each of these three scenario classes has two variants each, as described in Table 7.4.

Six major drivers were identified, along with the trend each driver would have to follow to cause the current state to transition into one of the six different future scenarios (Figure 7.23). In addition to these six drivers, the scenarios considered the belief systems that would lead to each future scenario (Figure 7.24). These beliefs are founded on the philosophical views of historical figures (Antecedents in Figure 7.24), which were condensed into easy-to-understand mottos. As was done in the IPCC SRES scenarios project, these storylines were quantified by translating qualitative drivers into quantifiable

Table 7.4 Variants of the Three Central Scenarios

Conventional World	
Market Forces	Competitive, open, integrated global markets drive world development
Policy Reform	Comprehensive, coordinated government action initiate for poverty reduction and sustainable development
Barbarization	
Breakdown	Conflict and crises spiral out of control, institutions collapse
Fortress World	Authoritarian response to breakdown divides the world into an elite minority living in interconnected, protected enclaves and an impoverished majority
Great Transition	
Eco-communalism	Bio-regionalism, localism, and face-to-face democracy abounds after the world has passed through a Barbarization phase
New Sustainability Paradigm	Rather than a retreat into localism, the character of global civilization as a whole changes toward unity and solidarity with cross-cultural fertilization, economic connectedness, and equity

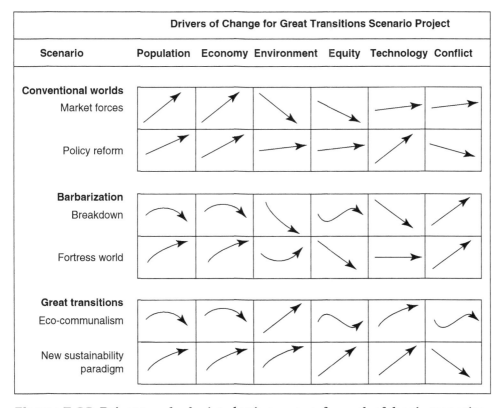

Figure 7.23 Driver trends, depicted using arrows, for each of the six scenarios.

| Belief Systems Shaping Great Transitions Scenarios ||||
Worldview	Antecedents	Philosophy	Motto
Conventional Worlds *Market*	Smith	Market optimism; hidden & enlightened hand	Don't worry, be happy
Policy Reform	Keynes Brundtland	Policy stewardship	Growth, environment, equity through better technology & management
Barbarization *Breakdown*	Malthus	Existential gloom; population/resource catastrophe	The end is coming
Fortress World	Hobbes	Social chaos; nasty nature of man	Order through strong leaders
Great Transitions *Eco-communalism*	Morris, social utopians, & Ghandi	Pastoral romance; human goodness; evil of industrialism	Small is beautiful
New Sustainability Paradigm	Mill	Sustainability as progressive global social evolution	Human solidarity, new values, the art of living

Figure 7.24 Different belief systems underlie each of the six scenarios.

proxy input variables for computer simulations. Future scenarios were constructed using model outputs, just as they were for the IPCC SRES project, where CO_2 emission rates (Figure 7.21) and cumulative CO_2 emissions (Figure 7.22) were the model outputs presented in this chapter (there were others not shown here). For the GSG project, model outputs were also presented and all scenarios were compared. Two examples of model outputs for the Conventional Worlds, Market Forces scenario are given in Figures 7.25 and 7.26. The interested reader should consult the *Great Transition* scenario project documents for more details on the meaning of the outputs and comparisons shown in these figures, as the full results will not be presented here.

The use of sustainability indicators, which was the topic of Chapter 4, in scenario analyses is illustrated by looking at variables shown on the y-axes of the graphs in Figure 7.26. The following variables are indicators used in this project and include indicators for all three pillars of sustainability: environment (e.g., CO_2 concentration for the climate graph, "billions of people" for the water stress graph), society (e.g., "conflict (>1000 deaths/year) for the peace graph, "gender equity" for the freedom graph), and economy (e.g., "GDP (trillions of $)" for economy graph). The GSG project indicators provide a more complete look at sustainability as compared to

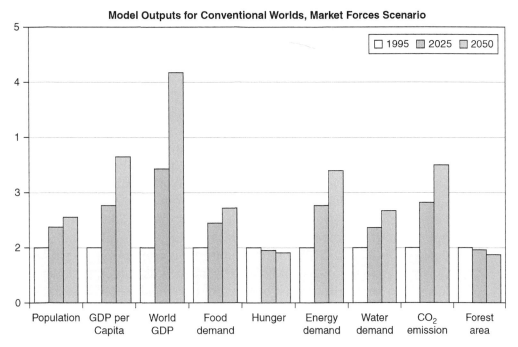

Figure 7.25 An example of quantitative model outputs for the Conventional Worlds, Market forces scenario.

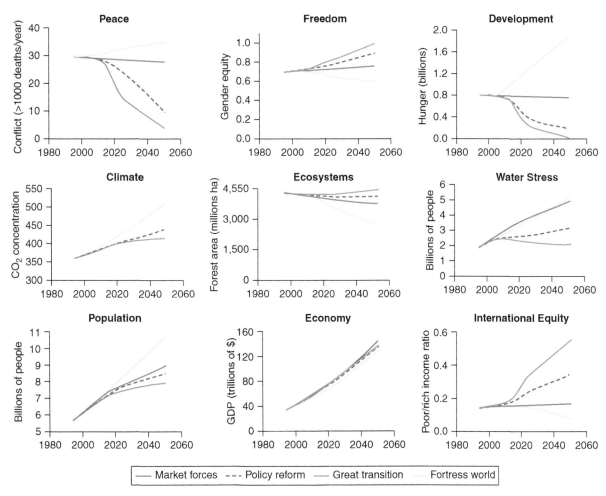

Figure 7.26 An example of model outputs for four of the six scenarios.

the IPCC SRES project, which used mostly indicators of environmental sustainability.

The final part of the *Great Transition* project, as a decision support scenario, was to suggest strategies, change agents, and value changes to inform decision-making processes. Specific strategies were suggested that might guide the current state toward a more sustainable future. The recommended strategies include both the market-led adaptations dominating Conventional Worlds, Market Forces and the government-led policy adjustments from Conventional Worlds, Policy Reform. However, they went beyond these traditional solution options to recommend fundamental societal value shifts that would decouple human well-being from material consumption. Specifically, they recommended that indirect drivers such as values, understanding, power, and culture must be altered. They made recommendations regarding specific actors who could make these changes in the world. Because world development is shaped by all people, they included actors from Conventional Worlds, Market Forces (global corporations, market-enabling governments, consumerist public), and Conventional Worlds, Policy Reform (governments), but also added critical actors who do not presently play a dominant role: civil society and engaged citizens. Specifically, these include intergovernmental organizations (e.g. World Bank), non-governmental organizations, non-profits, and the general public (especially youth). The GSG made more recommendations for a sustainability transformation, and the interested reader should consult the *Great Transition* project documents for more details.

This concludes the future scenarios section of this chapter. The case studies described show how future scenarios can be used to consider a variety of ways that a sustainability problem may develop into the future. All scenario examples produced a set of plausible futures through either an intuitive or a formal design process. In terms of project goals, the exploratory scenario projects (WBCSD and IPCC SRES) promoted awareness, learning, and capacity building by using an open-ended approach to anticipating the long-term future. In contrast, the decision support scenario projects (*Destino Colombia* and *Great Transitions*) produced scenarios intended for use in decision making about more tangible problems in the relative short-term. For the *WBCSD, IPCC SRES*, and *Great Transition* scenario projects, participants were largely experts who (with input from a broader audience) constructed the scenarios. For *Destino Colombia*, participants were much more diverse and collectively developed scenarios through a more participatory process. This type of participatory process is relevant to the next section of this chapter on visioning, which is a process by which stakeholders come together to collectively imagine a desired future.

Section 7.2: Visioning and Future Thinking

Core Question: How can we determine what a desirable and sustainable future should be?

No one can predict the future, but we can invent and make the future because we know how to do so and we have done so consistently through our history as a species. Inventing and making however mean thinking clearly about where we wish to go and then creating and devising the means to get there.

—Garry D. Brewer, Inventing the Future, 2007

[T]he future depends on people's ability to judge preferable futures, to understand the human values and goals that define conceptions of the good society. Knowing what is possible and probable helps to guide our actions, but we need to know more. We also need to know what is desirable. Which futures should we want to achieve? Which futures should we try to avoid? To choose to act one way or another involves an evaluation of the desirability of alternative futures. Of all that is possible, what do we want?"

—David Hicks, Lessons for the Future: The Missing Dimension in Education, 2004

Future scenarios are about assembling a range of plausible futures for a certain purpose (exploratory or decision support), using either an intuitive or formal process, that results in multiple scenarios (i.e., three to six for the case study examples in the last section). However, not all future scenarios are desirable, and this is where visioning comes in. Rather than answering "where are we headed?" using future scenarios, visioning is focused on considering "where do we want to go?" (Figure 7.1). A vision describes a desirable ideal future state where society, or a subgroup of stakeholders in society, would like to go. Visioning is a process by which stakeholders collectively develop a picture of the future they would like to see. However, a desirable future defined by a group of stakeholders based on their preferences, needs, and wants may not necessarily be a sustainable one. Thus, visions aimed at addressing sustainability problems must be based on both stakeholder preferences *and alignment with sustainability principles*. Visioning is important for moving purposefully toward the future. As alluded to in one of the opening quotes to this section, if we do not think clearly about where we want to go, then progress toward a sustainable future will not be efficient but rather difficult, slow, and frustrating.

The use of visioning to think about the future of sustainability problems is even younger than scenario analysis. Contemporary visioning schemes that integrate holistic systems thinking with participatory stakeholder engagement practices have been applied to energy futures, in corporate settings, and in local communities. In the 1980s, John Robinson pioneered a related method called backcasting, which will not be described in this textbook. It was not until 1996 with the work of systems thinking pioneer Donella Meadows that visioning was first applied to sustainability problems. Meadows' study was focused on a vision for putting an end to hunger worldwide. Since this time, visioning has been used in many different sustainability contexts. Despite this, a consistent methodology and widely accepted way of going about the visioning process does not currently exist. As a result, there is no existing typology with which to frame different visioning processes, as there is for future scenarios. However, Professor Arnim Wiek and graduate student David Iwaniec from the School of Sustainability at Arizona State University have recently developed quality criteria for distinguishing good visions, and visioning processes, from those not as useful for addressing sustainability problems. (Many of these quality criteria may also be used to evaluate future scenarios.) This section begins by providing a general overview of what visions are and how they can help us move toward sustainability. Then, an example of how one might go about the visioning process, and a few example visions, are presented. Finally, the quality criteria of Wiek and Iwaniec are presented as a guide to developing visions for sustainability.

Section 7.2.1—Visions for Sustainability

Key Concept 7.2.1—Visions promote movement toward sustainability by motivating and inspiring, guiding and directing change, building capacity and social capital, giving solutions "staying power," and promoting long-term thinking.

Visions describe desirable futures. They are not scenarios or predictions. Rather, they describe a target or ideal condition for the future that people would like to establish. As such, they provide a picture of the future, and include the notion of transitioning toward that future. Although visions provide idealistic pictures of the future, they are also based in reality. For example, a group of people might envision a future society living on another planet surrounded by healthy ecosystems similar to those on earth. However, while these images might provide fodder for science fiction films and novels, they are not based in the realm of present possibilities (Figure 7.27). As such, they are not useful for authentically moving today's society toward a more sustainable world.

Figure 7.27 Unrealistic visions, such as living on another planet surrounded by lakes and healthy ecosystems, might be desirable, but they are not useful for developing strategies for moving societies toward a realistic future.

Visions are often broad statements about the future, but must also be specific to the aspirations of a defined group of people living in a distinct place during a given time period to be useful. To achieve this, the visioning process should be participatory (similar to that described for the *Destino Colombia* scenario project) such that diverse stakeholders gather together to collectively agree on a shared vision for the future. This is because *the transformative power of a vision depends on the extent to which is it common to the majority of people*. As noted by the planning consultant Steve Ames, based on his work with institutions and communities in the state of Oregon during sustainability visioning processes: "A vision must in time catch fire with the people of a place—or it will not succeed." (Refer to Section 3.3.1 of Chapter 3 for more on deciding who should be included as a stakeholder, so that an equitable and accurate representation of diverse perspectives and agendas is including during the visioning process.)

If asked what a group of stakeholders would like the future to look like, they may come up with an ideal picture of the future according to their preferences. However, depending on who the stakeholders are, this future may not be sustainable or viable. Thus, developing visions for a sustainable future must not only be that which is desired by stakeholders, but also must be aligned with the goal of sustainability: to simultaneously promote healthy ecosystems, human well-being, and viable economies (Figure 7.28).

Visions, and the visioning process itself, promote movement toward sustainability in the following ways:

1. **Motivate and inspire.** Visioning can be very powerful because it removes the logistical hurdles of getting things done, for the moment, and gets people's emotions rolling and helps them feel

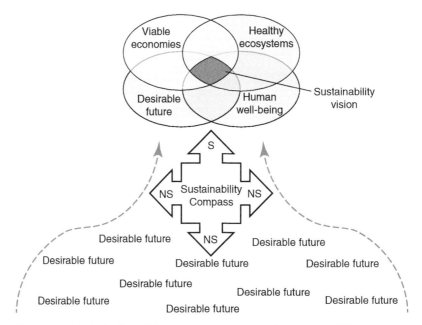

Figure 7.28 Stakeholders may come up with many different desirable futures, but a sustainable vision must be guided by the principles of sustainability as well as being desirable.

inspired. They can be catalysts for a sustainability transition by providing the political will, societal momentum, and inspiration needed for change. Part of the motivation and inspiration provided by visions comes from the fact that they are positive projections of the future, as opposed to the typical doom and gloom images of pending environmental and social disaster often portrayed in the media. These sensational headlines sell magazines, but they do little to inspire positive change, and can actually be harmful if people become paralyzed from action or apathetic in the face of seemingly overwhelming problems. Visions are forward-looking dreams about a better world that are pursued with passion and imagination. The importance of visions has been realized since biblical times with the saying "Without vision the people will perish" (Proverbs 29:18). Many organizations, companies, and cities have used visions to rouse sustainability. Today, political leaders employ visions to gain the support of voters by harnessing imagination. Many international declarations and visions of sustainability exist, such as the *Earth Charter* and *Agenda 21*. Generating political will and momentum at local and national levels for a more sustainable world is one of the most important undertakings presently facing humanity.

2. **Guide and direct change.** Visions also catalyze sustainability transitions by guiding future decisions and actions. A vision is a prerequisite to actually developing strategies for reaching an ideal future

state. Transition strategies to move from the current state to a sustainable future is the subject of the next chapter, but a vision is needed before strategies can be designed. To guide strategies, broad visions are translated into qualitative goals (e.g., stop anthropogenic climate change) or quantitative targets (e.g., atmospheric CO_2 concentrations of 350 ppm or less). Organizations, companies, and cities use these to guide action plans, investment decisions, and political tactics. Visions guide action by allowing prioritization and analysis of tradeoffs. Because of the limited time, money, and other resources needed to pursue visions, tradeoffs among different priorities must be made. For example, a student is getting ready to leave a university for the Christmas holiday and has many things to accomplish beforehand, she examines priorities and tradeoffs. She must prepare for final exams, arrange for time off from her part-time job in the student advising offices on campus, buy Christmas gifts for her friends, apply for several internships for the next semester, and find time to attend holiday parties in her department. She has a vision of her future: to be an excellent student, a loyal employ, a caring friend, an active participant in her university department, and to successfully get a job after graduation. With this vision in mind, and in the face of resource constraints, she prioritizes her actions based on tradeoffs. Tradeoffs include things such as more time spent studying versus shopping for Christmas gifts, which is really trading off being an excellent student against being a caring friend. Or if she spends more time at department Christmas parties than preparing her internship application, she may be trading off "active participation in her department" for "successful job acquisition following graduation." In the end, she sacrifices none of these, and instead makes a tradeoff with sleep! But she is not too worried, because she'll have a lot of time to sleep over the semester break. By considering priorities and tradeoffs, visions provide the basis for decisions and actions.

3. **Build capacity and social capital.** Developing a vision is much more complicated than an individual's internal battle for prioritization and making tradeoffs, because visions involve many different individuals with their own agendas and diverse perspectives. As with the *Destino Colombia* scenario project, engaging diverse stakeholders during visioning brings people together to develop relationships based on trust and respect. It bolsters stakeholder identity and solidarity toward a common purpose by developing a consistent language for talking about the future and nurturing partnerships. Also, as in *Destino Colombia*, the participatory aspect of visioning creates a safe space for talking honestly and openly. All of this builds the relationships and social capital needed for collectively dealing with the future. When making decisions, there is an overwhelming amount of information to consider and a vast

range of legitimate perspectives to reflect on. It is better to do this in a deliberate way, before a crisis occurs, so that stakeholders have a common direction and course of action beforehand. Crisis situations often provide short-lived opportunities for change. To take advantage of these opportunities, a vision and plan for action must already be well-defined and developed.

4. **Give solutions "staying power."** Involving representatives of all relevant perspectives, whether dominant or minority, from the start of the visioning process, gives solution strategies "staying power." This has to do with stakeholder ownership over ideas about future directions. By engaging the values and preferences of the community, a vision is more likely to be achieved. For example, imagine the difference in your excitement level in pursuing a college degree if your parents choose your degree for you instead of you choosing it yourself. No one likes to be told what to do, and you are more likely to pursue your degree with passion, persistence, and gusto if you get to choose a degree that you prefer and value. Even if your parents approve of your decision, the very act of choosing your own path gives you a sense of ownership over your life, and engenders much more enthusiasm from you than if it were chosen by your parents (Figure 7.29). Similarly, in visioning aimed at solving sustainability problems, allowing communities to choose their vision (rather than imposing a vision on them) allows them to choose their own goals and paths into the future. Although visions aimed at sustainability problems must be constrained by sustainability principles, presenting people with information about sustainability, and then stepping back to allow them to choose, often results in their choosing the "right thing" anyway. If it does not, visions must be gently and diplomatically guided toward sustainability if they

Figure 7.29 If your personal vision is to earn a college degree in sustainable agriculture to become an organic farmer (*left*), you will be much more enthusiastic about this choice, and more likely to complete your degree, than if your parents demanded that you pursue a business degree (*right*).

are to effectively address sustainability problems. Ownership, gained through a participatory process, also leads to accountability. People like to do what they say they are going to do. Thus, if a sustainability pledge is made in front of a large group of stakeholders during visioning, then it is more likely to come to fruition.

5. **Promote long-term thinking.** We naturally consider our needs and desires on a short-term, daily basis. Considering the long-term future does not come naturally. Visioning provides a way to do this. Considering the long-term future is a requirement for sustainability because we are concerned with the effect of our actions on future generations. The degree to which we are able to meet the needs of future generations is a direct result of our present-day vision of what a desirable future looks like for them. Although it can be difficult, if not impossible, to determine the needs and desires of people not yet born, it must be attempted. One way to handle this is to view visioning as an ongoing process, such that current visions are continually revised as conditions change.

Section 7.2.2—The Visioning Process

Key Concept 7.2.2—Vision development involves building consensus around questions related to defining human progress, needs, and ethics and the roles of technology and place in sustainability.

This section provides an example of the types of deep, almost philosophical, questions that are examined during visioning. These questions are derived from a chapter on visioning contained in a book called *Cities as Sustainable Ecosystems: Principles and Practices* written by Peter Newman and Isabella Jennings in 2008. As already mentioned, there is not one agreed-upon way to go about the visioning process. However, the following questions can guide vision development and are a good, easily digestible example for the reader who is very new to visions and visioning. Newman and Jennings propose five questions to help in formulating a vision:

1. **What do we value? How do we define human progress?** Clarifying stakeholder values and beliefs about the meaning of progress leads to more productive discussions. Before developing a shared vision, differences and agreements regarding what people value about the future, and how they define future progress, must be addressed. Recall from Section 3.3.2 in Chapter 3 that values are general principles and standards for defining what is worthwhile or important. They provide criteria for defining actions, and evaluating people and events, and give us a fundamental sense of what is

appropriate. People value things such as freedom, independence, equity, or a structured hierarchy, for example Values are not "right" or "wrong" and are generally difficult to change. Thus, when confronting sustainability problems, we should *engage* stakeholder values rather than try to change them. Table 7.5 shows a human value characterization developed by Dr. Shalom Schwartz based on ten human motivational goals, driven by three universal necessities of the human condition: meeting basic needs (food, water, and shelter), coordinating social interaction, and satisfying group needs (such as welfare and survival). Although there are many different ways to characterize values, the Schwartz categorization is based on value surveys in at least 67 nations around the world and is framed by human motivational goals relevant to sustainability. Thus, it is a good place to start.

The second question is about defining what is meant by human progress. This is based on beliefs about what progress means. Recall from Chapter 1 that the meaning of human progress has evolved through time. In general, the idea of progress is

Table 7.5 Basic Values According to the Motivational Value Theory of Schwartz

Motivational Goals	Values
Security. Safety, harmony, and stability of society, of relationships, and of self.	Social Order
Power. Social status and prestige, control or dominance over people and resources.	Authority, Wealth
Achievement. Personal success through demonstrating competence according to social standards.	Success, Ambition
Hedonism. Pleasure and sensuous gratification for oneself.	Pleasure
Stimulation. Excitement, novelty, and challenge in life.	Exciting Life
Self-Direction. Independent thought and action; choosing, creating, exploring.	Creativity, Freedom
Universalism. Understanding, appreciation, tolerance, and protection for the welfare of all people and for nature.	Social Justice, Equality
Benevolence. Preserving and enhancing the welfare of those with whom one is in frequent personal contact (the "in-group").	Helpfulness
Tradition. Preserving and enhancing the welfare of those with whom one is in frequent personal contact (the "in-group").	Humility, Devoutness
Conformity. Restraint of actions, inclinations, and impulses likely to upset or harm others and violate social expectations or norms.	Obedience

the notion that human societies will advance through time, and that this has driven development. Centuries ago, progress was defined as moral advancement toward the purpose of achieving salvation after death. Although this idea of progress is still alive in many cultures around the world, the definition today is dominated by the secular aim of a better material life on earth, and improvement of the human condition, by modernization and technological advances. Since the Industrial Revolution, economic growth, increased material wealth, and human domination over nature have been added to the list of what it means for a society to progress. This definition of progress has resulted in many benefits to humanity, but it has also resulted in costs, such as increased demand for natural resources, emission of harmful pollution into natural systems, and much social inequity and human suffering. The visioning process allows stakeholder groups to think about these issues in light of what they value for the future, and to possibly redefine progress. In sustainability, progress is centered on promoting the conditions necessary for the survival of prosperous human societies. The specific "conditions," and the meaning of the word "prosperous," will depend on the place and time in which a community lives.

2. **What are human needs?** As noted in the 1987 Brundtland Commission Report (*Our Common Future*), sustainability is concerned with "development that meets the needs of the present without compromising the ability of future generations to meet their own needs." Meeting needs of present and future generations is centered on ensuring human well-being. As mentioned in Chapter 1, human well-being is defined in many ways. In this book, human well-being is defined in general terms as vitality and integrity (**Figure 7.30**). Vitality includes basic needs for survival, such as food, water, and shelter, whereas integrity is concerned with human needs beyond the basics that make life worth living. These general categories are further divided into five major aspects of human well-being, in order to provide a more detailed framework for thinking about human needs: basic material for a good life, health, security, good social relations, and freedom of choice and action. These categories are vague, but there is a reason for this. Different people,

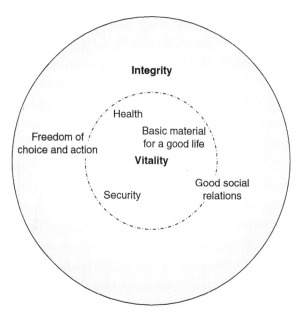

Figure 7.30 Vitality and integrity overlap with the five major aspects of human well-being, as defined by the Millenium Ecosystem Assessment.

living in different places, during distinct time periods, define needs differently. Vision statements developed by stakeholders should clarify and specify needs as related to their particular situation and context. Visioning for sustainability helps stakeholders reconcile how needs are presently met with the resource limitations of natural systems and the requirements of equity and economic viability. This provides the opportunity to uncover new means for meeting needs that are kinder to ecosystems and people.

3. **What are our ethics? How should we treat each other and the natural world?** Ethics is the study of right and wrong. It considers what should be done in certain situations and why things should be done this way. For much of human history, the judgment of right and wrong action in Western society has been centered on interactions between people. More recently, this has been expanded to the non-human world by considering ethical action toward nature and other species. In the United States, the development of these ethics began largely with the conservationist–preservationist debate of the late 19th century. Gifford Pinchot, the first head of the U.S. Forest Service, promoted a conservationist ethic that was utilitarian in nature: "The greatest good for the greatest number." As long as the "greatest good" to the human and non-human world was promoted by some action, then that action was considered "right." John Muir held a different idea about right and wrong, often referred to as the preservationist ethic or the wilderness ethic. He argued that wilderness should be left untouched by human activity so that it would remain in a "pristine" condition. With the rise of the scientific discipline of ecology, holistic systems concepts, such as interconnectivity and interdependence, were merged into these ethics to result in Aldo Leopold's land ethic: "[A] land ethic changes the role of Homo sapiens from conqueror of the land-community to plain member and citizen of it. It implies respect for his fellow-members, and also respect for the community as such" (*A Sand County Almanac*). "A thing is right when it tends to preserve the integrity, stability, and beauty of the biotic community. It is wrong when it tends otherwise." These are examples of ethics that have led us to a sustainability ethic, which might look something like this: "A thing is right when it preserves the conditions necessary for the survival of prosperous human societies by promoting healthy ecosystems, human well-being, and viable economies for people living today, and for those who will be living in the future. It is wrong when it tends otherwise." This is the "what" aspect of ethics. For the "why" of this statement, the case was made in Chapter 1 that if we do not address the challenges of natural system degradation and growing inequity, then prosperous human societies will not survive. During visioning, ethics are specified to a particular place and time.

The ethics that guide visioning can, and have, led to tangible outcomes in the real world (Figure 7.31). Gifford Pinchot's conservation ethic led to the creation of the National Forest Service, which manages public lands for many uses, including hiking, hunting, mining, and grazing. John Muir's wilderness ethic led to the National Park Service, which oversees public lands where human activities are very limited. Aldo Leopold's land ethic led to concepts related to sustainability, such as socioecological systems, where humans are viewed are a part of ecosystems rather than separate from them, and using wind and sun as renewable energy sources in an effort to minimize harm to the all living things. The sustainability ethic has brought human needs and economies to the forefront, along with ecosystem health, to guide the development of products that are beneficial to both people and nature.

4. **What is the role of technology?** The role of technology must be considered in any vision of the future because it has the potential to effect human well-being and natural system health. One influence of technology on human well-being is explained by Aidan Davison in his 2001 book *Technology and the Contested Meanings of Sustainability*:

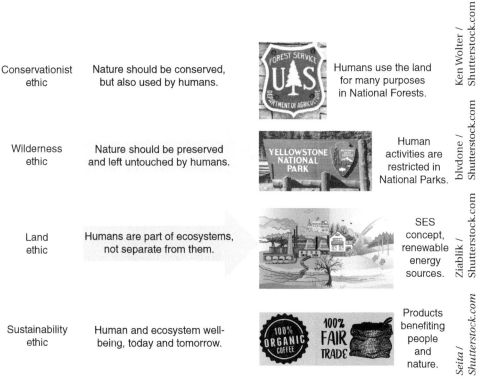

Figure 7.31 Ethics help guide decisions and actions, and have led to real changes in the world.

The fulfillment of the promise of liberation in everyday practice is problematic because we are simultaneously liberated from what burdens us and from what we care about.... Devices undermine our relationships to those things, places, and people we want to be free to be able to cherish. (p. 111)

What he is getting at here is the fact that technology brings us comfort, control, and security in the face of uncertainty. However, at the same time, it can isolate us from people and nature. This separation and isolation can harm well-being, both physically and psychologically, as noted by many researchers, and as summarized in a 2008 book by Richard Louv called *Last Child in the Woods: Saving Our Children from Nature Deficient Disorder*. Technology can also harm well-being by giving some people power and control over others. This affects social equity because those without power can become marginalized from the greater society, or dependent on those in control, which engenders more power. Technology can also empower people in positive ways, such as by organizing community action through Facebook, or warning of flooding events via cell phone in rural areas where people do not have access to weather forecasts. Finally, technology can harm natural systems through increased resource extraction and pollution generation. For example, current agricultural systems depend on massive inputs of energy via fossil fuels, fertilizer, pesticides, and water. Not only does this result in resource extraction, but it also generates pollution harmful to ecosystems by leading to problems such as climate change and eutrophication. On the other hand, the solution to some of these problems may lie in new technologies, such as wind power generation with turbines. Technology is a double-edged sword because it can provide benefits to human societies and ecosystems, but also inflicts costs on them. As a result, the role of technology in future visions of a sustainable world should be carefully considered.

5. **What is the role of place in sustainability?** The importance of connecting sustainability to place when developing visions has already been touched on in each of the four preceding questions by emphasizing that values and beliefs about progress, definitions of human needs, notions of ethics, and the role of technology in the future must all be made specific to a given stakeholder group, place, and time period. Doing this emphasizes distinct local cultures, which promotes stakeholder "buy in" to a sustainability vision, and gives it "staying power" in a local community. It also embeds a vision in the uniqueness of each place, such that specific assets and constraints can be examined when devising transition strategies (Chapter 8). Visions can be general and global, or local and more specific. An example of a global vision statement (Figure 7.32) and a vision statement for a state (Figure 7.33) are shown here. Visions for cities and local municipalities would get even more specific.

> We urgently need a shared vision of basic values to provide an ethical foundation for the emerging world community.
>
> **I. Respect and Care for the Community of Life**
> 1. Respect Earth and life in all its diversity.
> 2. Care for the community of life with understanding, compassion, and love.
> 3. Build democratic societies that are just, participatory, sustainable, and peaceful.
> 4. Secure Earth's bounty and beauty for present and future generations.
>
> **II. Ecological Integrity**
> 5. Protect and restore the integrity of Earth's ecological systems, with special concern for biological diversity and the natural processes that sustain life.
> 6. Prevent harm as the best method of environmental protection and, when knowledge is limited, apply a precautionary approach.
> 7. Adopt patterns of production, consumption, and reproduction that safeguard Earth's regenerative capacities, human rights, and community well-being.
> 8. Advance the study of ecological sustainability and promote the open exchange and wide application of the knowledge required.
>
> **III. Social and Economic Justice**
> 9. Eradicate poverty as an ethical, social, and environmental imperative.
> 10. Ensure that economic activities and institutions at all levels promote human development in an equitable and sustainable manner.
> 11. Affirm gender equality and equity as prerequisites to sustainable development and ensure universal access to education, health care, and economic opportunity.
> 12. Uphold the right of all, without discrimination, to a natural and social environment supportive of human dignity, bodily health, and spiritual well-being, with special attention to the rights of indigenous peoples and minorities.
>
> **IV. Democracy, Nonviolence, and Peace**
> 13. Strengthen democratic institutions at all levels, and provide transparency and accountability in governance, inclusive participation in decision making, and access to justice.
> 14. Integrate into formal education and life-long learning the knowledge, values, and skills needed for a sustainable way of life.
> 15. Treat all living beings with respect and consideration.
> 16. Promote a culture of tolerance, nonviolence, and peace.
>
> The Earth Charter, 2000. Reprinted with permission by Earth Charter Center for Education for Sustainable Development at UPEACE. www.earthcharter.org.

Figure 7.32 The Earth Charter provides a vision for the entire planet and all of humanity.

Before moving on to the quality criteria section, three final points will be made about visions and the visioning process. First, the way that future visions are presented varies widely. The products that result can be in the form of statements, charters, or declarations, as shown in Figures 7.32 and 7.33. These are sometimes also expanded into stories or narratives of a future world, similar to what was done with scenarios for the WBCSD Biotechnology project and *Destino Colombia*. Narratives are often accompanied by illustrations, graphics, or photographs, in order to engage the larger public outside of the core stakeholder group. This is done to get feedback, gain "buy in," and build momentum for achieving the vision. The more representative the core stakeholder group of society at large, the more aligned the vision statement will be with the general public as a whole. This brings up the second point of *who to engage* in visioning. Participants should represent both

> **Vision for Western Australia state, AUS**
>
> **Foundation Principles**
> - **Long-term economic health.** Sustainability recognizes the needs of current and future generations for long-term economic health, diversity, innovation, and productivity of the earth.
> - **Equity and human rights.** Sustainability recognizes that an environment needs to be created where all people can express their full potential and lead productive lives, and that significant gaps in sufficiency, safety, and opportunity endanger the earth.
> - **Biodiversity and ecological integrity.** Sustainability recognizes that all life has intrinsic value and is interconnected, and that biodiversity and ecological integrity are part of the irreplaceable life support systems upon which the earth depends.
> - **Settlement efficiency and quality of life.** Sustainability recognizes that settlements need to reduce their ecological footprint (i.e., less material and energy demands and reductions in waste), while they simultaneoulsy improving their quality of life (health, housing, employment, community . . .).
> - **Community, regions, "sense of place," and heritage.** Sustainability recognizes the reality and diversity of community and regions for the management of the earth, and the critical importance of "sense of place" and heritage (buildings, townscapes, landscapes, and culture) in any plans for the future.
> - **Net benefit from development.** Sustainability means that all development, and particularly development involving extraction of non-renewable resources, should strive to provide net environmental, social, and economic benefit for future generations.
> - **Common good.** Sustainability recognizes that planning for the common good requires equitable distribution of public resources (like air, water, and open space) so that ecosystem functions are maintained and a shared resource is available to all.
>
> **Process Principles**
> - **Integration.** Sustainability requires that economic, social, and environmental factors be integrated by simultaneous application of these principles, seeking mutually supportive benefits with minimal trade-offs.
> - **Accountability, transparency, and engagement.** Sustainability recognizes that people should have access to information on sustainability issues, that institutions should have triple bottom-line accountability, that regular sustainability audits of programs and policies should be conducted, and that public engagement lies at the heart of all sustainability principles.
> - **Precaution.** Sustainability requires caution, avoiding poorly understood risks of serious or irreversible damage to environmental, social, or economic capital, designing for surprise and managing for adaptation.
> - **Hope, vision, symbolic, and iterative change.** Sustainability recognizes that applying these principles as part of a broad strtegic vision for the earth can generate hope in the future, and thus it will involve symbolic change that is part of many successive steps over generations.
>
> Government of Western Australia 2003, *Hope for the Future: The Western Australian State Sustainability Strategy*, Department of Environment and Conservation (DEC). Reprinted by permission.

Figure 7.33 This vision statement is specific to a single state in Australia.

dominant and minority perspectives, and be drawn from key organizations and institutions in society, such as government, the private sector, non-profit and intergovernmental groups, and community leaders. In addition to the knowledge of these practitioners, expert academic knowledge should be part of the process as well so that discussions of complicated technologies (e.g., genetically modified organisms, nanotechnologies) and natural and social systems (e.g., ecosystems, institutions) can be handled. There are criteria and tools that can be used to evaluate who should be invited to attend visioning workshops, as described in Section 3.3.1 of Chapter 3. Another key workshop participant is a facilitator, which brings up the final point.

In addition to considering who to engage, figuring out *how to engage* those selected is also crucially important. As a result, professional facilitators are often hired. The quality of the visioning product depends on the ability of participants, often with diverse perspectives and agendas, to work productively together. Facilitators ensure that discussions are constructive and respectful by promoting good communication and listening, and helping to resolve conflicts that arise. In addition to this, the facilitator makes sure goals and objectives are clear upfront. This is especially important for visions of sustainability because all stakeholders must understand what a sustainable system means, and agree that this will be the outcome. Sometimes during this process, stakeholders must be challenged to go beyond their own perspectives and align them with sustainability. This can be a delicate process, and facilitators help. Appropriate media and events should be used to engage stakeholders in non-technical and informal ways. Although scientists and government officials are accustomed to reading formal documents, visions need to be communicated to the public by other means, such as through documentaries, movies, and popular magazines. Creative events, children's activities, and interviews are used to get input from groups not in attendance at formal meetings. Sometimes, visioning is closely preceded by scenario analysis, such that the scenarios provide stakeholders with plausible futures to think about, and decide what they would and would not like to see happen in the future.

Section 7.2.3—Quality Criteria for Visioning

Key Concept 7.2.3—Ten criteria can be used to guide visioning in order to distinguish efforts that will be effective for addressing sustainability problems from those that will not.

Since the 1980s, many different visioning projects have been carried out, and some were specific to sustainability. These specific projects will not be discussed here, but the interested reader should consult the Wiek and Iwaniec resource *Quality Criteria for Visions and Visioning in Sustainability Science*, found in this chapter's bibliography, for more information. As already mentioned, no consistent methodology for visioning currently exists. However, ten quality criteria developed by Wiek and Iwaniec in the School of Sustainability at Arizona State University can be used to guide visioning, and distinguish effective efforts from those that may not be as useful for addressing sustainability problems. These quality criteria are described next. They are not mutually exclusive and overlap in many ways.

Criterion 1: Visionary. Visions describe an ideal future that is desirable based on judgments of how things ought to be. These judgments are rooted in values, ideals of progress, definitions of needs, and ethics, rather than descriptive facts founded on science and empirical evidence. Descriptive facts describe *what is* and, in sustainability, can be used to assess the current state of an SES, and how it might plausibly evolve into the future. Visions are about how the current state *should be* developed into the future. Although visions must account for *what is* so that they are based in reality (see Criterion 5), they are fundamentally about imagining a future based on desires, ideals, ethics, and preferences.

Criterion 2: Sustainable. The vision should be aligned with the principle of sustainability. As such, the preferences and desires of stakeholders must be passed through a "sustainability filter," or be guided by a "sustainability compass" (Figure 7.28), so that the vision promotes a future that preserves the conditions necessary for the survival of prosperous human societies by promoting healthy ecosystems, human well-being, and viable economies for today and for future generations.

Criterion 3: Systemic. Visions must view the future holistically by considering all system components, system relationships, causal chains, feedbacks, and the overall behavior of complex systems. The tools for doing this were described in previous chapters, and are used to holistically describe the system for which the vision is being developed. Thus, a vision should not be a static list of goals for the future, but a dynamic description of a future system.

Criterion 4: Coherent. Vision coherence is aimed at fostering maximum compatibility and consistency among goals, and minimizing conflicts. If goals conflict too much, and visions contain grossly inconsistent elements, then when the vision is translated into real-world strategies and actions, it may fail to materialize. For example, a city seeking to reduce air pollution envisions that new technologies, such as electric cars, will help solve their air pollution problem. As a result of this vision, the city offers tax incentives for people that buy electric cars and sets up charging stations around the city. The citizens living in the city are concerned about air pollution, but they are equally concerned about the effect of their sedentary life style on their health. Instead of continuing to drive, they want to solve the air pollution problem by walking and biking. The vision of the city, which is focused on using technology to reduce air pollution, is not coherent with the vision of it's citizens (Figure 7.34). Citizens would like to see abundant bike lanes, frequent bus lines, and an extensive light rail network. They do not want to buy another car nor do they want to continue to sit in traffic jams. It is unlikely that the city's vision would be achieved with

Figure 7.34 A city's vision of using electric cars to address air pollution is not likely to be successful, in practice, if the citizens of the city would prefer to bike and walk to work.

these incompatibilities, inconsistencies, and conflicts. Coherence is especially important in developing systemic visions (Criterion 3) focused on sustainability problems, where solution options must be developed for several different parts of the system simultaneously.

Criterion 5: Plausible. The meaning of *plausible* is the same as in scenario analysis, such that plausible futures describe those that could possibly occur, but are not necessarily the most likely. Plausible futures can arise from unanticipated events and surprises due to the inherent behavior of complex systems. This criterion is centered on the idea that visions should be based in reality or *what is*. The realism of a vision can be judged by comparing it to historical cases in which a similar vision was successfully implemented, or to some present-day effort in another place. If neither is available, a consensus by experts and practitioners with appropriate knowledge to judge may be enough to deem a vision realistic.

Criterion 6: Tangible. If visions are too vague or abstract, they will not be understood by stakeholders, nor will they be useful for devising strategies to move from the current state to the future vision. Broad visions are translated into goals (e.g., rich biodiversity) or targets (using indicators, e.g., extinction rates below ten species per million species per year) that make them more tangible (more on this in Chapter 8). Recall the development of sustainability indicators by the Slovenian stakeholders using AMOEBA diagrams from Section 4.2.2 in Chapter 4. They used indicators of environment, society, and economy to determine the current state of their system. Such indicators make visions more tangible. For example, instead of stating "The future environment will be healthy," which is a very vague vision of environmental health, the specifics of public waste removal and households sewer systems can be included in the vision. However, visions should not be so specific that they do not allow for wiggle room in devising solutions. Thus, waste removal and sewage system capacity are probably too specific. A better vision might read: "The future environment will be free of garbage and sewage waste." This makes it clear that pollution is the issue of concern when it comes to environmental health for this region, but leaves solution strategies open.

Criterion 7: Relevant. Visions must be relevant to the people for which the desired future state is being imagined. For example, a vision of roadways dominated by electric cars fueled by clean energy does not mean much to the average person. Relevance is promoted by knowing how real people will use these vehicles (will it affect me?), who precisely will be using them (will I be expected to use them?), what new responsibilities and roles will evolve (will I have my own charging station at work or will they be available only on a first come, first served basis?), how governance systems will be involved (will the city install a charging station in my home or will I have to buy one?), and the primary aims and objectives of others (are car companies promoting this technology because it is more sustainable or more profitable, or both?). Such visions are not only more relevant, but also more tangible (Criterion 6), complex (Criterion 3), and offer a chance to examine issues of social equity in the vision (Criterion 2).

Criterion 8: Nuanced. When first developing a vision, many different stakeholder perspectives and agendas are brought to the table. As the vision matures, they must be prioritized, with tradeoffs examined, so that visions can effectively guide and direct change in the face of limited resources. A nuanced vision fosters the development of a coherent vision (Criterion 4) because evaluating tradeoffs will help bring to light incompatibilities, inconsistencies, and conflicts.

Criterion 9: Motivational. Visions are meant to catalyze change through inspiration and motivation. They must have certain qualities in order to do this. First, they must be believable (Criterion 5). If you do not believe that something can happen, then you are less likely to be motivated to try. Second, they must create "Ah-ha" moments that change the way we see the world when pushed to confront our assumptions in light of new information. Such perspective-shifting experiences are exciting and can inspire action. Third, motivations come from seeing how the vision is relevant to you or your family (Criterion 7). Finally, visions communicated in interesting ways such as by stories, plays, and movies are more motivating and inspiring than those presented in dry, rigid formal reports or analytical pieces.

Criterion 10: Shared. At the end of the day, in order to move society toward a more sustainable world in solidarity, a common direction and vision for the future must exist. However, consensus among diverse stakeholders does not spontaneously emerge from the visioning process nor is there necessarily ever unanimous agreement. Such is the nature of a sustainability problem. However, the very act of collectively laying diverse perspectives and agendas on the table builds the social capital important to negotiation processes. A diversity of viewpoints also brings key insights and information to the table that no one person could have compiled alone. Collectively building a vision also promotes stakeholder "buy in" and "staying power" due to a sense of ownership and accountability.

Future scenarios and visioning are concerned with thinking about the future. Transition strategies, the topic of the next chapter, are about actually steering the future direction of our world toward sustainability by devising action plans for change implemented on the ground today. As a result, knowledge derived from all of the tools presented thus far, that help us think about the present day and the future, is used to inform transition strategies.

Bibliography

Ames, S. C. 1997. Community visioning: planning for the future in Oregon's local communities. Web link to original source no longer available.

Bradfield, R., G. Wright, G. Burt, G. Cairns, and K. van der Heijen. 2005. "The Origins and Evolution of Scenario Techniques in Long Range Business Planning." *Futures* 37: 795–812.

Brewer, G. 2007. Inventing the future: scenarios, imagination, mastery and control. Sustainability Science 2: 159–177.

Carpenter, S. R., and C. Folke. 2006. "Ecology for Transformation." *Trends in Ecology and Evolution* 21 (6): 309–15.

Chermack, T. J., and S. A. Lynham. 2002. "Definitions and Outcome Variables of Scenario Planning." *Integrative Literature Review* 1 (3): 366–83.

Davison, A. 2001. Technology and the contested meanings of sustainability. Albany, N.Y.: State University of New York Press, 281 pp.

Destino Colombia. 1998. "A Scenario-Planning Process for the New Millennium." *Deeper News* 7 (1): 131.

Hicks, D. 2002. Lessons for the future: the missing dimension in education. New York, Routledge/Falmer, 145 pp.

Holdrege, C. 2008. "Understanding the unintended effects of genetic manipulation." *The Nature Institute*. Accessed May 26, 2017. http://natureinstitute.org/txt/ch/nontarget.php.

Johnson, K. A., G. Dana, N. R. Jordan, K. J. Draeger, A. Kapuscinski, L. K. Schmitt Olabisi, and P. B. Reich. 2012. "Using Participatory Scenarios to Stimulate Social Learning for Collaborative Sustainable Development." *Ecology and Society* 17 (2): 9–21.

Kahane, A. 2004. "Colombia: Speaking Up." *Development* 47 (4): 95–98.

Kahane, A. 2004. Solving tough problems: an open way of talking, listening, and creating new realities. San Francisco: Berrett-Koehler, 149 pp.

Lempert, R. J., S. W. Popper, and S. C. Bankes. 2004. *Shaping the Next One Hundred Years: New Methods for Quantitative Long-Term Policy Analysis*. Santa Monica, CA: RAND Corporation. Accessed July 23, 2012. http://www.rand.org/content/dam/rand/pubs/monograph_reports/2007/MR1626.pdf.

Leopold, A. 1949. A Sand County almanac, and sketches here and there. New York: Oxford University Press, 1989, c1949, 228 pp.

Louv, R. 2006. Last child in the woods: saving our children from nature-deficit disorder. Chapel Hill, N.C.: Algonquin Books of Chapel Hill, 334 pp.

Meadows, D. H. 1996. Envisioning a Sustainable World. Third Biennial Meeting of the International Society for Ecological Economics,

October 24–28, 1994, San Jose, Costa Rica. (Last accessed 6/3/2016: http://donellameadows.org/archives/envisioning-a-sustainable-world/)

Nakicenovic, N., O. Davidson, G. Davis, A. Grubler, T. Kram, E. Lebre La Rovere, B. Metz, T. Morita, W. Pepper, H. Pitcher, A. Sankovski, S. Priyadarshi, R. Swart, R. Watson, and Z. Dadi. 2000. *Intergovernmental Panel on Climate Change, Special Report on Emissions Scenarios*. Accessed September 26, 2012. www.ipcc.ch/pdf/special-reports/spm/sres-en.pdf.

Nakicenovic, N., and R. Swart, eds. 2000. *IPPC Special Report on Emissions Scenarios*. Cambridge, UK: Cambridge University Press.

Newman, P., and I. Jennings. 2008. *Cities as Sustainable Ecosystems: Principle and Practices*. Washington, DC: Island Press, 8–30.

Parris, T. M., and R. W. Kates. 2003. "Characterizing a Sustainability Transition: Goals, Targets, Trends, and Driving Forces." *PNAS* 100 (14): 8068–73.

Pielke, R., T. Wigley, Jr., and C. Green. 2008. "Dangerous Assumption." *Nature* 452 (3): 531–32.

Pojman, L.P. 2005. Environmental ethics: readings in theory and application. Belmont, CA: Thomson/Wadsworth, 678 pp.

Popper, S. W., R. J. Lempert, and S. C. Bankes. 2005. "Shaping the Future." *Scientific American* 292 (4): 66–71.

Proverbs 29:18. *The Holy Bible, King James Version*. Cambridge Edition: 1769; *King James Bible Online*, 2016. www.kingjamesbibleonline.org.

Raskin, P. et al. 2002. Great Transition: The Promise and Lure of the Times Ahead. Report of the Global Scenario Group. (Last accessed 6/3/2016: http://www.greattransition.org/documents/Great_Transition.pdf)

Raskin, P. D. 2005. "Global Scenarios: Background Review for the Millennium Ecosystem Assessment." *Ecosystems* 8: 133–42.

Raskin, P., F. Monks, T. Ribeiro, D. van Vuuren, M. Zurek, A. A. Concheiro, and C. Field. 2005. "Global Scenarios in Historical Perspective, *Ecosystems and Human Well-being: Scenarios*, Vol. 2 of *Millennium Ecosystem Assessment*, 35–44. AccessedSeptember 24, 2012. www.millenniumassessment.org/documents/document.771.aspx.pdf.

Robinson, J. 2003. "Future Subjunctive: Backcasting as Social Learning." *Futures* 35: 839–56.

Schwartz, S.H. 2012. An overview of the Schwartz theory of basic values. Online Readings in Psychology and Culture, International Association for Cross-Cultural Psychology. (Last accessed 6/6/2016: http://scholarworks.gvsu.edu/cgi/viewcontent.cgi?article=1116&context=orpc)

Selin, C. 2006. "Trust and the Illusive Force of Scenarios." *Futures* 38: 1–14.

Swart, R. J., P. Raskin, and J. Robinson. 2004. "The Problem of the Future: Sustainability Science and Scenario Analysis." *Global Environmental Change* 14: 137–46.

Thompson, J. R., A. Wiek, J. Swanson, S. R. Carpenter, N. Fresco, T. Hollingsworth, T. A. Spies, and D. R. Foster. 2012. "Scenario Studies as a Synthetic and Integrative Research Activity for Long-Term Ecological Research." *Bioscience* 62 (4): 367–76.

UNDP. 2000. *Destino Colombia 1997–2000: A Treasure to Be Revealed*. Working Paper, Civic Scenario/Civic Dialogue Workshop, Antigua, Guatemala, November 8–10, 2000. Accessed September 23, 2012. www.democraticdialoguenetwork.org/index.pl.

WBCSD. 1997. *Exploring Sustainable Development, World Business Council on Sustainable Development Global Scenarios 2000–2050, Summary Brochure*. Accessed September 25, 2012. www.wbcsd.org/pages/edocument/edocumentdetails.aspx?id=143&nosearchcontextkey=true.

WBCSD. 2000. *Biotechnology Scenarios: 2000–2050: Using the Future to Explore the Present*. Accessed September 25, 2012. www.wbcsd.ch/DocRoot/BPlWgkkSZJUWEulZJMlk/biotech-scenarios.pdf.

Wiek, A., C. R. Binder, and R. W. Scholz. 2006. "Functions of Scenarios in Transition Processes." *Futures* 38 (7): 740–66.

Wiek, A., and D. Iwaniec. 2014. "Quality Criteria for Visions and Visioning in Sustainability Science." *Sustainability Science* 9: 497–512.

Wiek, A., L. K. Withycombe, V. Schweizer, and D. J. Lang. 2013, in press. Plausibility indications in future scenarios. *International Journal of Foresight and Innovation Policy*.

Withycombe, L. K. 2010. "Anticipatory Competence as a Key Competence in Sustainability Education." Master's thesis, Arizona State University, Tempe, AZ.

World Commission on Environment and Development. 1987. *Our common future*. Oxford: Oxford.

End-of-Chapter Questions

General Questions

1. Determine whether each of the following hypothetical situations describes a probable future or a plausible future. Explain why you picked the answer that you did.

    a. Part of your overall plan for earning money to attend college is to invest in the stock market. You diversify your stock portfolio by investing your money in two different types of stock: low-risk stocks that accrue money slowly, and high-risk stocks that tend to lose all of your money quickly but on rare occasions will rapidly increase your earnings. During the summer between your sophomore and junior year, one of your high-risk stocks goes through the roof and you now have all of the money that you need to fully pay for your junior and senior years of college.

b. Amphibians, such as frogs, salamanders, and toads, are at great risk of extinction. This is thought to be one of the greatest threats to biodiversity worldwide. The exact cause of amphibian extinctions is not known, and is an active area of research, but it has been tied to habitat loss and fragmentation, stratospheric ozone layer depletion, pollution, and disease. Currently, the amphibian extinction rate is thought to be more than 200 times higher than natural background rates of extinction. If trends in habitat loss and fragmentation, stratospheric ozone layer depletion, pollution, and disease continue into the future, amphibian extinction rates are expected to remain the same, or even increase, into the future.

c. Population growth is a major concern around the world. This is especially true in developing countries, where the majority of future human population growth is expected to occur. Education rates, especially of women, are thought to be a central factor in decreasing population growth rates in many developing countries. As a result, education and literacy programs for women have been implemented in these countries by nonprofit organizations around the world. In the regions where these programs have been implemented, birth rates have been declining. If these programs continue to educate women into the future, then many believe birth rates and population growth will continue to decline into the future.

d. The Earth's climate changes over short to long time scales as a result of three general factors. First, Milankovitch cycles alter the orientation between the Earth and the sun and, therefore, the amount of sunlight that reaches the earth. Second, changes in greenhouse gas concentrations, such as CO_2, CH_4, and N_2O, in the earth's atmosphere hold outgoing infrared radiation within the atmosphere and this warms the earth. Third, changes in the albedo of the Earth's surface result in more or less sunlight being absorbed or reflected. If an asteroid were to smash into the Earth's surface today, it would spew tons of dust and other particles into the atmosphere. This would alter the earth's albedo and prevent much sunlight from reaching the earth's surface for a two- to five-year period, changing the Earth's global climate in the future.

2. Use one or more of the general SES characteristics, as described in Section 7.1.1, to explain why predictive modeling alone is not sufficient for addressing the issues described in the scenarios below.

   a. Ponderosa pine forest ecosystems, and the people and local economies dependent on them, are at risk. These ecosystems exist on the Colorado plateau in the western United States. Their health and, ultimately, their long-term survival are threatened by many factors, including forest fire, the mountain pine beetle (MPB), and climate change. Due to past misguided forest management practices, many of these relatively dry forest ecosystems are tinderboxes of dry dead organic matter waiting to be ignited by a natural lightning strike, or a cigarette carelessly thrown out the window of a vehicle. Recent management practices, such as forest thinning, seem to be helping. MPB outbreaks in these forests can result in the decimation of hundreds to thousands of trees within short periods of time. The role of climate change in ponderosa pine forest loss is uncertain, and an active research area, but it is thought to contribute in two general ways: (i) milder winters allow more MPB larvae to survive, whereas previously colder and harsher winters kept MPB populations in check, and (ii) drought conditions make it more difficult for forests affected by fire, the MPB, or both, to regrow because it is more difficult for ponderosa pine seedlings to establish in drier soils. All of this makes the future of ponderosa pine forest ecosystems difficult to predict.

   b. As described in Box 2.4 of Chapter 2, the precise cause of the 2007–2008 Food Crisis is hard to pinpoint and is the subject of much debate. The five major factors thought to have contributed to the spike in food prices during this crisis include increased demand for biofuels, rising oil prices, climate change, more demand for meat, and market speculation. Anticipating another food crisis in the future is even more difficult than understanding the causes of one that has already occurred. Using the information provided in Box 2.4 of Chapter 2 regarding the five major factors that are thought to have contributed to the 2007–2008 Food Crisis, explain why predicting a food crisis in the future would be difficult.

3. Use the 2003 van Notten scenario typology to determine whether each of the following hypothetical scenario projects is (i) exploratory or decision support and (ii) intuitive or formal.

   a. A community of urban farmers gathers together to think about the future of food production in their city neighborhoods and what this might look like 50 to 100 years into the future. Prior to this scenario development project, urban farmers competed with each other at farmers markets. Afterwards, they found new ways to work together and share information about consumer demand for locally produced food, and new city regulations affecting their farming operations. In addition to including farmers as part of the scenario development project sessions, organizers also invited representatives from the city government, non-farming citizens, and large grocery retailers. By including these other groups to collectively come up with scenarios for the future of urban farmers in the city, all groups learned about barriers and opportunities for a viable local urban food production system and how they might benefit from this type of future. After scenario construction was complete, three different possible future scenarios—"Food Deserts Continue," "Wealthy Are Healthy," and "Sustainable Food City"—were shared with the city's general public through a local comic book sold at newsstands, and a play put on by a local urban theatre group.

b. The board of directors for an electric car company needs to determine how many electric vehicles to produce at its manufacturing facilities for the next year based on projected consumer demand, economic growth, material costs, and profit maximization. Several scenarios, ranging from worst case to best case, are developed to inform future strategies for dealing with these four factors as they change over time. Because electric cars are so new, useful data on sales, economic activity, and material costs are sparse. Thus, in order to think about the future and create scenarios to help inform their business strategies, the company invites several groups to the scenarios development sessions. These include vehicle consumers with a range of product preferences, economists, policy analysts, and engineers. Together with these groups, the business develops four possible scenarios, in the form of story narratives accompanied by catchy illustrations. They share these scenarios with business strategists in the company so that they can determine production levels for the following year.

c. The elected representatives of a federal government are in a heated debate near the end of a budgetary legislative session. Due to budget cuts, before the session ends they must make a choice between reducing subsidies to farmers or to oil companies for the next fiscal year. In order to gather more information about which choice to make, they ask economic and political analysts to determine future economic growth, unemployment rates, and re-electability for three different possible futures: one where cuts are made to farmers' subsidies, another where cuts are made to oil company subsidies, and a final one where partial but smaller cuts are made to subsidies for both. After developing the three scenarios, the analysts present them to the representatives in the forms of charts, tables, and graphs for comparison.

    d. A group of social scientists are gathered together for a scenario project in order to learn more about how people living in a metropolitan area might respond to different demand-side water conservation measures, if they were implemented in the future. A few of the biggest uses of water in households, and where some of the greatest potential for water conservation lies, are toilets, car washing, and lawn watering. Thus, they decided to focus on these three aspects of water use, and use game theory models to determine possible future human behavioral responses to water conservation. Each model had slightly different assumptions, but one underlying assumption is that people always act in their own self-interest. The project resulted in many tables and graphs for the social scientists to explore. By the end of the scenario project, all scientists felt they had more knowledge about human behavior within the context of water conservation. They all planned to use this new knowledge to inform their research on solutions to the region's water crisis.

4. Use the ten criteria for effective sustainability visions, as described in Section 7.2.3, to determine the quality of the hypothetical visions and visioning processes described below.

    a. A city has major toxic waste management issues that pollute local wetlands and cause health issues for people exposed to the chemicals. The city feels very strongly, on the basis of social equity and biodiversity preservation, that the polluting companies should pay billions of dollars to clean up the river until it is completely contaminant free. In order to come up with this vision, the city works together with the people affected by pollution and species conservation groups. They do not include the polluting companies, however. After all, these companies caused the problem, and they should not have a say in how they make amends. The group comes up with a vision that lays out a list of five goals to be accomplished by the polluting companies over the next ten years: (1) contaminants in the water are cleaned up using solid sorbents dispersed throughout the river channel; (2) contaminants in river sediments are remediated by dredging sediments from the river to a depth of 5 meters; (3) compensation for species losses comes in the form of a wildlife rehabilitation center, funded by the company and run by the species conservation group; (4) any citizen harmed by contamination sends his or her medical bills to the contaminating company; and (5) local fishermen who have ingested contaminated fish undergo a medical toxicity assessment, paid for by the company. They hand these goals over to the company, making it clear that the goals are non-negotiable and unchangeable no matter what the future may bring. This type of mandated cleanup has never been accomplished in a city before. Thus, the visioning group feels they are being rather innovative and revolutionary. The vision was heavy handed, and many in the visioning group felt the company may have been treated unfairly, but they forged ahead with the vision creation nonetheless.

b. A coalition of local urban school districts from across a geographic region come together to form a vision for the future health of school children. This vision is based on recent statistical data showing an alarming increase in obesity of school children over the past ten years. Their goal is to create a future in which all children have access to healthy food. They decide to pursue this goal by starting a farming program where students grow their own food as part of an afterschool program. This type of program worked in a school district in a nearby rural area, so they figured that it should work here too. No one really knows where they will get the money to fund this program, but they strongly believe in the saying: "where there is a will, there is a way!" The school administrators and teachers who came up with the vision are so excited that they keep the vision a secret from the school children. They decide to reveal their vision in a surprise musical that the children will all attend at the beginning of the next school year.

Project Questions

1. *Scenario Development.* Use information on drivers and indicators for your system, which you developed as part of the Project Questions in previous chapters, to sketch out plausible alternative future scenarios for your system 25 years from now. The scenarios should include a "snapshot" photograph of what the future will look like 25 years from now, as well as a "film" or "series of snapshots" detailing the key events that took place to result in the current state becoming the "snapshot" of the future 25 years later.

   a. Develop three scenarios for your system: worst case, best case, and middle ground. To do this, think about the different ways that drivers might shape the future. Think about what the future might look like, when shaped by these drivers, for each of the societal sectors in the table below. This table is a tool to help you structure your scenario building around five different societal sectors, as was shown in Table 7.2. To help you think about what the future may be like, use your knowledge of the system from your current state analysis to identify significant trends within the system today. If these trends were to continue, what could they lead to? For example, in the political organization sector, you might note an increase in government surveillance and intrusion into citizens' privacy. This might lead you identify an overt authoritarian police state incrementally developing in response to successive "security events" in the worst-case scenario, but in the best-case scenario, perhaps, it leads to mass protests that result in a new generation of "clean" politicians and the emergence of a more transparent and accountable government.

In this exercise, the type of scenarios you are building are somewhere between exploratory and decision support, and on the scenario design continuum you are using an intuitive approach as opposed to formal. Remember, scenarios should be plausible and realistic! Also, the scenarios should be relevant to the context of your problem, but you should keep them at a fairly general level of detail.

b. Next, identify at least three indicators that could help you track whether the system is changing toward each of the different scenarios you developed in part (a). Go back and look at the indicators you have developed for Project Questions in previous chapters. For each indicator, state what it is, and which sector it is used to track. Describe how you would expect it to change for each scenario. An indicator may be useful for one scenario but not another.

Example: Indicator: number of salmon returning to river each year—ecosystem health

- In the worst-case scenario, salmon would be expected to decrease and eventually disappear as pollution levels in the river increase.
- In the middle-ground scenario, salmon would experience an initial decrease followed by a recovery to suboptimal levels, but to levels that will maintain a viable population in the river.
- In the best-case scenario, salmon would be expected to increase and then stabilize at a healthy level as pollution is eradicated and the river is cleaned up.

	Economic Structure	Political Organization	Social Policy	International Situation	Ecosystem Health
Significant Current Trends					
Scenario 1 (Worst Case)					
Scenario 2 (Middle Ground)					
Scenario 3 (Best Case)					

2. *Vision Development.* Create a vision of a desirable future for your system in 25 years time. Use the ten criteria for an effective vision, as described in Section 7.2.3, to guide your vision development. To help you include the perspectives of many different stakeholders, role playing may be helpful. To help you be specific in your vision, imagining what a "day in the life" would be for each of the stakeholders may also help. Start by writing a two or three sentence "day in the life" vignette for each stakeholder.

> **Example:** Fisherman Stakeholder. *"Four days a week I go out in my boat and, unlike before, I always find big fish and lots of crab that bring money back into my community, but more importantly into my household. No longer are the fish small, and the job has become very reliable for me even though it is hard work. Today, I don't fish and I am working on the community farm, which the government helped us to establish and showed us new farming methods that are good for the environment as well as for the community."*

After creating the vignettes for each stakeholder, synthesize the different stakeholder perspectives, and other ideas you might have, into a general description of your overall vision. This should be three or four paragraphs in length. You might want to make each paragraph describe a different aspect of the system, such as in the scenario-building exercise. Another approach could be to describe how the different sectors (economic, government, social, etc.) come together to affect different aspects of people's lives, such as working, getting around (transportation), recreation, and eating.

Chapter 8
SUSTAINABILITY TRANSITIONS

Students Making a Difference

Urban Forestry and Orlando's One Person, One Tree Program

After completing his Ph.D., Braden Kay was hired by the City of Orlando. The city has a sustainability plan, and he was hired to implement parts of that plan. One goal for the city is to have 25% tree cover by 2040, which would promote greater livability and walkability. When Braden started his job, the city was at 16% tree coverage. One of the first things he did was to identify assets: what already existed that was working well? Orlando had a street tree program, where city workers planted trees along the streets. As part of this program, people could obtain trees from the city to plant in their yards. However, people did not know about the program because it was not well advertised. This was a barrier to the program's success. To overcome this barrier, they pursued two solution options simultaneously: they worked on getting more people on board with the program and also making it easier for people to get trees.

To get more people on board, Braden and his team needed to execute and facilitate many disparate and diverse activities simultaneously: "We needed to figure out these different points of intervention where we could get some energy going." In one instance, they worked with young professionals who went door-to-door. For this project, about 50 trees were planted over several months. In another instance, in an African American neighborhood, they worked with a local church, which got excited about tree planting and then worked to get their parishioners on board. In a third instance, in an affluent neighborhood, they worked directly with a grandfather whose granddaughter had died of cancer. Just before she passed away, she and her grandfather had gone door-to-door and got about 50 trees planted. She wanted to have a lasting impact on her neighborhood, no matter what happened to her. In her legacy, her grandfather continued the project with her sister and eventually got about 250 more trees planted. The grandfather became a champion for this cause and went with Braden and his team to other neighborhoods to get people inspired and excited about tree planting. In the end, Braden found that they really needed to tailor their approach, in each instance, to each specific neighborhood.

They also created a website to make it easier for people to get trees. People could order a tree using this website, but they could do much more. There was information about tree maintenance and other educational resources, including videos. In partnership with the Arbor Day Foundation, they helped people figure out how much energy savings they could achieve for their home by planting a certain type of tree in a specific location on their property. The homeowner could then use this information to pick the type of tree that they wanted and the tree would be delivered to their home.

About 5,000 trees had been planted in the first year, but at this rate, the goal of 25% tree cover by 2040 would not be achieved and many challenges remained. People that had lost trees in storms didn't want to replace them because they feared property damage, such as a tree falling on their house. Those that did want trees preferred fruit trees, which provided less shade and could lead to problems with rodents and disease. Public utilities were also not 100% behind the project because trees knocked down power lines. Braden had to adjust their strategies as a result. He advises that adjustments are an inherent part of sustainability transitions: "The thing that it is impossible to say enough is that you are constantly reiterating your goal and you are constantly learning more about the dynamics of the problem that you've been working on. As you march along, and create new experiments and new programs and new pilot projects, you're learning more and more about your current state, and you're learning more and more about your vision, and you're learning more and more about how to keep on pushing it through. It is really important to know that [guiding transitions] is an iterative process. It is a learning process, and it is a process that, as sustainability professionals, we have to be the motor for. There is no question that transitions are always a work in progress and require a lot of relationship building, resource gathering, and creativity." In this chapter, you will learn about these aspects of transitions as well as tools used by sustainability scientists and professionals to help guide transitions.

Core Questions and Key Concepts

Section 8.1: Understanding Transitions

Core Question: How can sustainability transitions be understood by studying past transitions?

Key Concept 8.1.1—Processes contributing to transitions occur at three functional levels that are distinguished from each other based on their relative degrees of structural stability and how this influences the role of agency in transitions.

Key Concept 8.1.2—Socioecological systems undergo different types of transitions, depending on the timing and nature of interactions among niche, regime, and landscape levels and also the magnitude, rate, and extent of landscape pressures on existing regimes.

Key Concept 8.1.3—Drivers, indicators, stability landscapes, regime shifts, adaptive cycles, and resilience are all concepts previously covered in this book that are relevant to understanding transitions.

Section 8.2: Guiding Sustainability Transitions

Core Question: How might SES transitions occurring today be guided toward sustainability?

Key Concept 8.2.1—The Multi-Phase Concept describes the various phases, as defined by their different dynamics and innovation adoption rates, that an SES experiences over time as it undergoes a transition.

Key Concept 8.2.2—Transition strategy building involves defining fundamental changes, assets, and barriers and devising strategic actions, interim targets, and indicators appropriate for each of the four MPC transition phases.

Key Concept 8.2.3—Intervention points are places in SESs that have powerful transformative potential, such that a small strategic action taken at one point in the system leads to an overall system transition.

Key Terms

transition
sustainability transition
multi-level perspective (MLP)
structuration theory
structure
agency
coevolution
innovation
negative externality
transition management (TM)
multi-phase concept (MPC)
unintended consequence
fundamental changes
assets
barriers
strategic actions
targets
intervention points
incentives
disincentives
self-organization

"The means must reveal the ends."

—Gandhi

The last chapter ended with a section about visioning, which is a process that allows stakeholders to collectively imagine a desirable and sustainable future that is different from the unsustainable current state. The vision that is developed during this process is the *ends* that we want to achieve, as referred to in the opening quote by Gandhi. However, the *means* by which the desired *ends* is actually arrived at is another matter entirely and is the topic of this chapter. Figuring out the *means* to the *end* helps answer the question: *How do we get there?* (Figure 8.1). In this question, "there" is the vision. "We" is left purposefully vague to allow for adaptation of the TSR framework to specific contexts around the world. Depending on the specific sustainability problem, and the stakeholders involved, "we" could mean the members of a small village, residents of a city, all people in a nation, or even our global society and humanity as a whole. The material presented in this book so far has provided you with ways to think about the current state (Chapter 3) and future scenarios and visions (Chapter 7), tools for determining the current state and tracking movement of the system into the future (indicators, Chapter 4), and key ideas about socioecological system (SES) complexity (Chapters 5 and 6). In this chapter, all of this is brought together to think about transition strategies (Figure 8.1). In essence, transition strategies are aimed at guiding present-day SESs toward more desirable and sustainable futures, while avoiding plausible yet undesirable futures.

The first section of this chapter focuses on understanding how societal transitions occurred in the past. It is useful to study the past as a basis for understanding the present and future. Observing societal processes and outcomes of past transitions that already have occurred can help in understanding what is happening today and how sustainability

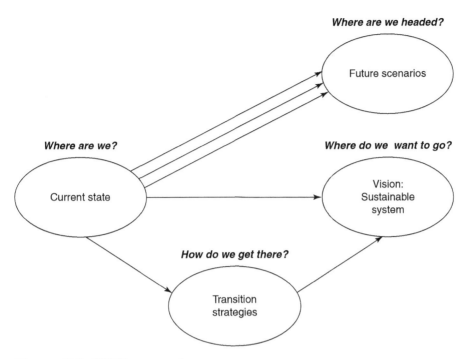

Figure 8.1 TSR Framework.

transitions might occur in the future. The second part of this chapter is about transition management, which is focused on present-day transitions and how SESs might be guided toward a more sustainable and desirable future. Much of the information in this chapter was drawn from a 2010 book titled *Transitions to Sustainable Development: New Directions in the Study of Long Term Transformative Change*. This book was a collaborative endeavor, incorporates ideas from a broad range of disciplines, and attempted to answer two questions: (1) How might transitions be understood? (2) How might transitions be shaped in the direction of sustainability? Section 8.1 provides options for answering the first question and Section 8.2 the second.

Section 8.1: Understanding Transitions

Core Question: How can sustainability transitions be understood by studying past transitions?

"If you stand back ... and look at the broad sweep of ... history, you see not a cumulating series of little fluctuations but just two great shifts: the agricultural revolution beginning around 10,000 years

ago, and the industrial revolution that began around 1750. This makes it more natural to think in terms of the preconditions for historically exceptional surges of innovation with progress coming in great storms rather than intermittent showers."

—Anonymous, The Economist,
July 8, 1999

> **transition** a profound change in the way an existing socioecological system functions to meet the needs of society.

> **sustainability transition** a profound change in the way an existing socioecological system functions to meets the needs of society *in ways that promote healthy ecosystems, human well-being, and viable economies.*

A **transition** is a profound change in the way an existing SES functions to meets its needs, in terms of both vitality and integrity (as described in Section 1.3.2). A **sustainability transition** aims to meet SES needs in ways that promote healthy ecosystems, human well-being, and viable economies. A transition can be thought of as a regime shift (as described in Chapter 5), such that when an SES shifts from one regime to another, the fundamental interactions among system components and the internal feedback structures change. Such changes result in an SES having different capacities to support societal needs after the shift than before. For example, recall the regime shifts described in Section 5.1: clear vegetated lake to turbid unvegetated lake, authoritarian government to democratic government, and grassland to desert shrubland. The clear vegetated lake supports biodiversity, provides food for people, and upholds economic activity. In contrast, once the regime shifts to a turbid unvegetated lake, it no longer supports these activities or supports them to a much lesser degree. The same is true when an authoritarian regime shifts to a democratic one, which better supports participatory decision making and equity, and when a grassland shifts to desert shrubland, which cannot support similar levels of cattle grazing and human well-being. The capacity of an SES to support certain needs arises from the interactions among system components and internal feedback structures, and is an emergent property of a societal system (as described in Section 6.1.2). SESs undergo transitions because they are adaptive (as described in Section 6.3) and respond to changing conditions, either in the external environment or internally.

How do SES transitions actually play out in the real world? This section's opening quote refers to two broad societal transitions from a preagricultural to an agricultural society and then from an agricultural to an industrial society. Historical transitions are important to study because it is the only way that an entire transition, from start to finish, can be examined. This section offers insight into how sustainability transitions might occur in the future by explaining how transitions occurred in the past. Did historical transitions occur as "exceptional surges" caused by "great storms" or were they more like "intermittent showers," as in the opening quote to this section? The general answer is: Probably a bit of both. A more specific answer is based in an understanding of the preconditions necessary for transitions, and the role played by innovation, which are both discussed next.

The information on transitions in this section was drawn from many disciplines, including sociology, evolutionary economics, and science and technology studies, and is collectively referred to as the sociotechnical approach to understanding transitions. The specific framework used for thinking about transitions, which comes out of this approach, is the **multi-level perspective (MLP)**, developed by Professor Frank Geels at the University of Manchester in the United Kingdom. The MLP claims that transitions result from interactions among processes occurring at three different functional levels: the niche, the regime, and the landscape. (The MLP transition landscape is not to be confused with a system's stability landscape, as described in Section 6.2.2). The three levels are part of a hierarchy (as described in Section 6.1.2 of Chapter 6) with a variety of niche subsystems composing or influencing a regime, and then many regime subsystems subsequently embedded within a landscape (Figure 8.2). The MLP is concerned with transitions occurring in regimes that meet specific societal needs, but are part of a broader societal landscape. For example, the regime subsystems in a society that meet specific needs include transportation, energy, waste, housing, health care, and food production (Figure 8.2). Today, these subsystems all operate under a landscape of globalization. In this section, the three functional levels are introduced, and then a description of different transition pathways

> **multi-level perspective (MLP)** a framework derived from the social sciences used for thinking about how transitions occur as a result of interactions occurring among processes operating at three different functional levels.

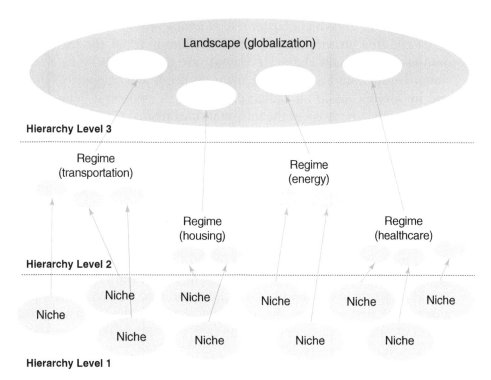

Figure 8.2 A landscape is made up of many different regimes, which are, in turn, composed of many different niches.

is given, along with an example of each. The pathways define a variety of patterns by which the three levels interact to ultimately result in a transition.

Structuration Theory. The MLP is rooted in a theory from the social sciences called **structuration theory**, which was proposed by British sociologist Professor Anthony Giddens in his 1984 book *The Constitution of Society*. Because this theory is the basis for many ideas about transitions presented in this section, a brief introduction is necessary before delving into the MLP framework. In the social sciences, there is a long-standing debate about the relative roles that **structure** and **agency** play during the establishment and propagation of societal systems. Structure refers to the factors that constrain human action, such as the formal and informal rules that are part of our social institutions. Agency refers to human action. The debate in the social sciences is concerned with the degree to which the action of an individual (agency) is carried out independently of constraining factors (structure). In other words: To what degree do humans have the ability to make their own choices, take actions free of constraints, and shape their own destinies within a given structure?

On one end of the spectrum are the structural determinists, who believe human action is fundamentally controlled and constrained by structure. From this perspective, outcomes of human action are predetermined by structural factors, such as formal or informal rules. For example, in the 2008 blockbuster film *Slumdog Millionaire*, a true story about children growing up in a Mumbai slum in India, the livelihood options for the main character (Jamal) are constrained by his social status. Throughout his life, Jamal meets his needs as a beggar, as a thief, and by serving tea at an Indian call center. A structural determinist would say that, because of the informal rules related to social status in Indian society, it is highly likely Jamal will continue to make a living in similar ways throughout his life and will remain at a low socioeconomic status.

At the other end of the spectrum are those who believe human action to be independent and free of constraints, such that any outcome of human action is possible regardless of existing rule structures. A social scientist with this view would point out that Jamal's actions, which included becoming a contestant on the Indian version of *Who Wants to Be a Millionaire?*, led him to a different future from one that would be predicted based on his social status, and related constraints, alone (Figure 8.3). Thus, the argument would be that human action (agency) operates independently of such constraints (structure) and that any outcome is possible.

The MLP takes the middle ground in this debate by giving preeminence to neither agency nor structure. Rather, it is based on the

---

structuration theory a social science theory that explains the establishment and propagation of societal systems as the continual and mutually influential interaction between structure and agency.

structure factors that constrain human action.

agency human action.

Figure 8.3 Is the future of a boy growing up in a Mumbai slum (*left*) pre-determined by structure or will he win a game show someday and become wealthy as a result of agency, as depicted by actor Dev Patel (*right*) in the movie *Who Wants to Be a Millionaire?*

notion that outcomes of human action result from continual interactions between structure and agency, such that each shapes the other over time through a process of coevolution. **Coevolution** means that structure and agency change together over time. Structural constraints are shaped by human actions and human actions are, in turn, guided by continually evolving constraints. In *Slumdog Millionaire*, Jamal correctly answered the game show host's questions on *Who Wants to Be a Millionaire?* Because of his social status, he was accused of cheating and questioned by the police. The suspicion of his cheating was based on informal expectations or rules related to his social status: People simply did not believe he was capable of knowing the answers, so they concluded he must be cheating. He eventually was allowed to continue on the game show and won a lot of money. This outcome may have changed structural constraints, such that they will be different in the future. For example, the outcome of his actions may have changed the expectations of Indian society regarding the capabilities of people from his social class. If another person who grew up in a Mumbai slum were to become a *Who Wants to Be a Millionaire?* contestant in the future and correctly answer questions, then that person might not be accused of cheating because expectations have changed. The MLP framework considers how structural constraints shape human actions at all three functional levels: niche, regime, and landscape (Figure 8.2). When human actions, or other significant events (e.g., natural disasters), result in outcomes that fundamentally transform structural constraints, a transition can occur.

> **coevolution** the continual process by which structure and agency change together and shape each other over time.

Section 8.1.1—Niches, Regimes, and Landscapes

Key Concept 8.1.1—Processes contributing to transitions occur at three functional levels that are distinguished from each other based on their relative degrees of structural stability and how this influences the role of agency in transitions.

The MLP is focused on how change happens through "innovation," as referred to in the opening quote to this section. The MLP is especially concerned with how innovation is inhibited or promoted by existing structural constraints. **Innovation** refers to development of new technologies. Technology is defined here as it was in Section 3.3.2, such that it denotes physical objects (e.g., tools, crafts) and also knowledge systems or ways of doing things (e.g., techniques, methods, systems of organization). Technology significantly affects the ability of human societies to meet their needs and adapt to changing environments. According to the MLP, the three different levels at which change processes happen during a transition, as SESs adapt, are niches, regimes, and landscapes. Niches are where innovation originates. If an innovation becomes well-established and incorporated into the existing regime, a transition may occur.

Agency and structure exist at all three levels, but to varying degrees. The role of agency is most prominent at the niche level, but it also exists at the regime and landscape levels. Structural constraints are stronger at the regime and landscapes levels, which are viewed as the overall constraints within which novel innovation originates and attempts to emerge from the niche level into a mainstream regime. Constraints at the regime and landscape levels, as well as the readiness of a niche-innovation for emergence, influence the success or failure of an innovation to become well-established. The three functional levels differ in terms of the stability of their structures (y-axis in **Figure 8.4**), with niches being the least stable, landscapes representing the most stable, and regimes falling somewhere in between.

It is important to emphasize that these levels are *functional*, rather than spatial or temporal as emphasized in other chapters. They are defined in terms of the function or role that each plays in a transition. Although functional levels do have spatial and temporal characteristics that are important to consider, these levels are used to understand transitions by paying attention to *their role or purpose in a transition*. Another way to think about this is that each level has a specific job as part of the overall transition. Each specific job contributes to overall SES functioning as the system does, or does not, undergo a transition. Thus, using the terminology of this book, each level contains distinct

---

innovation a new technological development, including both physical objects and knowledge systems.

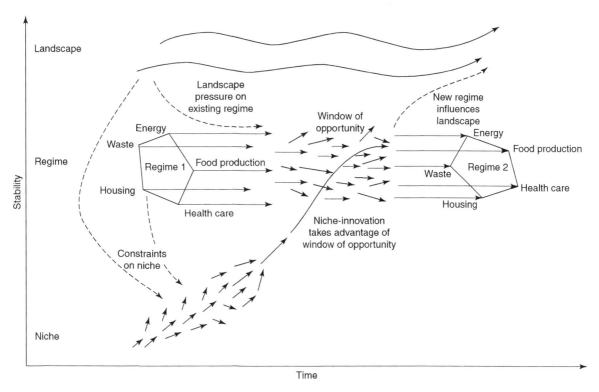

Figure 8.4 Landscapes, regimes, and niches change overtime and exhibit different degrees of stability.

system components that interact in different ways to contribute to the overall functioning of the system as it undergoes a transition (or not). The components of each functional level are analogous to the different organisms that contribute to overall functioning in an ecosystem through their varied roles (e.g., pollinators, decomposers, primary producers, secondary consumers) or to different people at a university with distinct jobs that support overall university functioning (e.g., students, teachers, administrators, researchers).

Niches. The least stable functional level is the niche, where structural constraints on human action are minimal. As a result, innovation is most likely to originate in niches. When conditions are right, fringe innovations become well established in existing regimes. This ultimately may culminate in a transition during which the innovation becomes an integral part of a new post-transition regime. Although innovation development at the niche level is a continuous process, only rarely does an innovation become well established in the dominant regime as part of a transition. The niche phase often lasts a long time, such as 10 to 30 years, until an innovation becomes stable and viable enough to compete with the better established components that dominate an existing regime. Even when an innovation is viable, stable, and ready to be introduced into the mainstream, the processes going on at the niche level must align with "windows of opportunity" that open at the regime level (Figure 8.4). These windows are often a result of pressure put on the regime from the landscape level, as shown in Figure 8.4, but there are other ways niche innovations become incorporated into regimes, and these are discussed in detail in Section 8.1.2.

Until a niche-innovation establishes in a dominant regime, it is inherently unstable. Niches consist of small, insecure, and uncertain social networks that link an array of actors keen on taking risks, such as innovators, entrepreneurs, and pioneers. Human actors that are part of such networks must exert great effort to sustain them. This is because the formal and informal structures that define interactions among actors within these networks, and promote stability, security, and certainty, are not well-established. This makes coordination among niche actors challenging and, therefore, it is more difficult for innovations to survive. However, certain processes operating at the niche level promote survival of emerging innovations through protective measures. Such measures arise when devoted actors within the niche allocate resources to support the innovation. This helps counter pressure from mainstream regime actors who may oppose an innovation, such as the fossil fuel industry opposing renewable energy. Another source of protection might be a government subsidy program providing support for a niche innovation, such as for renewable energy technology research. Such a subsidy comes from actors in the dominant regime who would like to see an innovation become well-established in an existing regime. Thus, protective measures promote innovation survival, whereas competitive pressures from the dominant regime can resist an innovation's establishment as part of the mainstream (Figure 8.5).

Despite their inherent instability, niches play a very important role in transitions by acting as the primary source of innovation. Continual development of innovations in niches is an ongoing process. It provides a diverse pool of new technologies from which an SES can draw to help it adapt to changing conditions. For example, if fossil fuels someday become prohibitively expensive for most people to use, then a diverse pool of renewable energy innovations will

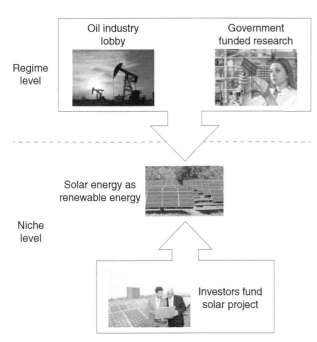

Figure 8.5 The success of an innovation introduced at the niche level, such as solar energy, depends on actions taken to support it, such as government-funded research into solar panels (regime level) and financial investments in solar energy projects (niche level), and also actions taken to oppose it, such a lobbying by the fossil fuel industry (regime level).

provide options for the future. (Recall from Chapter 5 that diversity bolsters adaptive capacity, or the ability to renew, reorganize, and learn when faced with change.) Innovations developed in niches could emerge to influence SES functioning in a way that meets new societal needs. To ensure adaptive capacity in SESs, niche innovation should be supported and maintained by promoting certain processes at the niche level. These include building protective social networks for niche-innovation, encouraging continual learning to bolster performance of new innovations, developing niche visions to guide coherent development, and continually working to draw attention and resources (e.g., funding, support) from within the niche, or from the dominant regime, to foster its survival and promotion.

Regimes. Relative to the niche and the landscape level, regimes are moderately stable. Along with landscapes, regimes constitute the structural constraints within which innovation happens. This is indicated by the dashed arrows in Figure 8.4. Compared to niches, constraints on human action in regimes are stronger as a result of broader, more established, and more secure social networks through which actors are linked. Less risk can be taken within regimes, and innovation is not as easily accomplished. When innovation does occur, it is more incremental and gradual in nature compared to radical niche-innovations. Unlike in niches, less effort is required to sustain the more stable social networks of regimes and protective measures are not as critical. This is because the formal and informal rules guiding relations among different actors are better established and foster a higher degree of security, stability, and certainty regarding outcomes of human action. Transitions researchers often refer to the factors contributing to regime stability, through factors influencing human actions, as cognitive, regulative, and normative rules. To maintain consistency in this book, structural constraints on human action are given as the same factors shaping stakeholder behavior as presented in Section 3.3: values and beliefs, informal social norms, formal rules and regulations, and available resources and technologies.

It is precisely the stability of regimes that make sustainability transitions difficult. Because regime structures are so stable, it is hard to make changes. For example, in the desert Southwest, rammed earth homes are more energy efficient and require fewer resources than homes constructed with traditional materials (Figure 8.6). If a person wanted to build a rammed earth home, he or she would need to have the building plans approved by their municipality to ensure that the minimal structural requirements for homes in the area are met (e.g., ability to withstand an earthquake, a fire, an ocean storm surge). An engineer would need to assess the structure and give his or her stamp of approval. Some engineers might not know enough about rammed

Figure 8.6 Rammed earth homes (*left*) are an innovation that makes sense in hot climates, but they are hard to build in practice due to the structural constraints present in the dominant regime, which is geared toward building homes made of traditional materials (*right*).

earth homes to assist with this process, or may not want to risk their reputation on a structure they are less familiar with if something goes wrong. These barriers could prevent the homebuilder from constructing a rammed earth home at all or add costs to the venture, such as time, effort, or money, in seeking out an engineer with the appropriate experience and willingness. Home construction materials and building designs have been standardized based on those commonly used in the past. Such regime structure stability can make it difficult for innovations, such as rammed earth homes, to survive and ultimately make transitions to new and different regimes challenging.

When the stable structural constraints of a regime break down, however, windows of opportunity for change open (Figure 8.4). Structures break down when some natural event or human action disrupts them in a fundamental way. At this time, innovations previously present only at the niche level can be incorporated into the new regime and become part of the new dominant structure. For example, green building concepts have been around for a long time, but only recently have become popular as "windows of opportunity" have opened. Using sunlight energy for electricity is a major idea in green building today, but this idea has been around for a while. During the Industrial Revolution, physicist Henri Becquerel first observed conversion of sunlight into electricity. In the late 19th and early 20th centuries, solar power was used for steam power generation. However, it did not become part of the dominant regime at the time, especially with the rise of energy-dense fossil fuels. People were not aware of what we realize today about fossil fuels: They are nonrenewable, and so would eventually become scarce, and they cause climate change and other forms of pollution (e.g., acid rain, photochemical smog). At the time, fossil fuels were viewed as a beneficial and highly concentrated energy source that could fuel industrial processes. With climate change, fossil fuel pollution, and scarcity looming on the horizon, solar power has crept into mainstream energy regime subsystems over the past several decades. During the 1970's oil crisis, scarcity was sharply perceived by the general public for the first time and solar panel use in green building started to take hold. Today, more government subsidies and support of other types, such as LEED (Leadership in Energy and Environmental Design) certifications, exist for developing solar power and green building capabilities.

Landscapes. Landscapes are the most stable of the three function levels and, like regimes, constitute the structural constraints within which niche-innovations occur (Figure 8.4). (The MLP landscape should not be confused with a system's stability landscape, as described in Section 6.2.2.) The MLP landscape represents the broad long-term processes external to a regime that influence it in a variety of ways. The allogenic drivers (Section 6.2.2) that regulate regime subsystems originate in the MLP landscape and influence a regime's stability landscapes over time in the form of either slow drivers (e.g., grassland cattle grazing, globalization) or fast drivers (e.g., meteor impact, nuclear war). This influence is not one-way from landscape to regime and niche, however, as indicated by the dashed arrow going from the regime to the MLP landscape in Figure 8.4. As already mentioned here and in Chapter 6, the conditions external to a system can be shaped by what is going on inside the system (dotted left-pointing arrow in Figure 6.12). The reason for the dotted left-pointing arrow in Figure 6.12 (rather than a solid arrow) is to emphasize that external system conditions (here, the MLP landscape) are not influenced by human actors in the short term, with some powerful exceptions such a nuclear war or a global financial crisis. This type of thinking about coevolution among landscape, regime, and niche levels is aligned with the middle ground between structural determinism and free agency taken by the MLP, as described at the beginning of this section.

Regimes and landscapes both constitute structural constraints on niche-innovation, so what distinguishes the two from each other? The MLP landscape represents the broad constraints on niche-innovation happening within any of the regime subsystems in the broader societal landscape. Transitions occur within specific regime subsystems, such as transportation, energy, waste, housing, health care, and food production (Figures 8.2 and 8.4). All of these subsystems are part of the broader landscape, so any major long-term trend in a landscape influences all the subsystems part of it. In the green building example, larger changes in the landscape, such as gradually changing public perceptions about fossil fuel use, influenced both the energy and housing subsystems. Regime subsystems embedded in the landscape are continually interacting with and influencing each other. The energy subsystem interacted with the housing subsystem to result in the increased use of household solar power. In Figure 8.4, each of the five corners of the two pentagons representing the two different regimes can be thought of as five different subsystems interacting and continually evolving together in a given regime through time.

The landscape is so broad, stable, and well-established that it is often taken for granted as the unchanging backdrop against which all human action takes place. For example, modern humans living today find it hard to imagine a much colder period in Earth history when glaciers covered the Earth's surface. This is because the last ice age ended more than 10,000 years ago. It is taken for granted that our current landscape of a warmer climate did not always exist (Figure 8.7). Similarly, most people living in developed countries today view electricity as a normal part of life and flip on light switches in the evening without a second thought. This was not the case in the past, and is still not the case in some less developed countries today. Despite the apparent long-term stability of landscapes, there are always periodic fluctuations, as indicated by the wavy lines depicting the landscape in Figure 8.4. When volcanoes, erupt, such as Mount Pinatubo in 1991, they temporarily cool global temperatures by small amounts. Power outages occasionally occur during hurricanes. Recessions sporadically occur within the landscape trend of continued economic growth.

Figure 8.7 Landscapes are so stable and well-established that we take them for granted, such as the fact that we presently live in an interglacial period and that much of our productive land, used to grow food and raise cattle today (*left*), was unproductive under huge sheets of ice in the past (*right*).

Section 8.1.2—Transition Pathways

Key Concept 8.1.2—Socioecological systems undergo different types of transitions, depending on the timing and nature of interactions among niche, regime, and landscape levels and also the magnitude, rate, and extent of landscape pressures on existing regimes.

This section describes four major pathways by which transitions occur. The pathways are idealized, since no transition will fit perfectly into any one. Nonetheless, they are useful tools for conceptualizing transitions. Each pathway is described generally and an historical example of each type is given in **Boxes 8.1** through **8.4**. The pathways are distinguished from one another based on the *timing of interactions* among the three functional levels, the *nature of the interactions* among those levels, and the *types of landscape changes* occurring at the time (**Table 8.1**). *Timing of interactions* is when landscape pressure on a regime occurs relative to the readiness of an innovation. Landscape pressure opens up "windows of opportunity" in regimes through which niche-innovations can enter to become well-established in the dominant regime. If an innovation is not fully developed when a "window

Table 8.1

	Pathway #1: Transformation	Pathway #2: De-, Re-Alignment	Pathway #3: Substitution	Pathway #4: Reconfiguration
Timing of interaction	Niche-innovation not ready	Niche-innovation not ready	Niche-innovation viable, stable, developed	Niche-innovation viable, stable, developed
Nature of interaction between levels	Pressure from outside groups and broader landscape, but not a window of opportunity; Symbiosis with niche innovations that conform to present regime; Strong regime structure stabilize	Sudden and large pressure opens window of opportunity; Competition among multiple niche innovations; Dominant innovation stabilized in new regime as additional regime subsystems form around it	Pressure opens window of opportunity; Niche-innovation competes with regime actors depending current system; New regime adjustments following substitution stabilize new regime	Symbiotic incorporation of niche-innovation into existing dominant regime; Competition among regime actors for best outcome as components of niche-innovations are blended with old regime components
Type of landscape change	Disruptive	Avalanche	Specific Shock, Disturbance, or Avalanche	Regular

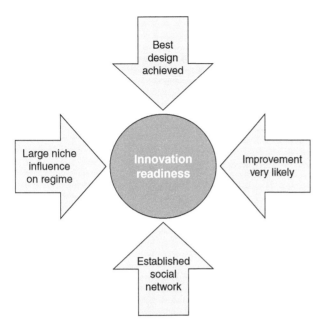

Figure 8.8 The arrows show four indicators that are used to assess whether an innovation developed at the niche level is ready for introduction into the dominant regime.

of opportunity" opens, it will not be able to take advantage of the open window that eventually might close.

Although regime actors and niche actors may disagree about whether an innovation is ready for introduction in the dominant regime, four general indicators have been proposed to assess readiness (**Figure 8.8**). First, consensus must exist among niche-innovation developers that the best possible design for an innovation has been achieved. Second, powerful actors who protect, fund, and provide other resources for promoting innovation survival must be part of the social network surrounding that innovation. Third, there are strong informal expectations for improving the innovation once it becomes part of the dominant regime, such as lower prices or improved efficiency of a renewable energy technology. Finally, the innovation is from a niche that has a large enough influence on the regime, such as renewable technologies having a greater than 5% market share.

The *nature of the interactions* among the three functional levels is concerned with how the different levels interact with each other: stabilizing, pressure, competition, or symbiosis. For example, if the broad societal landscape is dominated by public beliefs that climate change is not occurring, or is not a problem, then this landscape trend stabilizes regimes dominated by energy subsystems reliant on fossil fuels, transportation subsystems with fossil-fuel-powered vehicles, and industrial-scale mechanized food production systems run on fossil fuels. However, if public beliefs shift toward concern about climate change, then this landscape trend could exert pressure on regime subsystems that results in change. These changes could include incorporation of niche-innovations that favor energy subsystems focused on renewable energy, transportation subsystems with electric cars, and smaller-scale food production systems based in local communities. Depending on the situation, regime actors might resist niche-innovations through competition or form symbiotic relationships with innovation if they are open to or in need of them.

Four major *types of landscape changes* characterize the different transition pathways discussed (**Table 8.2**). *Regular* landscape changes do not exert intense pressure on regimes. Rather, regime changes occur gradually and are small in magnitude and extent. Landscape changes that are *specific shocks* are high magnitude and occur rapidly, but happen rarely and the effects dissipate over time. Thus, they do not

Table 8.2 Types of Landscape Change

	Regular	Specific Shock	Disruptive	Avalanche
Magnitude	Low	High	High	High
Rate	Slow	Fast	Slow	Fast
Influence	Low	Low	Low	High

continue to influence a system after the initial shock and the overall effect on the system is low. *Disruptive* landscape change is similar to specific shocks, except they occur more slowly over longer periods of time. Finally, *avalanche* landscape pressures result in the highest extent of influence, occur very rapidly, and are high magnitude. The transition pathways discussed next assume that some sort of pressure from landscapes is required for a transition and regime shift to occur.

Pathway 1: Transformation Pathway. For this transition pathway, disruptive landscape pressure influences a regime, but niche-innovations are not yet ready. Although niches do not influence the regime, because they are not ready, regime actors still respond to disruptive landscape pressure by adjusting the direction of their regime's development path and modifying their activities related to innovation. The magnitude of the disruption is high in this case because there are so many groups exerting pressure on the regime to change. Disruptive landscape pressure comes in the form of outsider groups bringing attention to the **negative externalities** generated by regime activity. Such groups include grassroots movements mobilizing public protests that compel governments to develop stricter regulations, expert groups (e.g., engineers, scientists, economists, sociologists) using their specialized knowledge to condemn malfunctioning technical features of dominant regimes and to suggest alternatives, and entrepreneurs and innovators proposing and developing alternatives to the status quo that cause dominant actors to rethink and reorient their current activities.

Although the size of the disruption is high, as a result of the numerous outside actors putting pressure on the regime, the rate at which regime change occurs is gradual and the overall effect is low. Gradual changes occur through a haze of conflict, disturbance, and struggles for power. Regime structures are changed in small ways by groups exerting pressure, such as social groups influencing governments to enact new rules and regulations. However, many small changes come from within, such that the overall regime structure is not altered dramatically or to a great extent. Instead, regime actors draw on their own adaptive capacity for dealing with change, as they adjust their activities to reorient the overall

> **negative externality** a cost incurred by an individual or group who did not agree to the action that caused the harm either because they did not have a choice or their interests were not taken into account.

development path of the regime. As such, the regime does not change dramatically, but a new one gradually grows out of the old one through small adjustments over time. New external actors do not replace current actors. Rather, regime actors persevere through the adjustments by slowly bringing in new knowledge that does not greatly conflict with their regime.

BOX 8.1 TRANSFORMATION PATHWAY: WASTE SYSTEM TRANSITION IN 19TH CENTURY NETHERLANDS

From the late 19th through early 20th century, the Netherlands went through a transition from unsanitary cesspool sewage systems to more modern sewage systems. Prior to the transition, people deposited human waste in household cesspools that leaked liquid waste into public water supplies and waterways, causing unsanitary conditions (Figure 8.9a). Private waste collection companies were contracted by households to empty cesspool tanks a few times a year. Human health was not the concern of city governments at the time. Instead, people were expected to look out for their own health. As such, public policies related to human health and waste disposal were nonexistent. Central regime actors included city governments, public works departments, citizens living in the cities, and private waste collection companies. The transition gradually began in the mid-19th century when doctors in the medical community first recognized that disease was highly correlated with cesspool waste and contaminated water. They acted as an expert group criticizing waste disposal practices using their specialized knowledge. However, medical theories based on microorganisms and disease had not yet been developed. Municipalities responded by implementing small changes, such as increasing water circulation in canals, dredging waste from canals more frequently, and even closing canals in some cases. The latter was met by resistance from some regime actors, such bargemen and traders, who relied on canals to do business. It was not until the late 19th century that several major events converged to add pressure to the regime: Increased urban populations made waste problems worse, the cholera outbreak of 1866–1867 led to the creation of a Dutch National Drinking Water Commission, public perceptions of waste were

(a) Unsanitary cesspool

(b) Modern sewage treatment plant

Figure 8.9 Sewage systems in the Netherlands gradually transitioned from unsanitary systems to modern treatment plants.

changed when Dr. Louis Pasteur discovered the link between disease and microorganisms in drinking water, a powerful coalition formed between doctors and engineers who saw new job opportunities with public works departments, and increasing concern for social equity was linked to health of the poor living under filthy conditions.

City governments in the Netherlands responded to this pressure in a few different ways. They continued to rely on the solution of improving water circulation. They also began to experiment with two different dry collection systems—the barrel system and the pneumatic Lierner system—but these systems still required regular emptying. Farmers and agriculturalists liked the barrel system because it provided fertilizer in the form of solid human waste, but medical experts and engineers opposed it because it was leaky and imprecise. The pneumatic Lierner system was cleaner, more precise, and still provided fertilizer, so it was supported by all groups. However, city government resisted it because it was complicated and expensive. In addition to dry collection systems, water-based sewer systems were discussed. However, at this point, there was not enough of a consensus for a waste system transition despite all the innovations available.

At the turn of the 20th century, waste problems worsened as urban population growth continued. Medical professionals and sanitation engineers continued protesting. Three cultural changes developing at the landscape level drove the regime even closer to transition. First, a cleanliness ideology emerged, such that the standard filth and stench of urban life was no longer viewed as an acceptable feature of a civilized, respectable, and virtuous society. Second, the movement for social equity and poverty reduction picked up momentum as public concern heightened as a result of the publication of several key novels, more abundant newspaper articles, and increasingly visible public protests. Third, democratic ideals developed and citizens now perceived an increased responsibility of governments to improve urban living and human well-being. In addition to these cultural changes, economic growth resulted in higher taxes on income, and sewer system projects became financially viable. New technologies in waste disposal (e.g., piped water systems made flushing easier) and agriculture (e.g., cheap artificial fertilizer eliminated the need to recover human waste for this purpose) converged with the more gradual landscape trends to finally result in a transition toward modern municipal sewage systems.

Pathway 2: Dealignment and Realignment Pathway. This type of transition is precipitated by a large and sudden avalanche-type landscape pressure with a wide impact. With this type of pressure, the informal and formal rule structures constraining human actions in an existing regime break down because of considerable internal difficulties, and the regime collapses. Regime actors do not remain part of the regime, because they have lost faith in the capacity of the regime to handle problems. This complete loss of dominant regime actors and constraining structures results in a void into which multiple niche-innovations enter. However, no one niche-innovation is viable or stable enough to become dominant in the new regime. As a result, there is an extended period of uncertainty as multiple niche-innovations coexist in an experimentation phase during which they compete for mainstream attention and available resources. Eventually, one niche-innovation is able to pick up momentum and establish dominance in the new regime. Once this happens, new actors and structures realign around the innovation and reinforce it, and it becomes a well-established feature of the stable new regime emerging from the transition.

BOX 8.2 DEALIGNMENT AND REALIGNMENT PATHWAY: TRANSPORTATION SYSTEM TRANSITION IN THE EARLY 20TH-CENTURY UNITED STATES

From the late 19th through early 20th century, transportation systems in the United States underwent a rapid transition from horse-drawn carriages to fossil fuel-powered automobiles (Figure 8.10a and 8.10b). Problems festering within the horse-based transportation system, including horse manure pollution, horse traffic congestion and safety issues, and rising costs, aligned with broader landscape developments to open a window of opportunity for new transportation options. These landscape developments included increasing public concern for health and horse manure sanitation issues, continued urban population growth compounded by rising immigrant populations living in filthy disease-laden slums, and a subsequent fleeing of the upper class to suburbs to escape these conditions. Transportation systems based on horses no longer met the needs of bigger cities and the need to travel longer distances to suburbs on their periphery. As such, they became very unstable and the transportation systems were primed for change. The window of opportunity was wide open. Despite the urgent need for new transportation options, no one niche-innovation was ready to replace horse-based systems. As a result, several contenders coexisted and competed with each other in an experimentation phase, until one eventually emerged as the new dominant transportation system: the automobile.

(a)
Library of Congress, Prints and Photographs, LC-USZ62-89691.

(b)
© Artens, 2013. Under license from Shutterstock, Inc.

Figure 8.10 When the innovation was finally developed, automobiles (b) quickly replaced the outdated and problematic horse-based transportation regime (a).

Early niche-level contenders were the bicycle and mechanically and electrically powered trams. Although originally used as toys for the wealthy in the 1830s, several decades later bicycles were widely used for recreation and practical purposes. Mechanical and electric tram innovations also existed, but the electric tram quickly came out ahead of the two: 16% of U.S. streets had electric trams in 1890 and, just 12 years later in 1902, 97% of streets had them. During that same time period, horse-based transportation dropped from 70% of all street transportation to less than 3%. The electric tram's fast speed, zero production of unsanitary waste, and support by powerful groups, such as tram companies that previously dominated horse-based transport markets, suburban real estate developers, electricity companies, and local government authorities, are the major factors that contributed to its success. Broad landscape developments also contributed to the success of electric trams. Immigration pressures led to more city slums and increased escape of the middle class to suburbs. Electric trams served suburbs much better than horses. The middle class itself also began to expand. Concern for public health and the unsanitary conditions created by horse manure continued. Electricity symbolized societal progress into a new age.

Together, the bicycle and the electric tram shaped the structure of a new post-horse-and-carriage regime, which laid the foundation for eventual automobile domination. The bike fostered a new public desire for independent transportation options and the novel practice of recreational bike touring. The Good Roads Movement arose to improve biking infrastructure, including smoother roads and traffic regulations for bikes that later morphed into automobile traffic regulations. The inexpensive mass transit system that emerged from electric tram popularity contributed to sprawl through even more suburbanization. Within cities, electric trams shifted public perceptions about the purpose of streets from public meeting places for socialization to arteries for transportation. People became accustomed to traveling at high speeds, as the average speed of the electric tram was twice that of a horse-drawn carriage. By 1920, the electric tram dominated urban transportation systems (Figure 8.11).

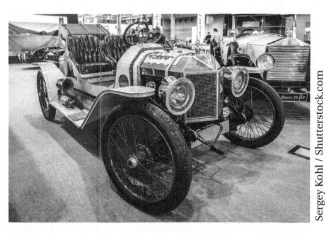

Figure 8.11 Electric trams, such as this old vintage tram in Lisbon, Portugal (*left*), dominated American streets for a short time in the early twentieth century, until cars powered by internal combustion engines took over, such as this Ford Model T (*right*).

(continued)

> However, the fate of transportation systems inside and outside of cities was yet to be determined. Automobiles emerged in many forms, including electric, internal combustion, and even steam-powered, and gained a strong foothold for specialized purposes such as taxis, luxury transport, racing, and touring. New cultural landscape trends related to increased leisure time, entertainment, and outdoor adventure led to preferences for internal combustion engines. This further propelled internal combustions autos toward market domination by 1905, as compared to electric and stream-driven cars. The affordable Model T pioneered by Henry Ford and the easier-to-use electric starter technology, as opposed to the unwieldy front-crank, pushed internal combustion autos even further into the mainstream. From 1910 through 1930, electric trams and automobiles competed for dominance. Several factors led to the automobile winning out in the end. The electric tram infrastructure became increasingly costly to maintain, which led to more crowded and less frequent trams. Government subsidies supported the automobile through publicly funded road improvements, whereas they taxed electric tram systems. Auto-based transportation was supported by policymakers as a means to increase suburbanization. Roads were improved and widened because of intense lobbying by the auto industry, whereas electric tram systems were left to further deteriorate. By 1930, the automobile was well-established as the new transportation regime subsystem and other subsystems in the new regime began to realign around it: Highway fast-food restaurant stops became part of food subsystems, drive-in movies were a new form of entertainment, and shopping mall complexes popped up at the edge of cities. Such structures and new actors further stabilized automobiles as the dominant transportation system and it continues to dominate U.S. systems today.

Pathway 3: Technological Substitution Pathway. As implied by the name of this transition pathway, a niche-innovation substitutes into the existing regime to replace what was previously dominant. Niche-innovations are capable of doing this because they are viable, stable, and developed enough to survive and overcome competition from the dominant regime (Figure 8.8). Prior to transition, the structures and human actors in the dominant regime are very stable and well-established. Small issues arise, but dominant actors do not pay much attention and address them through gradual innovation within the regime. They tend to ignore radical fringe innovations developed by actors in niches. Without any pressure from the landscape, business-as-usual continues to run its course.

Once pressure from the landscape arises, in the form of specific shock, disruptive, or avalanche, it is high enough in magnitude to result in critical tensions within a regime. These tensions open up windows of opportunity for viable and stable niche-innovations to enter. Diffusion of niche-innovations into the dominant regime increases as they gain increasingly larger and more established footholds. Rather than accepting and incorporating the innovation, dominant regime actors defend existing systems by investing in enhancements and improvements. As a result, competition and struggle ensue between dominant regime actors protecting the current regime and niche actors attempting to upend that regime with their innovations. Eventually, the niche-innovation wins out and substitutes into the old regime, old regime structures and actors experience a downfall from their dominant role, and a new regime forms around the niche-innovation made up of new dominant structures and actors.

BOX 8.3 TECHNOLOGICAL SUBSTITUTION PATHWAY: TRANSITION FROM SAILING TO STEAMSHIPS IN LATE 19TH-CENTURY GREAT BRITAIN

In the early 19th century, fleets of sailing ships dominated ocean transportation systems, for both freight shipment and passenger travel (Figure 8.12a). Sailing ships had problems, including slow and inconsistent speeds, uncertainty and unpredictability in arrival times because of high dependence on ocean currents and prevailing wind directions, and poorly coordinated trade over long-distance because of slow communication through letters sent by mail. Despite these problems, sailing ships continued to dominate travel and trade until the mid-19th century. At this time, new societal needs arose and a transition to steamships began (Figure 8.12b).

Although steamships were faster and more reliable than sailing ships, several problems still existed with steamships in the early 19th century. They required a lot of coal, which was expensive and reduced the cargo capacity. The paddle-wheel technology did not work well on rough ocean waters because it either submerged underwater or completely emerged out of water, which damaged engines and reduced wheel efficiency. The structure of wooden hulls warped under the massive weight of the boilers, condensers, and engines required for steam power. Despite these problems, steamships were pioneered in small niches within

(a)

(b)

Figure 8.12 Steamships (b) replaced sailing ships (a) when a window of opportunity finally opened.

(continued)

the dominant regime where they did work well, such as for inland waterway travel in the United States for westward settlement (such as on the Hudson River), as steam tugs pulling unwieldy sailing ships through crowded ports in Great Britain, and as a means to increase coordination of international trade within the British Empire by speeding up mail-based communications.

In the mid-19th century, a window of opportunity opened for steamships as landscape trends exerted pressure on the dominant sailing ship regime and improvements were made to the steamship. One major trend at the time was massive emigration to the United States because of disruptions and social hardships occurring in Europe, such as the Irish Potato Famine (1845–1849) and the 1848 political revolutions. This broadened ocean-based transportation markets for passengers. Another landscape trend was trade liberalization, which fostered competition among trading and shipping companies and also infrastructure changes, such as the Suez Canal in 1869. Sailing ships were not permitted to travel through the canal, but steamships were. The third major landscape trend was industrialization, with raw natural resources and manufactured products being shipped all over the world. Improvements in steamship technology, such as increased coal energy efficiency, paddle-wheel replacement by propellers, and iron rather than wooden hulls, merged with landscape trends to give steamships an advantage over sailing ships.

Between 1870 and 1880, steamships finally broke through to become the dominant form of ocean transportation. In addition to landscape pressure, shifts in regime structures also contributed to this breakthrough. Ports were deepened and enlarged to accommodate the larger steamships, port loading-unloading systems improved, and ship building was transformed with larger shipyards, novel machinery for working with iron, and new competencies of engineers and other skilled laborers. Although sailing ship companies attempted to defend themselves by improving and enhancing their product, including manufacturing bigger ships with higher cargo capacities and adding more sails to enhance speed, they experienced their downfall during this time and now remain only in niche markets. Sailing ships also failed to make the switch to iron, which, in the face of increasing wood scarcity for ship building, contributed to their decline.

Pathway 4: Reconfiguration Pathway. Similar to the Technological Substitution Pathway, niche-innovations in the Reconfiguration Pathway are stable, viable, and developed enough to be introduced into the dominant regime (Figure 8.8). However, unlike the Technological Substitution Pathway, there is no resistance from or competition with regime actors. Rather, innovations are readily accepted by regime actors as additions or replacements for some regime components. As such, the relationship between regime and niche is symbiotic and introduction of the innovation into the dominant regime is mutually beneficial for actors from both functional levels. Acceptance of innovations is primarily a result of economic drivers, such as the need for performance improvements or to resolve certain small issues within the existing regime. Thus, regime structures and actors remain largely unchanged when the innovation is first incorporated.

Over time, however, the adopted innovations lead to gradual changes in regime structure. Regime actors experiment with novel configurations among old dominant regime components and new innovations to figure out the best combinations. Thus, transition is not driven by emergence of a single new radical innovation through a window of opportunity, as in the Technological Substitution Pathway. Instead, an innovation is accepted into the new regime, and undergoes a series of renovations, as it is merged with and reworked alongside existing regime components, until it becomes a permanent part of the dominant regime. Regular landscape pressures drive this regime reconfiguration process. Eventually, a new regime emerges out of the old one, similar to the Transformation Pathway. However, unlike the Transformation Pathway, the old regime is reconfigured and restructured. Old regime actors survive in the new regime, but constraining structures are reworked and competition ensues among actors for developing the best combination of old and new as the system is reconfigured.

BOX 8.4 RECONFIGURATION PATHWAY: TRANSITION FROM TRADITIONAL FACTORIES TO MASS PRODUCTION IN LATE 19TH- AND EARLY 20TH-CENTURY AMERICA

The transition from traditional factories to mass production assembly lines came about through several bouts of innovation. The process started out by swapping various new innovations into the existing regime and culminated with a major system reconfiguration constituting the transition to a mass production regime. Such multistage change is characteristic of the Reconfiguration Pathway. In the 1850s and 1860s, traditional factory production was the status quo. Tools were general-purpose, machines were steam engine powered, and labor was largely unskilled. A certain type of factory building characterized this regime: multistoried, narrow, and very long. It was multistoried because technologies for continuous horizontal material movement, such as conveyor belts, were not yet invented and it was easier to move materials vertically with cranes. They were narrow because indoor electric lighting in buildings was not yet possible and natural lighting from windows was used instead. Line shafts distributed steam engine power all at once to the many machines in a factory because individual electricity-based plug-in technologies did not yet exist. Buildings needed to be long to accommodate line shafts (Figure 8.13). The setup of these factories presented several problems, such as poor lighting, inflexible setups with machines tied to line shafts, low energy efficiency, and substandard working conditions (e.g., safety, noise, dust).

(continued)

Figure 8.13 Before electrical plug in-style outlets existed, vertical line shafts distributed electricity to factory machines.

Several niche-innovations emerged in the 1870s that were incorporated symbiotically into factories to address these problems. To provide lighting and electricity for spaces too small to accommodate large steam engines, such as for lathes in jewelry workshops and drills in dentists' offices, small battery-driven electric motors were invented. In the canning and meat packing industries, conveyor belts emerged to allow for horizontal material movement in production lines. With the increased use of steel, tools for more specialized purposes and with greater precision arose. This sped up product assembly and the utility of horizontal conveyor belts increased. Also with steel came wider buildings with structures supported by longer horizontal beams made of stronger steel materials and concrete reinforced with steel bars. Such buildings could accommodate horizontal conveyor belts. Two landscape trends pressured incorporation of these new technologies into the old factory production system. First, production scales grew for many industries. As a result, the low energy efficiency and inflexibility of line shafts, which limited factory size and design, became urgent issues. Second, industrial engineers emerged as a new skilled labor group with a focus on production efficiency. This further fueled the movement toward factory redesign and incorporation of new technologies.

Between 1900 and 1910, small electric motor use really took off. They replaced traditional line shafts that, along with technological developments in conveyor belts and building construction and landscape trends, pushed traditional factory systems further toward systems of mass production. Individualized electric motors led to a reconfiguration of traditional power distribution systems to ones that were more flexible (for example, specialized tools could be swapped into and out of production lines as needed), had higher energy efficiency, and provided indoor electricity for lighting and fans for removing dust. The percentage of total factory power supported by electric motors in U.S. industries rose from 5% in 1900 to 25% in 1910. By the 1920s, industrial engineers had redesigned the size and layout of factories to maximize material flow with conveyor belts. The transition to mass production based on the assembly line model culminated with Henry Ford's fully integrated automobile production plant on the Rouge River in Dearborn, Michigan in 1920. This new mass production model merged niche-innovations, such as the electric motor, conveyor belts, and specialized steel tools, with the old factory model to meet new needs defined by landscape trends in industrial production efficiency and the rise of skilled professions (Figure 8.14).

Figure 8.14 Several bouts of innovation, and landscape pressures, eventually led to the dominance of mass production assembly lines in factories.

Section 8.1.3—Tying It All Together

Key Concept 8.1.3—Drivers, indicators, stability landscapes, regime shifts, adaptive cycles, and resilience are all concepts that are relevant to understanding transitions.

This brief section ties concepts from previous chapters to new ideas presented thus far in this chapter. Specifically, drivers (Chapter 3); indicators (Chapter 4); stability landscapes, regime shifts, adaptive cycles, and resilience (Chapter 5); and interactions between internal autogenic drivers with the allogenic drivers external to a system (Chapter 6) are all ideas relevant to thinking about transitions. When a transition occurs, the existing regime shifts to a new regime. You learned in Chapter 3 that drivers are the governing forces acting on an SES that either cause it to change or to remain in its current state. They are analogous to drivers of a car, which push the car forward, steer it in a certain direction, or keep it in the same place. Similarly, drivers push SESs toward sustainability, pull them away from it, or prevent systems from changing at all. In the MLP, which is used to understand how transitions occur, drivers are present at all three functional levels.

Allogenic and autogenic drivers were introduced in Chapter 6. When thinking about transitions, autogenic drivers can be thought of as operating within the existing dominant regime. They can be conceptualized as the momentum of an SES, which keeps it moving forward along a certain pathway of development. This momentum is governed by interactions and feedbacks among internal system components, such as those driving the forward movement of a forest along a certain development path during ecological succession. Similarly, the interactions and feedbacks among MLP regime components (e.g., structures and agency) govern the progression of a societal system down a certain development pathway. In the Netherlands waste system example (Box 8.1), prior to the transition people deposited waste in cesspool tanks, it was collected by private companies, and human health issues were not considered the purview of city governments. The interactions and feedbacks among structures and agency in this regime kept it on a certain development path. That is, until outside forces or allogenic drivers began influencing the system.

Recall that allogenic drivers are the forces external to an SES that influence its pathway of development. They can either set the system back along its development path or cause it to set off in an entirely new direction. In ecological succession (Section 6.2.2), a fire that burns newly established shrubs at the beginning of succession sets the forest back to an earlier successional stage. If a fire occurs and an invasive shrub is introduced after the fire that outcompetes native shurbs, then the forest development path could set off in a entirely new direction to become a completely different forest. In the Netherlands example (Box 8.1), the allogenic drivers that set the waste regime subsystem off in new directions toward a regime dominated by modern municipal sewage subsystems came from the landscape and niche level and also from other subsystems within the dominant regime. For example, the medical community, a subsystem in the dominant regime, discovered correlations between microorganisms and disease and used specialized knowledge to criticize waste disposal practices. Sanitation workers in the engineering community, another subsystem, formed alliances with doctors to advocate for more sanitary waste disposal practices. A cholera outbreak also occurred among the citizenry. From the niche level, new innovations emerged, such as piped water systems and cheap artificial fertilizer, making dry waste systems less desirable. Broader landscape trends contributed slow allogenic drivers that put pressure on the regime, such as population growth and increasingly filthy urban living conditions, changing public perceptions of waste and fostering a new cleanliness ideology, a social movement for increased equity, novel democratic ideals, and economic growth.

Allogenic drivers can change a regime subsystem's stability landscape over time, either gradually as slow drivers (e.g., grazing pressure in a grassland, Figure 6.23) or rapidly as fast drivers (e.g., a sudden destructive grassland fire, Figure 6.25). Recall from Section 6.2.2 that the shape of a system's stability landscape affects system resilience (Figure 6.21), which is its ability to absorb disturbances and still maintain the same fundamental structure. Therefore, if the resilience of a regime is weakened enough, a transition could occur. In Chapters 5 and 6, you learned that systems cross thresholds into new regimes as a result of landscape changes, fast disturbances, or some combination of both. Using the ball-and-basin model, and incorporating this into the MLP framework, Figures 8.15 and 8.16 show two examples of how stability landscape changes, and disturbances, might affect a regime subsystem during a transition to cause a new regime to emerge. Figure 8.15 shows how a regime subsystem can shift to a new one due

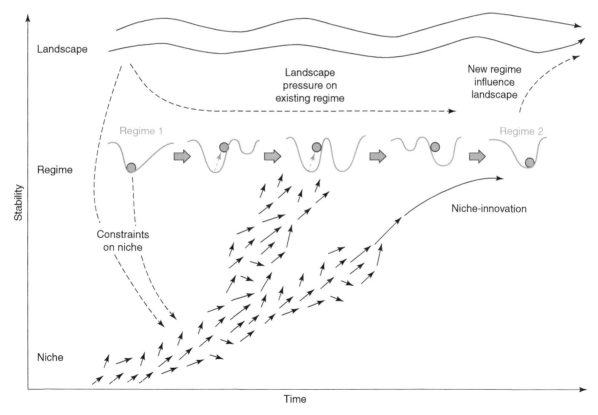

Figure 8.15 When the shape of a regime's stability landscape (curved blue lines) changes, and a new regime develops as a result, niche innovations can enter as the system (the ball) transitions to a new regime.

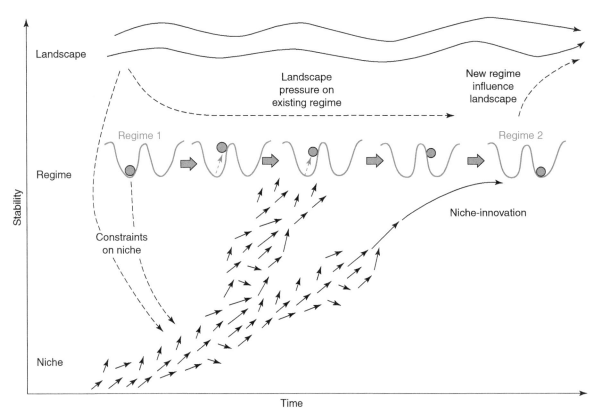

Figure 8.16 When a regime change occurs because of fast disturbances (orange arrows) to a system (the ball), niche innovations can enter as the system (the ball) moves into a new regime.

to stability landscape changes and how the stability of the new regime is reinforced by these changes. In this case, the niche-innovation enters as the stability landscape changes. Figure 8.16 shows less stability landscape alteration and, instead, a regime change that occurs as a result of fast disturbances to the system. In this case, the niche-innovation enters the system as it moves into a new regime.

Indicators are useful when studying historical transitions and are absolutely crucial to guiding present-day transitions to sustainability. Recall from Chapter 4 that indicators are tools for understanding where an SES currently stands with respect to sustainability, how far it must go to reach sustainability, and which way it is headed (toward or away from sustainability) or whether it is changing at all. In the historical transportation system transition example (Box 8.2), there was competition between horse-and-carriage systems and electric trams early on. Dominance of the tram over the horse-and-carriage led to structural developments that paved the way for later auto-based systems. Simple indicators of the direction that transportation systems were headed during that time were "% of streets with electric trams" and "% of horse-based transportation." As noted in Box 8.2, 16% of U.S. streets had electric trams in 1890 and 97% had them by 1902, whereas during that same time period horse-based transportation dropped from 70% to less than 3%. These were valuable indicators of the direction of transportation systems at the time. When thinking about guiding transitions of present-day systems toward sustainability, it is crucial to know which indictors provide relevant information, to diligently track these indicators to see how a given transition approach is working, and to make necessary adjustments to the approaches based on knowledge gained from indicators.

The final connection to be made between the MLP framework and concepts from previous chapters is focused on adaptive cycles (Section 5.3). Recall that systems are continuously passing through these cycles and that the cycle itself characterizes patterns of change that influence resilience, with some phases of the cycle being more resilient to disturbance than others. Thus, the phase that a given regime subsystem is in could determine the type of transition pathway taken. For example, in the reorganization phase (α), resilience is high, change is the rule rather than the exception, and system structure is flexible. Following a large, sudden, and widescale avalanche-style disruption, which is actually the release phase (Ω), a transition sets off along the Dealignment/Realignment pathway while the regime is in a reorganization phase. During the avalanche-style release phase, innovations enter the new regime easily and become well established during the reorganization phase without having to compete with dominant regime structures (which collapsed during the release phase). In contrast, in the conservation phase (K), systems are very rigid and have inflexible structures and tight interactions among components. Thus, a regime subsystem in this phase might be more likely to undergo a transition via the Reconfiguration Pathway, for example, where existing regime structures are well establised and niche innovations are incorporated into it symbiotically.

This ends the section on historical transitions, which was aimed at answering the question: How might transitions be understood? The MLP provides tools for answering this question based on three interacting functional levels and four possible transition pathways. The next section is focused on a different question: How might transitions be shaped in the direction of sustainability? This information is used to guide present-day SESs toward sustainability and is a process known as Transition Management.

Section 8.2: Guiding Sustainability Transitions

Core Question: How might SES transitions occurring today be guided toward sustainability?

"Transition management... is the attempt to influence the societal system into a more sustainable direction. [T]here are no ready-made solutions for [sustainability] problems, we can only explore promising future options and directions. Managing transitions therefore implies searching, learning and experimenting. As such, transition management is a quest, not a recipe for robust solutions.... We cannot answer unequivocally the question of whether transition management really works. And it might take another decade before we can answer this question. But the potential and positive effects of the transition management approach are clear and encouraging [and] ... we remain convinced that it is an attractive and useful model for governance towards sustainable development."

—Rotmans & Loorbach, 2010, Towards a Better Understanding of Transitions and Their Governance

transition management (TM)
a governance approaching used to guide socioecological systems toward sustainability.

Transition Management (TM) is a governance approach for guiding SESs toward sustainability. As alluded to in the opening quote, it is still quite young and largely in development. As such, although the transformative potential of TM and its positive influences on the world thus far in terms of fostering sustainability are encouraging, its absolute effectiveness in guiding SESs toward a more sustainable future remains uncertain. Because TM is young, different yet converging frameworks exist. The framework for developing a transition strategy presented in this section comes from recent transformational sustainability research done by sustainability scientists Dr. Braden Kay and Professor Arnim Wiek from Arizona State University, in collaboration with Dr. Derk Loorbach from the Dutch Research Institute For Transitions (DRIFT) located in the Netherlands. This section also continues to draw on ideas from the previously mentioned 2010 book titled *Transitions to Sustainable Development*, which attempts to answer the question: How might transitions be shaped in the direction of sustainability?

The word *management* implies precise control over an SES's behavior and future direction. In TM, such precision and control is not a part of the process because transitions are not determinative. As a result, the TM approach is often referred to as transition *governance*. It is not possible to know with certainty what a new SES regime will look like after a transition or which of the many different paths it will take to get there. Based on what was learned in previous chapters, you can guess that exerting precise control over SESs might be impossible because of their profound complexity and unpredictability. You would be correct. Therefore, in the context of TM, the word *management* does not imply precise control. Rather, it denotes guiding or influencing systems in ways that prod or coax them toward sustainability. Transition managers do not attempt to control human actions, and therefore the outcomes of these actions, in the same way one controls characters in a video game. Instead, they strive to shape the actions of individuals, and their interactions with regime structures, in ways that steer systems toward sustainability. As noted by Rotmans and Loorbach, TM is "a quest, not a recipe for robust solutions."

With these things in mind, when managing transitions it is important to continually observe, learn, and adjust the approach. As also noted by Rotmans and Loorbach, "searching, learning, and experimenting" are all key elements of the TM process. Contrary to what is accepted in many professional situations, in present-day problem solving arenas, one cannot approach transition challenges with complete confidence in the knowledge or strategies on which TM approaches are based. Rather, it is critical to approach TM with a humble and curious attitude grounded in an acceptance that mistakes will be made, a desire to learn from mistakes, and a willingness to adjust approaches in light of continual learning. Dr. Donella Meadows articulated these ideas well in her 2008 book *Thinking in Systems* when she wrote the following under a section titled "Stay Humble–Stay a Learner":

> *Systems thinking has taught me to trust my intuition more and my figuring-out rationality less, to lean on both as much as I can, but still to be prepared for surprises. Working with systems, on the computer, in nature, among people, in organizations, constantly reminds me of how incomplete my mental models are, how complex the world is, and how much I don't know. The thing to do, when you don't know, is not to bluff and not to freeze, but to learn. The way you learn is by experiment—or, as Buckminster Fuller put it by trial and error, error, error. In a world of complex systems, it is not appropriate to charge forward with rigid, undeviating directives. "Stay the course" is only a good idea if you're on course. Pretending you're in control even when you aren't is a recipe not only for mistakes, but for not learning from mistakes. What's appropriate when you're learning is small steps, constant monitoring, and willingness to change course as you find out more about where it's leading.*

In the next section (section 8.2.1), a new framework for thinking about transitions called the **multi-phase concept (MPC)** is presented. This framework is useful for understanding transitions in a way that

---

multi-phase concept (MPC) a framework derived from the social sciences used for thinking about how transitions occur, based on four different phases that a socioecological system passes through over time as it experiences a transition.

helps transition managers build strategies. A description of how to actually build a transition strategy is given in Section 8.2.2. Finally, Section 8.2.3 presents a list developed by systems thinker Dr. Meadows regarding where and how to intervene in complex systems to effect change. This list ranks, from least to most effective, possible intervention points that a transition manager might use to steer SESs toward sustainability.

Section 8.2.1—Multi-Phase Concept

Key Concept 8.2.1—The Multi-Phase Concept describes the various phases, as defined by their different dynamics and innovation adoption rates, that an SES experiences over time as it undergoes a transition.

Like the MLP, the multi-phase concept (MPC) is another framework for thinking about transitions. Instead of functional levels, the MPC focuses on four different phases that an SES goes through sequentially over long time periods as it experiences a transition: *predevelopment*, *take-off*, *acceleration*, and *stabilization* (Figure 8.17). Indicators (Chapter 4, y-axis in Figure 8.17) are used to determine where a system is at any given time. It should be emphasized again here that this framework is a tool intended for exploratory purposes only and not a deterministic predictive model, which alone cannot explain the behavior of SESs as they progress into the future (an issue discussed in Chapter 7, Section 7.1.1). Nonetheless, the MPC phases are useful when building transition strategies for guiding current systems toward sustainability. Each of the four phases has its own dynamics that are different from the other phases. Recall from Chapter 2 that a system's dynamics are the patterns of change it exhibits over time (Section 2.1.2). During some phases, change occurs slowly (i.e., predevelopment, stabilization) and during others it is faster (i.e., take-off, acceleration).

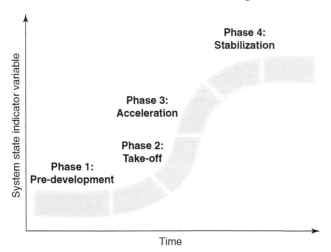

Figure 8.17 The Multi-Phase Concept (MPC) emphasizes four different phases that an SES experiences when going through a transition.

The dynamics of the four phases are based on the adoption rate of innovations into the dominant regime at each phase. Change in social systems can be characterized by such patterns of adoption. Adopters of innovations may be individuals or institutions, including businesses, cities, neighborhoods, schools, households, state governments, or citizens. The percentage of potential adopters that actually adopt an innovation varies with each transition phase. (The percentages given next are not fixed and could vary based on the specific system and innovation of concern.) Early on, during the *predevelopment phase*, only a small percentage of possible adopters (< 5%) have accepted an innovation. The system starts to undergo background changes, but they are not visible. In the *take-off phase*, a larger percentage (10% to 15%) adopt an innovation as the change process really starts to pick up. This phase is the transition ignition point, or the spark, that really gets the transition going. At this point, a critical mass of adopters has accepted an innovation and the transition process becomes self-sustaining. As the transition takes on a life of its own, it becomes more difficult to steer it in any direction other than the one in which it is headed. This momentum creates path dependence (Section 6.3.2), which is hard to oppose or resist and will influence the system's future course. During the *acceleration phase*, the adoption rate further picks up. When the system reaches the *stabilization phase*, adoption rates drop and only the stragglers continue to accept innovations.

The adoption rate during each phase is indicated by the slope of a line tangent to the MPC curve (Figure 8.18). During the *predevelopment* and *stabilization phases*, the slope of the tangent line is very small because the rate of adoption during these phases is small relative to the other phases. Although the *take-off phase* is the time during which a transition begins to unfold and take on a life of its own, adoption rates are still modest relative to the acceleration phase. The *acceleration phase* exhibits the highest adoption rate and, therefore, the steepest slope. Recall from Chapter 5 (Section 5.2.1) that change is governed by the shifting dominance of stabilizing and reinforcing feedbacks. Because change during the *predevelopment* and *stabilization phases* is minimal, stabilizing feedbacks are stronger than reinforcing feedbacks. On the other hand, during the *take-off* and *acceleration phases* when a lot of change is occurring, reinforcing feedbacks dominate and change is nonlinear. These concepts are illustrated with a waste management case study in **Box 8.5**.

Before leaving this section, it should be noted that not all transitions follow the idealized S-shaped

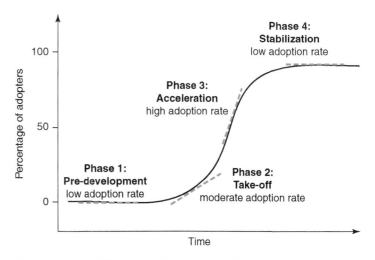

Figure 8.18 The rate of adoption of innovations varies with each phase of the MPC.

BOX 8.5 WASTE MANAGEMENT IN THE NETHERLANDS: SUSTAINABILITY TRANSITION FROM 1950S TO PRESENT

This case study is not about sewage waste, which was the focus of the previous Netherlands case study (Box 8.1). Instead, it is about other types of waste traditionally discarded as garbage into landfills, including paper, organics, metal, glass, wood, and other materials produced by many sectors of society (e.g., households, governments, industry). The transition described here is from a local, decentralized, and unsustainable waste management regime of the mid-20th century to a more centralized, better coordinated, and more sustainable system of the early 21st century. The description of this transition is broken down into the four MPC phases (Figure 8.21).

Predevelopment Phase. Following World War II, several landscape trends led to enormous amounts of waste production within the waste management regime that existed at the time. In addition to population growth, rapid economic growth, and the rise of democracy, individualism and consumerism increased the amount of waste produced and put pressure on the dominant regime. In the 1960s, the environmental impacts of waste were first noticed, which led to a new landscape trend of societal concern about these issues. This trend grew stronger into the 1970s, adding pressure on the existing regime to address the waste issue. Problems within the existing regime, and new niche-innovations, further pushed the system toward a critical point beyond which a transition occurred. Governments experienced landfill space shortages (Figure 8.19b) and sought out other means for waste disposal, such as garbage incinerators (Figure 8.19a). Although governments traditionally managed waste, private companies cropped up to meet disposal and collection challenges. However, they exported waste problems, rather than resolving them, as waste was moved by companies from the Netherlands to less dense areas outside of the country where landfill capacity still existed.

During the 1970s and 1980s, many innovations in waste management technologies and practices emerged. At local scales, glass and fabric recycling programs appeared. People donated materials to second-hand stores for resale, rather than throwing them away. Government regulations related to waste production and management were implemented and

Figure 8.19a Toxic, pollution-generating waste incinerator.

Figure 8.19b Overflowing landfill.

new institutions created (e.g., Ministry for the Environment) to promote sustainable practices. Technologies for waste collection and emission reductions were developed. Increasing landscape pressures for implementation of the reduce-reuse-recycle ethic forced private collectors to participate in programs that separated recyclables from landfill waste. Toxic legacies left in landfills, which affected both human and ecosystem health, became more of a concern. As a result, incineration became the preferred waste disposal method until the late 1980s, when the public pressured governments to reduce harmful toxic emissions (e.g., dioxins) from incineration plants. Although pressure for change and problems escalated the situation, adoption of innovations in waste management by the dominate regime was still quite low and had not yet reached a critical mass.

Take-off Phase. In early 1990s, the waste management subsystem was in full-blown crisis. Landfilling and incineration were both deemed unsustainable, and the situation reached a tipping point. (This tipping point is analogous to the threshold concept described in Chapter 5, Section 5.1.2). At this time, landscape trends pressuring for change finally converged with innovations in technology and waste management practices that had developed over the previous two decades. This convergence resulted in increased adoption rates of innovations by the dominant regime and tipped the scales toward more sustainable waste practices. At this point, the transition was ignited and movement toward more sustainable waste regime subsystems picked up momentum. The transition took on a life of its own.

Acceleration Phase. Throughout the 1990s, innovations and trends that had existed at smaller scales in the 1970s and 1980s accelerated and adoption rates increased even more than during the take-off phase. In just one decade, major aspects of waste management were transformed toward sustainability. Government institutions, such as the Waste Consultation Agency, were established and new regulations that centralized waste management were approved. Waste collection companies consolidated into international corporations, which helped streamline processes and increase efficiency. Business associations for sustainable waste management, such as the Association of Waste Companies, appeared. Markets for waste emerged and what was traditionally viewed as useless garbage was increasingly associated with profit. Physical infrastructure changed, including waste separation within households and industry alike. Separate collection of different wastes, such as recyclables and compost, became common. Emissions reductions technologies for waste incineration plants were developed. In addition to this, technologies for energy-recapture and conversion to useful electricity in these plants became popular (Figure 8.20). For waste that remained as garbage, landfilling operations implemented practices for protecting soils from contamination and for methane gas recovery. Methane gas is used for energy production in a similar way that natural gas is used. People also adopted more sustainable waste management practices on an individual basis and the overall culture of waste was transformed.

Stabilization Phase. The new regime that emerged from this transition was centralized waste management. New perceptions considered waste to be an economic good used to produce

(Continued)

Figure 8.20 Waste to energy plant in the Netherlands.

energy. At the turn of the 21st century, about 85% to 90% of all waste was kept from landfills through some combination of re-use, recycling, and incineration. Many industries and households have assumed best practices along these lines, founded on a new ethic of environmental sustainability. Although the stabilization phase of this transition is still in process, new waste management systems that emerged from this transition are quite firmly embedded in the broader societal landscape.

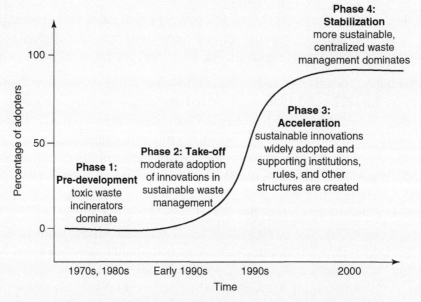

Figure 8.21 Over several decades, the waste management system in the Netherlands transitioned from a toxic, unsustainable regime to a more productive, efficient, and sustainable one.

curve depicted in Figures 8.17 and 8.18. This idealized curve depicts a transition during which the regime successfully adjusts to change. When such a transition plays out, adjustments occur in such a way that the newly emerged regime exhibits a greater degree of complexity as

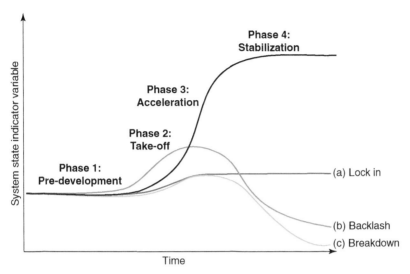

Figure 8.22 Sometimes, in the face of external pressures, a system does not undergo an idealized transition and, instead, experiences lock in (a), backlash (b), or breakdown (c).

compared to the pretransition regime. In other cases, when external pressures act on systems to change them, existing regimes do not always adjust successfully. When there is strong path dependence, such that past choices and events have excluded a subset of possible opportunities for future change, the deeply ingrained and unchangeable regime structures might result in lock-in (line (a) in Figure 8.22). Under these conditions, change is not possible and, despite external pressure, the regime continues on with its existing structure. Sometimes, a transition starts to take off, but then factors that don't respond to external pressures, such as inadequate information, lack of support for change, or deeply ingrained structures that don't respond to external pressure, cause enough resistance to create backlash against a transition already in progress (line (b) in Figure 8.22). The regime that results from backlash may be less desirable than the one that existed previously. Finally, a complete breakdown could occur (line (c) in Figure 8.22). In this case, the existing regime structures completely collapse and the system ceases to function or to meet societal needs in any way.

Whether presented as the idealized S-shape curve, or some other possible non-ideal pattern, the MPC is a tool for thinking about how transitions play out over time. This tool is used to figure out how to guide present-day SESs toward sustainability by developing transition strategies that will be effective at each MPC phase. How to build transition strategies using the MPC framework is the topic of the next section.

Section 8.2.2—Building a Transition Strategy

Key Concept 8.2.2—Transition strategy building involves defining fundamental changes, assets, and barriers and devising strategic actions, interim targets, and indicators appropriate for each of the four MPC transition phases.

As mentioned above, Dr. Kay and Professor Wiek at ASU's School of Sustainability developed a general framework that can be used to construct transition strategies. That framework is presented in this section. Many of the specific examples and conceptualizations of the TM approach presented in this section are based on lessons created by graduate student Nigel Forrest, who conducts transformational sustainability research with Professor Wiek. These lessons were created, in collaboration with the author of this textbook, for the introductory sustainability course taught at Arizona State University on which this book is based. This section first discusses how strategic actions are aligned with the different MPC phases. Then, it describes the process of developing a transition strategy, which includes an initial assessment of assets, barriers, strategic actions, interim targets, and indicators. A case study about a transportation subsystem transition that occurred in the city of Curitiba in Brazil is presented in **Box 8.6** to illustrate these concepts.

Aligning Strategic Action and MPC Phases. To facilitate transition strategy building, the four MPC phases are translated into four phases for strategic action (blue box in Figure 8.23). Each strategic action phase is aimed at different interim targets and has it's own characteristics because of the varied dynamics present at each phase of a transition. Strategic actions focus on stimulating innovation development and encouraging adoption, promotion, and stabilization of innovations within a regime. Knowledge about how an existing regime might interact with innovation, depending on which MPC transition phase it is in, is used to efficiently and intelligently deal with innovation during TM. For example, if TM is focused on an existing energy subsystem based on nonrenewable fossil fuels, and the desire is to transition to a system dominated by renewable energy sources, then strategies for managing renewable energy innovations are informed by the MCP phase that the energy system is in at any given time. If the system is in the predevelopment phase, for example, then the regime is not ready to accept innovations "with open arms." Thus, in the predevelopment phase, strategies should be focused on stimulating innovation development, which will ensure they are viable and stable when the chance for acceptance into the dominant regime arrives. In the take-off phase, during which a critical mass of adopters can be reached, strategies encouraging innovation adoption should be the focus.

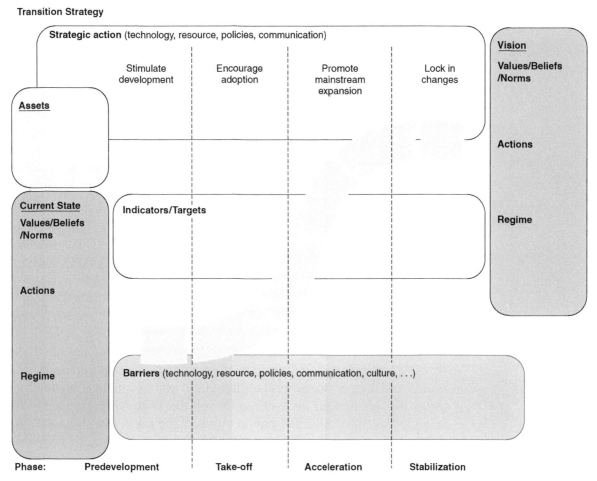

Figure 8.23 Adapted from Braden Kay, Arnim Wiek, Derek Loorbach. (2012). The concept of transition strategies toward sustainability. Working Paper. Sustainability Transition and Intervention Research Lab, School of Sustainability, Arizona State University. Tempe, AZ.

When an SES is in the predevelopment phase, strategies generally should be aimed at *Stimulating Development* of innovations (blue box in Figure 8.23). Before any actions are taken to stimulate innovation development, however, a good understanding of the system (current state), its possible futures (future scenarios), and a desirable future (vision) is needed (Figure 8.1). Because sustainability problems are complex, and spread over many different subsystems within a regime, transition strategies should focus on only one subsystem (e.g., waste, transportation, energy). This simplifies complexity and makes TM less daunting. Collecting information about the subsystem before devising strategies provides a detailed understanding of the current system, the actions to be avoided to prevent undesirable futures, and the desired future that is the goal. This knowledge is the basis for devising an overall transition strategy. It is also needed to inform strategic actions intended to stimulate innovation development during the predevelopment phase.

During the take-off phase, the focus of strategic actions is to *Encourage Adoption* of innovations by early adopters in the dominant regime (blue box in Figure 8.23). There are many mechanisms by which this is achieved, including raising awareness of problems faced by the current regime through educational efforts and building social support structures that will maintain the overall vision, innovations that support that vision, and plans for strategic action to propel innovations into the mainstream. Strategic actions during this phase also might be focused on organizing and directing projects to pilot innovations and test their viability on small scales. Such experimentation is crucial to learning that will inform future strategic actions, making them more effective. At some point, a critical adoption rate will be reached and the transition will take on a life of its own and become more difficult to steer in a given direction.

The main aim of strategic action in the acceleration phase is to *Promote Mainstream Expansion* of the innovation beyond the early adopters targeted during the take-off phase (blue box in Figure 8.23). Doing this may involve several different activities. Insights gained from pilot projects can be used to strengthen and grow supporting social structures that were initiated during the take-off phase. Innovation improvements, and deeper integration of these innovations into existing regime structures, also might be fostered during this time. Such activities are intended to gain continued buy-in and backing for innovations. At this point, both the innovations and the overall transition strategy have the basic political, financial, and societal sponsorship necessary for wide implementation of the strategy.

Finally, the objective of strategic actions during the stabilization phase is to *Lock in Changes* to ensure they are long-lasting (blue box in Figure 8.23). At this point, the new regime structures have become the accepted norm. The purpose of lock-in strategies is to ensure those structures survive in the long-term. Part of this process is about paying attention to undesirable **unintended consequences** and addressing them quickly enough so that the new regime structure does not "get a bad name." Recall from Chapter 2 (Section 2.1.1) that sustainability problems are wicked problems that have no endpoint, such that new problems are almost guaranteed to arise when any solution is implemented. Thus, it is important to stay on top of things and resolve new problems as they come up. If this is not done, opposing landscape trends eventually could develop that put pressure on the newly established regime to change. If the new regime is sustainable, and growing landscape trends drive the system in an unsustainable direction, this would be undesirable and should be avoided.

Developing a Transition Strategy. Several components should be included when devising an overall transition strategy. After selecting

---

unintended consequence an outcome that was not intended that can be beneficial or harmful and is typically not expected or anticipated because of the inherent complexity of socioecological systems.

the subsystem to focus on (e.g., waste, energy, food), the **fundamental changes** needed to transition the system from the current unsustainable state to a sustainable future state should be defined. This information is gleaned by comparing the system's current state and the vision, with specific focus on defining the regime, evaluating actions, and assessing the values, beliefs, and norms of the current system and also noting what these look like in a desired future (pink and purple boxes in Figure 8.23). During this assessment of fundamental changes, a transition manager should think about changes in values, beliefs, and norms that will be needed to drive a transition as part of a broader societal landscape pressure. Thoughts about actions focus on evaluating how individuals and organizations do things today and how they should do them differently in the future. Finally, defining the regime is about characterizing major structures and components that compose the current regime and those that would compose the desired future regime. For example, decentralized waste management was a major component of the unsustainable waste management system in the Netherlands (Box 8.5). By the early 21st century, the system had transitioned to a centralized, and more sustainable, form.

Assets (dark yellow box in Figure 8.23) are evaluated in order to understand what currently existing aspects of the system could be used to promote the desired transition and to get things done. Assets exist in many different forms, including technical (e.g., infrastructure, buildings, equipment), natural (e.g., natural resources, weather), human (e.g., skills, knowledge, capabilities, labor), financial (e.g., money, property), cultural, political, and institutional. They often are not recognized as valuable because the focus is usually on identifying problems, deficiencies, and what is lacking rather than on realizing what beneficial system aspects already exist. For example, U.S. society tends to view elderly people as a burden. As a result, these populations are commonly marginalized and excluded from major activities in society. However, elderly populations also could be considered underutilized assets. They have knowledge, experience, and time to contribute that younger populations might not possess. For example, a senior community member could serve as a volunteer who possesses the professional skills necessary to manage a project. Creativity often is required to identify assets that might be useful for contributing to a given transition.

Barriers are roadblocks to getting things done (orange box in Figure 8.23). If there were no barriers, transitions would be easy, so barriers are at the heart of the challenges faced during a transition. Like assets, barriers come in technical, natural, human, financial, cultural, political, and institutional forms. For example, three major barriers to a transition from vehicles fueled by nonrenewable fossil fuels to ones fueled by clean electricity could be cultural, political, and financial. *Range anxiety* is a culturally pervasive belief that electric cars cannot hold a charge long enough to travel to distant places, and therefore might leave the driver stranded in a remote location. Governments continue to subsidize, and promote

---

fundamental changes the essential transformations in values, beliefs, and norms, actions, and basic regime characteristics that must happen in a socioecological system in order for it to transition from the current state to the vision.

---

assets the beneficial and valuable aspects of a system that already exist and that can be used to "get things done" in ways that promote the desired transition.

---

barriers the roadblocks or challenges to "getting things done" that are at the heart of any transition.

through other incentives, fossil fuel-based vehicles. Electric cars are more expensive than traditional vehicles. Thus, even if someone wanted to buy an electric car, financial barriers might stop her. Barriers are specific to MPC phases. For example, during the predevelopment phase, battery technologies for electric cars might be barriers, whereas in the take-off phase high car prices might be barriers for early adopters.

Strategic actions (blue box in Figure 8.23) are designed with the intent that they will lead to a desired outcome. They are focused on moving a system toward the next transition phase by using assets or overcoming barriers. Ultimately, the desired outcome is expressed in the vision. Prior to using visions for devising transition strategies, specific goals must be operationalized into concrete targets. For example, one goal of a vision for a farming community might be that they have a stable climate that allows for productive crop systems. To operationalize this goal, a number might be used to set a specific target, such as 350 ppm as the maximum CO_2 concentration in the atmosphere. A vision contains many different goals. Thus, many different strategic actions will need to be carried out. These actions work together, and reinforce each other, over an extended time, as an existing unsustainable regime shifts toward the sustainability goals. Strategic actions are of many different forms that fall into the same categories as were listed for assets and barriers: they come in technical, natural, human, financial, cultural, political, and institutional forms. They can be focused on a diversity of projects, including developing technologies, building resources, devising policies, and communicating through education to raise awareness.

As already described, strategic actions are tailored to the different transition phases because some will be more effective in one phase as compared to another. As a result of this need to employ different strategies at different phases, interim **targets** (light yellow box in Figure 8.23) are set for each of the four transition phases. Interim targets are intermediate goals that eventually will allow longer-term goals to be attained. They break up the overall transition into more immediately achievable steps that are possible based on the system's state in each MPC phase. Interim targets have three characteristics. First, they are a measurable output of the system's activities. Second, they are focused on a value to be obtained (e.g., 350 ppm). Third, they include a time by which this value should be achieved (e.g., by the year 2050). Targets are needed for planning strategic actions. In practice, strategic actions and targets are developed together. Indicators (Chapter 4) allow progress toward interim targets to be tracked and the overall transition monitored. They go hand-in-hand with targets because they are in the same measurable units. For example, a target of "350 ppm atmospheric CO_2 concentration by 2050" would have a companion indicator of "CO_2 concentration in the atmosphere" that allows progress, toward or away from the target, to be followed.

---

strategic actions planned actions focused on specific interim targets and carried out at different phases of a transition that are designed to make use of assets and overcome barriers in ways that move the socioecological system continually forward toward the ultimate vision.

targets intermediate goals for each transition phase that guide a socioecological system toward the longer-term goals of the vision.

BOX 8.6 A TRANSPORTATION SYSTEM TRANSITION IN CURITIBA, BRAZIL

Curitiba is a city in southeastern Brazil located in Paraná state (Figure 8.24a). It has a large population on par with major U.S. cities, such as Philadelphia or Houston. As is characteristic of many developing nation cities, it has minimal resources for dealing with overwhelming issues such as poverty, illiteracy, unemployment, inequity, disease, corruption, pollution, and congestion. Unlike other cities in southern Brazil, however, Curitiba was able to creatively combine responsible city governance with entrepreneurship to resolve many of its problems.

Prior to the 1970s, before the transition to Curitiba's current state, traffic congestion and air quality plagued the city. People spent a lot of time traveling around the city to carry out daily tasks. Whether in their cars, or riding on an inefficient bus system, the large amount of time spent traveling each day left less time for friends, family, and generally enjoying life. Even worse, the existing transportation system led to social inequity. People living in poverty who needed access to amenities, such as medical clinics, daycare and schools, and supermarkets, had a hard time because no efficient bus system existed nor could they afford a car. This all began to change in the early 1970s.

In 1971, Jaime Lerner, an architect, urban planner, and engineer, was appointed major of Curitiba. His vision for transportation subsystems was to provide democratic access through public transit to meet the needs of society (especially the poor), instead of a particular mode of transportation, and be financially responsible by working with existing resources to avoid overspending. Based on these goals, Curitiba began to overhaul its transportation system by focusing on the bus system that already existed. Changes in routing, schedules, and boarding procedures allowed the city to inexpensively transform

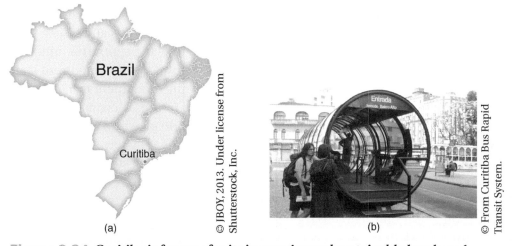

Figure 8.24 Curitiba is famous for its innovative and sustainable bus-based transportation system.

(Continued)

its current system toward one that met societal needs in a more sustainable way. Routing and scheduling was switched from a manual system to a more efficient computerized one. Boarding systems were changed by altering bus entrance ways from previously narrow ones meant to prevent people from evading busfare to wider, more efficient, and comfortable ones. Curitiba's famous tube stations (Figure 8.24b) made the bus system more like a subway, with turnstiles for efficient payment, clear route map displays, and easy handicapped access. Passengers enter the bus through a door on one side and exit through a door on the other. This efficiency led to rush hour bus frequencies of one bus *per minute*.

Although more like a subway in terms of efficiency, the bus system was 200 times less expensive and took much less time to build than the subway in Rio de Janeiro. It pays for itself and also provides a profit for private bus companies. Three-quarters of all Curitiba commuters use this system, which is the most heavily used bus system in Brazil and more so even than New York City's. Almost 30% of bus users do own cars, but they choose to ride the bus because it is so convenient and pleasant. Today, Curitiba has the best air quality of any city in Brazil and also the lowest rates of car commuting. On a per capita basis, Curitiba uses 25% less fuel than other cities in Brazil. It is often noted that the city is a victim of its own success. It is dealing with new population, waste management, and other problems resulting from masses of poor people migrating to this city renowned for its high quality of life. These are the undesirable unintended consequences that arose from solutions implemented to resolve transportation subsystem challenges. These problems have spilled over into other subsystems, such as housing and waste. Good or bad, this is the current state of Curitiba. But how might a transition to the current system have been managed? What might a transition strategy have looked like back in the early 1970s when it all began?

Using the original framework presented in Figure 8.23, Figure 8.25 is an example of what a transition strategy for Curitiba's transportation system may have looked like. It should be noted that this example is not based on an actual analysis and is not representative of what actually happened in Curitiba. It is merely used to illustrate concepts and offer a sampling of what a transition strategy might look like. Figure 8.25 shows the regime definition, evaluated actions, and values, beliefs, and norms clearly laid out for both the current state and the vision. Recall that all of this is done in order to assess the fundamental changes needed to transition the system from the present to a sustainable future. Then, assets and barriers are laid out. Notice that barriers are specific to each transition phase. In Curitiba, one major asset was the already existing road system and buses. Some may have viewed buses as a burden and instead pushed for a shiny new subway or light rail system. However, Curitiba viewed them as assets and used them as such to build its new system. Strategic actions, targets, and indicators are developed together for each phase of the transition. For example, for the take-off phase, one planned strategic action might have been to implement pilot bus routes to test their viability for widespread usage and allow space for learning to improve these innovations for larger-scale projects. A target for this phase may have been to have five main pilot routes by 1975. An indicator used to track progress toward this target would be "Number of Main Routes."

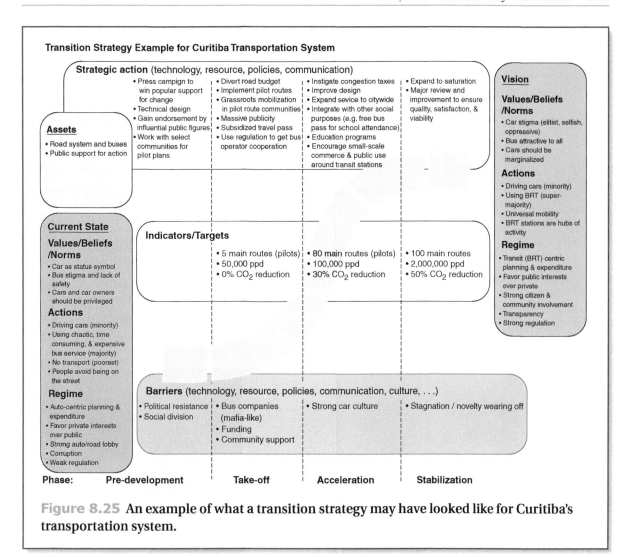

Figure 8.25 An example of what a transition strategy may have looked like for Curitiba's transportation system.

Section 8.2.3—Intervention Points

Key Concept 8.2.3—Intervention points are places in SESs that have powerful transformative potential, such that a small strategic action taken at one point in the system leads to an overall system transition.

> "Give me a place to stand and with a lever I will move the world."
> —Archimedes, 12th century

This chapter ends with a section focused on guidelines for intervening in SESs when attempting to steer them toward sustainability. **Intervention points**, also known as leverage points, are key elements

intervention point (aka. leverage point) an efficient place to intervene in a system because a relatively small change or effort ends up causing a large overall shift in the behavior of a system.

of transition strategies. However, they are presented as a separate section in this textbook for clarity. In reality, they should be fully integrated and developed along with transition strategies. The interested reader should consult the references at the end of this chapter to learn more about how to do this.

Intervention points are places in a system where a relatively small change ends up causing a large overall shift in system behavior. They can be thought of as levers of power, such that they can "move the world," as expressed by Archimedes in the 12th century (Figure 8.26a). In the context of sustainability transitions, intervention points are efficient and smart places to intervene in SESs, with the intent of influencing its future direction. If a high leverage point can be located in a SES, then only a relatively small problem-solving effort (force) is required to profoundly affect the behavior of the system (Figure 8.26b) and possibly cause a sustainability transition to occur.

In addition to causing a sustainability transition to occur, locating and using SES leverage points reduces the resources required to do so. However, locating these powerful points in SESs is far from straightforward,

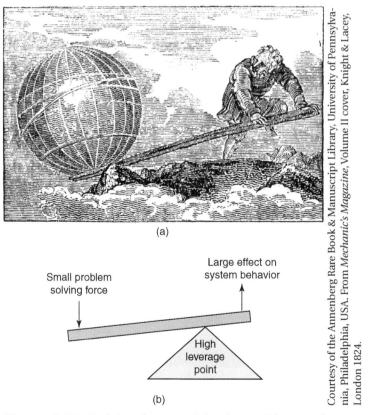

Figure 8.26 Archimedes noted that he could move the world with a lever (a), which allows a small force exerted to result in a large effect (b).

and there is no systematic way to locate them with absolute certainty. What is presented here is a general list of 12 intervention points offered by Dr. Meadows in *Thinking in Systems*. As mentioned in her book, this list should be considered a collection of insights based on years of experience, but still a work in progress rather than a hard-and-fast list. Also, although the list is ordered from least to most effective, the ordering is not definitive. Exceptions always can be found that move some items up or down on the list, depending on the specific system and sustainability problem of interest. The more effective the intervention point, the more resistance to change there will be from the system. Thus, magic intervention points are not easily manipulated, even if their location, and the direction they need to be pushed, is known. The list is presented here as a countdown from least to most effective.

Intervention Point 12: Numbers. Recall from Section 2.2.1 that systems, at a very basic level, are made up of components and interactions among those components. Components are the different parts of a system, such as trees in an ecosystem or consumers in an economic system. Interactions are the processes through which different system components associate with each other, such as photosynthesis (an interaction) in an ecosystem by which trees (Component 1) interact with CO_2 in the air (Component 2) or purchasing (an interaction) in an economic system by which consumers (Component 1) interact with producers (Component 2). The *magnitude* of an interaction depends on the size of the numbers. For example, purchasing rates of 200 electric cars per year is a bigger number than 100 cars per year. The larger rate represents a higher magnitude interaction between consumer and producer.

When resolving sustainability problems, it is sometimes desirable to speed up interaction rates and, at other times, to slow them down. For example, you are trying to transition a transportation subsystem from one dominated by fossil fuel-based vehicles to a more sustainable system of electric vehicles. To promote this transition, it is desirable to slow down purchasing rates of fossil fuel-based vehicles by consumers and speed up rates at which they purchase electric vehicles. There are several intervention points within the economic system that might be used to effect these changes. Government subsidies to electric car companies could be used to speed up purchasing rates of electric vehicles by making them less expensive for consumers to buy. Governments could impose a gas tax to deter consumers from purchasing gas-guzzling vehicles, which would be more expensive to own. A regulation could be passed requiring all car manufactures to convert 5% of their vehicles to electric by 2025, 10% by 2030, and 50% by 2050. This might make electric vehicles a more widespread option for consumers.

The **incentives** (e.g., subsidies) and **disincentives** (e.g., gas tax) described in this example are some of the most overused intervention

incentives tools and mechanisms that are used to encourage certain behaviors.

disincentives tools and mechanisms that are used to discourage certain behaviors.

points and much time, effort, and energy is spent on them, especially debating them in political arenas. This is because they are relatively easy to vary and manipulate, assuming the political will to do so exists. However, they are also the least effective in terms of causing overall shifts in system behavior and lasting transformative change. Changing consumer purchasing rates alone rarely changes the overall behavior of an economic system. That said, manipulating numbers can become a powerful intervention point *when changes in numbers set off reinforcing feedbacks* that drive systems across a threshold and into a new regime. This process will be discussed in more detail below under "Intervention Point 7: Reinforcing Feedbacks."

Intervention Point 11: Buffers. This intervention point is about changing the *magnitude* of the components themselves rather than the rates at which they interact with each other, as was the case for Intervention Point 12. The overarching idea is that, when the magnitude of some component is large, relative to the interaction rates, the system as a whole is more stable and resilient. In other words, the system is buffered, such that the effect of some disturbance is less than it would be if the components were small relative to the interaction rates. The stability and resilience offered by buffers is the reason that more money in your bank account is better than less. If have a good deal of money in your accounts and suddenly lose your job (a disturbance), then you will be able to pay rent, buy food, afford car insurance, and all of the other things that you need to survive while you seek out new employment. However, if you do not have a large enough buffer, your "system" is unstable and has low resilience. Before losing your job you existed in one regime (*working and living independently of your parents*), but if you lose your job and do not have a large enough financial buffer to get you to the next job, then your "system" might shift to a new regime (*unemployed and living in your parent's basement*).

An SES example is the climate system. The overall state of this system depends, in part, on atmospheric CO_2 concentrations. If concentrations increase or decrease too quickly, other systems dependent on climate (e.g., coral reef ecosystems, food production systems) might not have time to adjust their behavior and adapt to the changes. Thus, a stable global climate is important to the overall stability of many SESs. Two components of the climate system are trees and CO_2 in the atmosphere. Two major interactions between these components are photosynthesis, whereby carbon atoms in CO_2 move from the atmosphere into organic molecules in tree biomass, and respiration, whereby carbon atoms move out of biomass and back into the atmosphere as CO_2. The size of the tree component is important to the overall stability and resilience of the climate system. Deforestation, an interaction between forests and humans (who are a third system component), is a disturbance to the climate system because it affects the amount of

CO_2 removed from the atmosphere by trees. The same amount of deforestation disturbance will not have as large an effect on the climate system if forests cover vast areas, as compared to very small areas (Figure 8.27). Thus, it is the magnitude, or size, of the forest *relative to* the amount removed by deforestation that is key. At a given deforestation rate, the more forest present, the more resilient, stable, and buffered the climate system.

Theoretically, a system can be stabilized, or kept from shifting to an entirely new regime, by increasing the magnitude of components relative to the rates at which they are changed by interaction with other components. Increased stability of your finances or the climate system occurs by contributing money to your accounts or allowing forests to regrow after deforestation, respectively. However, these things take time. In general, SES component magnitudes are slow to change and, although a powerful influence on system behavior, this is why buffer interventions are low on the list relative to other intervention points. Finally, in addition to the time it takes to augment buffers to promote stability, a system with large buffering components, by it's very nature, is difficult to change.

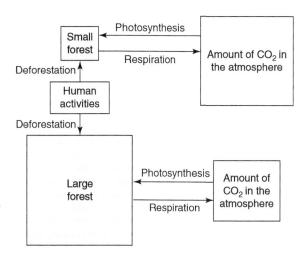

Figure 8.27 A planet with a large amount of forested area is more buffered from climate change effects because large areas of forest, compared to small areas, are not affected as much when disturbances, such as deforestation, effect small parts of them, simply due to the fact that there are more trees available for CO_2 removal in large forests.

Intervention Point 10: Physical Structures. System structure is concerned with the overall layout of an SES in terms of the *arrangement of components relative to each other*. This greatly influences how a system can be changed. For example, a city's physical layout influences transportation system possibilities. When managing an unsustainable transportation subsystem for a transition to sustainability, the physical structure of a city affects the types of sustainable transportation systems that are feasible. If a city has high population densities and mixed-use development, then transition to a light rail system would be a viable option. However, if a city had low population densities with households located very far from each other and from daily amenities, such as the bank, post office, and grocery store, then a light rail might not be a feasible option because of the miles of expensive track that would need to be built to span such great distances. In this case, a transportation system based on solar-powered vehicles might be more workable.

Although changing system structure has great transformative potential, it is low on the list because it is a very slow and expensive way to change a system. The low-density city could be made more dense, and mixed-use development implemented, but this would take a lot of time, money, and effort. Solar-powered vehicles also would take time to be realized because it would require restructuring other systems connected to the transportation systems. It could take decades for car manufacturing and distribution systems to transform from supplying vehicles run on fossil fuels to those run on sunlight. There also would be very high upfront costs for this industry. The bottom line is that, once a system is built in a certain way, it is hard to change, so smart system structural design in the first place is of utmost importance.

Beyond that, viewing certain system features as assets and working with the existing physical layout, rather than starting with an intent for complete restructuring, can be a surprisingly effective way to carry out a transition (as with the bus system in Curitiba, Box 8.6). That said, sometimes crises provide opportunities for rebuilding. For example, geographer Dr. William Solecki studies the impact of Hurricane Sandy in 2012 on sewage, power, water, transportation, and other New York City subsystems. He also studies how these subsystems might be transformed to more sustainable systems as they are rebuilt. According to Solecki, it appears that at least telecommunication systems are undergoing a post-Hurricane Sandy transition.

Intervention Point 9: Delays. Reducing system delays can influence their overall responsiveness to strategic action. Delays occur in systems both in terms *receiving information* needed for decisions and *responding* to a given situation that needs to be addressed. For example, a multinational electric car company needs to figure out how many cars to manufacture the next year. Such decisions often need to be made before all information is available. This is because there is a delay between the time cars are sold by dealers all around the world and the time that information reaches the company in a format that can be used as a basis for decision making. Forecasting models attempt to project future demand, but are not always accurate, and a company might end up producing more or fewer cars than are actually needed for the next year.

Delays also occur when responses take much longer than is desired. For example, building a power plant to generate electricity takes several years and, once built, will last for decades. A state government might decide that it will need five new power plants over the next 10 years to meet electricity demand and set forth a plan to build these new plants. While the plants are being built, actual electricity demand might exceed forecasted demand. When this happens, there is a delay in how fast a state government can respond to new demands because it takes time to plan and build additional plants. As a result, electricity production systems in this state might have to function at overcapacity until the additional plants are ready. Response delays are why giant organizations often respond to societal issues inefficiently. This is true whether the organization is a centrally planned government (e.g., former Soviet Union) or a very large corporation (e.g., General Motors).

Reducing delay length in systems is an effective intervention point because it can make systems more responsive to strategic actions intended to change them. However, they are low on the list because delays are very difficult to reduce. Some system behavior is simply slower than desired and that is just the way they operate. There is often not much that can be done to speed up the time it takes to build a new power plant or for information to reach an electric car company's headquarters in time for decisions to be made about the next year. It also should be noted that speeding up delays is not always a good thing. Sometimes it decreases system stability, as evidenced by decreased delays in money and information-transfer in increasingly unstable financial markets over past decades. Some delay is a good thing when it provides systems with the amount of stability required for them to function in ways that meet societal needs.

Intervention Point 8: Stabilizing Feedbacks. By moving into the topic of feedbacks, we are starting to move into the realm of effective intervention points. As you know from Chapter 5 (Section 5.2.1), when stabilizing feedbacks dominate over reinforcing ones, a system will remain in its current regime. Thus, if a given SES currently is sustainable, but on the pathway to being unsustainable, then strengthening stabilizing feedbacks are a powerful place to intervene. For example, democracy is a governance system put in place to ensure all citizens have an equal say in decisions that affect their lives. In a democracy, citizens have information about elected officials' activities and they can respond based on their approval or disapproval of these activities by voting them into or out of office. A reinforcing feedback between governments and wealthy private interests is driving governance systems away from *one that serves all citizens* (Regime 1) toward *one that favors private interests* (Regime 2). To prevent this shift, certain stabilizing feedbacks could be strengthened. Examples of strategic actions taken to do this include passing the Freedom of Information Act, so that citizens know what governments are doing, and protecting whistleblowers who expose corruption in government agencies. Stabilizing feedbacks must be managed in parallel with reinforcing feedbacks because, if the latter become stronger, then the former also must be strengthened.

Intervention Point 7: Reinforcing Feedbacks. Trying to weaken reinforcing feedbacks driving an SES toward an unsustainable situation is generally more effective than strengthening stabilizing

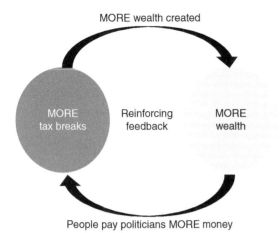

Figure 8.28. (a) A reinforcing feedback between wealthy citizens and politicians widens the income gap, and increases inequity.

Figure 8.28. (b) A stabilizing feedback between people who cannot afford to pay off politicians for favors and anti-poverty campaigns helps to keep people out of poverty.

feedbacks. This is because weakening reinforcing feedbacks gives the stabilizing ones a fighting chance to keep a system in a sustainable state. Inequity is increasing worldwide as that gap between rich and poor widens. This is driven by a reinforcing feedback: The wealthy pay politicians to decrease their taxes, leaving them with more money for their children's inheritances and educations (**Figure 8.28a**). People who are poor can't influence authorities with money or get an education that will land them a high-paying job, and college costs continue to rise. Antipoverty campaigns are stabilizing feedbacks that barely influence this strong reinforcing feedback (**Figure 8.28b**). Weakening the reinforcing feedback would be more effective, such as through progressive income and inheritance taxes on the wealthy and open access high-quality education for all.

Intervention Point 6: Creating Feedbacks. A major problem with many systems is one or more missing information flows. For example, for a time people thought resource scarcity would be prevented through information provided by price signals in markets. If prices got too high, people would stop exploiting a resource because it was too expensive. What actually happens, however, is that as a resource becomes scarcer, its price goes up. This drives more people to exploit it for a higher profit and, in the process, make it even scarcer. Thus, price cannot be the only information flow providing a stabilizing feedback to prevent resource scarcity. Other information about the resource system, such as the number of trees left in a forest or fish in an ocean, is needed to provide an effective stabilizing feedback, in the form of an information flow, that will prevent resource scarcity.

One of the reasons creating feedbacks is so important is the pervasive tendency for humans to evade accountability for their actions. What if a company had to locate its water intake pipes *downstream* from the pipe it uses to empty polluted effluent into a river? What if a politician responsible for declaring war or approving a nuclear power facility actually had to serve as a soldier in the war or store the waste from the nuclear facility in his or her backyard? Such feedbacks certainly would be effective in changing the way things are currently done. However, these feedback loops are still relatively low on the list because they can be difficult to create. The people in positions of power, who don't always want their actions accounted for, are the same people who have the power to put new feedbacks in place.

Intervention Point 5: Rules. Rules constrain and shape human action, such as the rules governing how a basketball player plays a game or how long a U.S. president can remain in office. Changing them can change human behavior. Not all rules can be changed, such as physical laws (e.g., second law of thermodynamics). However, rules created by society (e.g., constitutions, laws, incentives and disincentives, informal social norms) can be changed to powerfully shape changes in human behavior. Imagine what would happen if there were no fouls in basketball or no limit to how long a president remained in office. Under those rules, basketball players might become seriously injured during games and presidents might morph into dictators or monarchs. Changing the rules of the game is a powerful intervention point, although it is not always the easiest to manipulate. Some of the strongest rules, such as those defined in the U.S. Constitution, are also some of the most difficult to change.

Intervention Point 4: Self-Organization. At the end of Chapter 6 (Section 6.3), the ability of SES's to adapt their behavior over time, to promote their own survival in the face of changing conditions, was discussed. Because they do this without any outside assistance, it is often referred to as *self-organization*. Self-organizing systems create new structures and behaviors following a disturbance, such as ecological succession in a forest after a fire or recovery of an economic system after a recession. When systems do this, they change one or more of the aspects discussed thus far (Intervention Points 5 through 12) to varying degrees. Fostering the ability of a system to self-regulate in the direction of sustainability is a very powerful way to intervene. Doing this involves promoting a system's resilience so that it has the adaptive capacity to innovate, reorganize, and learn when faced with disturbance or change. Recall from Section 5.1.3 that diversity is the key to fostering adaptive capacity. Thus, ensuring survival of diversity—of species, of cultures, of technological innovations—bolsters the adaptive capacity for ecological, social, and economic systems alike. Promoting components of diversity that are sustainable, such as solar power rather than nuclear, encourages a system's self-regulation toward sustainability.

> **self-organization**
> the ability of a socioecological system to adjust and adapt its behavior over time in a way that promotes its survival in the face of changing conditions.

Intervention Point 3: Goals. Many of the intervention points discussed thus far are about managing various components or aspects of a system to alter some behavior. Establishing the right goals toward which system behavior is directed is a powerful intervention point. It directs

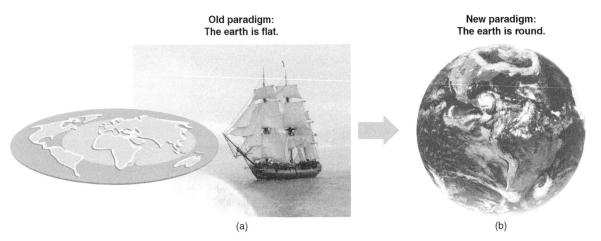

Figure 8.29 The paradigm of a round Earth is so deeply embedded in our society that it is hard to imagine that most people used to believe that it was flat.

8.19a(1): © Dawn Hudson, 2013. Under license from Shutterstock, Inc.
8.19a(1): © James Steidl, 2013. Under license from Shutterstock, Inc.
8.19b: © Loskutnikov, 2013. Under license from Shutterstock, Inc.

a system toward some *ends*, no matter what the specifics of its internal component structure and overall functioning might be. Rules (Intervention Point 5) are the conditions for "playing the game," but goals are about the meaning of the game. Did a basketball player join the National Basketball Association for satisfaction derived from a love of the game or is he there to become wealthy and famous? A system goal is essentially a vision for the future. If all players in a system have the goal of sustainability, then all the details of system functioning fall into place to meet that *end*. However, a common goal of sustainability is a difficult one to set given the diverse values, agendas, and perspectives of stakeholders in a system.

Intervention Point 2: Changing Paradigms. Paradigms are ideas shared by a society, such as beliefs or assumptions about how the world works. They often are taken for granted, or not realized explicitly, because they are deeply embedded and unquestioned. Paradigms are so obviously known by all that they remain implicit and unstated. One paradigm deeply held today is that the Earth is round. However, it was not always this way. Prior to about 500 BCE, most people believed the Earth was flat. Around 500 BCE, Pythagoras noticed that the view of ships returning to port, with the mast appearing first, followed by the hull, could not be explained by the flat Earth paradigm. Eventually, the flat Earth paradigm shifted to the round Earth one (**Figure 8.29**). This paradigm is so deeply embedded in society today that one would be shocked to find out that the Earth were square, for example. The reason paradigms are such powerful intervention points is that they determine what is possible to know, to be, to change, or to do (e.g., Is it safe to sail if the Earth is flat? Will the ship sail off the the edge of the Earth?, Figure 8.29a). When it comes to transitions, paradigms influence the details of all other intervention points discussed this far, such as systems goals, adaptive capacity, setting rules, and creating and influencing feedbacks. Changing paradigms is an important intervention point because old paradigms limit thinking, progress, and ultimately forward movement toward sustainability.

One major limitation to a sustainability transition was pointed out by ecological economist Herman Daly, who argues for a shift from the empty world to the full world paradigm. In an empty world, human populations are low and resources seemingly limitless. Under these conditions, environmental degradation and social equity are not pervasive problems. As discussed in Chapter 1 (Section 1.1), humans lived in an empty world for much of history, but conditions have changed. We now live in a full world, with high human populations and increasingly scarce resources. In this world, environmental degradation and social equity are serious problems. Yet, many continue to act as if we live in an empty world. This is understandable, as old paradigms and ways of thinking die hard. For example, in the 16th century, Copernicus first discovered that the Earth revolves around the Sun rather than vice versa. His heliocentric solar system model is now the basis of modern astronomy. However, the paradigm shift took a while. In the 17th century, astronomer Galileo was held under house arrest by the Inquisition for promoting this theory, which was considered heresy. Although the theory has been widely accepted by astronomers for some time, it was not until 1992 that the Catholic Church acknowledged Galileo's unjust treatment. Thus, as most other intervention points high on the list, paradigms are powerful but difficult to change.

Intervention Point 1: Transcending Paradigms. Even more effective than changing paradigms is promoting their transcendence. No one paradigm, or way of viewing the world, is completely and always accurate, including the one that defines your own personal assumptions and beliefs. For example, ecologists and economists approach the problem of human population growth with different paradigms and, therefore, come to different conclusions and suggest different solution options. Figure 8.30 depicts this as different lenses through which they view the world. Ecologists take the Malthusian view, which assumes that human population growth is controlled by environmental limits. If populations exceed these limits, then war, famine, disease, food shortages, and other disastrous outcomes will result. Economists have different paradigms for understanding human population growth. Danish economist Ester Boserup, in her 1965 book *The Conditions of Agricultural Growth*, claimed that growing human populations will meet rising demands for food by increasing agricultural production through technological innovation.

Figure 8.30 Everyone views the world with their own perspective, or with their own lens, such that no one person has a completely accurate assessment of reality.

Like the ecologist or the economist, we all view the world through our own lens, which filters out some aspects of reality and emphasizes others. Reality is some combination of all different viewpoints. As already stated, paradigms can limit what is possible to know, to do, to be, or to change. Thus, if they can be transcended, then many (but not all) limitations are removed. Some inevitable limitations will always remain because of the finite capacity of humans to fully comprehend the world around us. Nonetheless, not becoming attached to any one paradigm means the any one can be used to achieve some end. When it comes to managing transitions that end in a sustainable world with healthy ecosystems, social equity, and viable economies, it is necessary to resolve problems by transcending paradigms through transdisciplinary understanding. The last chapter of this book is focused on promoting a fuller picture of who should be involved in building these understandings. Specifically, solutions traditionally have been focused on market and government mechanisms, whereas community-based solutions have been neglected until relatively recently. It is the latter that is the topic of the last chapter.

Bibliography

Daly, H. E., and J. Farley. 2004. *Ecological Economics: Principles and Applications*. Washington, DC: Island Press.

De Haan, J., and J. Rotmans. 2011. "Patterns in Transitions: Understanding Complex Chains of Change." *Technological Forecasting and Social Change* 78: 90–102.

Geels, F. W., and J. Schot. 2007. "Typology of Sociotechnical Transition Pathways." *Research Policy* 37: 399–417.

Grin, J., J. Rotmans, and J. Schot. 2010. *Transitions to Sustainable Development: New Directions in the Study of Long Term Transformative Change*. London: Routledge Taylor and Francis Group.

Gunderson, L. H., and C. S. Holling. 2002. *Panarchy: Understanding Transformations in Human and Natural Systems*. Washington, DC: Island Press.

Kay, B., A. Wiek, and D. Loorbach. 2012. "The Concept of Transition Strategies Towards Sustainability." Working Paper, Sustainability Transition and Intervention Research Lab, School of Sustainability, Arizona State University, Tempe, AZ.

Loorbach, D. 2007. *Transition Management: New Mode of Governance for Sustainable Development*. Utrecht, The Netherlands: International Books.

Meadows, D. H. 2008. *Thinking in Systems: A Primer*. White River Junction, VT: Chelsea Green.

Rotmans, J., and D. Loorbach. 2010. "Towards a Better Understanding of Transitions and Their Governance: A Systemic and Reflexive Approach." In *Transitions to Sustainable Development: New Directions in the Study of Long Term Transformation Change*, edited by J. Grin, J. Rotmans, and J. Schot, 105–220. New York: Routledge Taylor and Francis Group.

Wiek, A. 2011. "Transformational Sustainability Research—Integrating Foresight, Backcasting, and Intervention Research." Working Paper, Sustainability Transition and Intervention Research Lab, School of Sustainability, Arizona State University, Tempe, AZ.

Wiek, A., Kay, B. 2012. "Strategies for Intentional Change Towards Sustainability—A Review of Key Paradigms." Working Paper, Sustainability Transition and Intervention Research Lab, School of Sustainability, Arizona State University, Tempe, AZ.

Wiek, A., and D. J. Lang. 2013. *Transformational Sustainability Research—From Problems to Solutions*. School of Sustainability, Arizona State University, Tempe, AZ.

Wiek, A., B. Ness, F. S. Brand, P. Schweizer-Ries, and F. Farioli. 2012. "From Complex Systems Analysis to Transformational Change:

A Comparative Appraisal of Sustainability Science Projects." *Sustainability Science* 7 (Supplement 1): 5–24.

Wiek, A., L. Withycombe, and C. L. Redman. 2011. "Key Competencies in Sustainability—A Reference Framework for Academic Program Development." *Sustainability Science* 6 (2): 203–18.

End-of-Chapter Questions

General Questions

1. In each of the following hypothetical transitions discussed below, define the niche, regime, and landscape levels for each specific scenario.

   a. Before the transition to widespread use of composting toilets, sewage systems nationwide involved using drinking water to flush human waste down the toilet. This water-waste mixture then was carried to a sewage treatment plant, where solids and organic matter were removed and waste effluent was released into rivers or into the ocean. The infrastructure for this dominant mode of waste disposal has been in existence and working effectively for decades. Parallel to this mainstream sewage waste disposal system, composting toilets have been in existence and working well for decades to centuries in remote locations that lack access to the dominant sewage disposal infrastructure, such as in campgrounds and national parks distributed all over the country. One broad trend that has prevented the widespread adoption of composting toilets by the mainstream has been the public perception that human waste should be taken "away" rather than remain near the place where it was deposited.

   Almost like clockwork, the old infrastructure of water distribution systems nationwide begins to break down within a five-year time period. The biggest problem is bursting pipes, which begin to leak sewage into waterways and into the streets of urban areas. Converging with this sudden crisis in water infrastructure breakdown is the rapidly increasing public perception of water shortages because of several years of recent drought. This perception has led to a push for water conservation. With budget cuts being the dominant trend in city, state, and federal governments in recent years, funds are simply not available to help companies rebuild the intricate infrastructure needed to continue with current water-sewage waste disposal systems. Many recognize that composting toilets are the answer, as they require minimal water use and little to no distribution infrastructure.

Despite this recognition, the private companies who manage the water-sewage waste disposal systems are determined to find a way to continue with their current practice. In order to raise the funds to repair and rebuild existing infrastructure, they work with public utilities to raise water rates. As a result, there is a public outcry that access to water is a basic right and, as such, it should not be extremely costly. After competition and struggle between private companies managing the old water-sewage waste disposal system and new companies promoting composting toilets, composting toilets become the human waste disposal system of the future. The old companies remain as a small market for the wealthy, who prefer to pay exorbitant amounts of money to retain their flush toilets.

b. Present-day transportation systems are dominated by vehicles fueled by gasoline derived from refined oil, which is a fossil fuel. Over the past 50 years, public perceptions about climate change have changed steadily, such that most consumers prefer transportation options based in renewable energy sources. In addition, with the reality of peak oil and the increasing cost of gasoline on the horizon, many in government and private industry recognize the need to transition systems to renewable fuels for transportation. They are actively searching for innovations that will allow them to do this. Several well-tested and feasible innovations have been in the works for many years, ranging from electric cars to hybrid gasoline-electric cars to those fueled by biodiesel. Changing consumer demand and peak oil have led many companies to accept these transportation innovations and slowly incorporate them into their existing gasoline-based models. After a time of experimentation to figure out the best combination of old and new technologies, two dominate models emerge: electric vehicles and vehicles run on biodiesel. Although both are good options, a trend in energy production systems eventually decides their fate: More than 50% of all grid power is still generated using dirty coal technologies. As a result, in addition to the advantage of creating a fuel resource from what was traditionally considered a waste product of food production systems, biodiesel vehicles win out and highways now smell like french fries. Companies that previously produced vehicles fueled by gasoline eventually refine their production systems to manufacture vehicles that can accept biodiesel.

c. A country with a governance system at odds with the rest of the world experiences a sudden embargo on fossil fuels, such that no other country in the world will engage with this country in trade or commerce of fossil fuels. This country lacks its own reserves of fossil fuels. As a result, the country has no access to fossil fuels, and the many systems dependent on this energy source rapidly breakdown. This includes housing, transportation, and food production. Because of emerging global trends focused on renewable energy use based on the widespread recognition that fossil fuels are nonrenewable resources, and the related localization movements for food production, many innovations exist in isolated geographical regions of the country. The country is located in a geologically active region, thus small villages have survived off the grid for decades using geothermal energy to generate household power and heat water. In other regions of the country, where geothermal energy is not as readily available, other small pockets of people have relied on wind energy as their primary source of power. In these same remote regions, innovations in transportation (such as using biodiesel from leftover cooking oils) and localized non-fossil fuel dependent food production systems have thrived for many years.

The complete collapse of fossil-fuel-dependent systems resulted in a period of experimentation, where many different private companies, government institutions, and citizen groups tried to figure out the best new model for several of the affected systems. From this period of experimentation, the most effective model emerged: localized food production systems dependent on community engagement in growing crops, rather than mechanized agriculture, and renewable energy produced from waste from these food production systems used to power transportation and households. Following the recognition of the "best" model, many traditional systems aligned with the new system. Food production systems began employing workers rather than mechanized methods, such as tractors and combines, to both harvest crops and convert wastes to usable energy forms. Companies adapted their manufacturing designs so that gas tanks in cars could now accept biodiesel rather than gasoline refined from oil. Developers also adapted housing complexes so that energy systems were more localized and distributed rather than relying on the traditional centralized grid system.

d. There is much pressure on the current health care system in a country to change. In this system, people are preyed upon by profit-driven insurance companies that are unregulated by the government. As a result, many low-income citizens, who cannot afford insurance premiums, suffer. Their low incomes necessitate the purchase of cheap, calorie-dense, low-nutrient foods produced by industrial agricultural systems. This is a vicious reinforcing feedback that results in exacerbated health problems and a need for even more costly health care, which leads to further reductions in income available for food and the purchase of more unhealthy foods. As a result, social welfare advocacy groups have pressured the federal government to provide adequate health care to low-income populations. Economists argue that the current health care system is not only ineffective for low-income populations, but it is also financially wasteful and irresponsible. In addition to these pressures, there is growing public opinion that people should be responsible for their own health by eating right and exercising. All of these pressures are occurring in a climate of massive budget cuts to federal programs and of increasing corporate social responsibility (CSR) for private for-profit organizations.

Start-up companies, think-tanks, and nonprofit organizations have been working on making the health care system more economically efficient and beneficial to low-income populations. These innovations include programs aimed at providing incentives, such as rebates for losing weight, lowering cholesterol levels, or buying healthy foods, and the ability to contribute to low-cost Health Savings Accounts (HSAs) matched by company funds. However, despite such efforts at innovation in health care by these institutions, none is quite ready to compete with the business models of mainstream health care companies. Recognizing the need to change, because of the conflicts with many stakeholders that they have experienced over the past decade, health care companies gradually begin to incorporate these innovations into their business models. Small yet effective changes in government regulations affecting these companies help speed up incorporation of these innovations. At the same time as health care companies are gradually changing, demand for healthier foods results in alternative food production models that begin to catch on in the mainstream. Because they have adapted gradually to the pressures for change exerted upon them, new health care companies and food production systems eventually grow out of the old models. Although not as immediately profitable as previous models, companies find that their new programs and business models result in healthier people and save money for all stakeholders in the medium-term, 3- to 5-year time frame.

2. Return to the hypothetical transitions described in Parts (a) through (d) of Question 1. Using the three criteria described in Section 8.1.2—*timing of interaction*, *nature of interaction*, and *type of landscape change*—determine which of the four idealized transition pathways most closely characterizes each hypothetical transition in (a) through (d). Use the table below to help you organize your thoughts.

Hypothetical Transition from Q1	Timing of Interaction	Nature of Interactions Between Levels	Type of Landscape Change	Idealized Transition Type
(a)				
(b)				
(c)				
(d)				

3. Use the Multi-Phase Concept (MPC), as described in Section 8.2.1, to characterize the four different phases of the hypothetical transitions described below.

   a. Only a few short decades ago, telephones connected to landlines with tape-based answering machines were the dominant way by which people communicated by speaking (unless talking in person). Now, cell phones with Internet connections are the dominant mode of communication, and many people no longer have phones connected to landlines in their homes. Initially, because of the high cost of cell phones, the effort required to make the switch, and the uncertainty associated with their effectiveness to communicate in an emergency, only a very small percentage of individuals and a few innovative businesses got rid of their landlines and changed their voice-based communication systems entirely to cellular phones. In addition to high costs, these new phones were bulky and unwieldy to carry around. After a few years, cell phone companies started to manufacture less expensive and smaller versions of these phones. The phones were now more appealing and a larger percentage of individuals and business switched over. Once this happened, a person almost could not survive without a cellular phone. Before the advent of cell phones, when meeting friends for dinner, all would agree to meet at one place and stick to that plan because there was no way to communicate about changes in dinner plans once leaving home. Now, dinner plans commonly change at the last minute, and those who do not have cell phones often are left in the dark at the original restaurant, with no way to find out where their friends have gone. Cell phone plans eventually become less expensive than owning a landline, especially for long distance calls to family members on the same plan. Situations such as these led to more people than ever switching to cell phones. By the end of this transition, only those very resistant to changing to cell phones still rely dominantly on landlines for communication. Even these curmudgeons eventually go out and buy a cell phone.

b. The general public attitude surrounding smoking cigarettes was supportive of smoking for quite some time. Not many worried about the health effects. In the early 20th century, the first strong scientific evidence that linked cigarette smoking to lung cancer appeared. At this point, the scientifically inclined and others "in the know" took notice and stopped smoking. However, this initial evidence did not affect a large percentage of the smoking population. As scientific evidence mounted, and nonprofit and government organizations mounted no-smoking campaigns for public health, an even larger percentage of people stopped smoking, and this seemed to drive most of the rest of society to stop. At this point, it became unfashionable to smoke in many social circles, as it was discouraged by friends and family who did not want to breathe in contaminated air. The habit became seen as generally unhealthy. Some companies looked to profit from this new widespread no-smoking mentality by selling gum and other products containing nicotine, meant to wean the smoker away from his or her addiction. Eventually, the no-smoking mentality became so widespread that smoking became forbidden in many public spaces, ranging from bars and restaurants to airports and courthouses. Only a few stragglers continued to smoke, taking breaks outside of buildings in areas where smoking was still permitted.

4. Using the 12 ranked criteria for effective intervention points described in Section 8.2.3, choose which of the two options in the examples given below would be a more *effective* intervention point. Explain why you picked the answer that you did. Then, determine which would likely be *easier* to achieve and explain why.

   a. Simultaneously solving problems of disposal of animal waste and synthetic fertilizer use by (1) fostering attitudes that support the goal of sustainable agriculture or (2) installing computer-automated systems that let farmers who need fertilizers for crops know immediately when fertilizer is available from a farmer raising pigs rather than relaying this information by writing it on a bulletin board at a weekly farmer's market in town.

   b. Preventing heart disease by (1) educating people so that they realize that healthy eating and exercise will make them feel good, which will result in even more healthy eating and more exercise, or (2) offering rebates on health care expenses for each drop in cholesterol levels.

   c. Ensuring that the economic livelihoods of farmers in a region survive by (1) bringing together many different people with a variety of beliefs, viewpoints, and perceptions to ensure that the most effective plan for long-term economic livelihood is implemented or (2) encouraging a diversity of livelihoods ranging from free-range cattle grazing to crop production to maple syrup extraction.

d. Preventing water shortages by (1) implementing a citywide ordinance that each household is limited to 80 gallons of water per person per day and, after this is exceeded, the water for that day is simply turned off or (2) increasing the amount of water in groundwater basins using infrastructure that enhances infiltration.

e. Creating local and sustainable food production systems by (1) rebuilding housing developments closer to urban farming sites so that people do not have to drive as far to get their food or (2) getting rid of government subsidies that favor large-scale fossil-fuel based agriculture, which results in profits used for campaign donations by the large companies that received these subsidies to elected politicians who will continue to promote such subsidies.

f. Reducing use of chemical pesticides and fertilizers on household lawns by (1) requiring that home owners who use these chemicals on their lawns funnel the runoff from these chemicals into their home's drinking water supply or (2) changing the deeply embedded social norm that all households must have a vibrant green lawn covered in grass.

Project Questions

1. *Devising a Transition Strategy.* Follow the step-by-step process below to devise a transition strategy for your system that will move it from its current unsustainable state toward the sustainable vision that you developed for your system in Question 2 of the Project Questions for Chapter 7.

    a. **Determine the main focus of the transition you are proposing for your system:** Sustainability problems are messy and spread over many different areas. As such, transitions may have multiple concurrent strands, each focusing on different parts of the system such as energy, transportation, or health. Pick only one strand to focus on. Otherwise, your transition strategy will become too complex. This is likely to be your original problem area, but may get more specific. If you can make it more specific, then do so.

    b. **Determine the fundamental changes that need to take place:** To do this, use your current state analysis and vision to roughly map out the starting and ending points of the transition in terms of the main social characteristics of the system. Transitions can be thought of as a regime shift, where a regime consists of the general set of influences or structures that determine the "rules of the game." For example, in Curitiba, one such regime shift was from a regime that favored automobiles, private property, and individualism to one the favors mass transit, collective benefits, and public interest. Use the table below to help you organize your thoughts.

	Pre-Transition (Current State)	Post-Transition (Vision)
What values, beliefs, and norms are prevalent?		
What actions do people and organizations take? In other words, what are the ways of doing things that characterize the system?		
How would you summarize the regime of the system? (See Project Question 1 from Chapter 5 to help you get started.)		

c. **Determine the available assets that may help or benefit the transition:** Use the table below to list at least four assets, describe whether the asset is technical, natural, human, financial, cultural, political, regulatory, institutional, or some other type, and describe how each will be used. Make the table bigger if you need more room or have more than four assets.

Asset	Asset Type	How will it be used?

d. Describe the barriers that may block or prevent the transition from happening: Use the table that follows to list at least four barriers, describe whether each barrier is technical, natural, human, financial, cultural, political, regulatory, institutional, or some other type, and describe how each barrier may block or prevent the transition from occurring. Make the table bigger if you need more room or have more than four barriers.

Example: *BedZED is a "zero energy" ecological housing complex in London that encountered numerous barriers making its sustainable wood-fuel combined heat and power system work. Barriers included: low reliability because it was untried and untested (technical); required a reliable source of sustainable wood chips (natural); lack of knowledge and skills needed to operate and repair it (human); expensive because it was new technology (financial); it was forced to shut down every night because of local "noise" restrictions (even though it was not noisy), which was a major cause of the technical problems (regulatory).*

Barrier	Barrier Type	How might it block the transition?

e. Determine the strategic actions that will be used to implement the transition: Sketch out the various strategic actions that you foresee could take you to your goals. It is likely that you would employ different strategies at different phases of the transition and would have interim targets that they would aim to achieve. You will define these targets in the next section. Strategies not only should take you to the next phase but might be required to overcome specific barriers. You may not have strategies for all of the types below or you may have other types.

Example: *In Curitiba, strategies may have been to gather internal political and external community support for change (political) in the predevelopment stage; stimulate BRT station and bus development (technical) in the take-off stage; force reorganization of the many existing bus companies (policies, political, institutional) in take-off stage; impose a car tax to raise money for*

BRT (policies, resources) in pre-development; implement a pilot BRT route (technical) in pre-development.

	Strategy	When would it be used (in what phase and for what purpose)?
Technologies		
Policies (rules, regulations etc)		
Communication (education, mass media, interpersonal etc.)		
What resources were needed?		
Institutional (organizational)?		
Other		

f. Determine what interim targets and indicators will guide the transition: You will have a suite of different indicators used to track whether interim targets, set for each of the four MCP transition phases, are met. As much as possible, try to devise interim targets using indicators that you devised in earlier project questions. However, not all of the indicators will be suitable or you may need to adapt them to make them work. You also may need or want to introduce new indicators that are more closely integrated with your strategies. Ultimately, for your previously defined indicator set, the interim target values should be within the upper and lower limits of the indicator.

Example: *In Curitiba, perhaps an indicator of CO_2 emissions was used to track progress toward ultimate goals. In predevelopment and take-off, you would not expect to see much difference*

532 *Sustainable World: Approaches to Analyzing and Resolving Wicked Problems*

in this indicator. In the acceleration phase, as BRT becomes widespread and people start to drive cars less, you would expect to see big changes. However, in addition to this CO_2 indicator, you also would want some other indicators, such as number of BRT routes and number of BRT passengers.

Indicator	Predevelopment Target	Take-off Target	Acceleration Target	Stabilization Target

g. Synthesis: Use the graphic below to synthesize and summarize the different aspects of your transition strategy.

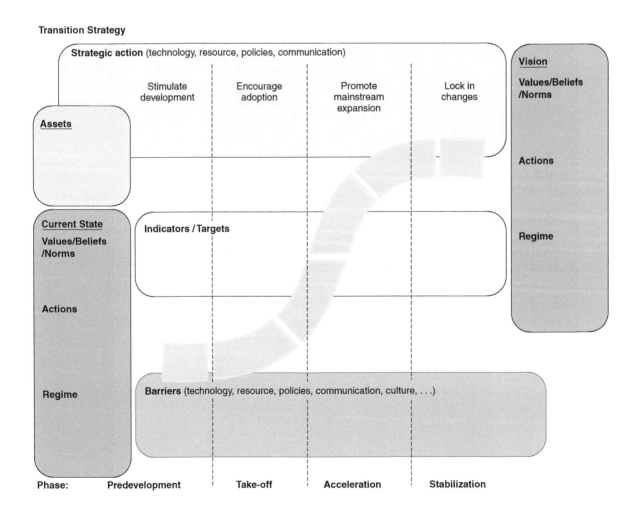

CHAPTER 9

GOVERNING THE COMMONS

Students Making a Difference

Water and Inequity in South Africa

Water is a big sustainability issue in South Africa. An historic drought has plagued the country for a decade or more. Although it has one of the most progressive constitutional rights to water in the world, on paper, in reality access is highly inequitable due to a legacy of apartheid-era discrimination and injustice. The water delivery infrastructure is not well developed. According to Chris Sanchez, an undergraduate student who set out to address this problem: "If you look at a map of the water infrastructure, it looks like a crazy spaghetti diagram of just pipes going everywhere." As part of a study abroad experience with Professor Michael Schoon at Arizona State University (ASU), Chris visited Kruger National Park and stayed with a household in a rural community located just outside the park. During his visit, he collected data for IMAGINE, a research project focused on water and equity in South Africa led by Dr. Melissa McHale, and made some important observations about what it takes to develop effective solutions to local sustainability problems.

The South African government recognizes the water issue and has an awareness campaign urging people to conserve water. However, according to Chris, not everyone is included in the problem definition: "[The water problem is] well defined in some ways, but there are smaller disenfranchised communities that aren't necessarily being included in that definition." Before leaving for South Africa, Chris spent a lot of time reading academic papers, and other literature written by experts, to understand the water access problem. When he arrived in South Africa and talked to local people, he realized that he didn't fully understand. As he went door-to-door conducting surveys in the Venda Homeland, a rural community outside the national park, he asked about certain issues that he thought local people were experiencing and they would look surprised because the thing that he'd mentioned wasn't a problem: "The big part. . .was realizing that there might be things that we as Americans, from the global North, can contribute to fixing that problem, but that we have to empower them to fix it themselves. If we just come in and say 'This is what I think the problem is,' then you're going to get situations where you just gloss over all of the nuance and implement solutions that don't actually sustainability fix anything. They [the local people] have to figure it out for themselves. We can help them do that, but the solutions and knowledge generation has to be from within."

Over the years, the South African government made these same realizations, after many attempts at solving the problem of lack of access to water for local communities. The country tried very top-down approaches, where the government managed all aspects of water governance and infrastructure. This did not work because water access problems vary on a village-to-village basis. Some have problems with nitrates in the water. Others have above-ground pipes that heat up water to unusably high temperatures during hot months. According to Chris: "It is really about getting really context-specific, local knowledge on the problem, and allowing those voices and that information to be represented, and then actually addressing the concerns that they are bringing." The South African government realized this and tried the approach of completely devolving management to local communities. This also failed because there was no organization or cohesion. The most recent approach has been an in-between approach of a Catchment Management Agency (CMA), which attempts to include local voices.

Equitable water access remains a problem in South Africa, as described by Chris: "The local knowledge is really important, but they have not developed the institutional capacity to incorporate that knowledge or those concerns, let alone address them and come up with the time or the funding to do anything." In this chapter, you will learn about different institutional arrangements for managing what are known as common pool resources, such as water, that involve the government, local communities, and markets, or some combination of them. You will see that there is no "one size fits all" solution to overexploitation of resources and that sustainability problems needs to be dealt with in a context-specific way.

Core Questions and Key Concepts

Section 9.1: Introduction to Tragedy of the Commons

Core Question: What types of resources are susceptible to the Tragedy of the Commons and how might this tragedy be addressed?

Key Concept 9.1.1—Socioecological systems likely to experience Tragedy of the Commons situations contain resources that are non-excludable and rival.

Key Concept 9.1.2—Community-based solutions to sustainability problems are just as viable as solutions based in market or government institutions.

Section 9.2: Characteristics of Successful Common Property Regimes

Core Question: Why do some communities self-organize to manage their natural resources and others do not?

Section 9.3: Evaluating the Tragedy of the Commons Using Essential Ingredients

Core Question: Do Hardin's conclusions regarding solutions to the Tragedy of the Commons hold up against the essential ingredients?

Key Terms

common-pool resource (CPR)
excludability
rivalry (aka. subtractability)
rival
nonrival
non-excludable
excludable
institutional arrangement
common property regime
community-based management (CBM)
outsiders
environmental generational amnesia
shifting baselines
substitutability
nested arrangements
devolution
co-management
biological community

> *"The knowledge of local people ... has a comparative strength with what is local and observable by eye, changes over time, and matters to people. It has been undervalued and neglected. But recognizing and empowering it should not lead to an opposite neglect of scientific knowledge.... The key is to know whether, where and how the two knowledges [sic] can be combined, with modern science as servant, not master, and serving not those who are central, rich and powerful, but those who are peripheral, poor and weak, so that all gain."*
>
> —Robert Chambers,
> Who's Reality Counts?
> Putting the First Last

This chapter describes the circumstances under which communities are likely to sustainably manage their own resources. The material presented in this chapter is not an explicit component of the problem-solving framework, which is what many of the chapters in this textbook have been focused on, but rather an umbrella covering the entire problem-solving framework. This chapter is aimed at answering questions regarding who should be involved in the process of resolving sustainability problems. Including a diversity of stakeholders is key. Doing this ensures that both local and expert knowledge are incorporated into the problem-solving process, and that solutions have "staying power." This chapter focuses on communities because their capacity to resolve sustainability problems has often been discounted and problem solving has been left to the "experts." However, as you will see, communities have enormous potential for contributing to the problem solving process *when certain conditions are met*. The first part of this chapter briefly introduces a pervasive problem in SESs known as the Tragedy of the Commons. This section also explains how it came to be that market and government solutions became the two widely accepted ways to address this problem. In the second section, a set of conditions under which communities are likely to sustainably manage their own resources, even in the absence of market and government institutions, are presented.

Section 9.1: Introduction to Tragedy of the Commons

Core Question: What types of resources are susceptible to the Tragedy of the Commons and how might this tragedy be addressed?

A Tragedy of the Commons (ToC) situation results when individuals acting in their own self-interest make choices that result in outcomes that are detrimental to society as a whole. In his influential 1968 article in *Science* magazine, Garrett Hardin used a herder example to illustrate the ToC situation. In this situation, the herders share a common pasture. It is in each individual herder's best interest to add another cow to the pasture. Over time, as each herder continues to act in his or her self-interest, the pasture commons become degraded and eventually destroyed by overgrazing. This is a bad outcome for all herders using the pasture.

common pool-resource (CPR) a type of resource that is susceptible to Tragedy of the Commons situations due to its non-excludability and rivalry.

Section 9.1.1—Excludability and Rivalry

Key Concept 9.1.1—Socioecological systems likely to experience Tragedy of the Commons situations contain resources that are non-excludable and rival.

ToC generally explains why natural resources become overexploited and eventually destroyed over time. In other words, it explains why resources are sometimes used unsustainably. The type of resource that is predisposed to ToC is known as a **common-pool resource (CPR)**. CPRs are defined by certain characteristics related to their **excludability** and **rivalry**. Non-excludability is used to describe how costly it is to prevent users from accessing a resource. Resources that are predisposed to ToC situations are **non-excludable**, which means that they are open for use by anyone, and it is costly to exclude resource users. For example, I own my bottle of shampoo. My shampoo is **excludable**. It is easy (not costly) for me to exclude you from using it. I buy it and lock it in my house so that you can't use it. In contrast, fish in the town lake where I fish are non-excludable. I may not want you to fish in the lake because you harvest too many fish, but it would be costly for me

excludability a measure of the cost that is required to prevent users from accessing a resource.

non-excludable a resource that has a high cost associated with excluding resource users from using it.

excludable a resource that has a low cost associated with excluding resource users from using it.

to prevent you from doing so. I could use physical force, buy the lake from the city, or try to get a law passed that requires fishing permits, but all of these things would cost me effort, time, or money. When classifying resources based on these properties, it is important to remember that non-excludability is a continuum, with non-excludable resources on one end and excludable ones on the other. Many resources fall in between. The closer a resource is to the non-excludable end of the continuum, the more susceptible it is to a ToC situation.

The same continuum holds for **rivalry, or subtractability**, which denotes the quantity of a resource left over after it has been consumed by a subset of resource users. **Rival** (also called *subtractable*) resources fall on one end of the continuum. When a resource is rival, it means that use of a resource by one person or group prevents use by another person or group. The fish in the town lake are an example of a rival resource. If I catch five fish from the lake, you cannot catch those same five fish. If there were 50 fish in the lake to start, there are now only 45. However, if I go to the town lake to enjoy the sunset, the view of the sunset is **non-rival**. My use of that resource does not prevent you from using it. If I enjoy the sunset, then there is the same amount of sunset there for you to enjoy. The closer a resource is to the rival end of the rivalry continuum, the more susceptible it is to a ToC situation.

Figure 9.1 shows the excludability and rivalry continuums as two intersecting axes. It also shows where resources susceptible to ToC situations (CPRs) fall along these axes. Examples of CPRs relevant to sustainability include community forests, irrigation systems, groundwater basins, open-ocean fisheries, the air in the atmosphere, and common grazing lands. It is difficult to prevent resource users from overusing or degrading the resources extracted from these CPRs because, in many instances, each person can access the resource without limit and use is rival. Figure 9.1 also shows three other general categories of resources—*toll goods*, *private goods*, and *public goods*—and provides examples of each. These three other types of goods will not be discussed in detail in this chapter. For our purposes, it is enough to note that four general types of goods, or resources, exist. This recognition combats the popular notion that resources should either be privately owned and regulated by markets or publicly owned and managed by a central government. This type of thinking has led to a focus on market or government solutions to ToCs as the only options. It is not until recently that management of resources by communities has been recognized as an equally viable option. A major take-home message from this chapter should be that there are many possible **institutional arrangements** for effective and sustainable resource management, which often includes some combination of markets, governments, and communities. Until relatively recently, communities have been excluded from this solution option mixture.

---

rivalry (aka. subtractability) a measure of the quantity of a resource left over after it has been consumed by a subset of resource users.

rival a resource whose use by one person or group prevents use by another person or group.

nonrival a resource whose use by one person or group does not prevent use by another person or group.

institutional arrangement some combination of formal and informal institutions used together for effective and sustainable resource management.

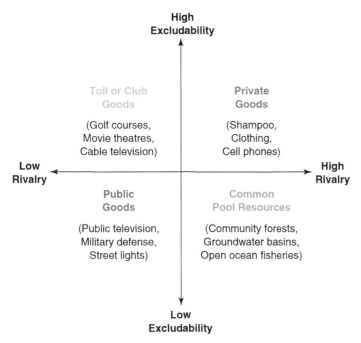

Figure 9.1 The excludability and rivalry continuums as two intersecting axes.

Section 9.1.2—The Market, the State, or Communities?

Key Concept 9.1.2—Community-based solutions to sustainability problems are just as viable as solutions based in market or government institutions.

How did this pervasive preference toward market and government solutions come to be? In his 1968 *Science* article, Hardin concluded that only these two types of institutional arrangements could successfully exclude resource users from CPRs to ensure sustainable resource use and prevent a ToC situation. He also concluded that the resource users were helpless to solve the ToC dilemma by themselves, which is why he concluded that resources need to be managed by governments as public goods, or owned by individuals and firms in markets as private goods. Policy and decision making aimed at sustainably managing resources have been guided by Hardin's conclusions for decades. Despite his arguments, private ownership and government regulation do not always work to solve ToC dilemmas. However, in certain situations, they do work and it is important to understand when they do. In addition to market and government institutions, there is a third general way CPRs can be managed sustainably, and this is through community management in common property regimes. In order to devise effective solutions, it is important to pay attention to the specific SES in which

common property regime an institutional arrangement by which a common resource is jointly owned by a community of resource users.

community-based management (CBM) the practice of resource management by a community of resource users.

a sustainability problem is embedded, in order to understand which combination of market, government, and community institutions can effectively resolve the problem.

In a **common property regime**, the resource is called a commons because it is jointly owned by a community of resource users. The practice of management by the community of resource users is often referred to as **community-based management (CBM)**. Often, the institutions, and accompanying rules, for management of the commons are created and adhered to voluntarily by the community. Rules frequently evolve into informal social norms that are shared by a community, and that resource users eventually internalize. As a result, the costly activities of monitoring, to make sure that resource users obey rules, and imposing penalties on those who violate the rules, are often unnecessary. For example, when buying food at a grocery store, the informal shared social norm in many countries is that you must wait in line for your turn to pay (**Figure 9.2**). There is no government regulation that states you must do this. Because this social norm is internalized, we do not need grocery store police patrolling the cash registers to make sure everyone stays in line, and to punish those who do not with a penalty or fine.

In the same way, a community of people in a common property regime may develop informal social norms regarding resource use. In traditional societies, these informal social norms are often supported by religious beliefs. For example, in the Palau District of Micronesia, there is a small island named Ngerur. According to traditional beliefs, the god of this island owns the island's turtles. Thus, turtles and turtle eggs are not allowed to be harvested by islanders while on this island. This

Figure 9.2 In most places around the world, there is an unspoken and seldom-questioned rule, a social norm, that you wait in line to buy products in a grocery store.

informal social norm, in the form of a religious belief, is well accepted by Palauans, and protects the long-term interests of their society over short-term individual interests. As a result, turtle populations that thrive on this island replenish populations on other islands where turtles are hunted, and turtle eggs are collected (Figure 9.3). Just as you simply don't cut to the front of the checkout line at the grocery store, if you are Palauan, you simply don't harvest turtles or turtle eggs from Ngerur Island. The informal social norm is internalized, so you don't think twice. You just carry out the internalized behavior. Sometimes in CBM, rules are more formal, but it is still the local community creating and enforcing rules, rather than a central government imposing them from outside of the community. This gives the community ownership and control over rules.

Contrary to Hardin's conclusion, private ownership and government regulation are not the only two ways to prevent ToC situations. As you will read about in the case studies in this chapter, local communities have successfully created, and adaptively maintained, self-governed common property regimes that have avoided ToC through the sustainable use of resources. These types of institutional arrangements have existed in the past (e.g., grazing pasture commons in medieval England) and still exist in parts of the world today (e.g., pastures used by nomadic herders in parts of Africa). For the remainder of this chapter, the characteristics of successful common property regimes will be explored in an attempt to answer the question:

Why do some communities self-organize to manage their natural resources from generation to generation, while others do not?

By the end of this chapter, you will have the beginnings of an answer to this question.

Figure 9.3 Religious beliefs of some people living on the islands of Palau forbid them from harvesting sea turtles and turtle eggs from Ngerur Island, essentially creating a nature preserve for this species.

Section 9.2: Characteristics of Successful Common Property Regimes

Core Question: Why do some communities self-organize to manage their natural resources and others do not?

All long-enduring common property regimes share fundamental similarities (**Table 9.1**). These were first developed by researcher Dr. Elinor Ostrom in her 1990 booked called *Governing the Commons*. In 2009, Ostrom became the first women to receive the Nobel Prize in Economics for demonstrating how common-pool resources could be successfully managed by local resource users. This idea challenges the mainstream belief, based on Hardin's conclusions in his 1968 *Science* article, that common-pool resources require either privatization or government regulation. It is this mainstream belief that informs many natural resource policies today. By understanding the characteristics of successful common property regimes, we will be able to make better future policy decisions regarding common pool resource use.

Through a variety of case studies, the essential ingredients (Table 9.1) of long-enduring common property regimes are described and illustrated in this section. By long-enduring, it is meant that the natural resource system is sustained by a local community from generation to generation, on time scales ranging from decades to millennia. The case study approach is essential here because Ostrom's theory of community resource management is still in development. In order to identify the essential ingredients discussed below, Ostrom and her colleagues spent decades combing through hundreds of individual case studies to identify commonalities among successful common property regimes in an effort to build a general theory. To this day, researchers continue to fully develop this theory, and it remains a work in progress.

When a local community self-organizes to sustainably manage a resource, the costs of management must not outweigh the benefits derived from the resource. Hardin concluded that resource users could not solve ToC themselves because they would not have any incentive to invest the effort and time necessary to do so and, therefore, government regulation or privatization were required. However, local communities *do* organize to manage the resources that they use and they *do* invest the necessary effort and time. The essential ingredients discussed below define situations in which communities are likely to invest time and energy to ensure sustainable resource use, because the benefits derived from the resource are greater than the costs of managing the resource.

Table 9.1 Essential Ingredients of Successful Common Property Regimes. Adapted from Table 1 in Ostrom's 2009 *Science* article titled "A General Framework for Analyzing Sustainability of Socio-Ecological Systems".

Essential Ingredient	Case Study Example
Characteristics of the Resource	

1. Size of the Resource System
    a. Defining boundaries
    b. Monitoring use patterns
    c. Accurate ecological knowledge
	
	Brazilian rainforests and logging
	Brazilian rainforests and logging
	Nova Scotia inshore fisheries

2. Productivity of the Resource System	Los Angeles groundwater overpumping
3. Predictability of the Resource System Dynamics	Los Angeles groundwater overpumping
4. Resource Unit Mobility	Gulf of Maine lobster and tuna fisheries
Characteristics of the Resource Users	
5. Number of Users	International fisheries, Maine lobsters
6. Presence of Leadership and Entrepreneurship	Canal irrigation systems in India
7. Shared Social Norms	Fishing cooperatives in southern Turkey
8. Knowledge of the Resource System	Cree Amerindians and caribou hunting
9. Importance of the Resource	Peak phosphorus, global forest regrowth
Characteristics of the Governance System	
10. Full Autonomy to Create and Enforce Rules	Forest communities in Nepal, Rice cultivation in Bali

1. **Size of the Resource System**

    Moderately sized commons territories are most conducive to community management. Areas that are too small simply do not generate enough of a desired resource to justify the costs of management. For example, the apple tree in my neighborhood park produces two bushels of apples (about 250 medium-sized apples) each year. All 50 families living in my neighborhood can harvest apples from the tree. However, this would provide only five apples per family, per year. In the eyes of the families in my neighborhood, the benefit derived from this apple tree is too small to bother with the costs of self-organizing to sustainably manage this apple

outsiders resource users that are not part of the community of users that compose a common property regime.

tree. Self-organizing as a community might involve monitoring the resource use through mandatory attendance at one-hour weekly meetings, at which each family reports the number of apples harvested. It might also involve protecting the resource by paying a guard to patrol the area to exclude non-neighborhood members (**outsiders**) from illegally harvesting apples. The time, effort, and financial costs of organizing are large relative to the benefits derived from this resource. When a commons is too large, such as the open ocean or a large forested region, community members are also unlikely to self-organize to sustainably manage a resource because the costs of defining the resource boundaries, monitoring resource use patterns, and gaining accurate ecological knowledge about the resource are too high. These three aspects of resource system size are discussed below.

(a) Defining boundaries. It is costly to delineate the boundaries of a resource that spans a large area. Without well-defined boundaries, it is hard to exclude outsiders and prevent them from overusing the resource. For example, in the rainforest of the Brazilian Amazon, logging in indigenous protected areas is illegal. However, it happens all the time because it is hard to keep illegal loggers out. Putting a fence around the protected area to keep people out would be time intensive and expensive to both install and maintain. Because of the high costs of protecting such a large resource, an indigenous community is unlikely to expend the effort and time necessary to do so, even if they derive substantial benefits from the resource (Figure 9.4).

Figure 9.4 Indigenous protected areas in the Amazon rainforest cover such a large area that it is costly to exclude illegal loggers.

(b) Monitoring use patterns. The second reason that community-based management is unlikely to occur in a very large commons territory is that it is costly to monitor resource use patterns. Monitoring resource use involves knowing who is using the resource, and how much of the resource is being used. This type of knowledge is necessary for sustainable management because the community needs to know that the resource is not being used illegally by outsiders, and that it is being used sustainably by legitimate resource users. For example, in order to monitor logging in the Brazilian Amazon, the indigenous community would need to hire guards to patrol the vast rainforest region (Figure 9.4). The guards would monitor resource use rates by each person to prevent overharvesting. They would also check permits to be sure resource users are part of the community, and impose penalties on outsiders who are illegally harvesting. This would be costly and require resources that the indigenous community likely does not have.

(c) Accurate ecological knowledge. If a resource system is too large, it is difficult to gain the appropriate ecological knowledge needed to inform management. Local community members use resources day after day, over the course of many years, and even from generation to generation. Thus, they have a thorough working knowledge of resource dynamics for a specific time and place. Local knowledge and careful observation over time is often the most appropriate information for both creating rules for resource use, and modifying those rules as conditions change. Central governments assume, like Hardin, that local communities do not have the capacity to manage their own resources, and often attempt to apply a set of one-size-fits-all regulations to a large and diverse area. With government regulation, time- and place-specific rules, devised by local communities, often go out the window. Government oversight often leads to degradation rather than conservation of a resource system that local communities had previously managed sustainably. The importance of accurate ecological knowledge is illustrated by the following case study on Nova Scotia inshore fisheries.

For hundreds of miles along the eastern coast of Canada, people have lived in small fishing villages for generations. Fishing for species such as cod, halibut, herring, mackerel, and lobster has been their primary livelihood. Over the years, these fishers have accumulated knowledge about which technologies are best to use in which locations, and during which times of the year. Fishers in villages along the coast of Nova Scotia have self-organized to create their own rules for sustainable management of their inshore fisheries (Figure 9.5). These involve rules dictating who can enter a fishing area and where they can fish based on the technology they are using

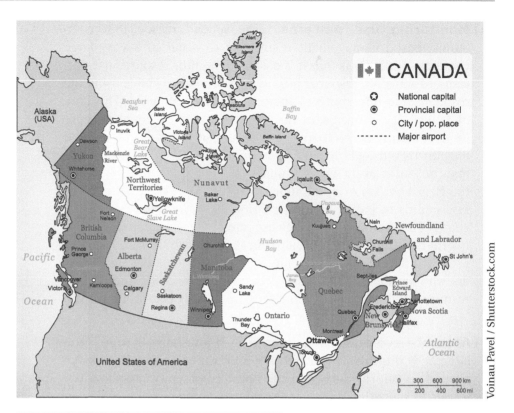

Figure 9.5 In small coastal villages of Canada's eastern province of Nova Scotia, fishing has been a way of life for generations.

(e.g., gill net, wooden trap). These rules are not recognized by the Canadian Department of Fisheries and Oceans (DFO).

In the mid-19th century, exploitation of deep sea offshore fisheries by foreign commercial fishing fleets resulted in severe depletion of fish populations in eastern Canadian coastal waters. In an attempt to remedy this problem, the Canadian government claimed jurisdiction over its exclusive economic zone (EEZ), which extended the country's borders out into the open ocean. (In response to increases in commercial fishing fleets and near-shore mining activities in the mid-19th century, countries with coastlines claimed ownership over resources in their EEZs.) As a result, the DFO began to regulate both the inshore and offshore fisheries located within its EEZ, located along the entire length of Canada's east coast. Local rules devised by fishing villages over generations were deemed null and void, and the DFO developed a single policy for all inshore and offshore fisheries. Local fishers felt the DFO rules related to licenses for fishing vessels and various technologies were arbitrary and not conducive to local conditions. In the words of one local fisher from Port Lameron, Nova Scotia:

> *What do they know about what we do? Fisheries Officers are only around here now and then. How do they know what's best for us? We've fished here for a long time and we know what's best for our ground. We know what it can take.*

The economically valuable cod fishery did, in fact, collapse in 2002. When the territory of a resource system is too large, it is difficult to gain the accurate ecological knowledge necessary to effectively manage a commons. Given this, it is not surprising that the DFO's management of Nova Scotian fisheries was less effective. Whether the managers are a local community, a government, or a private firm, detailed time- and place-specific information is difficult to obtain for a large region. This type of information is necessary for creating effective rules, and for adapting those rules to changing conditions over time.

2. **Productivity of the Resource System**

A community needs to be concerned about the degradation, and eventual loss, of a valuable resource before it will devote time, effort, and other assets to self-organize for sustainable management. In order for this concern to arise, resource users need to witness some degree of resource scarcity. If users perceive the resource as abundant, they will not take actions to conserve it because they do not see a need for management. For example, in the case study discussed in the next section, regarding groundwater extraction from the West Basin near Los Angeles, resource users did not take action to conserve their groundwater, or evaluate other potential water supplies for their region, until they realized that their groundwater resource was threatened by overuse.

In the case of groundwater, the users cannot actually see the amount of water being stored under the ground without the aid of geological survey techniques and certain technologies. The inability to observe scarcity is not the only factor that contributes to the level of community concern about degradation and the loss of a valuable resource. Perception, or the resource users' interpretation of a situation, is also a factor. Ideas such as **environmental generational amnesia** from the discipline of psychology, or **shifting baselines** from ecology, convey that

environmental generational amnesia a concept from psychology that explains why people often do not realize the full extent of environmental degradation that has occurred due to the tendency to measure environmental degradation against the time scale of their own lifetimes rather than degradation that has occurred slowly over the hundreds of years before they were born.

shifting baselines a concept from ecology that denotes the continual, slow, almost imperceptible changes in baseline reference conditions that occur over long periods of time.

people measure environmental degradation against their own experiences and what they encounter during their early lifetimes, rather than the actual degradation that may have taken place slowly over the hundreds of years before they were born. This results in resource users perceiving less degradation of a resource than actually exists. In this case, resource users may not perceive scarcity, but in reality the resource on which they depend may be threatened. A good example of this is changing climate and glacial water supply. For someone born 20 years ago, it may not appear that the size of glaciers, and the accompanying water supply, are shrinking. However, for someone born 80 years ago, the shrinking of glaciers is apparent.

Another reason that a community might not be concerned about resource degradation is a belief in **substitutability**. This is the viewpoint of neoclassical economists who believe that there are substitutes for all natural resources, whereas ecological economists, on the other end of the spectrum, believe that some natural resources have no substitutes. A neoclassical economist might not take action to conserve a resource because he or she believes that technology will provide a substitute. An ecological economist, on the other hand, will observe resource scarcity and take action because he or she believes that there are few substitutes for natural resources and services. Whatever the reason, resource users must witness scarcity in order to be willing to bear the costs of organizing to manage resources sustainably.

3. Predictability of Resource System Dynamics

The dynamics of a resource system need to be predictable enough so that a community can foresee the influence of its actions on resource productivity. Resource system dynamics are largely dependent on resource type. To illustrate how the dynamics of different resource systems influence a community's ability to sustainably manage that resource, three resources systems are contrasted here: forests, groundwater basins, and ocean fisheries (**Figure 9.6**). Of the three, forests are usually the most predictable. Trees remain in a defined area for their entire lifespan and they do not move around. This makes determining the growth rate of trees in a forest and monitoring the use of forest resources by local communities relatively straightforward. Therefore, it can be fairly easy for a local community to determine the influence of their actions on a forest resource system.

Groundwater resource systems are relatively more difficult for a local community to manage sustainably. Groundwater can be destroyed if the rate at which resource users extract water is greater than the rate at which the groundwater is replenished by infiltration and percolation. However, the groundwater is underground

---

substitutability refers to a situation in which a resource can be replaced by a man-made technology.

Chapter 9 *Governing the Commons* **549**

Figure 9.6 It is easier to predict resource dynamics for a forest (*top left*), where the effects of actions such as logging can be easily observed, than for a groundwater resource (*right*), which is out-of-plain sight under the ground, or for the ocean (*bottom left*), which appears as a vast expanse of blue water when viewed from the surface.

and, therefore, difficult to observe directly (Figure 9.6). As a result, local resource users might not know that they are using the resource too much, and management efforts might come too late. These points can be illustrated by the case of groundwater use in the West Basin near Los Angeles, California in the early- to mid-20th century. The West Basin is a groundwater basin with a surface area of roughly 170 mi^2 located next to the Pacific Ocean. Groundwater overuse by residents of the West Basin was rampant in the early 20th century. It was not until the early 1940s that residents started to notice that the water flowing out of their taps was becoming increasingly salty. Resource users disagreed on whether the salty water was a temporary problem that would go away, or a more persistent problem that threatened the future of their groundwater. It was not until the local community commissioned a joint study of the salt problem, by the U.S. Geological Survey and the Los Angeles County Flood Control District, that residents learned that their unsustainable pumping rates threatened the future of their water supply. The freshwater stored in the West Basin was being invaded by seawater from the Pacific Ocean, which is a process known as saltwater intrusion. It is caused by overpumping of groundwater basins near an ocean. At this point, the community decided to do something about it and reduce the rate at which they pumped groundwater. Thus, it was not until the community was able to foresee the influence of their actions on resource productivity that they self-organized for sustainable resource management.

Ocean fisheries are one of the least predictable resource systems (Figure 9.6). In order for resource users to foresee the influence of their harvesting activities on fisheries, they need to know how their fishing efforts effect fish population sizes. Theoretically, they could count the number of fish available for harvest at any given time, just as the number of trees in a forest can be counted. In reality, this is extremely difficult to do. Unlike trees in a forest, or water in a defined underground basin, fish populations move throughout a large ocean, and this makes accurate monitoring of population sizes difficult. Like water stored beneath the earth's surface in groundwater resource systems, fish populations exist beneath the water's surface. Therefore, unlike the easily observable trees in forests, water under the ground and fish in the ocean are difficult to observe (Figure 9.6). However, scientists and engineers have devised ways to accurately measure the amount of water in a groundwater basin and monitor use rates with technologies such as piezometers, seismic surveys, and, more recently, satellite remote sensing. Similarly accurate direct measurements of fish population sizes are not available, and ocean fishery managers rely largely on mathematical models. However, fish population dynamics are extremely chaotic and unpredictable. This makes them difficult to describe mathematically. In most developed countries, the basis for government-based fisheries management is the stock-recruitment model. This model accounts for the effect of fishing on the population size of a given fish species. In reality, fish population sizes also depend on **biological community** interactions such as predation, competition for resources, and even cannibalism. These dynamic interactions are difficult to capture in models. Using current models, fish population sizes can be predicted to within about 30–50% of their actual sizes. Researchers continue to improve these models, but fisheries are still one of the most unpredictable resource systems.

biological community multiple populations of different species living in the same area and interacting with each other in a variety of ways.

4. Resource Unit Mobility

Resources that are mobile, such as ocean fish and terrestrial wildlife, do not stay within defined boundaries, making monitoring difficult. As a result, there are high costs associated with observation and management. Resources that are stationary, such as trees or fish in a small lake, are more easily observed and managed. Therefore, stationary resources are more likely to be sustainably managed through community self-organization. These ideas can be illustrated with two fisheries case studies focused on two marine species harvested from the ocean in the Gulf of Maine: American lobster (*Homarus americanus*) and Atlantic bluefin tuna (*Thunnus thynnus*).

The American lobster is found in North American coastal waters extending from Newfoundland to the Carolinas. However, lobster fisheries harvesting this species within its geographical range are not equally productive. Local lobster fisheries in the Gulf of Maine have consistently produced more lobsters than any other. Their success is attributed mostly to community management of their fisheries (Figure 9.7). These common property regimes were first studied by anthropologist Dr. James Acheson at the University of Maine in the 1970s. The American lobster is a relatively stationary species. Its only migration goes from very close to the shoreline during the summer molting season into warmer, deeper waters located three to ten miles offshore during the winter. Lobster fishers follow this seasonal migration by relocating their lobster traps (called "pots") to deeper waters in the winter, and back to shallower waters in the summer. The state of Maine is involved in managing this fishery by issuing licenses to lobster fishermen, and restricting the sex and size of lobsters allowed to be caught. Technically, the state also enforces these regulations by monitoring lobster fishing activity. In reality, the local lobster fishers themselves carry out much of the monitoring and enforcement of these regulations. They also have their own informal regulations. According to the state,

Figure 9.7 Fishermen living in towns along the coast of Maine (*right*) have sustainably managed their stationary lobster populations (*right*), which live within the same area, about 3–10 miles off the coast, year round.

anyone with a lobstering license can fish anywhere. However, local lobster fishers are very exclusive of outsiders through their own enforcement mechanisms. Lobster fishers tend to claim territories, often near their homes along the coast, and exclude outsiders from their territories using verbal threats and physical destruction of lobster pots when necessary. Because of state licensing regulations, each pot is marked by the owner's license number, and this aids local lobster fishers in identifying who is fishing in their territory. The effect of local, informal monitoring and regulation has been to prevent overharvesting of lobsters. This type of community-based management is possible because lobsters are relatively stationary and remain within the territories defined by lobster fishers.

Another species historically harvested from the Gulf of Maine is the Atlantic bluefin tuna. Unlike the American lobster, the bluefin tuna is a highly mobile species that crosses international boundaries during its annual migration (Figure 9.8). Populations of bluefin tuna exist in the Atlantic, Pacific, and Southern Oceans, and in the Mediterranean Sea. All are threatened by extinction due to overharvesting to fulfill a high demand for the fish, especially the fatty and buttery belly meat used in sushi. Dr. Barbara Block and her team of researchers at Stanford University's Tuna Research and Conservation Center have detected two populations of Atlantic bluefin tuna using electronic tagging techniques. Both populations occupy foraging grounds in the North Atlantic Ocean, but the populations migrate to separate regions to reproduce: one migrates to the Gulf of Mexico, while the other heads to the Mediterranean Sea. Juveniles and adults that forage in the Gulf of Maine migrate to the Gulf of Mexico from April until June, where they are subject to fishing pressures, oil spills, and other hazards during their critical spawning period. The health of Atlantic bluefin tuna in the Gulf of Maine, as indicated by their fat content, has declined since the 1990s, and the reasons for this are unclear. Whatever the cause, it is not possible for local fishing communities in the Gulf of Maine alone to self-organize to sustainably manage the Atlantic bluefin tuna caught in its waters because it is such a mobile species. Management of this species requires cooperation among all countries that influence tuna populations, throughout their entire international migration.

5. **Number of Users**

Group size is almost always a factor in determining whether communities self-organize to manage resources. However, group size can either discourage or encourage self-organization, and this will depend on the specific situation. In some situations, if a group is too small, fixing a problem may require access to certain financial or labor resources that small groups can't manage to muster, but that larger groups can. Alternatively, the costs of bringing a large

Figure 9.8 Blue fin tuna (*left*) are very mobile, with migration routes that cross three of the five major global oceans (*right*), and are difficult to manage sustainably as a result.

number of community resource users together to share information about a resource, and reach agreement regarding actions to be taken, can be very high. In this case, communities are not likely to organize. For example, management of open ocean fishery resource systems by a global community of users is a particularly difficult challenge. No one country owns the vast open ocean, and it is open to all for fishing without formal regulation or enforcement. The United Nations Convention on International Trade in Endangered Species (CITES) is an attempt at international regulation. Member countries meet every two years to review the status of commercially desirable marine species ranging from cod and sea urchins to whales. CITES member countries ban or control the trade of species considered threatened with extinction. However, this is a tricky process that brings together nations from around the world, with diverse cultures, norms, beliefs, value systems, and commercial interests, to agree on what to do. Thus, it is a costly undertaking.

In the previous section, the Gulf of Maine lobster fishery case study was used to illustrate how resource mobility is an important factor in determining whether a community of users will self-organize for sustainable management. The American lobster is considered a stationary species and so, in contrast to the Atlantic bluefin tuna, it is easier to deal with because it remains within the boundaries of fishing territories defined by the local community. There is another reason why the Maine lobster fishery has been successful at sustainably managing its resource over the years, and that has to do with group size. Because the number of resource users is small, it is not costly to gather together to share information or discuss actions that need to be taken. In these small Maine fishing communities, lobster fishers might meet informally at a local pub after a day of fishing, at the harbor while deploying or docking their lobstering vessels, or in the town park during a Saturday picnic with their families. It is during these informal meetings that much discussion occurs. If discussion and collective agreement on some issue by all lobster fishers in a certain harbor were required, then it would not be difficult to gather together in the local town hall for a formal meeting.

Large group size does not preclude sustainable management of resources by communities if effective **nested arrangements** are used. In these arrangements, all three types of institutions for resource management—markets, governments, and local communities—are often involved. In the case of the West Basin in Los Angeles, there were about 500 groups of resource users involved, including residents from local cities and towns, public and private water utilities, and private industries. With so many resource users, spanning such a large geographical region, discussions at local town hall meetings were not an option. Instead, the resource users created the West Basin Water Association (WBWA) as a forum for discussing problems, deciding on actions through negotiation, and keeping all parties up to date by disseminating current information in a weekly newsletter. With the creation of the WBWA, groundwater users in the region were able to come together as a community to collectively devise strategies and carry them out. As a result, the 500 resource user groups cut back on water use and enhanced local supplies. They did this without state or federal government regulation.

On the other hand, a large number of resource users promote community organization in some cases. For example, if a large

---

nested arrangements institutional arrangements used to promote effective, bottom-up community management of resources when large group size is an issue by nesting smaller management units within larger units.

number of people are needed to monitor a large territory, then group size will be beneficial, as long as all resource users are willing to bear the costs of individual monitoring. A successful example of this type of situation is large community forests in southern India.

6. **Presence of Leadership Potential**
Self-organization is more likely when some members of a community possess leadership potential. This can be in the form of entrepreneurial skills that contribute to a community's capacity for innovation, or influential individuals who are well-connected and respected both inside and outside of a community. Dr. Ruth Meinzen-Dick of the International Food Policy Institute in Washington, DC studied the factors affecting the likelihood of local farmer participation, during the devolution of canal irrigation systems in the Karnataka and Rajasthan states in India (Figure 9.9). Devolution, or management-transfer, involves government agencies relinquishing control over resource systems to local users, and has become widespread in countries around the world. This practice is driven by a realization of the limited capacity of governments to create and enforce rules that are appropriate for local resource management, budget cuts, and movement toward democracy and collective participation.

Despite policy trends toward devolution, it cannot be assumed during policy development that local resource users will want to absorb the costs of management. Dr. Meinzen-Dick recognized the

devolution a transfer of management by which a government relinquishes control over a resource system to a local community.

Figure 9.9 In canal systems located in dry regions of India, where access to water is crucial to local food production, community leadership was important for successful sustainable management once the Indian government returned this responsibility to local communities.

importance of understanding the factors affecting the likelihood that local resource users would be willing to take on these costs. In the Karnataka and Rajasthan states in India, water is scare and irrigation systems are critical to food production. Meinzen-Dick studied 48 different irrigation systems, and determined that local community leadership potential is an important factor in predicting the likelihood of farmer participation. Management of these systems involves both the physical maintenance of irrigation canals and interaction with government officials, who remain partially involved. This is generally known as **co-management**, which occurs when a resource system is managed by some combination of government regulation, privatization, and local community control.

In Indian canal irrigation communities, leadership potential came in the form of college graduates and influential leaders respected by the community. College graduates possessed the entrepreneurial skills necessary for innovation and were usually pivotal in canal maintenance activities. They also had the education, skills, and ideas required to deal with formal government organizations, and to set up a formal local management organization. Influential persons were well-connected within the community and had contact with officials outside of the community. This facilitated the creation of a local organization for irrigation and maintenance of the system through appeals to government agencies for better service.

7. Shared Social Norms

When people in a community have shared social norms, they have similar expectations about how to behave in resource use situations. Oftentimes, these informal rules are internalized (as in the examples of the grocery store checkout line and the Palauan turtles described in Section 9.1). It is easier for groups with shared social norms to establish common goals based on trust and understanding. They are also more likely to exhibit reciprocity. Reciprocity is when a positive action by one member of the group produces a sense of obligation for the receiving group member to return the favor. Community members often genuinely care about other community members when there are shared social norms. Therefore, they view rule-following as "the right thing to do", or as a responsibility.

When a community has shared social norms, it is likely that members will self-organize to collectively manage a resource because the costs for reaching an agreement on rules, monitoring resource use, and enforcing rules are low. Deliberation costs for reaching agreements on rules are low because people who have shared social norms have similar expectations about appropriate behavior, and often have similar values and belief systems.

---

co-management
a situation in which a resource system is managed by some combination of government, market, and community institutions.

Resource users in a community with shared social norms are more likely to follow rules for resource use, because they trust that others in their community will keep their promise to follow the rules as well. They are likely to follow rules even if no one is watching, because a rule violator will be viewed in a negative light by other community members. This provides an incentive not to violate rules, and lowers the costs of monitoring and enforcement. On the flipside, when a large number of outsiders infiltrate a community in a short time period, they may not accept the social norms of the resident community. This lack of shared social norms often results in resource degradation. These ideas can be illustrated by contrasting resource exploitation in two Mediterranean fishing cooperatives studied in the 1970s by Dr. Fikret Berkes at the University of Manitoba in Canada.

The two fishing cooperatives described in this case study are the villages of Alanya and Bodrum, which are both located along the southern coast of Turkey (Figure 9.10). The small, local

Figure 9.10 In Alanya, a fishing community closed to outsiders in the 1970s, social norms strongly influence the behavior of fisherman, but further west along Turkey's southern coast in Bodrum, where the local fishing community fished alongside outsiders, such as commercial fisher and tourists, social norms did not contribute as strongly to sustainability fishery management.

fishing areas are owned by the government, but leased and managed by local cooperatives. In Alanya, fishers are part of a local community where informal rules regulate access to the fishing resource. At the beginning of the fishing season each September, each fisherman is randomly assigned a fishing area based on predefined and agreed-upon fishing locations. As the season progresses, fishers rotate through all possible fishing locations to ensure that each gets an opportunity to fish in the more productive sites. Fisherman who are part of this community all view these rules as fair and tend to follow them. Violators are penalized verbally, and sometimes by use of physical violence, when the local fishermen meet in the village coffee house at the end of the day. As a result of social pressure for acceptance by the community, most people have an incentive to follow the rules. This is analogous to the discomfort you might feel if you cut to the front of the line at a grocery store, and then the people, whom you have just cut in front of, start complaining. In contrast, locals in Bodrum fish alongside commercial fishers and fishing boats chartered by tourists who are not members of the local community, and do not accept their informal local rules. These outsiders have no incentive to follow rules because they do not respond to community social pressures. The idea of shared social norms helps explain why international ToC situations, such as climate change and overexploitation of open ocean fisheries, are so challenging. In these instances, a variety of cultural social norms are at play.

8. Knowledge of the Resource System

When community members have accurate knowledge about the resource system that they are using, the costs of initially creating rules, and adaptively advising those rules over time, will be low. Therefore, they are more likely to self-organize for sustainable management of a resource system that brings benefits. The importance of accurate time and place knowledge in resource system management was mentioned earlier in the context of resource system size. Recall that, in response to increased competition from foreign fishing fleets in deep offshore fisheries, the Canadian government claimed jurisdiction over its exclusive economic zone (EEZ). The EEZ extends Canada's territory 200 miles out into the open ocean. Thus, by assuming responsibility for hundreds to thousands of miles of coastal ocean, extending 200 miles offshore of the provinces of Nova Scotia and Newfoundland, the government had to devise a fisheries management system for an area on the order of 1,000,000 mi^2. They had to do this without the detailed knowledge of local communities, who had been fishing in these waters for generations. However, even in smaller territories, it can be difficult for anyone other than local communities to acquire the long-term, real-time information necessary for adaptive management of changing resource systems.

There are two general reasons that knowledge of a resource is important. First, appropriate information is needed to initially design a sustainable management system. Second, as resource systems change over time, due to either natural cycles or human influences, it is necessary to adapt the management scheme to changing conditions. Thus, knowledge of changes occurring in a resource system is also critical. These ideas can be illustrated through a case study of the Chisasibi Cree Amerindians of the Canadian subarctic, where a combination of local indigenous knowledge, rituals, and general guidelines have informed sustainable management practices of wildlife resources (caribou, beaver, fish, and waterfowl) for thousands of years. This can be viewed in contrast to government wildlife management, which often relies on remote scientific data collection over shorter time scales of years to decades.

The Chisasibi live in a lichen-woodland transition zone of a subarctic ecosystem, characterized by a cold climate (mean annual temperatures of $-3°$ C) and low biological productivity. Their livelihood historically consisted of hunting, trapping, fishing, and gathering activities. Their hunting territory, which borders James Bay in central western Québec, is relatively small (about 100,000 mi^2). Chaotic, unpredictable environmental conditions are typical of the region in general. Wildlife populations, upon which the Chisasibi depend, experience large natural variations in size from year to year. This is especially true for caribou, which temporarily move into Chisasibi hunting territory each year during their annual migration, and have been a subsistence food resource for generations (Figure 9.11). Migrating caribou population sizes fluctuate from year to year. To this day, wildlife ecologists cannot accurately predict these annual population increases and declines because they do not understand the underlying reasons. In contrast, indigenous peoples seem to have the knowledge necessary to foresee caribou population fluctuations. The Inuit, who live further north in arctic ecosystems, believe there is an 80-year natural population cycle. For the Chisasibi, this type of knowledge is derived not from scientific data, but from stories passed down through generations by elders based on the hunters' day-to-day observations.

The Chisasibi used this "data" to hunt and sustainably manage caribou populations for millennia. Then, a change in the caribou resource system dynamics occurred in the 1910s when rifles became available to hunters. With this new technology, the Chisasibi decimated migrating caribou populations. In their fervor of overexploitation, they easily disregarded their traditional practices. They

Figure 9.11 For generations, the Chisasibi Cree hunted caribou primarily as a food source and, as a result, their culture holds valuable knowledge about the behavior of migrating caribou populations.

also failed to observe another traditional practice that involved properly disposing of animal carcasses by burial after slaughtering, which resulted in a river polluted with waste. The caribou did not return the next year and the Chisasibi believed this was because they had disrespected traditional hunting ethics. The caribou slaughter of the 1910s became part of their oral history and greatly influenced the next generation of young hunters. The Chisasibi did not see caribou again until the 1980s. By this time, the Chisasibi Cree Trappers Association (CTA) had been created by the local community to monitor the yearly hunt and ensure it was conducted in a responsible manner, and in accordance with traditional standards. As a result of CTA regulation, the caribou were only lightly hunted this time and returned each year into the future. Not only did they return, but caribou started to re-establish viable populations in areas of the Chisasibi hunting territory in which they had not existed in the past few decades. The Chisasibi believed this to be a result of the restoration of proper hunting ethics, and have continued to adhere to these practices.

Wildlife ecologists studying population dynamics have found that, in some species, light hunting can actually increase population size. When individuals are lost to hunting, there is reduced competition for food, shelter, and other resources among the surviving individuals in the population. This can increase survival and reproduction rates, and ultimately stimulate population growth beyond what it would be if hunting did not occur. However, scientific data for caribou populations migrating through the Chisasibi hunting territory are neither abundant, nor extensive enough, to determine if this hunting effect exists in this species of caribou. Therefore, there are no data available to inform wildlife management by the Canadian government. In contrast, the local Chisasibi "data" seems to have helped them figure out, over time, that light hunting can be beneficial, whereas overexploitation (such as in the 1910s) threatens their resource system. Although the Chisasibi collected their "data" by trial and error, rather than the scientific method, the detailed knowledge accumulated over long time periods has allowed them to sustainably and adaptively manage their caribou resource system for many generations.

Scientific knowledge definitely plays a critical role in certain situations, such as in groundwater and atmospheric resource systems. For example, in the case of the West Basin in Los Angeles (first described under the section on predictability of resource system dynamics), groundwater users could not have known the reason for increased salt in their groundwater, nor that groundwater supplies were being severely depleted without the USGS survey data. Without this knowledge, they would not have been able to devise an effective management plan. Another resource arena where scientific knowledge plays an important role is in addressing climate change. If Svante Arrhenius, a Swedish physical chemist, had not called attention to the role of CO_2 in warming the surface temperature of the Earth through the greenhouse effect in 1896, then we might be wondering today why global mean temperatures are rising.

Finally, it is important to realize that information gathering and rule revision time scales must be compatible. Real-time information about caribou populations allowed the Chisasibi to change hunting rules on a day-to-day basis, if necessary. This can be contrasted with a politically gridlocked central government legislature that, even if accurate information on the situation were available, could not change rules fast enough to keep up with the changes in the dynamic resource systems. This is one reason that it is important for local communities to have full autonomy in creating and enforcing their own rules (see Essential Ingredient 10).

9. **Importance of the Resource System**
If resource users are highly dependent on a resource system for subsistence, then they will be more willing to bear the costs of self-organizing for sustainable management. When people have a high dependence on a resource system, it may be because they do not have the ability to access larger markets that can provide them with similar resources. This can be true in remote regions where local communities are not well connected with larger markets; although these instances are becoming less and less common with globalization. If people have other opportunities available for their

economic livelihood outside of a community, then they have less of an incentive to expend the necessary effort to conserve a resource, because they can simply leave the community when the resource becomes scarce and move to a place where the resource is abundant. However, if community livelihoods are highly dependent on resource availability, then they have stronger incentives to protect the resource.

Another reason for high dependence on a resource may be that substitutes for a resource may not exist. A good example is the relatively recent attention given to peak phosphorus (Figure 9.12). Phosphorus is a chemical element that is needed by plants for photosynthesis, and by other organisms as a key ingredient of DNA and for many metabolic processes. Phosphorus is mined from rocks to provide human societies with fertilizer for crops, and without it, current crop yield would drop drastically. Although controversial, the world phosphorus production from rock sources is expected to peak around 2030. Phosphorus is a special case because, unlike other nonrenewable resources, such as oil, where renewable alternatives exist, there is no known substitute for phosphorus. In anticipation of this scarcity, much effort is being put into developing technologies to increase efficiency of phosphorus use. For example, technologies are being developed for removal of phosphorus from wastewater for use as organic fertilizer. Previously, phosphorus was left in wastewater to be released into ecosystems where it is not only of no use for crop fertilization, but can also cause harm, such as the eutrophication of freshwater systems.

Figure 9.12 The supply of phosphate, an intensively mined resource that his crucial to the success of our present-day food production systems, is expected to reach peak production in the year 2030.

Resource dependence can also be characterized by high commercial value, but sustainable management of this type of resource must be tempered with high levels of rule enforcement. If resource users can sell the resource for cash income, then overexploitation is likely. Therefore, enforcement of rules developed for sustainable resource extraction is necessary to prevent overuse. Researchers studied forest commons in countries around the world using a 15-year dataset created by the International Forestry Resources and Institutions Research Program. They found that forest regrowth was more likely to occur in forests yielding products with high commercial value, but only when enforcement rates were high. When enforcement rates were low, these commercially valued forests actually became more degraded.

10. Full Autonomy to Create and Enforce Rules

When resource users have full autonomy to create and enforce their own rules, the costs of sustainably managing a resource are low. In order for this to be the case, centralized governments need to recognize the effectiveness of rules that local communities use to manage their own resources. In situations in which central governments exert control over resources that were previously managed successfully by local communities, degradation often accelerates rather than slowing down. This is illustrated by a case study of forest nationalization in the Himalayan foothills of Nepal.

Due to increasing human populations, the accompanying growing demand for forest products, and observed forest degradation, the government of Nepal took control of all forest land in 1957 (Figure 9.13). There were several problems with this approach, and it eventually led to accelerated deforestation rather than forest conservation, which was the intent of nationalization. The first major problem was the difficulty of implementing and enforcing government regulations in thousands of extremely inaccessible and remote forests across a mountainous landscape. Firewood harvesting required government permits, which households had to acquire from forest rangers located more than a day's walk from their village. As a result, it was easier for people to illegally harvest wood, especially since enforcement by forest rangers was next to impossible. On top of that, local communities now viewed forests as government property rather than their own. Instead of protecting their forests from outsiders, using traditional community-based management systems, they now had no incentive to protect forests that they no longer owned. Overall, forest degradation accelerated after nationalization.

By 1978, the Nepalese government recognized this increase in degradation and returned control of the forests to local communities. As a

Figure 9.13 When forest degradation in the Himalayan foothills increases in the mid-twentieth century, the national government took control of these forests, which made forest degradation even worse.

result, forest degradation slowed. The government of Nepal still owns all forest land, but under three different tenure regimes based on management practices: (1) *national forests* managed by the government, (2) *community forests* managed by local communities, and (3) *leasehold forestry programs*, where highly degraded land is leased to a small number of poor households for subsistence living. In 2007, Dr. Harini Nagendra of the Center for the Study of Institutions, Population, and Environmental Change at Indiana University compared 55 different Nepalese forests and found that forest density was increasing in community forests and leasehold forestry programs, whereas national forests continued to be degraded. **Box 9.1** describes a similar case study focused on management of rice paddies in Bali.

BOX 9.1 SUSTAINABLE AGRICULTURE AND RESOURCE GOVERNANCE IN BALI

By Ramanuj Mitra, MS Candidate, School of Sustainability, Arizona State University

The island of Bali in Indonesia is blessed with fertile volcanic soils. This makes it ideal for rice cultivation, which has been the primary occupation of the inhabitants of Bali for centuries. A large number of rivers and streams run down the hill sides, cutting deep gullies into the land on the island. This makes the land less tractable. Hence, farmers cultivate rice by steppe farming and irrigate their fields through a network of canals (**Figure 9.14**). The areas serviced by these canals are known as Subaks. In Bali, agricultural practices are deeply tied to religion and rituals. The elements of nature are worshipped as Gods and Goddesses. Each Subak has a temple, where farmers gather to discuss irrigation, planting schedules, fallow periods, and resolve disputes. Planting and irrigation schedules are democratically organized in order to optimize yields, prevent infestation by pests, and make irrigation water available to the farthest reaches of the system downhill. A group of Subak temples are governed by a larger water temple, and every water temple falls under the mother temple on Lake Batour, which is considered the source of all water on the island (Figure 9.15). Temple priests govern the rituals and act as mediators to resolve disputes regarding irrigation and planting. This ancient system granted perfect autonomy to Bali in organizing its rice cultivation. Unfortunately, things changed drastically with outside intervention.

Figure 9.14 Steep hillsides, cut by gullies, make rice cultivation on stepped fields necessary.

Food shortage hit Indonesia in the 1960s and 1970s. Chaos and discontentment led to military intervention. The old government was overthrown and the new government focused its resources on Bali, which was

Figure 9.15 This temple on Lake Batour, which is a popular tourist attraction, governs all Subak temples on the island of Bali.

considered the "rice bowl" of Indonesia. Measures were implemented to increase rice yield. Around the same time, the International Rice Research Institute in the Philippines developed a high yielding variety of seed that would double the yield of rice, but was dependent on heavy doses of fertilizers and water. The Asian Development Bank (ADB) spearheaded a project to fund "agriculture packets," consisting of seeds and fertilizers that all farmers were required to use. Modern management techniques were imposed on the Subaks, while the role of the temples in water governance was ignored as a ritualistic and superstitious practice. For the next two decades, the well-organized modern system imposed by the government fell into total chaos. The new seed variety proved to be highly susceptible to pests,

Contributed by Ramanuj Mitra. Copyright © Kendall Hunt Publishing Company.

(Continued)

and each new version of genetically modified seed was vulnerable to different factors. Pest attacks increased manifold due to the absence of synchronized fallow periods. The irrigation system failed without the temples to oversee democratic use of water and planned planting schedules. Use of fertilizers and pesticides rose steadily until the authorities had to use planes to spray pesticides on the island. Overuse of phosphates and other chemicals in the fertilizers led to problems downstream, such as formation of algae, contamination of water sources, and destruction of coral reefs in the surrounding ocean. Farmers began selling their land to developers, endangering the whole integrated system. Luckily, before the whole system fell apart, Dr. Stephen Lansing and his team at the University of Arizona were able to use their research, which focused on ecological modeling of the Subaks, to convince the ADB of the superiority of the traditional system. Eventually, management was handed back to the Subaks. In 2012, the Subaks were declared as a UNESCO World Heritage site.

Section 9.3: Evaluating the Tragedy of the Commons Using Essential Ingredients

Core Question: Do Hardin's conclusions regarding solutions to the Tragedy of the Commons hold up against the essential ingredients?

The essential ingredients described in the last section are based on the eight design principles first identified by Ostrom in her 1990 book *Governing the Commons*. They are described in Section 9.2, as laid out by Ostrom and others in a 2009 *Science* article describing characteristics of communities that are likely to self-organize to manage their natural resources over the long term. However, there are many more essential ingredients of successful common property regimes, and those described in the last section are only a small subset of the total set of variables that researchers have identified. At least 43 additional variables have been identified as important, for a grand total of 53 variables, and these are listed in **Table 9.2**. The purpose of this chapter is to provide a taste of the complexity and diversity of community-based management schemes. Therefore, we will not go any further into descriptions of the other factors important for predicting when a CBM situation will result in sustainable management, and when it will not. However, the essential ingredients listed in Table 9.2 will be used to illustrate why Hardin's conclusion, that government regulation or privatization are the only means for preventing ToC, is a vast oversimplification of reality. For

Table 9.2 Essential Ingredients of Successful Common Property Regimes. Adapted from Table 1 in Ostrom's 2009 *Science* article titled "A General Framework for Analyzing Sustainability of Socio-Ecological Systems." Ingredients marked with an asterisk appeared in Table 1 and were discussed in Section 2 of this chapter. Superscripted numbers correspond to the numbering system in Table 1. † symbols indicate ingredients considered by Hardin in his herder scenario.

Essential Ingredients	
Characteristics of the Resource System (RS)	*Characteristics of the Governance System (GS)*
RS1 – Type (e.g. water, forests, pasture, fishery)†	GS1 – Government organizations
RS2 – Clarity of system boundaries	GS2 – Non-government organizations
RS3 – Size of the resource system*(1)†	GS3 – Network structure
RS4 – Human-constructed facilities	GS4 – Property-rights system
RS5 – Productivity of the resource system*(2)†	GS5 – Operational rules
RS6 – Equilibrium properties	GS6 – Full autonomy to create and enforce*(10)
RS7 – Predictability of the dynamics*(3)	GS7 – Constitutional rules
RS8 – Storage characteristics	GS8 – Monitoring and sanctioning processes
RS9 – Location	
Characteristics of the Resource Units (RU)	*Characteristics of Interactions*
RU1 – Resource unit mobility*(4)†	I1 – Harvesting levels of diverse users†
RU2 – Growth or replacement rate	I2 – Information sharing among users
RU3 – Interaction among resource units	I3 – Deliberation processes
RU4 – Economic value†	I4 – Conflicts among users
RU5 – Size	I5 – Investment activities
RU6 – Distinctive markings†	I6 – Lobbying activities
RU7 – Spatial and temporal distribution	I7 – Self-organizing activities
	I8 – Network activities
Characteristics of the Resource Users (U)	*Characteristics of Outcomes*
U1 – Number of users*(5)†	O1 – Social performance measures
U2 – Socioeconomic attributes of users	O2 – Ecological performance measures†
U3 – History of use	O3 – Externalities to other SESs
U4 – Location	*Characteristics of Related Ecosystems*
U5 – Leadership and entrepreneurship*(6)	ECO1 – Climate patterns
U6 – Shared social norms*(7)	ECO2 – Pollution patterns
U7 – Knowledge of the resource system*(8)†	ECO3 – Flows into and out of focal SES
U8 – Importance of the resource*(9)	
U9 – Technology used	

the last part of this chapter, the herder scenario is assessed within the framework of the 53 essential ingredients listed in Table 9.2.

The essential ingredients are factors that help distinguish between situations that are likely to lead to sustainable community management of a resource, and situations that are likely to end in resource loss and, consequently, tragedies for the societies depending on that resource. In Hardin's scenario, herders acting in their individual self-interest add cattle to a pasture for a full positive benefit to themselves, but only a partial negative cost. Over time, the pasture resource system eventually becomes degraded; this is a bad outcome for the herders as a whole. In order to come to this conclusion, Hardin sets up the scenario as follows, according to the variables shown in Table 9.2 (variables considered by Hardin are marked with a † symbol in the table):

- The resource system type is a pasture (RS1).
- There is no governance system associated with the pasture (no GS variable).
- The cows are mobile, but on a stationary grass resource that composes the pasture (RU1).
- Cows have distinctive markers (perhaps they are branded) so that they can be identified on the common pasture system by the owner (RU6) and have market value such that they can be sold for a profit after a period of grazing and fattening (RU4).
- The size of the pasture is small (RS3) relative to the number of resource users (U1) and, as a result, the resource users are clearly using the pasture in a way that threatens its long-term productivity (RS5).

In Hardin's scenario, resource users do not communicate with each other, but rather independently make decisions about resource use that maximize their own short-term gains (U7). This leads to an interaction between the resource users and the resource, with high harvesting rates of the pasture grasses by all herders' cows (I1). The outcome is overuse and ecological degradation of the pasture resource system (O2).

Now, let us return to the real-world case study of the Maine lobster fishers described in Section 9.2, put it into the same framework, and compare it to Hardin's herder scenario. The lobster fisher scenario is similar to Hardin's in the following ways:

- With the exception of seasonal migrations to deeper waters, the lobsters are a relatively stationary resource (like the cows in the pasture) that are easy to keep track of (RU1).
- Lobsters that could be legally sold have distinctive marks (RU6). In Maine fisheries, females of reproductive age are marked with V-notches to prevent pregnant females from being harvested and sold. The marks disappear after two to three molts.

- The lobsters have market value such that they can be sold for a profit after catching them (RU4),
- The size of the fishing area is small (RS3) relative to the possible number of resource users able to obtain a lobstering permit (U1).
- If resource users were to overharvest lobsters, there would be a clear indication (decreased yield) that overuse was occurring and threatening the long-term productivity of the lobster fishery (RS5).

The Maine lobster fishery differs from Hardin's herder scenario in some ways, and there are additional essential ingredients relevant to the lobster fisher scenario that were not even considered by Hardin. There is no governance system associated with Hardin's pasture (no GS variable); whereas state licensing regulations in Maine (GS1) require that all lobster fishers mark their lobster pots with a license number. This allows local lobster fishers to identify outsiders fishing in their territory (GS8), thereby keeping harvesting rates low (I1) by limiting the number of resource users (U1). Hardin's herders did not communicate with each other, but the lobster fishers are part of fishing communities with regular face-to-face interaction that encourages discussion about resource use, builds informal social norms (U6), and fosters cooperative interaction that avoids conflicts among users (I4). The nature of the small lobstering communities also results in low costs for sharing information about the resource (I2), as well as for getting together as a group for deliberation when decisions about resource use need to be made (I3). Finally, lobster fishers have lived in these communities for generations (U3), and as a result have a strong sense of place and deep roots in their communities (U4), have local leadership (U5), and have accurate real-time knowledge about their resource system that can be used to adapt rules over time (U7). This all results in a situation where the lobster fishery has been used sustainably for generations (O2).

In the lobster fisher case study, 20 essential ingredients come into play. This can be contrasted with Hardin's herder scenario, where only 10 were considered. It is easy to imagine a real-world pasture commons situation in which herders might exhibit some of the additional essential ingredients included in the lobster fishery case study. In fact, with the exception of commercial fishing in international waters of the unregulated open ocean, most real-world commons situations are much more complex than Hardin's herder scenario, and many do not involve tragedies of resource overuse. Therefore, it can be concluded that Hardin's influential work was an oversimplification of what, in reality, is often a complex commons situation.

Hardin's work has greatly influenced resource management policy, as it has imbued decision makers with the underlying assumption that only government regulation or privatization can solve ToC dilemmas. However, as illustrated in this chapter, communities *can* and *do* sustainability manage their own resource systems. The trick is determining in which instances communities are likely to self-organize for sustainable management, and in which instances some degree of government regulation or privatization might need to be part of the institutional arrangement. This knowledge would help us to better inform local, state, national, and international policy aimed at natural resource management. It is important to realize that there is no panacea, or a one size fits all solution, to ToC dilemmas. Resource systems in the real world are complex, variable, and diverse. Thus, we need a diversity of institutional arrangements—governments, markets, and communities—that can match the diversity of resource management situations that exist.

Bibliography

Acheson, J. M. 1975. "The Lobster Fiefs: Economic and Ecological Effects of Territoriality in the Maine Lobster Industry." *Human Ecology* 3 (3): 183–207.

Acheson, J. M. 1987. "The lobster fiefs, revisited: Economic and Ecological Effects of Territoriality in the Maine Lobster Industry". In *The Question of the Commons*, edited by B. McCay and J. Acheson, 37–65. Tucson: University of Arizona Press.

Acheson, J. M., J. A. Wilson, and R. S. Steneck. 1998. "Managing Chaotic Fisheries In *Linking Social and Ecological Systems: Management Practices and Social Mechanisms for Building Resilience*, edited by F. Berkes, C. Folke, and J. Colding, 390–413. New York: Cambridge University Press.

Agrawal, A. 2001. "Common Property Institutions and Sustainable Governance of Resources." *World Development* 29 (10): 1649–72.

Arnold, J. E. M., and J. G. Campbell. 1986. "Collective Management of Hill Forests in Nepal: The Community Forestry Development Project." In *Proceedings of the Conference on Common Property Resource Management*, 425–54. Washington, DC: National Research Council, National Academy Press.

Arrhenius, S. A. 1896. "On the Influence of Carbonic Acid in the Air upon the Temperature of the Ground." *Philosophical Magazine* 41: 237–76.

Baland, J-M., and J-P. Platteau. 1996. *Halting Degradation of Natural Resources: Is there There a Role for Rural Communities?* Food and Agriculture Organization of the United Nations. Accessed October 31, 2011. Available online from Indiana University at http://dlc.dlib.indiana.edu/dlc/bitstream/handle/10535/21/Halting_degradation_of_natural_resources.pdf?sequence.

Banfield, A. W. F., and J. S. Tener. 1958. "A Preliminary Study of the Ungava Caribou." *Journal of Mammology* 39: 560–73.

Berkes, F. 1986. "Local-Level Management and the Commons Problem: A Comparative Study of Turkish Coastal Fisheries." *Marine Policy* 10 (3): 215–29.

Berkes, F. 1986. "Marine Inshore Fishery Management in Turkey." In *Proceedings of the Conference on Common Property Resource Management*, 63–84. Washington, DC: National Research Council, National Academy Press.

Berkes, F. 1998. "Indigenous Knowledge and Resource Management Systems in the Canadian Subarctic." In *Linking Social and Ecological Systems: Management Practices and Social Mechanisms for Building Resilience*, edited by F. Berkes, C. Folke, and J. Colding, 98–128. New York: Cambridge University Press.

Block, B. A., S. L. H. Teo, A. Walli, A. Boustany, M. J. W. Stokesbury, C. J. Farwell, K. C. Weng, H. Dewar, and T. D. Williams. 2005. "Electronic Tagging and Population Structure of Atlantic Bluefin Tuna." *Nature* 434: 1121–27.

Chambers, R. 1997. Whose Reality Counts? Putting the First Last. Institute of Development Studies. Intermediate Technology Publications.

Chhatr, A., and A. Agrawal. 2008. "Forest Commons and Local Enforcement." *Proceedings of the National Academy of Science of the United States* 105 (36): 13286–91.

Cushing, D. H. 1977. "The Study of Stock and Recruitment." In *Fish Population Dynamics*, edited by J. A. Gulland, 105–28. London: John Wiley and Sons.

Cunningham, B. 2012. "Sustainable Agriculture in Bali." Accessed June 16, 2017. https://www.youtube.com/watch?v=dGOQLX86gEE.

Fora TV. 2013. "A Perfect Order: A Thousand Years in Bali, J. S. Lansing." Accessed ... https://www.youtube.com/watch?v=Ir92cnpX7Pw]

Hardin, G. 1968. "The Tragedy of the Commons." *Science* 162: 1243–48.

Johannes, R. E. 1981. *Words of the Lagoon: Fishing and Marine Lore in the Palau District of Micronesia*. Los Angeles: University of California Press.

Meinzen-Dick, R. 2002. "What Affects Organization and Collective Action for Managing Resources? Evidence from Canal Irrigation Systems in India." *World Development* 30 (4): 649–66.

Nagendra, H. 2007. "Drivers of Reforestation in Human-Dominated Forests." *Proceedings of the National Academy of Science of the United States* 104 (39): 15218–23.

National Research Council. 2002. *The Drama of the Commons*. Washington, DC: National Academy Press.

Ostrom, E. 1990. *Governing the Commons: The Evolution of Institutions for Collective Action*. New York: Cambridge University Press.

Ostrom, E. 2005. *Understanding Institutional Diversity*. Princeton, NJ: Princeton University Press.

Ostrom, E. 2009. "A General Framework for Analyzing Sustainability of Socio-Ecological Systems." Science 325: 419–22.

Ostrom, E., R. Garder, and J. Walker. 1994. *Rules, Games, and Common-Pool Resources. Ann Arbor:* University of Michigan Press.

Wade, R. 1988. *Village Republics: Economic Conditions for Collective Action in South India*. New York: Cambridge University Press.

End-of-Chapter Questions

General Questions

1. Determine whether each of the resources described in the examples below has (i) high or low excludability and (ii) high or low rivalry. Then, use Figure 9.1 to determine whether the resource is a toll or club good, a private good, a public good, or a common-pool resource. Finally, determine whether the resource is susceptible to a tragedy of the commons situation. Why or why not?
   a. Clean air in an urban area
   b. Money in your bank account
   c. National public radio
   d. City botanical gardens

2. Use the ten essential ingredients to come up with reasons to explain why the hypothetical communities described below will or will not sustainably manage their own resources.
   a. People living on a small island in a remote region of the South Pacific Ocean have collectively managed their fisheries for generations. They are isolated from global food markets and dependent on their fisheries for food. At the start of each fishing season, based on the fish catches observed in the previous season, rules are developed to define catch limits for each family. Their local religion includes a god of the ocean. Based on their religious beliefs, they must avoid taking too many fish from the ocean at once. If they do, they know he will bring lower fish populations to their island the next year.

b. Villagers in a small, remote, dryland region of Northeast India have collectively managed an irrigation system for generations. They depend on the system to water their crops and for drinking water. At the start of each planting season, they use observations from the previous season to determine rules for water that year. Their religion includes a goddess of irrigation systems. They know that they must avoid contaminating irrigation waters with human and animal waste. If they do, they know she will bring a grave illness to their family.

3. Watch this short video, called *The Solution to Pollution is Life*, on community engagement in soil remediation for urban farming, which is occurring today in the city of Philadelphia: http://ourland.tv/?p=336

   Based on this video, which of the ten essential ingredients, needed for successful community-based management of this resource system, are present? Which are still needed?

Project Questions

1. Use the excludability and rivalry framework, as presented in Figure 9.1, to characterize the resource system(s) of concern to your sustainability problem. Determine if this resource system(s) is subject to a tragedy of the commons situation, explain why or why not.

2. Use the ten essential ingredients, as defined in Section 9.2 of this chapter, to analyze your system. Based on your analysis, how likely is it that communities will self-organize to sustainability manage the resource systems of concern in the system that you defined in Question 1 of this section?

GLOSSARY

active stakeholders   a class of stakeholders who have direct or indirect influence over an issue.

adaptation   actions aimed at adjusting human or natural systems to minimize the harmful impacts of hazards, such as by changing agricultural practices in ways that will allow food production systems to continue to function in a different climate.

adaptive capacity   the ability of a system to innovate, reorganize, and learn when faced with change.

adaptive cycle   four different phases of long-term change characteristic of many complex systems.

agency   human action.

albedo   the fraction of solar radiation that is reflected from a surface rather than absorbed.

allogenic drivers   the external forces that disturb the directional development pathway of a system by either setting it back to an earlier stage or causing it to move in a new direction.

alternative stable state   a term often used in place of regime.

assets   the beneficial and valuable aspects of a system that already exist and that can be used to "get things done" in ways that promote the desired transition.

autogenic drivers   the internal forces that regulate a system's behavior and cause it to move forward along a certain directional pathway of development.

barriers   the roadblocks or challenges to "getting things done" that are at the heart of any transition.

beliefs (aka. worldview)   a sense of how the world works that is susceptible to knowledge and, therefore, can be fundamentally correct or incorrect.

biological community   multiple populations of different species living in the same area and interacting with each other in a variety of ways.

biological population   all individuals of the same species that live in the same geographical location and are capable of producing offspring.

boundary   an imaginary line drawn around a system to conceptually separate the internal components and interactions from the environment external to the system.

carrying capacity   the maximum number of individuals of a given species that a certain environment can continue to support with the available resource base.

cascading effects a chain of events set off by an action in one part of a system that results in relatively larger and typically unpredictable changes in the rest of the system.

causal chain analysis a tool for analyzing socioecological systems by classifying key drivers, establishing the relationships among drivers, and determining their relative influences on a system.

change agent an individual who uses his or her knowledge, skills, and attitudes to directly or indirectly serve as a catalyst for change.

chaotic the type of unpredictable behavior exhibited by a system that lacks strong internal feedbacks such that the state of a system at any given time is determined largely by random external disturbances.

closed systems systems in which energy, materials, and information cannot move across system boundaries from the external environment into the system nor vice versa.

co-management a situation in which a resource system is managed by some combination of government, market, and community institutions.

coproduction a process through which knowledge about a situation is generated by blending the specialized knowledge of experts with the practical knowledge of local communities.

coevolution the continual process by which structure and agency change together and shape each other over time.

common-pool resource (CPR) a type of resource that is susceptible to Tragedy of the Commons situations due to its non-excludability and rivalry.

common property regime an institutional arrangement by which a common resource is jointly owned by a community of resource users.

community-based management (CBM) the practice of resource management by a community of resource users.

complex adaptive system (CAS) a system capable of evolving over time in a manner that helps it adjust to changing conditions and typically in ways that promote its survival.

complex system a system in which the many parts that compose the system are interconnected in an irreducible way, such that the whole system is greater than the sum of the parts.

component a single element in a system that plays a specific role as part of the overall system.

critical transition a nonlinear threshold-based regime shift that is hard to reverse and often difficult to anticipate in socioecological systems.

cultural eutrophication response of aquatic ecosystems to the addition of excess nutrients that are a by-product of human activities.

current state analysis a process used to evaluate and understand the present-day situation of a given socioecological system, and for a certain sustainability problem facing that system, with the aim of producing information and insight for thinking about the future of that system and devising strategies for the transition of that system to a more sustainable future.

desertification a phenomenon induced by human activity that occurs in dryland systems and results in the loss or decline of economic or biological yield as a dryland system shifts to a less productive system.

desirable futures futures that are preferred.

devolution a transfer of management by which a government relinquishes control over a resource system to a local community.

direct drivers drivers that clearly and unequivocally influence the behavior of a system.

disincentives tools and mechanisms that are used to discourage certain behaviors.

dynamics patterns of change exhibited by a system over time.

economic development a process by which the quality of life is improved for everyone without damaging the resources and services provided by natural systems.

ecosystem services the benefits afforded by ecosystems to human systems that make human life both possible and worth living, including goods such as food, fiber, fuel, and fresh water and services such as soil formation, nutrient cycling, crop pollination, climate regulation, and spiritual values and inspiration.

emergent property a sophisticated feature that emerges from a system and cannot be explained from the behavior of the system's individual components nor the simple interactions among them.

environmental generational amnesia a concept from psychology that explains why people often do not realize the full extent of environmental degradation that has occurred due to the tendency to measure environmental degradation against the time scale of their own lifetimes rather than degradation that has occurred slowly over the hundreds of years before they were born.

eutrophication the process by which a body of freshwater or saltwater, such as a lake or coastal area, receives excessive quantities of nutrients that lead to low oxygen conditions and toxic algae blooms.

excludability a measure of the cost that is required to prevent users from accessing a resource.

excludable a resource that has a low cost associated with excluding resource users from using it.

fast driver a type of driver that rapidly influences a system over very short time periods.

feedback a specific type of interaction among system components, or with the external environment, such that the outcome of some process returns to affect the factor that originally initiated the process.

flow a system interaction that represents the rate at which some quantity of energy, material, or information is transferred among system components or between the system and its external environment.

fluctuations periodic changes in the state of a system within a given regime that do not constitute a regime shift.

functional diversity the ecological processes of energy flow and material cycling that are required for the success of populations of species, biological communities, and ecosystems.

fundamental changes the essential transformations in values, beliefs, and norms, actions, and basic regime characteristics that must happen in a socioecological system in order for it to transition from the current state to the vision.

Gross Domestic Product (GDP) the total market value of all goods and services produced in a certain period of time, usually a year.

hierarchy different levels of organization by which items are arranged as a variety of levels relative to each other.

holistic thinking   a point of view that claims systems *cannot* be understood by studying each individual component and the interactions among those components in isolation from the overall system.

human agency   the ability of humans to make decisions that influence the behavior of socioecological systems over time.

human capital   the physical labor and mental talents of people in economic systems.

incentives   tools and mechanisms that are used to encourage certain behaviors.

indicators   quantitative or qualitative data used to determine the current state of a system, how far it is from some ultimate goal, and which way it is changing, or whether it is even changing at all.

indirect drivers   drivers that influence the behavior of a system in a more diffuse way by altering one or more direct drivers.

inertia   the momentum of a system that keeps it moving in a certain direction and is difficult to resist when attempting to stop the system or steer it in a different direction.

infiltration   the downward passage of water through the soil.

inherited trait   a characteristic that is controlled by genes and is passed from parents to offspring.

innovation   a new technological development, including both physical objects and knowledge systems.

input variables   information about a specific system that is measured in some way and used as data for models.

institutional arrangement   some combination of formal and informal institutions used together for effective and sustainable resource management.

integrity   human desires beyond basic needs that make life worth living.

interaction   a process through which a subset of system components relate to each other.

intergenerational equity   social equity among people living today and those who will live in the future.

Intergovernmental Panel on Climate Change (IPCC)   a scientific body established in 1988 by the World Meteorological Organization and the United Nations Environmental Programme to regularly review the most recent work of thousands of climate scientists from 195 countries around the world and assess the potential environmental, social, and economic impacts of climate change based on this review in order to provide the most up-to-date scientific information for decision makers.

intervention point (aka. leverage point)   an efficient place to intervene in a system because a relatively small change or effort ends up causing a large overall shift in the behavior of an entire system.

intragenerational equity   social equity among all people living on Earth today.

key stakeholders   a critical class of stakeholders, who can be primary or secondary stakeholders or neither, and who have the power to significantly influence or change a given situation.

latitude (L)   a measure of the maximum amount of change a system can undergo before crossing a threshold.

livelihood   the continuous means by which a person adequately secures life's necessities to meet the needs of the self and household in a dignified way, such as through earning wages or self-employment.

manufactured capital   the factories, equipment, and other types of infrastructure in economic systems.

megafauna   a word used, especially by paleontologists and archeologists, to denote any large mammal with a weight in excess of 100 pounds.

Millennium Ecosystem Assessment (MEA)   a five-year project (2001–2005) based on the work of more than 1,300 international experts that assessed the present conditions and future trends of ecosystem services worldwide, the effects of changes in the conditions of these services on human well-being, and the options for action that would promote the conservation and sustainable use of these services.

mitigation   actions intended to permanently eliminate or lessen long-term hazards or risks to human systems, such as by ceasing or reducing anthropogenic CO_2 emissions contributing to climate change.

multi-level perspective (MLP)   a framework derived from the social sciences used for thinking about how transitions occur as a result of interactions occurring among processes operating at three different functional levels.

multi-phase concept (MPC)   a framework derived from the social sciences used for thinking about how transitions occur based on four different phases that a socioecological system passes through over time as it experiences a transition.

natural capital   the resources and services provided by natural systems that support and sustain human activities.

negative externality   a cost incurred by an individual or group who did not agree to the action that caused the harm either because they did not have a choice or their interests were not taken into account.

nested arrangements   institutional arrangements used to promote effective, bottom-up community management of resources when large group size is an issue by nesting smaller management units within larger units.

niche   describes the "way of life" of a given species as defined by the total uses made by the species of the abiotic and biotic resources available in the environment in which it lives.

non-excludable   a resource that has a high cost associated with excluding resource users from using it.

nonlinear   an interaction defined by the fact that the magnitude of the factor causing some interaction in a system and the actual outcome of that interaction are not proportional.

open systems   systems in which energy, materials, and information can move freely across system boundaries from the external environment into the system and vice versa.

order of magnitude   an expression of the approximate magnitude of something based on a power of 10 difference between that quantity relative to another.

outsiders   resource users that are not part of the community of users that compose a common property regime.

participatory approach   a democratic problem-solving process that incorporates both the specialized knowledge of experts and the practical knowledge of local communities as a basis for defining and understanding problems and devising solution options.

passive stakeholders   a class of stakeholders who are affected by the actions of active stakeholders, but who do not necessarily influence the problem.

path dependence   a condition under which past events and present actions lay the groundwork for socioecological system behavior in the future by constraining possible development pathways and influencing present-day decisions.

per capita   a term that literally means "by head" in Latin and is used to denote the average per person, such as the quantity of resources used or pollution generated by the average person in a certain context.

percolation   the general movement of any liquid, including water, through porous material such as sediment, gravel, or soil.

plausible futures   futures that could possibly occur but have not been deemed the most likely because they arise from unanticipated events.

precariousness (P)   a measure of the nearness of a system to a threshold.

pressure indicators (aka. control indicators)   a class of indicators focused on the cause of a problem by tracking the processes regulating a system's behavior that lead to certain conditions within a system.

primary stakeholders   a class of stakeholders who strongly affect other stakeholders by their actions or are profoundly affected by the actions of other stakeholders.

probable futures   futures that are deemed likely to happen based on knowledge of past systems.

proxy variable   a quantifiable variable used in a computer model in place of a qualitative variable that cannot be directly represented in a model.

Q_{10} factor   a dimensionless quantity used in both biology and chemistry to define how reaction rates change with every 10°C change in temperature, which is typically by a factor of about 2 to 3 times.

reductionist thinking   a point of view that claims systems can be understood by studying each individual component and the interactions among those components as separate from the overall system.

reflexivity   a quality unique to the human components of socioecological systems, as opposed to their other biophysical components, that results in the actual decisions made and actions taken being influenced by the very act of *thinking about* the future impacts of decisions and actions.

regime (aka. alternative stable state)   the dynamic and constantly changing yet characteristic pattern of conditions under which a system can exist and by which it supports certain functions or purposes.

regime shift   a typically nonlinear change process by which a system transitions from an existing regime to a new regime under which the system supports different functions or purposes than before.

reinforcing feedback (aka. positive feedback or amplifying feedback)   a general feedback type that causes a system to change and ultimately shift to a new regime.

resilience   the ability of a system to absorb impacts caused by disturbances and remain in its current regime with the same conditions, structure, function, and identity.

resistance (R)   a measure of how easy or difficult it is to change a system.

response diversity   the number of different species in the same functional group in an ecosystem that have varied responses to different disturbances.

response indicators   a class of indicators focused on evaluating solution strategy effectiveness by assessing the stakeholder actions that were taken in response to a problem and its cause.

rivalry (aka. subtractability)   a measure of the quantity of a resource left over after it has been consumed by a subset of resource users.

runoff   the portion of total precipitation that is not absorbed into the soil and instead flows across the land surface.

scales   a reference by which to classify, arrange, and understand internal system components and their interactions in relation to their external environment.

scenario   a carefully constructed quantitative model or qualitative story about one of many plausible alternative pathways a system might take into the future and what its future state might look like.

secondary stakeholders   a class of intermediary stakeholders who indirectly affect other stakeholders or who are only slightly affected by the actions of other stakeholders.

self-organization   the ability of a socioecological system to adjust and adapt its behavior over time in a way that promotes its survival in the face of changing conditions.

shifting baselines   a concept from ecology that denotes the continual, slow, almost imperceptible changes in baseline reference conditions that occur over long periods of time.

shifting dominance   a constant interplay between the two general types of feedback in a system that compete with each other for dominance and that determines the system's overall behavior at any given time.

simple system   a system in which components interact with each other to serve some purpose, but are not connected in an irreducible way as in a complex system.

slow driver   a type of driver that influences systems slowly and over relatively long periods of time.

social capital   the productive social relationships grounded in mutual trust and understanding by which diverse groups of people have the capacity to communicate with one another, find common ground, and work together to resolve sustainability problems.

social equity   equal opportunity for all people through fair access to education, livelihoods, and other resources, democratic participation, and self-determination.

social norms   defines what is approved or disapproved of within a society and, as a result, tends to shape or guide people's actions.

socioecological systems (SESs)   an integrated conceptual model for understanding how energy, matter, and information are transferred among the different components of natural and human systems and how these exchanges impact the long-term development of both systems.

stability landscape   the character of a system's internal feedback structure at a given point in time as determined in part by the shifting dominance among stabilizing and reinforcing feedbacks.

stabilizing feedback (aka. negative feedback) a general feedback type that keeps a system in its current regime.

stakeholders individuals, organizations, or other entities that benefit or are harmed by other's actions, or who carry out the actions causing the benefits or harms, and as a result have an interest in some policy, conflict, organizational goal, or other issue that will influence their future actions.

state the conditions under which a system exists at any given time.

state indicators a class of indicators focused on the problem outcomes by tracking undesirable conditions within a system, or other indirect factors that shape these conditions, and render the system unsustainable.

stock (aka. reservoir) a system component that reflects some quantity of energy, material, or information contained in a system.

strategic actions planned actions focused on specific interim targets and carried out at different phases of a transition that are designed to make use of assets and overcome barriers in ways that move the socioecological system continually forward toward the ultimate vision.

strong sustainability a belief that the substitutability of natural system components by technologies developed within human systems is limited and that tradeoffs among the three pillars of sustainability are also limited.

structuration theory a social science theory that explains the establishment and propagation of societal systems as the continual and mutually influential interaction between structure and agency.

structure factors that constrain human action.

substitutability refers to a situation in which a resource can be replaced by a man-made technology.

sustainability transition a profound change in the way an existing socioecological system functions to meets the needs of society *in ways that promote healthy ecosystems, human well-being, and viable economies.*

system a set of components and the interactions among them that function together as a whole to accomplish or serve some purpose.

system relationships interactions among different system components that are used in quantitative models when they can be defined mathematically, such as in equations.

tame problems relatively simple problems that can be solved using traditional approaches, such as scientific and technological advances.

targets intermediate goals for each transition phase that guide a socioecological system toward the longer-term goals of the vision.

threshold (aka. tipping point) the point beyond which a system shifts nonlinearly to a different regime.

tradeoff a situation in which one or more aspects of a pillar of sustainability are lost in exchange for gaining one or more aspects of the other sustainability pillars.

transdisciplinary knowledge and understanding based on both the specialized expertise of academic experts and the practical real-world know-how of communities outside of a university.

Transformational Sustainability Research (TSR) a general problem-solving framework that can be used to understand and resolve sustainability problems.

transition a profound change in the way an existing socioecological system functions to meet the needs of society.

transpiration the process by which plants absorb water in liquid form, typically through their roots, and release it into the atmosphere as water vapor, primarily through openings on their leaves called stomata.

troposphere the layer of the atmosphere closest to the earth's surface that extends from 4 miles elevation at the poles to 11 miles at the equator and is characterized by temperature decreases with increases in altitude of about 6.5°C per kilometer.

unintended consequence an outcome that was not intended that can be beneficial or harmful and is typically not expected or anticipated due to the inherent complexity of socioecological systems.

value a general principle or standard that defines what is important and provides criteria for action based on a fundamental sense of what is worthwhile and what is not; cannot be deemed correct or incorrect.

vision a desirable future ideal envisioned by a society or a subset of society.

visioning the process by which a vision is collectively created by a society or a subset of society.

vitality basic needs for survival that make life possible.

vulnerability science an emerging field that seeks to understand socioecological systems with a focus on lessening or eliminating harm to human systems and, to a lesser extent, the supporting natural systems.

weak sustainability a belief that the resources and services provided by natural systems can be substituted for by technologies developed by human systems and that extensive tradeoffs can be made among components of the three pillars of sustainability.

wicked problems difficult problems that cannot be addressed using only traditional approaches, such as scientific and technological advances, and that require continuous attention because they can never be completely resolved.

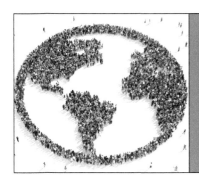

INDEX

A

Active stakeholders, 165
Activism, sustainability, 1–2
Adaptation, 7
   biological evolution, natural selection, 361–362
   cultural evolution by learning, 362–366
   and path dependence in Netherlands, 364–366
   process, 360
Adaptive capacity, 257
Adaptive cycles, 235, 285–286
   application
      ecological succession in forest, 287
      fostering resilience, 294–296
      progression of American automobile industry, 289–290, 294
      role of feedbacks in ecological succession, 287–289
   and climate change on Yucatan Peninsula, 291–293
   conservation phase, 286
   exploitation phase, 286
   release phase, 285
   reorganization phase, 286
African villages, playpumps and water in small rural, 169–173
Agency, 466
   role of, 468
Agricultural system
   in Bali, 318–320
   during Industrial Revolution, 78–79
Air conditioning
   and integrity, 36
   and vitality, 35
Alaskan halibut fishery, 47–48
Albedo, 95, 260
Algae-dominated reefs, 94

Allogenic drivers, 330, 488, 489
   fast, 335–338
   slow, 334–335
   slow drivers *vs.* fast drivers, 338–340
Alternative stable state, 93, 236
American automobile industry, 289–290, 294
Anthropogenic climate change, 32–34
A Sand County Almanac (Leopold), 41
Assets, 503
Autogenic drivers, 330, 488

B

Ball-and-basin model, 251
   resilience using, 249–250
Banff National Park, Canada, 10
Barriers, 503
Becquerel, Henri, 472
Beliefs, 167
Biodiversity loss, direct drivers of, 152, 153
Biofuels, 89
   Sweden's car culture from gasoline to, 277–279
Biogeochemical cycle alteration, 14–17
Biological community, 312, 548
Biological population, 312
Biomass production, 131–132
Biotechnology Scenarios: 2000–2050 Using the Future to Explore the Present, 395
Black Sea, 129, 130
Blessed Unrest (Hawken), 54
Boreal forests, 137
Bottom-up approach, 107
Boundary, 81, 82
Brundtland Commission, 51
1987 Brundtland Commission Report, 437
Bubonic plague, 150, 151

585

Buffers, 510–511
Bureau of Land Management (BLM), 269
Butterfly effect, 86

C

Canadian Wildlife Service (CWS), 109, 110
Carbon capture and sequestration (CCS), 11
Carbon cycle, 10
Carrying capacity, 316
Carson, Rachel, 42
CAS. *See* Complex adaptive system (CAS)
Cascading effects, 86–90
Cassava, 12
Catchment Management Authorities (CMAs), 284, 532
Causal chain analysis (CCA), 123, 146–147, 156, 159, 163
    assessing causality, 147–151
    extent of influence and current trends, 152–154
CBD. *See* Convention on Biological Diversity (CBD)
CCA. *See* Causal chain analysis (CCA)
CCS. *See* Carbon capture and sequestration (CCS)
Change agent, 57
Chaotic behavior, 337
Charismatic megafauna, 21, 22
Chicago heat wave (1995), air conditioning during, 35–36
Cities as Sustainable Ecosystems: Principles and Practices, 435
Clean Air Act, 42
Clean Water Act, 42
Climate change, 10–12
    anthropogenic, 32–34
    global, 88
    and U.S. Military, 231–232
    on Yucatan Peninsula, 291–293
Climate system, competing feedbacks in, 267
Closed system, 81
Clothing, social and environmental costs of cheap, 140–146
CMAs. *See* Catchment Management Authorities (CMAs)
Coastal dead zones, 126
Coastal eutrophication, 131, 132, 135
CO_2 emissions, 193
Coevolution, 467
Collapse: How Societies Choose to Fail or Succeed (Diamond), 96
Collective intelligence, 310–311
Co-management, 554
Commission on Sustainable Development (CSD), 53
Common-pool resource (CPR), 535
Common property regime, 538
    characteristics of, 540–564
    ecological knowledge, 543–545
    essential ingredients of, 541
    full autonomy to create and enforce rules, 561–564
    leadership potential, 553–554
    monitoring use patterns, 543
    number of users, 550–553
    resource system
        boundaries of, 542
        dynamics predictability, 546–548
        importance of, 559–561
        knowledge of, 556–559
        productivity, 545–546
        size, 541–542
        unit mobility, 548–550
    shared social norms, 554–556
Community-based management (CBM), 538
Comoros, 160, 161
Competing feedbacks, climate system, 267
Complex adaptive system (CAS), 98, 284, 309
    adaptation
        biological evolution, natural selection, 361–362
        cultural evolution by learning, 362–366
        and path dependence in Netherlands, 364–366
        process, 360
    emergent feature and behaviors
        emergent properties *vs.* collective properties, 314–320
        sophisticated properties, simple individual interactions, 310–313
    system *vs.* external conditions interactions
        behavior, 321
        disturbance and patterns of fluctuation, 322–330
        disturbance, internal dynamics, 324–331
        fast allogenic drivers, 335–338
        forest fire suppression in United States, 323–324
        resilience, 331–334
        Scheffer ball-and-basin stability landscape model, 342–345
        slow allogenic drivers, 334–335

slow drivers *vs.* fast drivers, 338–340
spruce-fir forest ecosystems, 345–346
stability landscape changes, in three dimensions, 340–341
stability landscapes, 331–334
Complexity: A Guided Tour (Mitchell), 98, 309
Complex system, 67, 90–97
feedback in, 265
Component, 79
Contentiousness, 221
Control indicators, 201
Convention on Biological Diversity (CBD), 53
Convention on International Trade in Endangered Species (CITES), 551–552
Coproduction, 107
Coral reefs, 160
changes in, 94
Corporate and social responsibility (CSR), 55
Coupled human-environmental systems. *See* Socioecological system (SES)
Critical transitions, 135, 241, 243
in grasslands, 248
CSD. *See* Commission on Sustainable Development (CSD)
CSR. *See* Corporate and social responsibility (CSR)
Cultural drivers, 138
Cultural eutrophication, 125, 126
Current state analysis, 100
CWS. *See* Canadian Wildlife Service (CWS)

D

Danube Delta Wildlife Reserve, 135
Dealignment and realignment pathway, 479–480
Degradation, environmental, 50–51
Delays, 512–513
Demographic drivers, 138
Dependency theory, 44
Desertification, 243, 386
grassland, 245–248
Desert shrubland
grassland to, 251–253
Desirable future, 383
Devolution, 553
Deworming, 203–205
Direct drivers, 123, 139
of biodiversity loss, 152–153
Western Indian Ocean Case Study, 156–159
Disincentives, 509

Disturbance
internal dynamics, 324–331
and patterns of fluctuation, 322–330
Diversity
functional, 250
and resilience, 250, 257–258
response, 257
Doughnut model, 244
Drivers, 124, 125
allogenic (*see* Allogenic drivers)
and causal chains, Western Indian Ocean case study, 155–161
demographic, 138
direct, 123, 139, 140, 152–153, 156–159
economic, 138
importance of context, 154–155
indirect, 123, 139, 140, 156–159
influence classification, 138–146
scale classification
spatial, 131–137
temporal, 134–137
Drylands
patterns of change in, 245–248
people living in, 246
vegetative productivity of, 245
Dynamics, 90

E

Earth Charter, 441
Earth Summit, 53, 56
East China Sea, 128, 130
eutrophication in, 140
Ecological succession
in forest, 287
role of feedbacks in, 287–289
Economic development, 43
Economic drivers, 138
Economic regime change, fostering resilience, 253–257
Economic well-being, 218
Ecosystem, human-dominated, 6
Ecosystem service, 6, 7
resilience of, 295
Ecosystem transformations, 6
by human activities, 7–20
biogeochemical cycle alteration, 14–17
climate change, 10–12
habitat degradation, 7–10
invasive species, 12–14
overexploitation, 17–20

EEZ. *See* Exclusive Economic Zone (EEZ)
Emergent feature and behaviors
   emergent properties *vs.* collective properties, 314–320
   sophisticated properties, simple individual interactions, 310–313
Emergent property, 314
Emissions, CO_2, 193
Endangered Species Act, 42
Environmental degradation, 41, 50–51
Environmental generational amnesia, 545–546
Environmental movement
   in India, 43
   in US, 41–43
Environmental Protection Agency (EPA), US, 48
Environmental Sustainability Index (ESI), 207, 211, 213, 214
   aggregate indicators, 215
   indicators used to calculate, 212
Environmental well-being, 218
Environment and natural capital, 40–43
Environment-society-economy, 40
Equity
   intergenerational, 45
   intragenerational, 45
   social, 44, 46, 56
ESI. *See* Environmental Sustainability Index (ESI)
Essential ingredients, 564–568
Ethics, 438–439
Eutrophication, 16, 17, 127
   coastal, 131, 132, 135
   of coastal waters, 125
   cultural, 125, 126
   in East China Sea, 140
   factors affecting, 127
   lake, 237–238
Evaporation, 14
Evolution, long-necked giraffe, 312–313
Excludability and rivalry, 535–537
Excludable, 535
Exclusive Economic Zone (EEZ), 158
Extinction
   of megafauna, 22
   rates, 19

F
Fast allogenic drivers, 335–338
Fast fashion, 140
   social and environmental costs of, 145
Feasibility, 221
Feedbacks, 91, 93, 235
   amplifying, 95
   in complex system, 265
   in ecological succession, 287–289
   in global climate system, 265–267
   in human body, 273–275
   intervention points
      creating, 514–515
      reinforcing, 513–514
      stabilizing, 513
   in markets, 271–273
   in political systems, 269–271
   positive, 95
   reinforcing, 263, 268, 269, 271–273, 275, 276, 281, 282
   runaway, 259–263
   in social systems, 268–269
   stabilizing, 263–266, 268, 270–276, 280, 282
   system fluctuations and, 279–284
Fertilizers, 128
Fish conservation and reduce pollution, 47–49
Fishery
   Alaskan halibut, 47–48
   overexploitation of, 18, 19
Flow, 91
Fluctuation, 93, 236
Food crisis (2007–2008), 89–90
Food security in Senegal, 185–186
Forest
   ecological succession in, 287
   fire suppression in United States, 323–324
   habitat, 9
Fossil fuels, 32, 78
Fostering resilience, 294–296
   through economic regime change, 253–257
Fuels, fossil, 32, 78
Functional diversity, 250
Fundamental changes, 503
Future thinking
   challenges to
      CO_2 in beaker, 385–386
      complexity results, 386–387
      human actors, 387–388
      long-term challenges, 388–389
      unanticipated events, 383–384
      unavailable information, 384–386
   formal–decision support, 423–428
   global climate, uncertainty and, 419–420
   scenario analysis
      1987 Brundtland Commission report, 391
      International Centre for Integrative Studies (ICIS), 392

present and past, 390
SES characteristics, 389
WWII, 391
scenarios and, 381–383
scenario typology
classification criteria, 392
complex, 392
exploratory and decision support, 393
intuitive–exploratory, 394–395
process design spectrum, 394
project goal and process design, 393
types of, 393
WBCSD Biotechnology scenarios, 394–397
socioecological systems characteristics, 383
visioning and
process, 435–443
quality criteria, 443–447
stakeholder preferences, 429
for sustainability, 430–435

G
Ganges, 143, 144
Garment factory workers, physical harm to, 148–149
GDP. *See* Gross domestic product (GDP)
Global climate system, feedbacks in, 265–267
Global ecological state, 245
Globalization, 156
Global oceans and rising CO_2 concentrations, 385–386
GNP. *See* Gross national product (GNP)
Goals, 515–516
Governance systems, regime shifts in, 237–239
Grassland
critical transitions in, 248
desertification, 245–248
to desert shrubland, 251–253
resilience of, 251
Greenhouses gas, 10, 11
Gross domestic product (GDP), 30, 56, 57, 191–193
Gross national product (GNP), 191, 192

H
Haber-Bosch process, 17
Habitat
degradation, 7–10
forest, 9
fragmentation, 8, 9
terrestrial, 8, 9

Happy Planet Index (HPI), 56
Hardin, Garrett, 535, 537
HDI. *See* Human Development Index (HDI)
HDSI. *See* Human Sustainable Development Index (HSDI)
Hierarchy, 314
in Boreal forests, 317–318
Holistic thinking, 85–86, 89–90
House temperature control in, 276
Human agency, 362
Human body, feedbacks in, 273–275
Human Development Index (HDI), 56, 207, 214
aggregate indicators, 215
Human-dominated ecosystem, 6
Human interaction, natural system, 21–23
Human Sustainable Development Index (HSDI), 207, 214
aggregate indicators, 215
Human system, regimes for, 237
Human transformation, natural system, 7–20
using I = PAT model, 25–31
Human well-being, 218
affected by natural system transformations, 36–39
definition, 34, 36
natural system and, 31–32
Hydrologic cycle
disruption of, 32
human impacts on, 15

I
Incentives, 509
Index of Sustainable Economic Welfare (ISEW), 192
Indicators, 100, 123, 188. *See also* Sustainability indicators
pressure, 201
response, 202
state, 200
of student success, 203–205
for sustainable society index, 216, 217
used to calculate ESI, 212
Indirect drivers, 123, 139
Western Indian Ocean Case Study, 156–159
Individual behavior and invisible hand of markets, 310–311
Individual selection and biological evolution, 312
Industrial Revolution, agricultural system during, 78–79
Inequity

Inertia, 96, 97
Infiltration, 132
Inherited trait, 312
Innovation, 468, 470
Input variable, 383
Institutional arrangements, 536
Integrity, 34
   air conditioning and, 36
Interaction, 80
Intergenerational equity, 45
Intergovernmental Panel on Climate Change (IPCC), 6, 56
Intergovernmental Panel on Climate Change-*Special Report Emissions Scenarios* (IPCC-SRES), 10
International Centre for Integrative Studies (ICIS), 392
International Union for the Conservation of Nature (IUCN), 50
Intervention points, 87, 507
   buffers, 510–511
   changing paradigms, 516–517
   delays, 512–513
   feedbacks
      creating, 514–515
      reinforcing, 513–514
      stabilizing, 513
   goals, 515–516
   numbers, 509
   physical structures, 511–512
   rules, 515
   self-organization, 515
   transcending paradigms, 517–518
Intragenerational equity, 45
Invasive species, 12–14
I = PAT model, 23
   application, 23–25
   human transformations, 25–26, 30–31
      affluence, 28–30
      population, 26–28
      technology, 30
IPCC. *See* Intergovernmental Panel on Climate Change (IPCC)
ISEW. *See* Index of Sustainable Economic Welfare (ISEW)
IUCN. *See* International Union for the Conservation of Nature (IUCN)

K
Key stakeholders, 164

L
Lakes
   eutrophication, 237–238
   Laguna, 38
   Laguna Basin, 37, 38
   regime shifts in, 237–239
Landscapes, 469, 473–474
   avalanche, 477
   disruptive, 477
   regular, 476
   specific shocks, 476
   types of, 476–477
Latitude (L), 332
Leverage points, 87, 102. *See also* Intervention points
Lionfish in Indian and South Pacific Oceans, 13, 14
Livelihood, 37
Long-necked giraffe, evolution, 312–313

M
Madagascar, 160, 161
Mangroves, climate change in, 103–105
Markets, feedbacks in, 271–273
Marx, Karl, 46
Mauritius, 160
MEA. *See* Millennium ecosystem assessment (MEA)
Megafauna, 21, 22
Mexican farmers and worst global coffee crisis in history, 83–84
Millennium ecosystem assessment (MEA), 6, 39, 124, 138, 139, 154
Mitigation, 7
Modernization theory, 44
Mortality rate, 316
Motivational value theory of Schwartz, 436
Muir, John, 42
Multi-level perspective (MLP), 465
Multi-phase concept (MPC), 493–495
Multiple regimes, thresholds and, 258–259
Multi-scalar socioecological systems, 87

N
Narain, Sunita, 43
Natural capital, 41
   environment and, 40–43
Natural system
   human interaction with, 21–23

human transformation of, 7–20
   aspects of, 36–39
     using I = PAT model, 25–31
   and human well-being, 31–32
   regimes for, 237
Nature of interactions, 476
Negative externality, 477
Negative feedback. *See* Stabilizing feedback
Nested arrangements, 552
NGOs. *See* Nongovernmental organizations (NGOs)
Niches, 287, 469
Nitrogen cycle, 18
Nitrous oxide gas, 17
Non-excludable, 535
Nongovernmental organizations (NGOs), 51, 53
Nonlinear, 91, 92
   change, 240
Non-rival, 536
Norms, 68
North American Caribou, protection, 108–111
Numbers, 509

O

Ocean fisheries, 548
OC RMDZ. *See* Orange County Recycling Market Development Zone (OC RMDZ)
Open systems, 81
Orange County Recycling Market Development Zone (OC RMDZ), 253–257
Order of magnitude, 136
Ostrom, Elinor, 540
Outsiders, 542
Overexploitation, 17–20
Oyster farming, 16

P

Paradigms, intervention point
   changing, 516–517
   transcending, 517–518
Participatory approach, 106
Passive stakeholders, 166
Path dependence, 362
Patron Saint of the Environment, 41
PCBs. *See* Polychlorinated biphenyls (PCBs)
Per capita, 23
Percolation, 14, 132
Peruvian Anchovy fishery management (1970), 207–210
Phosphorus, 16
Photosynthesis, 126
Physical structures, 511–512
Pinchot, Gifford, 42
Plague, bubonic, 150, 151
Plausible futures, 383
Polio, 150, 151
Political systems, feedbacks in, 269–271
Pollution
   fish conservation and reduce, 47–49
   solid waste, 157, 158
Polycentric governance system, 295
Polychlorinated biphenyls (PCBs), 156
The Population Bomb (Ehrlich), 42
Precariousness (P), 332
Precipitation, 14
Pressure indicators, 201
Primary stakeholders, 164
Primorska Region of Slovenia, case study, 220–226
Probable futures, 383

Q

Q_{10} factor, 131
Quality criteria, visioning and future thinking
   coherent, 444–445
   motivational, 447
   nuanced, 447
   plausible, 445–446
   relevant, 446
   shared, 447
   sustainable, 444
   systemic, 444
   tangible, 446
   visionary, 443–444

R

Range anxiety, 503–504
Ream National Park, Cambodia, 104
Reconfiguration pathway, 484–487
Recycling Market Development Zone (RMDZ) program, 253–257
Reductionist thinking, 85
REEF. *See* Reef Environmental Education Foundation (REEF)
Reef Environmental Education Foundation (REEF), 13
Regimes, 93, 235, 471–472
   shift, 93, 236
     in lakes and governance systems, 237–239
     and resilience, 251–253
     thresholds and, 239–248

Reinforcing feedback, 95, 263, 266–269, 271–273, 275, 276, 281, 282
Religious drivers, 138
Reservoir, 91
Resilience, 235, 248–249, 331–334
    diversity and, 250, 257–258
    of ecosystem services, 295
    fostering, 294–296
    of grasslands, 251
    in human systems, 257
    regime shifts and, 251–253
    through economic regime change fostering, 253–257
    using ball-and-basin model, 249–250
Resistance (R), 332
Resources, categories of, 536
Resource system, common property regimes
    boundaries of, 542
    characteristics of, 540–564
    dynamics predictability, 546–548
    ecological knowledge, 543–545
    essential ingredients of, 541
    full autonomy to create and enforce rules, 561–564
    importance of, 559–561
    knowledge of, 556–559
    leadership potential, 553–554
    monitoring use patterns, 543
    number of users, 550–553
    productivity, 545–546
    shared social norms, 554–556
    size, 541–542
    unit mobility, 548–550
Response diversity, 257
Response indicators, 202
Riparian zones, 133
Rival, 536
Rivalry, excludability and, 535–537
Rules, 515
Runaway feedback, 259–263

S
Saltwater intrusion, 548
San Cristobal River basin, 38
Scales, 87
    multiple, 89–90
Scenario, 101
Scheffer ball-and-basin stability landscape model, 342–345
    critical transitions using, 347–352

SDGs. *See* Sustainable Development Goals (SDGs)
Sea walls, climate change in, 103–105
Secondary stakeholders, 164
Self-organization, 515
SESs. *See* Socioecological system (SES)
Sewage systems in Netherlands, 478–479
Seychelles, 160
Shifting baselines, 545–546
Shifting dominance, 93, 263
Silent Spring (Carson), 42
Simple system, 79–82
Slow allogenic drivers, 334–335
Slumdog Millionaire, 467
Smith, Adam, 45, 46
SNA. *See* Social network analysis (SNA)
Social capital, 38, 57
Social equity, 44, 46, 56
Social network analysis (SNA), 173, 174
Social systems, feedbacks in, 268–269
Socioecological system (SES), 5, 75, 235, 282–284
    capacity, 464
    cascading effects, 86–90
    holistic thinking, 85–86, 89–90
    interactions between human and natural systems, 76–77
    interventions in, 87
    multi-scalar, 87
    scales, 87
    transitions, 464
Sociopolitical drivers, 138
Solid waste pollution, 157, 158
Southeast Asia, climate change in, 103–105
Spruce-fir forest ecosystems, 345–346
Stability landscapes, 331–334
    changes in three dimensions, 340–341
    Scheffer ball-and-basin model, 342–345
    spruce-fir forest ecosystems, 345–346
    in traditional and modern-day Shimshal society, 352–360
Stabilizing feedback, 94, 263–266, 268, 270–276, 280, 282
Stakeholders, 36, 164
    active, 165
    analysis, 161–162
    behavior, interest, and influence, 165–173
    categories, 165
    identifying, 162–165
    key, 164
    passive, 166
    primary, 164

relationships, investigation, 173–174
secondary, 164
State, 93, 236
State indicators, 200
Statement of Forest Principles, 53
Stock, 91
Strategic actions, 504
Strong sustainability, 102, 105
Structuration theory, 466
Structure, 466
Substitutability, 546
Subtractability. *See* Rivalry
Sulfur cycle, 16
Sustainability, 5, 39–40, 440–441
  birth of, 50–57
  challenges for, 56, 66
  economy and development, 43–44
  environment and natural capital, 40–43
  environment pillar of, 41
  global movement toward, 54–55
  making tradeoffs for, 102–106
  SSI rankings for pillars of, 218
  strong, 102, 105
  through activism, 1–2
  United Nations meetings related to, 52
  visioning and future thinking, 430
    build capacity and social capital, 433–434
    guide and direct change, 432–433
    motivate and inspire, 431–432
    promote long-term thinking, 435
    staying power, 434–435
  weak, 102
Sustainability indicators, 189–190
  development, participatory process, 220–226
  effective characteristics
    collectively determined by stakeholders, 202–207
    current state, drivers, and responses, 200–202
    easy to understand, even by nonexperts, 199–200
    realistically available data, 196–199
    representative of real-world complexity, 207–219
    use appropriate time and space scales, 194–196
  vs. traditional indicators, 190–193
Sustainability Indicators: Measuring the Immeasurable? (Bell and Morse), 194
Sustainability transitions, 464
  guiding, 492–494
  from 1950s to present
    acceleration phase, 497
    predevelopment phase, 496–497
    stabilization phase, 497–498
    take-off phase, 497
Sustainable Development Goals (SDGs), 219
Sustainable Society Foundation, 216
Sustainable Society Index (SSI), 216
  indicators for, 216, 217
  rankings, 218
Sweden's car culture, from gasoline to biofuels, 277–279
Syrian civil war, 32–34
System, 79
  complex, 90–97
  definition, 125–131
  fluctuations and feedbacks, 279–284
  relationships, 383
  simple, 79–82
System *vs.* external conditions interactions
  allogenic drivers
    fast, 335–338
    slow, 334–335
    slow drivers *vs.* fast drivers, 338–340
  behavior, 321
  disturbance
    internal dynamics, 324–331
    and patterns of fluctuation, 322–330
  forest fire suppression in United States, 323–324
  resilience, 331–334
  spruce-fir forest ecosystems, 345–346
  stability landscapes, 331–334
    changes in three dimensions, 340–341
    Scheffer ball-and-basin model, 342–345
    spruce-fir forest ecosystems, 345–346
    in traditional and modern-day Shimshal society, 352–360

T
Tame problems, 66, 67
  vs. wicked problems, 68–69, 73–74
Targets, 504
Technological substitution pathway, 482–484
Temperature control, in house, 276
Terrestrial habitat, 8, 9
Thinking
  holistic, 85–86, 89–90
  reductionist, 85
Threshold, 93, 235
  and multiple regimes, 258–259
  and regime shifts, 239–259
Tipping point. *See* Threshold

Tonle Sap Lake, Cambodia, 11
Top-down approach, 107
Tradeoffs, 102
  for resilience to climate change, 103–105
Traditional indicators vs. sustainability indicators, 190–193
Tragedy of the Commons (ToC) situation, 535
  common property regimes, characteristics of, 540–564
  essential ingredients, 564–568
  excludability and rivalry, 535–537
Transcendentalist movement, 41
Transdisciplinary knowledge, 107, 108
Transformational Sustainability Research (TSR), 98–100
  framework, 102, 103
Transformation pathway, 477–478
  in 19th century netherlands, 478–479
Transition, 101, 464
Transition management (TM), 492–494
Transition pathways, 475–477
Transition strategy
  aligning strategic action and MPC phases, 500–502
  development, 502–504
  transportation system transition in Curitiba, Brazil, 505–507
Transpiration, 15
Transportation system transition, 480–482
Troposphere, 17
TSR. *See* Transformational Sustainability Research (TSR)
TSR framework, 463

U
UNCED, 53
UN Environment Program (UNEP), 50
UNEP. *See* UN Environment Program (UNEP)
UNEP Global International Waters Assessment project, 152
Unequivocal, 138
UNFCCC. *See* UN Framework Convention on Climate Change (UNFCCC)
UN Framework Convention on Climate Change (UNFCCC), 53
Unintended consequences, 502
United Nations Convention to Combat Desertification (UNCCD), 247

United States
  environmental movement in, 41–43
  Environmental Protection Agency (EPA), 42, 48
Utility death spiral, 260–263

V
Values, 167
Vision, 101
  statement, 442
Visioning, 101
  process, 435
    1987 Brundtland Commission Report, 437
    Earth Charter, 441
    ethics, 438–439
    formulating, 435
    motivational value theory of Schwartz, 436
    sustainability, 440–441
    technology, 439
    vision statement, 442
  quality criteria
    coherent, 444–445
    motivational, 447
    nuanced, 447
    plausible, 445–446
    relevant, 446
    shared, 447
    sustainable, 444
    systemic, 444
    tangible, 446
    visionary, 443–444
  stakeholder preferences, 429
  for sustainability, 430
    build capacity and social capital, 433–434
    guide and direct change, 432–433
    motivate and inspire, 431–432
    promote long-term thinking, 435
    staying power, 434–435
Vitality, 34
  air conditioning and, 35
Vulnerability science, 50

W
Waste management in Netherlands, 496–498
Water
Watershed boundaries, 127
WCED. *See* World Commission on Environment and Development (WCED)

Weak sustainability, 102
The Wealth of Nations (Smith), 45
West Basin Water Association (WBWA), 552
Western Indian Ocean case study, 155–161
Wicked problems, 66
    freeing females from twenty-first century oppression, 70–73
    irreversible, 69, 72–73
    no end point, 69, 72
    vs. tame problems, 68–69, 73–74
    undefined solution, 68, 70–72
    unique, 73
    urgent, 69, 73
    vague problem definition, 68, 70
Wiek, Arnim, 166
Wildlife corridors, 9
World Business Council on Sustainable Development (WBCSD) Biotechnology Scenarios, 55, 394–395
    BBI niche market, 404
    biotrust, 404–409
    Blame Biotech, 398
    Domino effect, 397–398
    feeding the world, 402–403
    hare and tortoise scenario, 400–401
    health consciousness, 401–402
    Intergovernmental Panel on Climate Change (IPCC) special report emissions scenarios, 415–419
    intuitive–decision support, Destino Colombia, 409–414
    revolution, 398–399
World Commission on Environment and Development (WCED), 51
World Development Indicators (WDI), 219
World Trade Organization (WTO), 55
World views, 167
World Wildlife Fund (WWF), 50
WSSD, 53
WWF. *See* World Wildlife Fund (WWF)

Y

Yangtze River, 127–129, 133
Yucatan Peninsula, adaptive cycles and climate change on, 291–293